Academic and Professional Skills:

Quantitative Methods

Fourth Edition

PEARSON

We work with leading authors to develop the strongest learning experiences, bringing cutting-edge thinking and best learning practice to a global market. We craft our print and digital resources to do more to help learners not only understand their content, but to see it in action and apply what they learn, whether studying or at work.

Pearson is the world's leading learning company. Our portfolio includes Penguin, Dorling Kindersley, the Financial Times and our educational business, Pearson International. We are also a leading provider of electronic learning programmes and of test development, processing and scoring services to educational institutions, corporations and professional bodies around the world.

Pearson Custom Publishing enables our customers to access a wide and expanding range of market-leading content from world-renowned authors and develop their own tailor-made book. You choose the content that meets your needs and Pearson Custom Publishing produces a high-quality printed book.

Every day our work helps learning flourish, and wherever learning flourishes, so do people.

To learn more please visit us at: www.pearsoncustom.co.uk

PEARSON CUSTOM PUBLISHING

Academic and Professional Skills:

Quantitative Methods

Fourth Edition

Compiled from:

Foundation Maths
Fifth Edition
by Anthony Croft and Robert Davison

Statistics for Economics, Accounting and Business Studies
Fifth Edition
by Michael Barrow

Harlow, England • London • New York • Boston • San Francisco • Toronto • Sydney • Auckland • Singapore • Hong Kong
Tokyo • Seoul • Taipei • New Delhi • Cape Town • Sao Paulo • Mexico City • Madrid • Amsterdam • Munich • Paris • Milan

Pearson Education Limited
Edinburgh Gate
Harlow
Essex CM20 2JE

And associated companies throughout the world

Visit us on the World Wide Web at:
www.pearsoned.co.uk

Compiled from:

Foundation Maths
Fifth Edition
by Anthony Croft and Robert Davison
ISBN 978 0 273 72940 2

Statistics for Economics, Accounting and Business Studies
Fifth Edition
by Michael Barrow
ISBN 978 0 273 71794 2

ISBN 978 1 78236 114 5

Printed and bound in Great Britain by Henry Ling Limited at the Dorset Press, Dorchester, DT1 1HD

Contents

Preface

This book has been customised by Pearson to match the structure of the Quantitative Methods part of the core Academic and Professional Skills module taken by all HUBS students in their first year. It draws on Michael Barrow's Statistics for Economics, Accounting and Business Studies and Anthony Croft and Robert Davison's Foundation Maths. We have included most of the Barrow book, which is obviously aimed at business school students, but have been more selective with Croft and Davison, which is a general introduction to basic maths; you may be relieved to hear that we have not included the trigonometry chapters here, for example!

The QM teaching within APS runs through both semesters. We are aware that students start their degree programmes with very different levels of maths knowledge, so we start with some "Refresher" lectures and classes to bring everyone up to speed on the basics (to something like GCSE level) before we move onto the more advanced topics. Attendance at some of these early sessions is optional, depending on whether you can handle the material confidently already. You should approach the early chapters of this book in the same way. For most people they will be a mixture of things you know well, things you vaguely remember learning once but can't quite remember now, and probably a few things you have never heard of. By all means skip over the chapters for your personal first category, but make sure you work through the material for the other two (as well as attending the relevant lectures and classes, and completing the associated exercises in MyMathLab). Steady work throughout the module is what is needed to do well, as later material will build on what has been done already.

Beyond the basic material there are a couple of lectures on algebra, but after that the course is mainly concerned with introductory statistics and calculus. From here on you need to be working through all the chapters in this book carefully, unless you have a strong (A-level or equivalent) knowledge of that particular topic already. Even then it will be worth your while to do at least some of the exercises included in this text to check that you really are on top of the material. You should also complete the regular MyMathLab exercises. The algebra and statistics are important for all students, and make up the bulk of the exam at the end of the year. The calculus lectures and classes are aimed principally at students who are registered for any degree including 'economics' in its title, or who think that they might wish to transfer to such a degree next year, but everyone else is of course welcome too. We hope you find the QM lectures and classes both useful and enjoyable, and assure you that the skills you will learn there and from this text will be valuable to you, both in the rest of your degree course and in whatever you go on to do after you graduate from Hull. Good luck!

Section Title 1

ALGEBRA

1 Arithmetic of whole numbers

Objectives: This chapter:

- explains the rules for adding, subtracting, multiplying and dividing positive and negative numbers
- explains what is meant by an integer
- explains what is meant by a prime number
- explains what is meant by a factor
- explains how to prime factorise an integer
- explains the terms 'highest common factor' and 'lowest common multiple'

1.1 Addition, subtraction, multiplication and division

Arithmetic is the study of numbers and their manipulation. A clear and firm understanding of the rules of arithmetic is essential for tackling everyday calculations. Arithmetic also serves as a springboard for tackling more abstract mathematics such as algebra and calculus.

The calculations in this chapter will involve mainly whole numbers, or **integers** as they are often called. The **positive integers** are the numbers

$$1, 2, 3, 4, 5 \ldots$$

and the **negative integers** are the numbers

$$\ldots -5, -4, -3, -2, -1$$

The dots (...) indicate that this sequence of numbers continues indefinitely. The number 0 is also an integer but is neither positive nor negative.

To find the **sum** of two or more numbers, the numbers are added together. To find the **difference** of two numbers, the second is subtracted from the first. The **product** of two numbers is found by multiplying the numbers together. Finally, the **quotient** of two numbers is found by dividing the first number by the second.

WORKED EXAMPLE

1.1

(a) Find the sum of 3, 6 and 4.

(b) Find the difference of 6 and 4.

(c) Find the product of 7 and 2.

(d) Find the quotient of 20 and 4.

Solution

(a) The sum of 3, 6 and 4 is

$$3 + 6 + 4 = 13$$

(b) The difference of 6 and 4 is

$$6 - 4 = 2$$

(c) The product of 7 and 2 is

$$7 \times 2 = 14$$

(d) The quotient of 20 and 4 is $\frac{20}{4}$, that is 5.

When writing products we sometimes replace the sign $\times$ by '$\cdot$' or even omit it completely. For example, $3 \times 6 \times 9$ could be written as $3 \cdot 6 \cdot 9$ or $(3)(6)(9)$.

On occasions it is necessary to perform calculations involving negative numbers. To understand how these are added and subtracted consider Figure 1.1, which shows a number line.

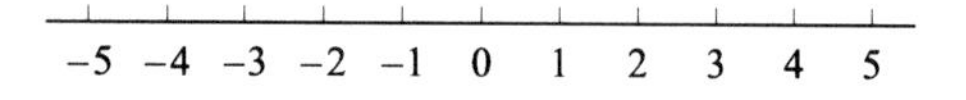

Figure 1.1
The number line

Any number can be represented by a point on the line. Positive numbers are on the right-hand side of the line and negative numbers are on the left. From any given point on the line, we can add a positive number by moving that number of places to the right. For example, to find the sum $5 + 3$, start at the point 5 and move 3 places to the right, to arrive at 8. This is shown in Figure 1.2.

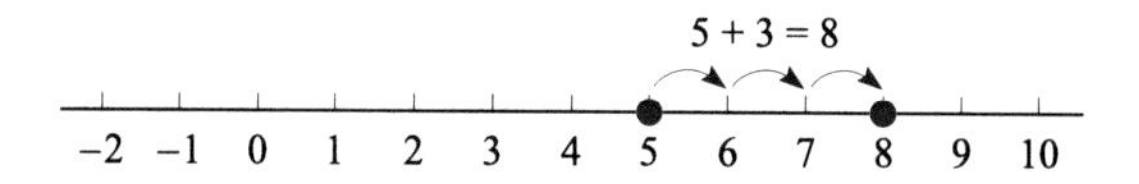

Figure 1.2
To add a positive number, move that number of places to the right

To subtract a positive number, we move that number of places to the left. For example, to find the difference $5 - 7$, start at the point 5 and move 7 places to the left to arrive at -2. Thus $5 - 7 = -2$. This is shown in Figure 1.3. The result of finding $-3 - 4$ is also shown to be -7.

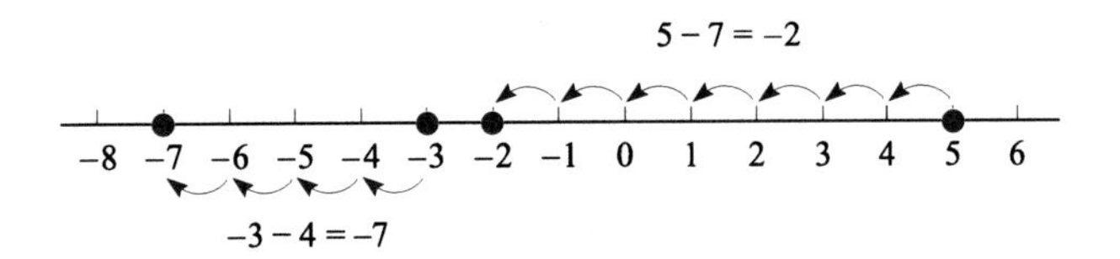

Figure 1.3
To subtract a positive number, move that number of places to the left

To add or subtract a negative number, the motions just described are reversed. So, to add a negative number, we move to the left. To subtract a negative number we move to the right. The result of finding $2 + (-3)$ is shown in Figure 1.4.

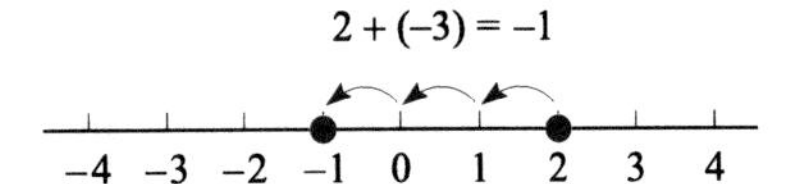

Figure 1.4
Adding a negative number involves moving to the left

We see that $2 + (-3) = -1$. Note that this is the same as the result of finding $2 - 3$, so that adding a negative number is equivalent to subtracting a positive number.

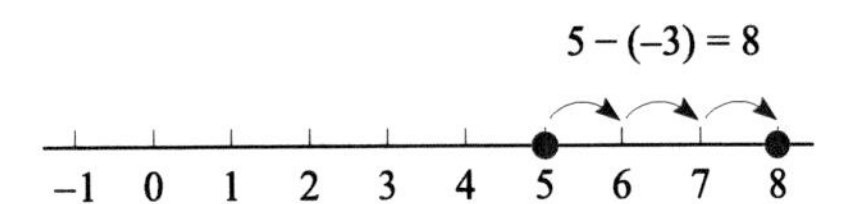

Figure 1.5
Subtracting a negative number involves moving to the right

The result of finding $5 - (-3)$ is shown in Figure 1.5.

We see that $5 - (-3) = 8$. This is the same as the result of finding $5 + 3$, so subtracting a negative number is equivalent to adding a positive number.

Key point

Adding a negative number is equivalent to subtracting a positive number. Subtracting a negative number is equivalent to adding a positive number.

WORKED EXAMPLE

1.2 Evaluate (a) $8 + (-4)$, (b) $-15 + (-3)$, (c) $-15 - (-4)$.

Solution (a) $8 + (-4)$ is equivalent to $8 - 4$, that is 4.

(b) Because adding a negative number is equivalent to subtracting a positive number we find $-15 + (-3)$ is equivalent to $-15 - 3$, that is -18.

(c) $-15 - (-4)$ is equivalent to $-15 + 4$, that is -11.

When we need to multiply or divide negative numbers, care must be taken with the **sign** of the answer; that is, whether the result is positive or negative. The following rules apply for determining the sign of the answer when multiplying or dividing positive and negative numbers.

Key point

(positive) × (positive) = positive
(positive) × (negative) = negative
(negative) × (positive) = negative
(negative) × (negative) = positive

and

$$\frac{\text{positive}}{\text{positive}} = \text{positive}$$

$$\frac{\text{positive}}{\text{negative}} = \text{negative}$$

$$\frac{\text{negative}}{\text{positive}} = \text{negative}$$

$$\frac{\text{negative}}{\text{negative}} = \text{positive}$$

WORKED EXAMPLE

1.3 Evaluate

(a) $3 \times (-2)$ (b) $(-1) \times 7$ (c) $(-2) \times (-4)$ (d) $\frac{12}{(-4)}$ (e) $\frac{-8}{4}$ (f) $\frac{-6}{-2}$

Solution

(a) We have a positive number, 3, multiplied by a negative number, -2, and so the result will be negative:

$$3 \times (-2) = -6$$

(b) $(-1) \times 7 = -7$

(c) Here we have two negative numbers being multiplied and so the result will be positive:

$$(-2) \times (-4) = 8$$

(d) A positive number, 12, divided by a negative number, -4, gives a negative result:

$$\frac{12}{-4} = -3$$

(e) A negative number, −8, divided by a positive number, 4, gives a negative result:

$$\frac{-8}{4} = -2$$

(f) A negative number, −6, divided by a negative number, −2, gives a positive result:

$$\frac{-6}{-2} = 3$$

Self-assessment questions 1.1

1. Explain what is meant by an integer, a positive integer and a negative integer.
2. Explain the terms sum, difference, product and quotient.
3. State the sign of the result obtained after performing the following calculations:
   (a) $(-5) \times (-3)$ (b) $(-4) \times 2$ (c) $\frac{7}{-2}$ (d) $\frac{-8}{-4}$.

Exercise 1.1

MyMathLab Global

1. Without using a calculator, evaluate each of the following:
   (a) $6 + (-3)$ (b) $6 - (-3)$
   (c) $16 + (-5)$ (d) $16 - (-5)$
   (e) $27 - (-3)$ (f) $27 - (-29)$
   (g) $-16 + 3$ (h) $-16 + (-3)$
   (i) $-16 - 3$ (j) $-16 - (-3)$
   (k) $-23 + 52$ (l) $-23 + (-52)$
   (m) $-23 - 52$ (n) $-23 - (-52)$
2. Without using a calculator, evaluate
   (a) $3 \times (-8)$ (b) $(-4) \times 8$ (c) $15 \times (-2)$
   (d) $(-2) \times (-8)$ (e) $14 \times (-3)$
3. Without using a calculator, evaluate
   (a) $\frac{15}{-3}$ (b) $\frac{21}{7}$ (c) $\frac{-21}{7}$ (d) $\frac{-21}{-7}$ (e) $\frac{21}{-7}$
   (f) $\frac{-12}{2}$ (g) $\frac{-12}{-2}$ (h) $\frac{12}{-2}$
4. Find the sum and product of (a) 3 and 6, (b) 10 and 7, (c) 2, 3 and 6.
5. Find the difference and quotient of (a) 18 and 9, (b) 20 and 5, (c) 100 and 20.

1.2 The BODMAS rule

When evaluating numerical expressions we need to know the order in which addition, subtraction, multiplication and division are carried out. As a simple example, consider evaluating $2 + 3 \times 4$. If the addition is carried out first we get $2+3\times4=5\times4=20$. If the multiplication is carried out first we get $2+3\times4=2+12=14$. Clearly the order of carrying out numerical operations is important. The BODMAS rule tells us the order in which we must carry out the operations of addition, subtraction, multiplication and division.

Key point

BODMAS stands for

Brackets ( )	First priority
Of $\times$	Second priority
Division $\div$	Second priority
Multiplication $\times$	Second priority
Addition $+$	Third priority
Subtraction $-$	Third priority

This is the order of carrying out arithmetical operations, with bracketed expressions having highest priority and subtraction and addition having the lowest priority. Note that 'Of', 'Division' and 'Multiplication' have equal priority, as do 'Addition' and 'Subtraction'. 'Of' is used to show multiplication when dealing with fractions: for example, find $\frac{1}{2}$ of 6 means $\frac{1}{2}\times6$.

If an expression contains only multiplication and division, we evaluate by working from left to right. Similarly, if an expression contains only addition and subtraction, we also evaluate by working from left to right.

WORKED EXAMPLES

1.4 Evaluate

(a) $2+3\times4$ (b) $(2+3)\times4$

Solution

(a) Using the BODMAS rule we see that multiplication is carried out first. So

$$2+3\times4=2+12=14$$

(b) Using the BODMAS rule we see that the bracketed expression takes priority over all else. Hence

$$(2+3)\times4=5\times4=20$$

1.5 Evaluate

(a) $4-2\div2$ (b) $1-3+2\times2$

Solution (a) Division is carried out before subtraction, and so

$$4 - 2 \div 2 = 4 - \frac{2}{2} = 3$$

(b) Multiplication is carried out before subtraction or addition:

$$1 - 3 + 2 \times 2 = 1 - 3 + 4 = 2$$

1.6 Evaluate

(a) $(12 \div 4) \times 3$ (b) $12 \div (4 \times 3)$

Solution Recall that bracketed expressions are evaluated first.

(a) $(12 \div 4) \times 3 = \left(\frac{12}{4}\right) \times 3 = 3 \times 3 = 9$

(b) $12 \div (4 \times 3) = 12 \div 12 = 1$

Example 1.6 shows the importance of the position of brackets in an expression.

Self-assessment questions 1.2

1. State the BODMAS rule used to evaluate expressions.
2. The position of brackets in an expression is unimportant. True or false?

Exercise 1.2

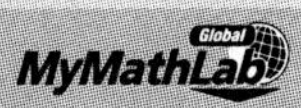

1. Evaluate the following expressions:
   (a) $6 - 2 \times 2$ (b) $(6 - 2) \times 2$
   (c) $6 \div 2 - 2$ (d) $(6 \div 2) - 2$
   (e) $6 - 2 + 3 \times 2$ (f) $6 - (2 + 3) \times 2$
   (g) $(6 - 2) + 3 \times 2$ (h) $\frac{16}{-2}$ (i) $\frac{-24}{-3}$
   (j) $(-6) \times (-2)$ (k) $(-2)(-3)(-4)$

2. Place brackets in the following expressions to make them correct:
   (a) $6 \times 12 - 3 + 1 = 55$
   (b) $6 \times 12 - 3 + 1 = 68$
   (c) $6 \times 12 - 3 + 1 = 60$
   (d) $5 \times 4 - 3 + 2 = 7$
   (e) $5 \times 4 - 3 + 2 = 15$
   (f) $5 \times 4 - 3 + 2 = -5$

1.3 Prime numbers and factorisation

A **prime number** is a positive integer, larger than 1, which cannot be expressed as the product of two smaller positive integers. To put it another way, a prime number is one that can be divided exactly only by 1 and itself.

For example, $6 = 2 \times 3$, so 6 can be expressed as a product of smaller numbers and hence 6 is not a prime number. However, 7 is prime. Examples of prime numbers are 2, 3, 5, 7, 11, 13, 17, 19, 23. Note that 2 is the only even prime.

Factorise means 'write as a product'. By writing 12 as 3×4 we have factorised 12. We say 3 is a **factor** of 12 and 4 is also a factor of 12. The way in which a number is factorised is not unique: for example, 12 may be expressed as 3×4 or 2×6. Note that 2 and 6 are also factors of 12.

When a number is written as a product of prime numbers we say the number has been **prime factorised**.

To prime factorise a number, consider the technique used in the following examples.

WORKED EXAMPLES

1.7 Prime factorise the following numbers:

(a) 12 (b) 42 (c) 40 (d) 70

Solution

(a) We begin with 2 and see whether this is a factor of 12. Clearly it is, so we write

$$12 = 2 \times 6$$

Now we consider 6. Again 2 is a factor so we write

$$12 = 2 \times 2 \times 3$$

All the factors are now prime, that is the prime factorisation of 12 is $2 \times 2 \times 3$.

(b) We begin with 2 and see whether this is a factor of 42. Clearly it is and so we can write

$$42 = 2 \times 21$$

Now we consider 21. Now 2 is not a factor of 21, so we examine the next prime, 3. Clearly 3 is a factor of 21 and so we can write

$$42 = 2 \times 3 \times 7$$

All the factors are now prime, and so the prime factorisation of 42 is $2 \times 3 \times 7$.

(c) Clearly 2 is a factor of 40,

$$40 = 2 \times 20$$

Clearly 2 is a factor of 20,

$$40 = 2 \times 2 \times 10$$

Again 2 is a factor of 10,

$$40 = 2 \times 2 \times 2 \times 5$$

All the factors are now prime. The prime factorisation of 40 is $2 \times 2 \times 2 \times 5$.

(d) Clearly 2 is a factor of 70,

$$70 = 2 \times 35$$

We consider 35: 2 is not a factor, 3 is not a factor, but 5 is:

$$70 = 2 \times 5 \times 7$$

All the factors are prime. The prime factorisation of 70 is $2 \times 5 \times 7$.

1.8 Prime factorise 2299.

Solution We note that 2 is not a factor and so we try 3. Again 3 is not a factor and so we try 5. This process continues until we find the first prime factor. It is 11:

$$2299 = 11 \times 209$$

We now consider 209. The first prime factor is 11:

$$2299 = 11 \times 11 \times 19$$

All the factors are prime. The prime factorisation of 2299 is $11 \times 11 \times 19$.

Self-assessment questions 1.3

1. Explain what is meant by a prime number.
2. List the first 10 prime numbers.
3. Explain why all even numbers other than 2 cannot be prime.

Exercise 1.3

MyMathLab

1. State which of the following numbers are prime numbers:
   (a) 13 (b) 1000 (c) 2 (c) 29 (d) $\frac{1}{2}$
2. Prime factorise the following numbers:
   (a) 26 (b) 100 (c) 27 (d) 71 (e) 64 (f) 87 (g) 437 (h) 899
3. Prime factorise the two numbers 30 and 42. List any prime factors which are common to both numbers.

1.4 Highest common factor and lowest common multiple

Highest common factor

Suppose we prime factorise 12. This gives $12 = 2 \times 2 \times 3$. From this prime factorisation we can deduce all the factors of 12:

2 is a factor of 12
3 is a factor of 12
$2 \times 2 = 4$ is a factor of 12
$2 \times 3 = 6$ is a factor of 12

Hence 12 has factors 2, 3, 4 and 6, in addition to the obvious factors of 1 and 12.

Similarly we could prime factorise 18 to obtain $18 = 2 \times 3 \times 3$. From this we can list the factors of 18:

2 is a factor of 18
3 is a factor of 18
$2 \times 3 = 6$ is a factor of 18
$3 \times 3 = 9$ is a factor of 18

The factors of 18 are 1, 2, 3, 6, 9 and 18. Some factors are common to both 12 and 18. These are 2, 3 and 6. These are **common factors** of 12 and 18. The highest common factor of 12 and 18 is 6.

The highest common factor of 12 and 18 can be obtained directly from their prime factorisation. We simply note all the primes common to both factorisations:

$$12 = 2 \times 2 \times 3 \qquad 18 = 2 \times 3 \times 3$$

Common to both is 2×3. Thus the highest common factor is $2 \times 3 = 6$. Thus 6 is the highest number that divides exactly into both 12 and 18.

Key point

Given two or more numbers the **highest common factor** (h.c.f.) is the largest (highest) number that is a factor of all the given numbers.
The highest common factor is also referred to as the **greatest common divisor** (g.c.d).

WORKED EXAMPLES

1.9 Find the h.c.f. of 12 and 27.

Solution We prime factorise 12 and 27:

$$12 = 2 \times 2 \times 3 \qquad 27 = 3 \times 3 \times 3$$

Common to both is 3. Thus 3 is the h.c.f. of 12 and 27. This means that 3 is the highest number that divides both 12 and 27.

1.10 Find the h.c.f. of 28 and 210.

Solution The numbers are prime factorised:

$$28 = 2 \times 2 \times 7$$
$$210 = 2 \times 3 \times 5 \times 7$$

The factors that are common are identified: a 2 is common to both and a 7 is common to both. Hence both numbers are divisible by $2 \times 7 = 14$. Since this number contains all the common factors it is the highest common factor.

1.11 Find the h.c.f. of 90 and 108.

Solution The numbers are prime factorised:

$$90 = 2 \times 3 \times 3 \times 5$$
$$108 = 2 \times 2 \times 3 \times 3 \times 3$$

The common factors are 2, 3 and 3 and so the h.c.f. is $2 \times 3 \times 3$, that is 18. This is the highest number that divides both 90 and 108.

1.12 Find the h.c.f. of 12, 18 and 20.

Solution Prime factorisation yields

$$12 = 2 \times 2 \times 3 \qquad 18 = 2 \times 3 \times 3 \qquad 20 = 2 \times 2 \times 5$$

There is only one factor common to all three numbers: it is 2. Hence 2 is the h.c.f. of 12, 18 and 20.

Lowest common multiple

Suppose we are given two or more numbers and wish to find numbers into which all the given numbers will divide. For example, given 4 and 6 we see that they both divide exactly into 12, 24, 36, 48, 60 and so on. The smallest number into which they both divide is 12. We say 12 is the **lowest common multiple** of 4 and 6.

Key point The lowest common multiple (l.c.m.) of a set of numbers is the smallest (lowest) number into which all the given numbers will divide exactly.

WORKED EXAMPLE

1.13 Find the l.c.m. of 6 and 10.

Solution We seek the smallest number into which both 6 and 10 will divide exactly. There are many numbers into which 6 and 10 will divide, for example 60,

120, 600, but we are seeking the smallest such number. By inspection, the smallest such number is 30. Thus the l.c.m. of 6 and 10 is 30.

A more systematic method of finding the l.c.m. involves the use of prime factorisation.

WORKED EXAMPLES

1.14 Find the l.c.m. of 15 and 20.

Solution As a first step, the numbers are prime factorised:

$$15 = 3 \times 5 \qquad 20 = 2 \times 2 \times 5$$

Since 15 must divide into the l.c.m., then the l.c.m. must contain the factors of 15, that is 3×5. Similarly, as 20 must divide into the l.c.m., then the l.c.m. must also contain the factors of 20, that is $2 \times 2 \times 5$. The l.c.m. is the smallest number that contains both of these sets of factors. Note that the l.c.m. will contain only 2s, 3s and 5s as its prime factors. We now need to determine how many of these particular factors are needed.

To determine the l.c.m. we ask 'How many factors of 2 are required?', 'How many factors of 3 are required?', 'How many factors of 5 are required?'

The highest number of 2s occurs in the factorisation of 20. Hence the l.c.m. requires two factors of 2. Consider the number of 3s required. The highest number of 3s occurs in the factorisation of 15. Hence the l.c.m. requires one factor of 3. Consider the number of 5s required. The highest number of 5s is 1 and so the l.c.m. requires one factor of 5. Hence the l.c.m. is $2 \times 2 \times 3 \times 5 = 60$.

Hence 60 is the smallest number into which both 15 and 20 will divide exactly.

1.15 Find the l.c.m. of 20, 24 and 25.

Solution The numbers are prime factorised:

$$20 = 2 \times 2 \times 5 \qquad 24 = 2 \times 2 \times 2 \times 3 \qquad 25 = 5 \times 5$$

By considering the prime factorisations of 20, 24 and 25 we see that the only primes involved are 2, 3 and 5. Hence the l.c.m. will contain only 2s, 3s and 5s.

Consider the number of 2s required. The highest number of 2s required is three from factorising 24. The highest number of 3s required is one, again from factorising 24. The highest number of 5s required is two, found from factorising 25. Hence the l.c.m. is given by

$$\text{l.c.m.} = 2 \times 2 \times 2 \times 3 \times 5 \times 5 = 600$$

Hence 600 is the smallest number into which 20, 24 and 25 will all divide exactly.

Self-assessment questions 1.4

1. Explain what is meant by the h.c.f. of a set of numbers.
2. Explain what is meant by the l.c.m. of a set of numbers.

Exercise 1.4

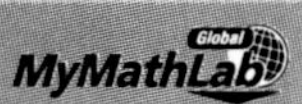

1. Calculate the h.c.f. of the following sets of numbers:
   (a) 12, 15, 21 (b) 16, 24, 40 (c) 28, 70, 120, 160 (d) 35, 38, 42 (e) 96, 120, 144
2. Calculate the l.c.m. of the following sets of numbers:
   (a) 5, 6, 8 (b) 20, 30 (c) 7, 9, 12 (d) 100, 150, 235 (e) 96, 120, 144

Test and assignment exercises 1

1. Evaluate
   (a) $6 \div 2 + 1$ (b) $6 \div (2 + 1)$ (c) $12 + 4 \div 4$ (d) $(12 + 4) \div 4$
   (e) $3 \times 2 + 1$ (f) $3 \times (2 + 1)$ (g) $6 - 2 + 4 \div 2$ (h) $(6 - 2 + 4) \div 2$
   (i) $6 - (2 + 4 \div 2)$ (j) $6 - (2 + 4) \div 2$ (k) $2 \times 4 - 1$ (l) $2 \times (4 - 1)$
   (m) $2 \times 6 \div (3 - 1)$ (n) $2 \times (6 \div 3) - 1$ (o) $2 \times (6 \div 3 - 1)$
2. Prime factorise (a) 56, (b) 39, (c) 74.
3. Find the h.c.f. of
   (a) 8, 12, 14 (b) 18, 42, 66 (c) 20, 24, 30 (d) 16, 24, 32, 160
4. Find the l.c.m. of
   (a) 10, 15 (b) 11, 13 (c) 8, 14, 16 (d) 15, 24, 30

Fractions

2

Objectives: This chapter:

- explains what is meant by a fraction
- defines the terms 'improper fraction', 'proper fraction' and 'mixed fraction'
- explains how to write fractions in different but equivalent forms
- explains how to simplify fractions by cancelling common factors
- explains how to add, subtract, multiply and divide fractions

2.1 Introduction

The arithmetic of fractions is very important groundwork that must be mastered before topics in algebra such as formulae and equations can be understood. The same techniques that are used to manipulate fractions are used in these more advanced topics. You should use this chapter to ensure that you are confident at handling fractions before moving on to algebra. In all the examples and exercises it is important that you should carry out the calculations without the use of a calculator.

Fractions are numbers such as $\frac{1}{2}$, $\frac{3}{4}$, $\frac{11}{8}$ and so on. In general a fraction is a number of the form $\frac{p}{q}$, where the letters p and q represent whole numbers or integers. The integer q can never be zero because it is never possible to divide by zero.

In any fraction $\frac{p}{q}$ the number p is called the **numerator** and the number q is called the **denominator**.

Key point

$$\text{fraction} = \frac{\text{numerator}}{\text{denominator}} = \frac{p}{q}$$

Suppose that p and q are both positive numbers. If p is less than q, the fraction is said to be a **proper fraction**. So $\frac{1}{2}$ and $\frac{3}{4}$ are proper fractions since the numerator is less than the denominator. If p is greater than or equal to q, the fraction is said to be **improper**. So $\frac{11}{8}$, $\frac{7}{4}$ and $\frac{3}{3}$ are all improper fractions.

If either of p or q is negative, we simply ignore the negative sign when determining whether the fraction is proper or improper. So $-\frac{3}{5}$, $\frac{-7}{21}$ and $\frac{4}{-21}$ are proper fractions, but $\frac{3}{-3}$, $\frac{-8}{2}$ and $-\frac{11}{2}$ are improper.

Note that all proper fractions have a value less than 1.

The denominator of a fraction can take the value 1, as in $\frac{3}{1}$ and $\frac{7}{1}$. In these cases the result is a whole number, 3 and 7.

Self-assessment questions 2.1

1. Explain the terms (a) fraction, (b) improper fraction, (c) proper fraction. In each case give an example of your own.
2. Explain the terms (a) numerator, (b) denominator.

Exercise 2.1

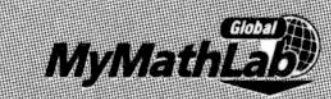

1. Classify each of the following as proper or improper:
   (a) $\frac{9}{17}$ (b) $\frac{-9}{17}$ (c) $\frac{8}{8}$ (d) $-\frac{7}{8}$ (e) $\frac{110}{77}$

2.2 Expressing a fraction in equivalent forms

Given a fraction, we may be able to express it in a different form. For example, you will know that $\frac{1}{2}$ is equivalent to $\frac{2}{4}$. Note that multiplying both numerator and denominator by the same number leaves the value of the fraction unchanged. So, for example,

$$\frac{1}{2} = \frac{1 \times 2}{2 \times 2} = \frac{2}{4}$$

We say that $\frac{1}{2}$ and $\frac{2}{4}$ are **equivalent fractions**. Although they might look different, they have the same value.

Similarly, given the fraction $\frac{8}{12}$ we can divide both numerator and denominator by 4 to obtain

$$\frac{8}{12} = \frac{8/4}{12/4} = \frac{2}{3}$$

so $\frac{8}{12}$ and $\frac{2}{3}$ have the same value and are equivalent fractions.

Key point

Multiplying or dividing both numerator and denominator of a fraction by the same number produces a fraction having the same value, called an equivalent fraction.

A fraction is in its **simplest form** when there are no factors common to both numerator and denominator. For example, $\frac{5}{12}$ is in its simplest form, but $\frac{3}{6}$ is not since 3 is a factor common to both numerator and denominator. Its simplest form is the equivalent fraction $\frac{1}{2}$.

To express a fraction in its simplest form we look for factors that are common to both the numerator and denominator. This is done by prime factorising both of these. Dividing both the numerator and denominator by any common factors removes them but leaves an equivalent fraction. This is equivalent to cancelling any common factors. For example, to simplify $\frac{4}{6}$ we prime factorise to produce

$$\frac{4}{6} = \frac{2 \times 2}{2 \times 3}$$

Dividing both numerator and denominator by 2 leaves $\frac{2}{3}$. This is equivalent to cancelling the common factor of 2.

WORKED EXAMPLES

2.1 Express $\frac{24}{36}$ in its simplest form.

Solution We seek factors common to both numerator and denominator. To do this we prime factorise 24 and 36:

Prime factorisation has been described in §1.3.

$$24 = 2 \times 2 \times 2 \times 3 \qquad 36 = 2 \times 2 \times 3 \times 3$$

The factors $2 \times 2 \times 3$ are common to both 24 and 36 and so these may be cancelled. Note that only common factors may be cancelled when simplifying a fraction. Hence

Finding the highest common factor (h.c.f.) of two numbers is detailed in §1.4.

$$\frac{24}{36} = \frac{\not{2} \times \not{2} \times 2 \times \not{3}}{\not{2} \times \not{2} \times \not{3} \times 3} = \frac{2}{3}$$

In its simplest form $\frac{24}{36}$ is $\frac{2}{3}$. In effect we have divided 24 and 36 by 12, which is their h.c.f.

2.2 Express $\frac{49}{21}$ in its simplest form.

Solution Prime factorising 49 and 21 gives

$$49 = 7 \times 7 \qquad 21 = 3 \times 7$$

Their h.c.f. is 7. Dividing 49 and 21 by 7 gives

$$\frac{49}{21} = \frac{7}{3}$$

Hence the simplest form of $\frac{49}{21}$ is $\frac{7}{3}$.

Before we can start to add and subtract fractions it is necessary to be able to convert fractions into a variety of equivalent forms. Work through the following examples.

WORKED EXAMPLES

2.3 Express $\frac{3}{4}$ as an equivalent fraction having a denominator of 20.

Solution To achieve a denominator of 20, the existing denominator must be multiplied by 5. To produce an equivalent fraction both numerator and denominator must be multiplied by 5, so

$$\frac{3}{4} = \frac{3 \times 5}{4 \times 5} = \frac{15}{20}$$

2.4 Express 7 as an equivalent fraction with a denominator of 3.

Solution Note that 7 is the same as the fraction $\frac{7}{1}$. To achieve a denominator of 3, the existing denominator must be multiplied by 3. To produce an equivalent fraction both numerator and denominator must be multiplied by 3, so

$$7 = \frac{7}{1} = \frac{7 \times 3}{1 \times 3} = \frac{21}{3}$$

Self-assessment questions 2.2

1. All integers can be thought of as fractions. True or false?
2. Explain the use of h.c.f. in the simplification of fractions.
3. Give an example of three fractions that are equivalent.

Exercise 2.2

MyMathLab

1. Express the following fractions in their simplest form:
(a) $\frac{18}{27}$ (b) $\frac{12}{20}$ (c) $\frac{15}{45}$ (d) $\frac{25}{80}$ (e) $\frac{15}{60}$
(f) $\frac{90}{200}$ (g) $\frac{15}{20}$ (h) $\frac{2}{18}$ (i) $\frac{16}{24}$ (j) $\frac{30}{65}$
(k) $\frac{12}{21}$ (l) $\frac{100}{45}$ (m) $\frac{6}{9}$ (n) $\frac{12}{16}$ (o) $\frac{13}{42}$
(p) $\frac{13}{39}$ (q) $\frac{11}{33}$ (r) $\frac{14}{30}$ (s) $-\frac{12}{16}$ (t) $\frac{11}{-33}$
(u) $\frac{-14}{-30}$

2. Express $\frac{3}{4}$ as an equivalent fraction having a denominator of 28.

3. Express 4 as an equivalent fraction with a denominator of 5.

4. Express $\frac{5}{12}$ as an equivalent fraction having a denominator of 36.

5. Express 2 as an equivalent fraction with a denominator of 4.

6. Express 6 as an equivalent fraction with a denominator of 3.

7. Express each of the fractions $\frac{2}{3}$, $\frac{5}{4}$ and $\frac{5}{6}$ as an equivalent fraction with a denominator of 12.

8. Express each of the fractions $\frac{4}{9}$, $\frac{1}{2}$ and $\frac{5}{6}$ as an equivalent fraction with a denominator of 18.

9. Express each of the following numbers as an equivalent fraction with a denominator of 12:
(a) $\frac{1}{2}$ (b) $\frac{3}{4}$ (c) $\frac{5}{2}$ (d) 5 (e) 4 (f) 12

2.3 Addition and subtraction of fractions

To add and subtract fractions we first rewrite each fraction so that they all have the same denominator. This is known as the **common denominator**. The denominator is chosen to be the lowest common multiple of the original denominators. Then the numerators only are added or subtracted as appropriate, and the result is divided by the common denominator.

WORKED EXAMPLES

2.5 Find $\frac{2}{3}+\frac{5}{4}$.

Solution The denominators are 3 and 4. The l.c.m. of 3 and 4 is 12. We need to express both fractions with a denominator of 12.

Finding the lowest common multiple (l.c.m.) is detailed in §1.4.

To express $\frac{2}{3}$ with a denominator of 12 we multiply both numerator and denominator by 4. Hence $\frac{2}{3}$ is the same as $\frac{8}{12}$. To express $\frac{5}{4}$ with a denominator of 12 we multiply both numerator and denominator by 3. Hence $\frac{5}{4}$ is the same as $\frac{15}{12}$. So

$$\frac{2}{3}+\frac{5}{4}=\frac{8}{12}+\frac{15}{12}=\frac{8+15}{12}=\frac{23}{12}$$

2.6 Find $\frac{4}{9} - \frac{1}{2} + \frac{5}{6}$.

Solution The denominators are 9, 2 and 6. Their l.c.m. is 18. Each fraction is expressed with 18 as the denominator:

$$\frac{4}{9} = \frac{8}{18} \qquad \frac{1}{2} = \frac{9}{18} \qquad \frac{5}{6} = \frac{15}{18}$$

Then

$$\frac{4}{9} - \frac{1}{2} + \frac{5}{6} = \frac{8}{18} - \frac{9}{18} + \frac{15}{18} = \frac{8 - 9 + 15}{18} = \frac{14}{18}$$

The fraction $\frac{14}{18}$ can be simplified to $\frac{7}{9}$. Hence

$$\frac{4}{9} - \frac{1}{2} + \frac{5}{6} = \frac{7}{9}$$

2.7 Find $\frac{1}{4} - \frac{5}{9}$.

Solution The l.c.m. of 4 and 9 is 36. Each fraction is expressed with a denominator of 36. Thus

$$\frac{1}{4} = \frac{9}{36} \quad \text{and} \quad \frac{5}{9} = \frac{20}{36}$$

Then

$$\frac{1}{4} - \frac{5}{9} = \frac{9}{36} - \frac{20}{36}$$

$$= \frac{9 - 20}{36}$$

$$= \frac{-11}{36}$$

$$= -\frac{11}{36}$$

Consider the number $2\frac{3}{4}$. This is referred to as a **mixed fraction** because it contains a whole number part, 2, and a fractional part, $\frac{3}{4}$. We can convert this mixed fraction into an improper fraction as follows. Recognise that 2 is equivalent to $\frac{8}{4}$, and so $2\frac{3}{4}$ is $\frac{8}{4} + \frac{3}{4} = \frac{11}{4}$.

The reverse of this process is to convert an improper fraction into a mixed fraction. Consider the improper fraction $\frac{11}{4}$. Now 4 divides into 11 twice leaving a remainder of 3; so $\frac{11}{4} = 2$ remainder 3, which we write as $2\frac{3}{4}$.

WORKED EXAMPLE

2.8

(a) Express $4\frac{2}{5}$ as an improper fraction.

(b) Find $4\frac{2}{5}+\frac{1}{3}$.

Solution

(a) $4\frac{2}{5}$ is a mixed fraction. Note that $4\frac{2}{5}$ is equal to $4+\frac{2}{5}$. We can write 4 as the equivalent fraction $\frac{20}{5}$. Therefore

$$4\frac{2}{5}=\frac{20}{5}+\frac{2}{5}$$

$$=\frac{22}{5}$$

(b) $$4\frac{2}{5}+\frac{1}{3}=\frac{22}{5}+\frac{1}{3}$$

$$=\frac{66}{15}+\frac{5}{15}$$

$$=\frac{71}{15}$$

Self-assessment question 2.3

1. Explain the use of l.c.m. when adding and subtracting fractions.

Exercise 2.3

MyMathLab Global

1. Find
   (a) $\frac{1}{4}+\frac{2}{3}$ (b) $\frac{3}{5}+\frac{5}{3}$ (c) $\frac{12}{14}-\frac{2}{7}$
   (d) $\frac{3}{7}-\frac{1}{2}+\frac{2}{21}$ (e) $1\frac{1}{2}+\frac{4}{9}$
   (f) $2\frac{1}{4}-1\frac{1}{3}+\frac{1}{2}$ (g) $\frac{10}{15}-1\frac{2}{5}+\frac{8}{3}$
   (h) $\frac{9}{10}-\frac{7}{16}+\frac{1}{2}-\frac{2}{5}$

2. Find
   (a) $\frac{7}{8}+\frac{1}{3}$ (b) $\frac{1}{2}-\frac{3}{4}$ (c) $\frac{3}{5}+\frac{2}{3}+\frac{1}{2}$
   (d) $\frac{3}{8}+\frac{1}{3}+\frac{1}{4}$ (e) $\frac{2}{3}-\frac{4}{7}$
   (f) $\frac{1}{11}-\frac{1}{2}$ (g) $\frac{3}{11}-\frac{5}{8}$

3. Express as improper fractions:
(a) $2\frac{1}{2}$ (b) $3\frac{2}{3}$ (c) $10\frac{1}{4}$ (d) $5\frac{2}{7}$
(e) $6\frac{2}{9}$ (f) $11\frac{1}{3}$ (g) $15\frac{1}{2}$ (h) $13\frac{3}{4}$
(i) $12\frac{1}{11}$ (j) $13\frac{2}{3}$ (k) $56\frac{1}{2}$

4. Without using a calculator express these improper fractions as mixed fractions:
(a) $\frac{10}{3}$ (b) $\frac{7}{2}$ (c) $\frac{15}{4}$ (d) $\frac{25}{6}$

2.4 Multiplication of fractions

The product of two or more fractions is found by multiplying their numerators to form a new numerator, and then multiplying their denominators to form a new denominator.

WORKED EXAMPLES

2.9 Find $\frac{4}{9} \times \frac{3}{8}$.

Solution The numerators are multiplied: $4 \times 3 = 12$. The denominators are multiplied: $9 \times 8 = 72$. Hence

$$\frac{4}{9} \times \frac{3}{8} = \frac{12}{72}$$

This may now be expressed in its simplest form:

$$\frac{12}{72} = \frac{1}{6}$$

Hence

$$\frac{4}{9} \times \frac{3}{8} = \frac{1}{6}$$

An alternative, but equivalent, method is to cancel any factors common to both numerator and denominator at the outset:

$$\frac{4}{9} \times \frac{3}{8} = \frac{4 \times 3}{9 \times 8}$$

A factor of 4 is common to the 4 and the 8. Hence

$$\frac{4 \times 3}{9 \times 8} = \frac{1 \times 3}{9 \times 2}$$

A factor of 3 is common to the 3 and the 9. Hence

$$\frac{1 \times 3}{9 \times 2} = \frac{1 \times 1}{3 \times 2} = \frac{1}{6}$$

2.10 Find $\frac{12}{25} \times \frac{2}{7} \times \frac{10}{9}$.

Solution We cancel factors common to both numerator and denominator. A factor of 5 is common to 10 and 25. Cancelling this gives

$$\frac{12}{25} \times \frac{2}{7} \times \frac{10}{9} = \frac{12}{5} \times \frac{2}{7} \times \frac{2}{9}$$

A factor of 3 is common to 12 and 9. Cancelling this gives

$$\frac{12}{5} \times \frac{2}{7} \times \frac{2}{9} = \frac{4}{5} \times \frac{2}{7} \times \frac{2}{3}$$

There are no more common factors. Hence

$$\frac{12}{25} \times \frac{2}{7} \times \frac{10}{9} = \frac{4}{5} \times \frac{2}{7} \times \frac{2}{3} = \frac{16}{105}$$

2.11 Find $\frac{3}{4}$ of $\frac{5}{9}$.

Recall that 'of' means multiply.

Solution $\frac{3}{4}$ of $\frac{5}{9}$ is the same as $\frac{3}{4} \times \frac{5}{9}$. Cancelling a factor of 3 from numerator and denominator gives $\frac{1}{4} \times \frac{5}{3}$, that is $\frac{5}{12}$. Hence $\frac{3}{4}$ of $\frac{5}{9}$ is $\frac{5}{12}$.

2.12 Find $\frac{5}{6}$ of 70.

Solution We can write 70 as $\frac{70}{1}$. So

$$\frac{5}{6} \text{ of } 70 = \frac{5}{6} \times \frac{70}{1} = \frac{5}{3} \times \frac{35}{1} = \frac{175}{3} = 58\frac{1}{3}$$

2.13 Find $2\frac{7}{8} \times \frac{2}{3}$.

Solution In this example the first fraction is a mixed fraction. We convert it to an improper fraction before performing the multiplication. Note that $2\frac{7}{8} = \frac{23}{8}$. Then

$$\frac{23}{8} \times \frac{2}{3} = \frac{23}{4} \times \frac{1}{3}$$

$$= \frac{23}{12}$$

$$= 1\,\frac{11}{12}$$

Self-assessment question 2.4

1. Describe how to multiply fractions together.

Exercise 2.4

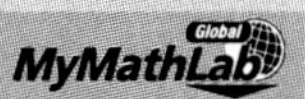

1. Evaluate

   (a) $\frac{2}{3} \times \frac{6}{7}$ (b) $\frac{8}{15} \times \frac{25}{32}$ (c) $\frac{1}{4} \times \frac{8}{9}$

   (d) $\frac{16}{17} \times \frac{34}{48}$ (e) $2 \times \frac{3}{5} \times \frac{5}{12}$

   (f) $2\,\frac{1}{3} \times 1\,\frac{1}{4}$ (g) $1\,\frac{3}{4} \times 2\,\frac{1}{2}$

   (h) $\frac{3}{4} \times 1\,\frac{1}{2} \times 3\,\frac{1}{2}$

2. Evaluate

   (a) $\frac{2}{3}$ of $\frac{3}{4}$ (b) $\frac{4}{7}$ of $\frac{21}{30}$

   (c) $\frac{9}{10}$ of 80 (d) $\frac{6}{7}$ of 42

3. Is $\frac{3}{4}$ of $\frac{12}{15}$ the same as $\frac{12}{15}$ of $\frac{3}{4}$?

4. Find

   (a) $-\frac{1}{3} \times \frac{5}{7}$ (b) $\frac{3}{4} \times -\frac{1}{2}$

   (c) $\left(-\frac{5}{8}\right) \times \frac{8}{11}$ (d) $\left(-\frac{2}{3}\right) \times \left(-\frac{15}{7}\right)$

5. Find

   (a) $5\,\frac{1}{2} \times \frac{1}{2}$ (b) $3\,\frac{3}{4} \times \frac{1}{3}$

   (c) $\frac{2}{3} \times 5\,\frac{1}{9}$ (d) $\frac{3}{4} \times 11\,\frac{1}{2}$

6. Find

   (a) $\frac{3}{5}$ of $11\,\frac{1}{4}$ (b) $\frac{2}{3}$ of $15\,\frac{1}{2}$

   (c) $\frac{1}{4}$ of $-8\,\frac{1}{3}$

2.5 Division by a fraction

To divide one fraction by another fraction, we invert the second fraction and then multiply. When we invert a fraction we interchange the numerator and denominator.

WORKED EXAMPLES

2.14 Find $\frac{6}{25} \div \frac{2}{5}$.

Solution We invert $\frac{2}{5}$ to obtain $\frac{5}{2}$. Multiplication is then performed. So

$$\frac{6}{25} \div \frac{2}{5} = \frac{6}{25} \times \frac{5}{2} = \frac{3}{25} \times \frac{5}{1} = \frac{3}{5} \times \frac{1}{1} = \frac{3}{5}$$

2.15 Evaluate (a) $1\frac{1}{3} \div \frac{8}{3}$, (b) $\frac{20}{21} \div \frac{5}{7}$.

Solution (a) First we express $1\frac{1}{3}$ as an improper fraction:

$$1\frac{1}{3} = 1 + \frac{1}{3} = \frac{3}{3} + \frac{1}{3} = \frac{4}{3}$$

So we calculate

$$\frac{4}{3} \div \frac{8}{3} = \frac{4}{3} \times \frac{3}{8} = \frac{4}{8} = \frac{1}{2}$$

Hence

$$1\frac{1}{3} \div \frac{8}{3} = \frac{1}{2}$$

(b) $$\frac{20}{21} \div \frac{5}{7} = \frac{20}{21} \times \frac{7}{5} = \frac{4}{21} \times \frac{7}{1} = \frac{4}{3}$$

Self-assessment question 2.5

1. Explain the process of division by a fraction.

Exercise 2.5

MyMathLab

1. Evaluate

(a) $\frac{3}{4} \div \frac{1}{8}$ (b) $\frac{8}{9} \div \frac{4}{3}$ (c) $\frac{-2}{7} \div \frac{4}{21}$ (d) $\frac{9}{4} \div 1\frac{1}{2}$

(e) $\frac{5}{6} \div \frac{5}{12}$ (f) $\frac{99}{100} \div 1\frac{4}{5}$ (g) $3\frac{1}{4} \div 1\frac{1}{8}$ (h) $\left(2\frac{1}{4} \div \frac{3}{4}\right) \times 2$

(i) $2\frac{1}{4} \div \left(\frac{3}{4} \times 2\right)$ (j) $6\frac{1}{4} \div 2\frac{1}{2} + 5$ (k) $6\frac{1}{4} \div \left(2\frac{1}{2} + 5\right)$

Test and assignment exercises 2

1. Evaluate

(a) $\frac{3}{4} + \frac{1}{6}$ (b) $\frac{2}{3} + \frac{3}{5} - \frac{1}{6}$ (c) $\frac{5}{7} - \frac{2}{3}$ (d) $2\frac{1}{3} - \frac{9}{10}$

(e) $5\frac{1}{4} + 3\frac{1}{6}$ (f) $\frac{9}{8} - \frac{7}{6} + 1$ (g) $\frac{5}{6} - \frac{5}{3} + \frac{5}{4}$ (h) $\frac{4}{5} + \frac{1}{3} - \frac{3}{4}$

2. Evaluate

(a) $\frac{4}{7} \times \frac{21}{32}$ (b) $\frac{5}{6} \times \frac{8}{15}$ (c) $\frac{3}{11} \times \frac{20}{21}$ (d) $\frac{9}{14} \times \frac{8}{18}$

(e) $\frac{5}{4} \div \frac{10}{13}$ (f) $\frac{7}{16} \div \frac{21}{32}$ (g) $\frac{-24}{25} \div \frac{51}{50}$ (h) $\frac{45}{81} \div \frac{25}{27}$

3. Evaluate the following expressions using the BODMAS rule:

(a) $\frac{1}{2} + \frac{1}{3} \times 2$ (b) $\frac{3}{4} \times \frac{2}{3} + \frac{1}{4}$ (c) $\frac{5}{6} \div \frac{2}{3} + \frac{3}{4}$ (d) $\left(\frac{2}{3} + \frac{1}{4}\right) \div 4 + \frac{3}{5}$

(e) $\left(\frac{4}{3} - \frac{2}{5} \times \frac{1}{3}\right) \times \frac{1}{4} + \frac{1}{2}$ (f) $\frac{3}{4}$ of $\left(1 + \frac{2}{3}\right)$ (g) $\frac{2}{3}$ of $\frac{1}{2} + 1$ (h) $\frac{1}{5} \times \frac{2}{3} + \frac{2}{5} \div \frac{4}{5}$

4. Express in their simplest form:

(a) $\frac{21}{84}$ (b) $\frac{6}{80}$ (c) $\frac{34}{85}$ (d) $\frac{22}{143}$ (e) $\frac{69}{253}$

3 Decimal fractions

Objectives: This chapter:

- revises the decimal number system
- shows how to write a number to a given number of significant figures
- shows how to write a number to a given number of decimal places

3.1 Decimal numbers

Consider the whole number 478. We can regard it as the sum

$$400 + 70 + 8$$

In this way we see that, in the number 478, the 8 represents eight ones, or 8 units, the 7 represents seven tens, or 70, and the number 4 represents four hundreds or 400. Thus we have the system of hundreds, tens and units familiar from early years in school. All whole numbers can be thought of in this way.

When we wish to deal with proper fractions and mixed fractions, we extend the hundreds, tens and units system as follows. A **decimal point**, '.', marks the end of the whole number part, and the numbers that follow it, to the right, form the fractional part.

A number immediately to the right of the decimal point, that is in the **first decimal place**, represents tenths, so

$$0.1 = \frac{1}{10}$$

$$0.2 = \frac{2}{10} \quad \text{or} \quad \frac{1}{5}$$

$$0.3 = \frac{3}{10} \quad \text{and so on}$$

Note that when there are no whole numbers involved it is usual to write a zero in front of the decimal point, thus, .2 would be written 0.2.

WORKED EXAMPLE

3.1 Express the following decimal numbers as proper fractions in their simplest form

(a) 0.4 (b) 0.5 (c) 0.6

Solution The first number after the decimal point represents tenths.

(a) $0.4 = \frac{4}{10}$, which simplifies to $\frac{2}{5}$

(b) $0.5 = \frac{5}{10}$ or simply $\frac{1}{2}$

(c) $0.6 = \frac{6}{10} = \frac{3}{5}$

Frequently we will deal with numbers having a whole number part and a fractional part. Thus

$$\begin{aligned} 5.2 &= 5 \text{ units} + 2 \text{ tenths} \\ &= 5 + \frac{2}{10} \\ &= 5 + \frac{1}{5} \\ &= 5\frac{1}{5} \end{aligned}$$

Similarly,

$$175.8 = 175\frac{8}{10} = 175\frac{4}{5}$$

Numbers in the second position after the decimal point, or the **second decimal place**, represent hundredths, so

$$0.01 = \frac{1}{100}$$

$$0.02 = \frac{2}{100} \quad \text{or} \quad \frac{1}{50}$$

$$0.03 = \frac{3}{100} \quad \text{and so on}$$

Consider 0.25. We can think of this as

$$\begin{aligned} 0.25 &= 0.2 + 0.05 \\ &= \frac{2}{10} + \frac{5}{100} \\ &= \frac{25}{100} \end{aligned}$$

We see that 0.25 is equivalent to $\frac{25}{100}$, which in its simplest form is $\frac{1}{4}$.

In fact we can regard any numbers occupying the first two decimal places as hundredths, so that

$$0.25 = \frac{25}{100} \quad \text{or simply } \frac{1}{4}$$

$$0.50 = \frac{50}{100} \quad \text{or } \frac{1}{2}$$

$$0.75 = \frac{75}{100} = \frac{3}{4}$$

WORKED EXAMPLES

3.2 Express the following decimal numbers as proper fractions in their simplest form:

(a) 0.35 (b) 0.56 (c) 0.68

Solution The first two decimal places represent hundredths:

(a) $0.35 = \frac{35}{100} = \frac{7}{20}$

(b) $0.56 = \frac{56}{100} = \frac{14}{25}$

(c) $0.68 = \frac{68}{100} = \frac{17}{25}$

3.3 Express 37.25 as a mixed fraction in its simplest form.

Solution

$$\begin{aligned} 37.25 &= 37 + 0.25 \\ &= 37 + \frac{25}{100} \\ &= 37 + \frac{1}{4} \\ &= 37\frac{1}{4} \end{aligned}$$

Numbers in the third position after the decimal point, or **third decimal place**, represent thousandths, so

$$0.001 = \frac{1}{1000}$$

$$0.002 = \frac{2}{1000} \quad \text{or} \quad \frac{1}{500}$$

$$0.003 = \frac{3}{1000} \quad \text{and so on}$$

In fact we can regard any numbers occupying the first three positions after the decimal point as thousandths, so that

$$0.356 = \frac{356}{1000} \quad \text{or} \quad \frac{89}{250}$$

$$0.015 = \frac{15}{1000} \quad \text{or} \quad \frac{3}{200}$$

$$0.075 = \frac{75}{1000} = \frac{3}{40}$$

WORKED EXAMPLE

3.4 Write each of the following as a decimal number:

(a) $\frac{3}{10} + \frac{7}{100}$ (b) $\frac{8}{10} + \frac{3}{1000}$

Solution

(a) $\frac{3}{10} + \frac{7}{100} = 0.3 + 0.07 = 0.37$
(b) $\frac{8}{10} + \frac{3}{1000} = 0.8 + 0.003 = 0.803$

You will normally use a calculator to add, subtract, multiply and divide decimal numbers. Generally the more decimal places used, the more accurately we can state a number. This idea is developed in the next section.

Self-assessment questions 3.1

1. State which is the largest and which is the smallest of the following numbers:
23.001, 23.0, 23.00001, 23.0008, 23.01
2. Which is the largest of the following numbers?
0.1, 0.02, 0.003, 0.0004, 0.00005

Exercise 3.1

1. Express the following decimal numbers as proper fractions in their simplest form:
   (a) 0.7 (b) 0.8 (c) 0.9
2. Express the following decimal numbers as proper fractions in their simplest form:
   (a) 0.55 (b) 0.158 (c) 0.98
   (d) 0.099
3. Express each of the following as a mixed fraction in its simplest form:
   (a) 4.6 (b) 5.2 (c) 8.05 (d) 11.59
   (e) 121.09
4. Write each of the following as a decimal number:
   (a) $\frac{6}{10}+\frac{9}{100}+\frac{7}{1000}$ (b) $\frac{8}{100}+\frac{3}{1000}$
   (c) $\frac{17}{1000}+\frac{5}{10}$

3.2 Significant figures and decimal places

The accuracy with which we state a number often depends upon the context in which the number is being used. The volume of a petrol tank is usually given to the nearest litre. It is of no practical use to give such a volume to the nearest cubic centimetre.

When writing a number we often give the accuracy by stating the **number of significant figures** or the **number of decimal places** used. These terms are now explained.

Significant figures

Suppose we are asked to write down the number nearest to 857 using at most two non-zero digits, or numbers. We would write 860. This number is nearer to 857 than any other number with two non-zero digits. We say that 857 to 2 **significant figures** is 860. The words 'significant figures' are usually abbreviated to s.f. Because 860 is larger than 857 we say that the 857 has been **rounded up** to 860.

To write a number to three significant figures we can use no more than three non-zero digits. For example, the number closest to 1784 which has no more than three non-zero digits is 1780. We say that 1784 to 3 significant figures is 1780. In this case, because 1780 is less than 1784 we say that 1784 has been **rounded down** to 1780.

WORKED EXAMPLES

3.5 Write down the number nearest to 86 using only one non-zero digit. Has 86 been rounded up or down?

Solution The number 86 written to one significant figure is 90. This number is nearer to 86 than any other number having only one non-zero digit. 86 has been rounded up to 90.

3.6 Write down the number nearest to 999 which uses only one non-zero digit.

Solution The number 999 to one significant figure is 1000. This number is nearer to 999 than any other number having only one non-zero digit.

We now explain the process of writing to a given number of significant figures.

When asked to write a number to, say, three significant figures, 3 s.f., the first step is to look at the first four digits. If asked to write a number to two significant figures we look at the first three digits and so on. We always look at one more digit than the number of significant figures required.

For example, to write 6543.19 to 2 s.f. we would consider the number 6540.00; the digits 3, 1 and 9 are effectively ignored. The next step is to round up or down. If the final digit is a 5 or more then we round up by increasing the previous digit by 1. If the final digit is 4 or less we round down by leaving the previous digit unchanged. Hence when considering 6543.19 to 2 s.f., the 4 in the third place means that we round down to 6500.

To write 23865 to 3 s.f. we would consider the number 23860. The next step is to increase the 8 to a 9. Thus 23865 is rounded up to 23900.

Zeros at the beginning of a number are ignored. To write 0.004693 to 2 s.f. we would first consider the number 0.00469. Note that the zeros at the beginning of the number have not been counted. We then round the 6 to a 7, producing 0.0047.

The following examples illustrate the process.

WORKED EXAMPLES

3.7 Write 36.482 to 3 s.f.

Solution We consider the first four digits, that is 36.48. The final digit is 8 and so we round up 36.48 to 36.5. To 3 s.f. 36.482 is 36.5.

3.8 Write 1.0049 to 4 s.f.

Solution To write to 4 s.f. we consider the first five digits, that is 1.0049. The final digit is a 9 and so 1.0049 is rounded up to 1.005.

3.9 Write 695.3 to 2 s.f.

Solution We consider 695. The final digit is a 5 and so we round up. We cannot round up the 9 to a 10 and so the 69 is rounded up to 70. Hence to 2 s.f. the number is 700.

3.10 Write 0.0473 to 1 s.f.

Solution We do not count the initial zeros and consider 0.047. The final digit tells us to round up. Hence to 1 s.f. we have 0.05.

3.11 A number is given to 2 s.f. as 67.

(a) What is the maximum value the number could have?
(b) What is the minimum value the number could have?

Solution

(a) To 2 s.f. 67.5 is 68. Any number just below 67.5, for example 67.49 or 67.499, to 2 s.f. is 67. Hence the maximum value of the number is 67.4999....
(b) To 2 s.f. 66.4999... is 66. However, 66.5 to 2 s.f. is 67. The minimum value of the number is thus 66.5.

Decimal places

When asked to write a number to 3 decimal places (3 d.p.) we consider the first 4 decimal places, that is numbers after the decimal point. If asked to write to 2 d.p. we consider the first 3 decimal places and so on. If the final digit is 5 or more we round up, otherwise we round down.

WORKED EXAMPLES

3.12 Write 63.4261 to 2 d.p.

Solution We consider the number to 3 d.p., that is 63.426. The final digit is 6 and so we round up 63.426 to 63.43. Hence 63.4261 to 2 d.p. is 63.43.

3.13 Write 1.97 to 1 d.p.

Solution In order to write to 1 d.p. we consider the number to 2 d.p., that is we consider 1.97. The final digit is a 7 and so we round up. The 9 cannot be rounded up and so we look at 1.9. This can be rounded up to 2.0. Hence 1.97 to 1 d.p. is 2.0. Note that it is crucial to write 2.0 and not simply 2, as this shows that the number is written to 1 d.p.

3.14 Write −6.0439 to 2 d.p.

Solution We consider −6.043. As the final digit is a 3 the number is rounded down to −6.04.

Self-assessment questions 3.2

1. Explain the meaning of 'significant figures'.
2. Explain the process of writing a number to so many decimal places.

Exercise 3.2

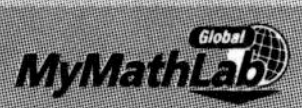

1. Write to 3 s.f.
   (a) 6962 (b) 70.406 (c) 0.0123
   (d) 0.010991 (e) 45.607 (f) 2345
2. Write 65.999 to
   (a) 4 s.f. (b) 3 s.f. (c) 2 s.f.
   (d) 1 s.f. (e) 2 d.p. (f) 1 d.p.
3. Write 9.99 to
   (a) 1 s.f. (b) 1 d.p.
4. Write 65.4555 to
   (a) 3 d.p. (b) 2 d.p. (c) 1 d.p.
   (d) 5 s.f. (e) 4 s.f. (f) 3 s.f. (g) 2 s.f.
   (h) 1 s.f.

Test and assignment exercises 3

1. Express the following numbers as proper fractions in their simplest form:
   (a) 0.74 (b) 0.96 (c) 0.05 (d) 0.25
2. Express each of the following as a mixed fraction in its simplest form:
   (a) 2.5 (b) 3.25 (c) 3.125 (d) 6.875
3. Write each of the following as a decimal number:
   (a) $\frac{3}{10}+\frac{1}{100}+\frac{7}{1000}$ (b) $\frac{5}{1000}+\frac{9}{100}$ (c) $\frac{4}{1000}+\frac{9}{10}$
4. Write 0.09846 to (a) 1 d.p, (b) 2 s.f., (c) 1 s.f.
5. Write 9.513 to (a) 3 s.f., (b) 2 s.f., (c) 1 s.f.
6. Write 19.96 to (a) 1 d.p., (b) 2 s.f., (c) 1 s.f.

Percentage and ratio

4

Objectives: This chapter:

- explains the terms 'percentage' and 'ratio'
- shows how to perform calculations using percentages and ratios
- explains how to calculate the percentage change in a quantity

4.1 Percentage

In everyday life we come across percentages regularly. During sales periods shops offer discounts – for example, we might hear expressions like 'everything reduced by 50%'. Students often receive examination marks in the form of percentages – for example, to achieve a pass grade in a university examination, a student may be required to score at least 40%. Banks and building societies charge interest on loans, and the interest rate quoted is usually given as a percentage, for example 4.75%. Percentages also provide a way of comparing two or more quantities. For example, suppose we want to know which is the better mark: 40 out of 70, or 125 out of 200? By expressing these marks as percentages we will be able to answer this question.

Consequently an understanding of what a percentage is, and an ability to perform calculations involving percentages, are not only useful in mathematical applications, but also essential life skills.

Most calculators have a percentage button and we will illustrate the use of this later in the chapter. However, be aware that different calculators work in different and often confusing ways. Misleading results can be obtained if you do not know how to use your calculator correctly. So it is better if you are not over-reliant on your calculator and instead understand the principles behind percentage calculations.

Fundamentally, a **percentage** is a fraction whose denominator is 100. In fact you can think of the phrase 'per cent' meaning 'out of 100'. We use the symbol % to represent a percentage, as earlier. The following three fractions all have a denominator of 100, and are expressed as percentages as shown:

$\frac{17}{100}$ may be expressed as 17%

$\frac{50}{100}$ may be expressed as 50%

$\frac{3}{100}$ may be expressed as 3%

WORKED EXAMPLE

4.1 Express $\frac{19}{100}$, $\frac{35}{100}$ and $\frac{17.5}{100}$ as percentages.

Solution All of these fractions have a denominator of 100. So it is straightforward to write down their percentage form:

$$\frac{19}{100} = 19\% \qquad \frac{35}{100} = 35\% \qquad \frac{17.5}{100} = 17.5\%$$

Sometimes it is necessary to convert a fraction whose denominator is not 100, for example $\frac{2}{5}$, into a percentage. This could be done by expressing the fraction as an equivalent fraction with denominator 100, as was explained in Section 2.2 on page 15. However, with calculators readily available, the calculation can be done as follows.

We can use the calculator to divide the numerator of the fraction by the denominator. The answer is then multiplied by 100. The resulting number is the required percentage. So, to convert $\frac{2}{5}$ we perform the following key strokes:

$$2 \div 5 \times 100 = 40$$

and so $\frac{2}{5} = 40\%$. You should check this now using your own calculator,

Key point To convert a fraction to a percentage, divide the numerator by the denominator, multiply by 100 and then label the result as a percentage.

WORKED EXAMPLES

4.2 Convert $\frac{5}{8}$ into a percentage.

Solution Using the method described above we find

$$5 \div 8 \times 100 = 62.5$$

Labelling the answer as a percentage, we see that $\frac{5}{8}$ is equivalent to 62.5%.

4.3 Bill scores $\frac{13}{17}$ in a test. In a different test, Mary scores $\frac{14}{19}$. Express the scores as percentages, and thereby make a comparison of the two marks.

Solution Use your calculator to perform the division and then multiply the result by 100.

Bill's score: $13 \div 17 \times 100 = 76.5$ (1 d.p.)

Mary's score: $14 \div 19 \times 100 = 73.7$ (1 d.p.)

So we see that Bill scores 76.5% and Mary scores 73.7%. Notice that in these percentage forms it is easy to compare the two marks. We see that Bill has achieved the higher score. Making easy comparisons like this is one of the reasons why percentages are used so frequently.

We have seen that percentages are fractions with a denominator of 100, so that, for example, $\frac{19}{100} = 19\%$. Sometimes a fraction may be given not as a numerator divided by a denominator, but in its decimal form. For example, the decimal form of $\frac{19}{100}$ is 0.19. To convert a decimal fraction into a percentage we simply multiply by 100. So

$$0.19 = 0.19 \times 100\% = 19\%$$

Key point

To convert a decimal fraction to a percentage, multiply by 100 and then label the result as a percentage.

We may also want to reverse the process. Frequently in business calculations involving formulae for interest it is necessary to express a percentage in its decimal form. To convert a percentage to its equivalent decimal form we divide the percentage by 100. Alternatively, using a calculator, input the percentage and press the % button, to convert the percentage to its decimal form.

WORKED EXAMPLE

4.4 Express 50% as a decimal.

Solution We divide the percentage by 100:

$$50 \div 100 = 0.5$$

So 50% is equivalent to 0.5. To see why this is the case, remember that 'per cent' literally means 'out of 100' so 50% means 50 out of 100, or $\frac{50}{100}$, or in its simplest form 0.5.

Alternatively, using a calculator, the key strokes

50 %

should give 0.5. Check whether you can do this on your calculator.

Key point

To convert a percentage to its equivalent decimal fraction form, divide by 100.

WORKED EXAMPLE

4.5 Express 17.5% as a decimal.

Solution We divide the percentage by 100:

$17.5 \div 100 = 0.175$

So 17.5% is equivalent to 0.175. Now check you can obtain the same result using the percentage button on your calculator.

Some percentages appear so frequently in everyday life that it is useful to learn their fraction and decimal fraction equivalent forms.

Key point

$10\% = 0.1 = \frac{1}{10}$ $\quad 25\% = 0.25 = \frac{1}{4}$

$50\% = 0.5 = \frac{1}{2}$ $\quad 75\% = 0.75 = \frac{3}{4}$ $\quad 100\% = 1$

Recall from Section 1.2 that 'of' means multiply.

We are often asked to calculate a percentage of a quantity: for example, find 17.5% of 160 or 10% of 95. Such calculations arise when finding discounts on prices. Since $17.5\% = \frac{17.5}{100}$ we find

$$17.5\% \text{ of } 160 = \frac{17.5}{100} \times 160 = 28$$

and since $10\% = 0.1$ we may write

$$10\% \text{ of } 95 = 0.1 \times 95 = 9.5$$

Alternatively, the percentage button on a calculator can be used: check you can use your calculator correctly by verifying

Because finding 10% of a quantity is equivalent to dividing by 10, it is easy to find 10% by moving the decimal point one place to the left.

17.5 % $\times\ 160 = 28$

10 % $\times\ 95 = 9.5$

WORKED EXAMPLES

4.6 Calculate 27% of 90.

Solution Using a calculator

$27\ \boxed{\%}\ \times 90 = 24.3$

4.7 Calculate 100% of 6.

Solution

$100\ \boxed{\%}\ \times 6 = 6$

Observe that 100% of a number is simply the number itself.

4.8 A deposit of £750 increases by 9%. Calculate the resulting deposit.

Solution We use a calculator to find 9% of 750. This is the amount by which the deposit has increased. Then

$9\ \boxed{\%}\ \times 750 = 67.50$

The deposit has increased by £67.50. The resulting deposit is therefore $750 + 67.5 = £817.50$.

Alternatively we may perform the calculation as follows. The original deposit represents 100%. The deposit increases by 9% to 109% of the original. So the resulting deposit is 109% of £750:

$109\ \boxed{\%}\ \times 750 = £817.50$

4.9 A television set is advertised at £315. The retailer offers a 10% discount. How much do you pay for the television?

Solution

10% of 315 = 31.50

The discount is £31.50 and so the cost is $315 - 31.5 = £283.50$.

Alternatively we can note that since the discount is 10%, then the selling price is 90% of the advertised price:

$90\ \boxed{\%}\ \times 315 = 283.50$

Performing the calculation in the two ways will increase your understanding of percentages and serve as a check.

When a quantity changes, it is sometimes useful to calculate the **percentage change**. For example, suppose a worker earns £14,500 in the current year, and last year earned £13,650. The actual amount earned has changed by 14,500 − 13,650 = £850. The percentage change is calculated from the formula:

Key point

$$\text{percentage change} = \frac{\text{change}}{\text{original value}} \times 100 = \frac{\text{new value} - \text{original value}}{\text{original value}} \times 100$$

If the change is positive, then there has been an increase in the measured quantity. If the change is negative, then there has been a decrease in the quantity.

WORKED EXAMPLES

4.10 A worker's earnings increase from £13,650 to £14,500. Calculate the percentage change.

Solution

$$\begin{aligned}\text{percentage change} &= \frac{\text{new value} - \text{original value}}{\text{original value}} \times 100 \\ &= \frac{14,500 - 13,650}{13,650} \times 100 \\ &= 6.23\end{aligned}$$

The worker's earnings increased by 6.23%.

4.11 A microwave oven is reduced in price from £149.95 to £135. Calculate the percentage change in price.

Solution

$$\begin{aligned}\text{percentage change} &= \frac{\text{new value} - \text{original value}}{\text{original value}} \times 100 \\ &= \frac{135 - 149.95}{149.95} \times 100 \\ &= -9.97\end{aligned}$$

The negative result is indicative of the price decrease. The percentage change in price is approximately −10%.

Self-assessment question 4.1

1. Give one reason why it is sometimes useful to express fractions as percentages.

Exercise 4.1

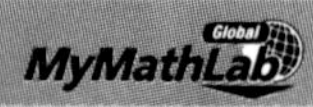

1. Calculate 23% of 124.
2. Express the following as percentages:

   (a) $\frac{9}{11}$ (b) $\frac{15}{20}$ (c) $\frac{9}{10}$ (d) $\frac{45}{50}$ (e) $\frac{75}{90}$
3. Express $\frac{13}{12}$ as a percentage.
4. Calculate 217% of 500.
5. A worker earns £400 a week. She receives a 6% increase. Calculate her new weekly wage.
6. A debt of £1200 is decreased by 17%. Calculate the remaining debt.
7. Express the following percentages as decimals:

   (a) 50% (b) 36% (c) 75%
   (d) 100% (e) 12.5%
8. A compact disc player normally priced at £256 is reduced in a sale by 20%. Calculate the sale price.
9. A bank deposit earns 7.5% interest in one year. Calculate the interest earned on a deposit of £15, 000.
10. The cost of a car is increased from £6950 to £7495. Calculate the percentage change in price.
11. During a sale, a washing machine is reduced in price from £525 to £399. Calculate the percentage change in price.

4.2 Ratio

Ratios are simply an alternative way of expressing fractions. Consider the problem of dividing £200 between two people, Ann and Bill, in the ratio of 7 : 3. This means that Ann receives £7 for every £3 that Bill receives. So every £10 is divided as £7 to Ann and £3 to Bill. Thus Ann receives $\frac{7}{10}$ of the money. Now $\frac{7}{10}$ of £200 is $\frac{7}{10} \times 200 = 140$. So Ann receives £140 and Bill receives £60.

WORKED EXAMPLE

4.12 Divide 170 in the ratio 3 : 2.

Solution A ratio of 3 : 2 means that every 5 parts are split as 3 and 2. That is, the first number is $\frac{3}{5}$ of the total; the second number is $\frac{2}{5}$ of the total. So

$$\frac{3}{5} \text{ of } 170 = \frac{3}{5} \times 170 = 102$$

$$\frac{2}{5} \text{ of } 170 = \frac{2}{5} \times 170 = 68$$

The number is divided into 102 and 68.

Note from Worked Example 4.12 that to split a number in a given ratio we first find the total number of parts. The total number of parts is found by adding the numbers in the ratio. For example, if the ratio is given as $m:n$, the total number of parts is $m+n$. Then these $m+n$ parts are split into two with the first number being $\frac{m}{m+n}$ of the total, and the second number being $\frac{n}{m+n}$ of the total. Compare this with Worked Example 4.12.

WORKED EXAMPLE

4.13 Divide 250 cm in the ratio 1 : 3 : 4.

Solution Every 8 cm is divided into 1 cm, 3 cm and 4 cm. Thus the first length is $\frac{1}{8}$ of the total, the second length is $\frac{3}{8}$ of the total, and the third length is $\frac{4}{8}$ of the total:

$$\frac{1}{8} \text{ of } 250 = \frac{1}{8} \times 250 = 31.25$$

$$\frac{3}{8} \text{ of } 250 = \frac{3}{8} \times 250 = 93.75$$

$$\frac{4}{8} \text{ of } 250 = \frac{4}{8} \times 250 = 125$$

The 250 cm length is divided into 31.25 cm, 93.75 cm and 125 cm.

Ratios can be written in different ways. The ratio 3 : 2 can also be written as 6 : 4. This is clear if we note that 6 : 4 is a total of 10 parts split as $\frac{6}{10}$ and $\frac{4}{10}$ of the total. Since $\frac{6}{10}$ is equivalent to $\frac{3}{5}$, and $\frac{4}{10}$ is equivalent to $\frac{2}{5}$, we see that 6 : 4 is equivalent to 3 : 2.

Generally, any ratio can be expressed as an equivalent ratio by multiplying or dividing each term in the ratio by the same number. So, for example,

5 : 3 is equivalent to 15 : 9

and

$\frac{3}{4}$: 2 is equivalent to 3 : 8

WORKED EXAMPLES

4.14 Divide a mass of 380 kg in the ratio $\frac{3}{4} : \frac{1}{5}$.

Solution It is simpler to work with whole numbers, so first of all we produce an equivalent ratio by multiplying each term, first by 4, and then by 5, to give

$$\frac{3}{4} : \frac{1}{5} = 3 : \frac{4}{5} = 15 : 4$$

Note that this is equivalent to multiplying through by the lowest common multiple of 4 and 5.

So dividing 380 kg in the ratio $\frac{3}{4} : \frac{1}{5}$ is equivalent to dividing it in the ratio 15 : 4.

Now the total number of parts is 19 and so we split the 380 kg mass as

$$\frac{15}{19} \times 380 = 300$$

and

$$\frac{4}{19} \times 380 = 80$$

The total mass is split into 300 kg and 80 kg.

4.15 Bell metal, which is a form of bronze, is used for casting bells. It is an alloy of copper and tin. To manufacture bell metal requires 17 parts of copper to every 3 parts of tin.

(a) Express this requirement as a ratio.

(b) Express the amount of tin required as a percentage of the total.

(c) If the total amount of tin in a particular casting is 150 kg, find the amount of copper.

Solution

(a) Copper and tin are needed in the ratio 17 : 3.

(b) $\frac{3}{20}$ of the alloy is tin. Since $\frac{3}{20} = 15\%$ we find that 15% of the alloy is tin.

(c) A mass of 150 kg of tin makes up 15% of the total. So 1% of the total would have a mass of 10 kg. Copper, which makes up 85%, will have a mass of 850 kg.

Self-assessment question 4.2

1. Dividing a number in the ratio 2:3 is the same as dividing it in the ratio 10:15. True or false?

Exercise 4.2

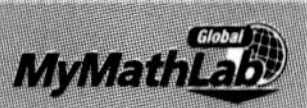

1. Divide 180 in the ratio 8:1:3.
2. Divide 930 cm in the ratio 1:1:3.
3. A 6 m length of wood is cut in the ratio 2:3:4. Calculate the length of each piece.
4. Divide 1200 in the ratio 1:2:3:4.
5. A sum of £2600 is divided between Alan, Bill and Claire in the ratio of $2\frac{3}{4} : 1\frac{1}{2} : 2\frac{1}{4}$. Calculate the amount that each receives.
6. A mass of 40 kg is divided into three portions in the ratio 3:4:8. Calculate the mass of each portion.
7. Express the following ratios in their simplest forms:
   (a) 12:24 (b) 3:6 (c) 3:6:12
   (d) $\frac{1}{3} : 7$
8. A box contains two sizes of nails. The ratio of long nails to short nails is 2:7. Calculate the number of each type if the total number of nails is 108.

Test and assignment exercises 4

1. Express as decimals
   (a) 8% (b) 18% (c) 65%
2. Express as percentages
   (a) $\frac{3}{8}$ (b) $\frac{79}{100}$ (c) $\frac{56}{118}$
3. Calculate 27.3% of 1496.
4. Calculate 125% of 125.
5. Calculate 85% of 0.25.
6. Divide 0.5 in the ratio 2:4:9.
7. A bill totals £234.5 to which is added tax at 17.5%. Calculate the amount of tax to be paid.
8. An inheritance is divided between three people in the ratio 4:7:2. If the least amount received is £2300 calculate how much the other two people received.

9. Divide 70 in the ratio of 0.5 : 1.3 : 2.1.
10. Divide 50% in the ratio 2 : 3.
11. The temperature of a liquid is reduced from 39 °C to 35 °C. Calculate the percentage change in temperature.
12. A jacket priced at £120 is reduced by 30% in a sale. Calculate the sale price of the jacket.
13. The price of a car is reduced from £7250 to £6450. Calculate the percentage change in price.
14. The population of a small town increases from 17296 to 19437 over a five-year period. Calculate the percentage change in population.
15. A number, X, is increased by 20% to form a new number Y. Y is then decreased by 20% to form a third number Z. Express Z in terms of X.

5 Algebra

Objectives This chapter:

- explains what is meant by 'algebra'
- introduces important algebraic notations
- explains what is meant by a 'power' or 'index'
- illustrates how to evaluate an expression
- explains what is meant by a 'formula'

5.1 What is algebra?

In order to extend the techniques of arithmetic so that they can be more useful in applications we introduce letters or **symbols** to represent quantities of interest. For example, we may choose the capital letter I to stand for the *interest rate* in a business calculation, or the lower case letter t to stand for the *time* in a scientific calculation, and so on. The choice of which letter to use for which quantity is largely up to the user, although some conventions have been developed. Very often the letters x and y are used to stand for arbitrary quantities. **Algebra** is the body of mathematical knowledge that has been developed to manipulate symbols. Some symbols take fixed and unchanging values, and these are known as **constants**. For example, suppose we let the symbol b stand for the boiling point of water. This is fixed at 100 °C and so b is a constant. Some symbols represent quantities that can vary, and these are called **variables**. For example, the velocity of a car might be represented by the symbol v, and might vary from 0 to 100 kilometres per hour.

Algebraic notation

In algebraic work particular attention must be paid to the type of symbol used, so that, for example, the symbol T is quite different from the symbol t.

Table 5.1
The Greek alphabet

A	α	alpha	I	ι	iota	P	ρ	rho
B	β	beta	K	κ	kappa	Σ	σ	sigma
Γ	γ	gamma	Λ	λ	lambda	T	τ	tau
Δ	δ	delta	M	μ	mu	Y	υ	upsilon
E	ε	epsilon	N	ν	nu	Φ	ϕ	phi
Z	ζ	zeta	Ξ	ξ	xi	X	χ	chi
H	η	eta	O	o	omicron	Ψ	ψ	psi
Θ	θ	theta	Π	π	pi	Ω	ω	omega

Your scientific calculator is pre-programmed with the value of π. Check that you can use it.

Usually the symbols chosen are letters from the English alphabet although we frequently meet Greek letters. You may already be aware that the Greek letter 'pi', which has the symbol π, is used in the formula for the area of a circle, and is equal to the constant 3.14159.... In many calculations π can be approximated by $\frac{22}{7}$. For reference the full Greek alphabet is given in Table 5.1.

Another important feature is the position of a symbol in relation to other symbols. As we shall see in this chapter, the quantities xy, x^y, y^x and x_y all can mean quite different things. When a symbol is placed to the right and slightly higher than another symbol it is referred to as a **superscript**. So the quantity x^y contains the superscript y. Likewise, if a symbol is placed to the right and slightly lower than another symbol it is called a **subscript**. The quantity x_1 contains the subscript 1.

The arithmetic of symbols

Addition (+) If the letters x and y stand for two numbers, their **sum** is written as $x + y$. Note that $x + y$ is the same as $y + x$ just as $4 + 7$ is the same as $7 + 4$.

Subtraction (−) The quantity $x - y$ is called the **difference** of x and y, and means the number y subtracted from the number x. Note that $x - y$ is not the same as $y - x$, in the same way that $5 - 3$ is different from $3 - 5$.

Multiplication (×) Five times the number x is written $5 \times x$, although when multiplying the × sign is sometimes replaced with '·', or is even left out altogether. This means that $5 \times x$, $5 \cdot x$ and $5x$ all mean five times the number x. Similarly $x \times y$ can be written $x \cdot y$ or simply xy. When multiplying, the order of the symbols is not important, so that xy is the same as yx just as 5×4 is the same as 4×5. The quantity xy is also known as the **product** of x and y.

Division (÷) $x \div y$ means the number x divided by the number y. This is also written x/y. Here the order is important and x/y is quite different from y/x. An expression involving one symbol divided by another is known as an **algebraic fraction**. The top line is called the **numerator** and the bottom line is called the **denominator**. The quantity x/y is known as the **quotient** of x and y.

A quantity made up of symbols together with $+$, $-$, $\times$ or $\div$ is called an **algebraic expression**. When evaluating an algebraic expression the BODMAS rule given in Chapter 1 applies. This rule reminds us of the correct order in which to evaluate an expression.

Self-assessment questions 5.1

1. Explain what you understand by the term 'algebra'.
2. If m and n are two numbers, explain what is meant by mn.
3. What is an algebraic fraction? Explain the meaning of the terms 'numerator' and 'denominator'.
4. What is the distinction between a superscript and a subscript?
5. What is the distinction between a variable and a constant?

5.2 Powers or indices

Frequently we shall need to multiply a number by itself several times, for example $3 \times 3 \times 3$, or $a \times a \times a \times a$.

To abbreviate such quantities a new notation is introduced. $a \times a \times a$ is written a^3, pronounced 'a cubed'. The superscript 3 is called a **power** or **index** and the letter a is called the **base**. Similarly $a \times a$ is written a^2, pronounced 'a squared' or 'a raised to the power 2'.

The calculator button x^y is used to find powers of numbers.

Most calculators have a button marked x^y, which can be used to evaluate expressions such as 2^8, 3^{11} and so on. Check to see whether your calculator can do these by verifying that $2^8 = 256$ and $3^{11} = 177147$. Note that the plural of index is **indices**.

As a^2 means $a \times a$, and a^3 means $a \times a \times a$, then we interpret a^1 as simply a. That is, any number raised to the power 1 is itself.

Key point

Any number raised to the power 1 is itself, that is $a^1 = a$.

WORKED EXAMPLES

5.1 In the expression 3^8 identify the index and the base.

Solution In the expression 3^8, the index is 8 and the base is 3.

5.2 Explain what is meant by y^5.

Solution y^5 means $y \times y \times y \times y \times y$.

5.3 Explain what is meant by x^2y^3.

Solution x^2 means $x \times x$; y^3 means $y \times y \times y$. Therefore x^2y^3 means $x \times x \times y \times y \times y$.

5.4 Evaluate 2^3 and 3^4.

Solution 2^3 means $2 \times 2 \times 2$, that is 8. Similarly 3^4 means $3 \times 3 \times 3 \times 3$, that is 81.

5.5 Explain what is meant by 7^1.

Solution Any number to the power 1 is itself, that is 7^1 is simply 7.

5.6 Evaluate 10^2 and 10^3.

Solution 10^2 means 10×10 or 100. Similarly 10^3 means $10 \times 10 \times 10$ or 1000.

5.7 Use indices to write the expression $a \times a \times b \times b \times b$ more compactly.

Solution $a \times a$ can be written a^2; $b \times b \times b$ can be written b^3. Therefore $a \times a \times b \times b \times b$ can be written as $a^2 \times b^3$ or simply a^2b^3.

5.8 Write out fully z^3y^2.

Solution z^3y^2 means $z \times z \times z \times y \times y$. Note that we could also write this as $zzzyy$.

We now consider how to deal with expressions involving not only powers but other operations as well. Recall from §5.1 that the BODMAS rule tells us the order in which operations should be carried out, but the rule makes no reference to powers. In fact, powers should be given higher priority than any other operation and evaluated first. Consider the expression -4^2. Because the power must be evaluated first -4^2 is equal to -16. On the other hand $(-4)^2$ means $(-4) \times (-4)$ which is equal to $+16$.

WORKED EXAMPLES

5.9 Simplify (a) -5^2, (b) $(-5)^2$.

Solution

(a) The power is evaluated first. Noting that $5^2 = 25$, we see that $-5^2 = -25$.

(b) $(-5)^2$ means $(-5) \times (-5) = +25$.

Recall that when a negative number is multiplied by another negative number the result is positive.

Note how the brackets can significantly change the meaning of an expression.

5.10 Explain the meanings of $-x^2$ and $(-x)^2$. Are these different?

Solution In the expression $-x^2$ it is the quantity x that is squared, so that $-x^2 = -(x \times x)$. On the other hand $(-x)^2$ means $(-x) \times (-x)$, which equals $+x^2$. The two expressions are not the same.

Following the previous two examples we emphasise again the importance of the position of brackets in an expression.

Self-assessment questions 5.2

1. Explain the meaning of the terms 'power' and 'base'.
2. What is meant by an index?
3. Explain the distinction between $(xyz)^2$ and xyz^2.
4. Explain the distinction between $(-3)^4$ and -3^4.

Exercise 5.2

MyMathLab Global

1. Evaluate the following without using a calculator: 2^4, $(\frac{1}{2})^2$, 1^8, 3^5 and 0^3.
2. Evaluate 10^4, 10^5 and 10^6 without using a calculator.
3. Use a calculator to evaluate 11^4, 16^8, 39^4 and 1.5^7.
4. Write out fully (a) a^4b^2c and (b) xy^2z^4.
5. Write the following expressions compactly using indices:
   (a) $xxxyyx$ (b) $xxyyzzz$
   (c) $xyzxyz$ (d) $abccba$
6. Using a calculator, evaluate
   (a) 7^4 (b) 7^5 (c) $7^4 \times 7^5$ (d) 7^9
   (e) 8^3 (f) 8^7 (g) $8^3 \times 8^7$ (h) 8^{10}
   Can you spot a rule for multiplying numbers with powers?
7. Without using a calculator, find $(-3)^3$, $(-2)^2$, $(-1)^7$ and $(-1)^4$.
8. Use a calculator to find $(-16.5)^3$, $(-18)^2$ and $(-0.5)^5$.
9. Without using a calculator find
   (a) $(-6)^2$ (b) $(-3)^2$ (c) $(-4)^3$
   (d) $(-2)^3$
   Carefully compare your answers with the results of finding -6^2, -3^2, -4^3 and -2^3.

5.3 Substitution and formulae

Substitution means replacing letters by actual numerical values.

WORKED EXAMPLES

5.11 Find the value of a^4 when $a = 3$.

Solution a^4 means $a \times a \times a \times a$. When we **substitute** the number 3 in place of the letter a we find 3^4 or $3 \times 3 \times 3 \times 3$, that is 81.

5.12 Find the value of $a + 7b + 3c$ when $a = 1$, $b = 2$ and $c = 3$.

Solution Letting $b = 2$ we note that $7b = 14$. Letting $c = 3$ we note that $3c = 9$. Therefore, with $a = 1$,

$$a + 7b + 3c = 1 + 14 + 9 = 24$$

5.13 If $x = 4$, find the value of (a) $8x^3$ and (b) $(8x)^3$.

Solution
(a) Substituting $x = 4$ into $8x^3$ we find $8 \times 4^3 = 8 \times 64 = 512$.
(b) Substituting $x = 4$ into $(8x)^3$ we obtain $(32)^3 = 32768$. Note that the use of brackets makes a significant difference to the result.

5.14 Evaluate mk, mn and nk when $m = 5$, $n = -4$ and $k = 3$.

Solution $mk = 5 \times 3 = 15$. Similarly $mn = 5 \times (-4) = -20$ and $nk = (-4) \times 3 = -12$.

5.15 Find the value of $-7x$ when (a) $x = 2$ and (b) $x = -2$.

Solution
(a) Substituting $x = 2$ into $-7x$ we find -7×2, which equals -14.
(b) Substituting $x = -2$ into $-7x$ we find -7×-2, which equals 14.

5.16 Find the value of x^2 when $x = -3$.

Solution Because x^2 means $x \times x$, its value when $x = -3$ is -3×-3, that is $+9$.

5.17 Find the value of $-x^2$ when $x = -3$.

Solution Recall that a power is evaluated first. So $-x^2$ means $-(x \times x)$. When $x = -3$, this evaluates to $-(-3 \times -3) = -9$.

5.18 Find the value of $x^2 + 3x$ when (a) $x = 2$, (b) $x = -2$.

Solution
(a) Letting $x = 2$ we find

$$x^2 + 3x = (2)^2 + 3(2) = 4 + 6 = 10$$

(b) Letting $x = -2$ we find

$$x^2 + 3x = (-2)^2 + 3(-2) = 4 - 6 = -2$$

5.19 Find the value of $\frac{3x^2}{4} + 5x$ when $x = 2$.

Solution Letting $x = 2$ we find

$$\frac{3x^2}{4} + 5x = \frac{3(2)^2}{4} + 5(2)$$

$$= \frac{12}{4} + 10$$

$$= 13$$

5.20 Find the value of $\frac{x^3}{4}$ when $x = 0.5$.

Solution When $x = 0.5$ we find

$$\frac{x^3}{4} = \frac{0.5^3}{4} = 0.03125$$

A **formula** is used to relate two or more quantities. You may already be familiar with the common formula used to find the area of a rectangle:

area = length × breadth

In symbols, writing A for area, l for length and b for breadth we have

$A = l \times b$ or simply $A = lb$

If we are now given particular numerical values for l and b we can use this formula to find A.

WORKED EXAMPLES

5.21 Use the formula $A = lb$ to find A when $l = 10$ and $b = 2.5$.

Solution Substituting the values $l = 10$ and $b = 2.5$ into the formula $A = lb$ we find $A = 10 \times 2.5 = 25$.

5.22 The formula $V = IR$ is used by electrical engineers. Find the value of V when $I = 12$ and $R = 7$.

Solution Substituting $I = 12$ and $R = 7$ in $V = IR$ we find $V = 12 \times 7 = 84$.

5.23 Use the formula $y = x^2 + 3x + 4$ to find y when $x = -2$.

Solution Substituting $x = -2$ into the formula gives

$$y = (-2)^2 + 3(-2) + 4 = 4 - 6 + 4 = 2$$

Self-assessment question 5.3

1. What is the distinction between an algebraic expression and a formula?

Exercise 5.3

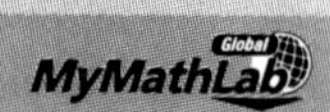

1. Evaluate $3x^2y$ when $x = 2$ and $y = 5$.
2. Evaluate $8x + 17y - 2z$ when $x = 6$, $y = 1$ and $z = -2$.
3. The area A of a circle is found from the formula $A = \pi r^2$, where r is the length of the radius. Taking π to be 3.142 find the areas of the circles whose radii, in centimetres, are (a) $r = 10$, (b) $r = 3$, (c) $r = 0.2$.
4. Evaluate $3x^2$ and $(3x)^2$ when $x = 4$.
5. Evaluate $5x^2$ and $(5x)^2$ when $x = -2$.
6. If $y = 4.85$ find
   (a) $7y$ (b) y^2 (c) $5y + 2.5$
   (d) $y^3 - y$
7. If $a = 12.8$, $b = 3.6$ and $c = 9.1$ find
   (a) $a + b + c$ (b) ab (c) bc (d) abc
8. If $C = \frac{5}{9}(F - 32)$, find C when $F = 100$.
9. Evaluate (a) x^2, (b) $-x^2$ and (c) $(-x)^2$, when $x = 7$.
10. Evaluate the following when $x = -2$:
    (a) x^2 (b) $(-x)^2$ (c) $-x^2$
    (d) $3x^2$ (e) $-3x^2$ (f) $(-3x)^2$
11. Evaluate the following when $x = -3$:
    (a) $\frac{x^2}{3}$ (b) $(-x)^2$ (c) $-(\frac{x}{3})^2$
    (d) $4x^2$ (e) $-4x^2$ (f) $(-4x)^2$
12. Evaluate $x^2 - 7x + 2$ when $x = -9$.
13. Evaluate $2x^2 + 3x - 11$ when $x = -3$. VIDEO
14. Evaluate $-x^2 + 3x - 5$ when $x = -1$.
15. Evaluate $-9x^2 + 2x$ when $x = 0$.
16. Evaluate $5x^2 + x + 1$ when (a) $x = 3$, (b) $x = -3$, (c) $x = 0$, (d) $x = -1$.
17. Evaluate $\frac{2x^2}{3} - \frac{x}{2}$ when
    (a) $x = 6$ (b) $x = -6$ (c) $x = 0$
    (d) $x = 1$
18. Evaluate $\frac{4x^2}{5} + 3$ when
    (a) $x = 0$ (b) $x = 1$ (c) $x = 5$
    (d) $x = -5$
19. Evaluate $\frac{x^3}{2}$ when
    (a) $x = -1$ (b) $x = 2$ (c) $x = 4$
20. Use the formula $y = \frac{x^3}{2} + 3x^2$ to find y when
    (a) $x = 0$ (b) $x = 2$ (c) $x = 3$
    (d) $x = -1$
21. If $g = 2t^2 - 1$, find g when
    (a) $t = 3$ (b) $t = 0.5$ (c) $t = -2$
22. In business calculations, the simple interest earned on an investment, I, is calculated from the formula $I = Prn$, where P is the amount invested, r is the interest rate and n is the number of time periods. Evaluate I when
    (a) $P = 15000$, $r = 0.08$ and $n = 5$
    (b) $P = 12500$, $r = 0.075$ and $n = 3$.
23. An investment earning 'compound interest' has a value, S, given by $S = P(1 + r)^n$, where P is the amount invested, r is the interest rate and n is the number of time periods. Calculate S when
    (a) $P = 8250$, $r = 0.05$ and $n = 15$
    (b) $P = 125000$, $r = 0.075$ and $n = 11$.

Test and assignment exercises 5

1. Using a calculator, evaluate 44^3, 0.44^2 and 32.5^3.
2. Write the following compactly using indices:
   (a) $xxxyyyyy$ (b) $\dfrac{xxx}{yyyy}$ (c) a^2baab
3. Evaluate the expression $4x^3yz^2$ when $x = 2$, $y = 5$ and $z = 3$.
4. The circumference C of a circle that has a radius of length r is given by the formula $C = 2\pi r$. Find the circumference of the circle with radius 0.5 cm. Take $\pi = 3.142$.
5. Find (a) $21^2 - 16^2$, (b) $(21 - 16)^2$. Comment upon the result.
6. If $x = 4$ and $y = -3$, evaluate
   (a) xy (b) $\dfrac{x}{y}$ (c) $\dfrac{x^2}{y^2}$ (d) $\left(\dfrac{x}{y}\right)^2$
7. Evaluate $2x(x + 4)$ when $x = 7$.
8. Evaluate $4x^2 + 7x$ when $x = 9$.
9. Evaluate $3x^2 - 7x + 12$ when $x = -2$.
10. Evaluate $-x^2 - 11x + 1$ when $x = -3$.
11. The formula $I = V/R$ is used by engineers. Find I when $V = 10$ and $R = 0.01$.
12. Given the formula $A = 1/x$, find A when (a) $x = 1$, (b) $x = 2$, (c) $x = 3$.
13. From the formula $y = 1/(x^2 + x)$ find y when (a) $x = 1$, (b) $x = -1$, (c) $x = 3$.
14. Find the value of $(-1)^n$ (a) when n is an even natural number and (b) when n is an odd natural number. (A natural number is a positive whole number.)
15. Find the value of $(-1)^{n+1}$ (a) when n is an even natural number and (b) when n is an odd natural number.

6 Indices

Objectives: This chapter:

- states three laws used for manipulating indices
- shows how expressions involving indices can be simplified using the three laws
- explains the use of negative powers
- explains square roots, cube roots and fractional powers
- revises multiplication and division by powers of 10
- explains 'scientific notation' for representing very large and very small numbers

6.1 The laws of indices

Recall from Chapter 5 that an index is simply a power and that the plural of index is indices. Expressions involving indices can often be simplified if use is made of the **laws of indices**.

The first law

$$a^m \times a^n = a^{m+n}$$

In words, this states that if two numbers involving the same base but possibly different indices are to be multiplied together, their indices are added. Note that this law can be applied only if both bases are the same.

Key point The first law: $a^m \times a^n = a^{m+n}$.

WORKED EXAMPLES

6.1 Use the first law of indices to simplify $a^4 \times a^3$.

Solution Using the first law we have $a^4 \times a^3 = a^{4+3} = a^7$. Note that the same result could be obtained by actually writing out all the terms:

$$a^4 \times a^3 = (a \times a \times a \times a) \times (a \times a \times a) = a^7$$

6.2 Use the first law of indices to simplify $3^4 \times 3^5$.

Solution From the first law $3^4 \times 3^5 = 3^{4+5} = 3^9$.

6.3 Simplify $a^4a^7b^2b^4$.

Solution $a^4a^7b^2b^4 = a^{4+7}b^{2+4} = a^{11}b^6$. Note that only those quantities with the same base can be combined using the first law.

The second law

$$\frac{a^m}{a^n} = a^{m-n}$$

In words, this states that if two numbers involving the same base but possibly different indices are to be divided, their indices are subtracted.

Key point

The second law: $\dfrac{a^m}{a^n} = a^{m-n}$.

WORKED EXAMPLES

6.4 Use the second law of indices to simplify $\frac{a^5}{a^3}$.

Solution The second law states that we subtract the indices, that is

$$\frac{a^5}{a^3} = a^{5-3} = a^2$$

6.5 Use the second law of indices to simplify $\frac{3^7}{3^4}$.

Solution From the second law, $\frac{3^7}{3^4} = 3^{7-4} = 3^3$.

6.6 Using the second law of indices, simplify $\frac{x^3}{x^3}$.

Solution Using the second law of indices we have $\frac{x^3}{x^3} = x^{3-3} = x^0$. However, note that any expression divided by itself equals 1, and so $\frac{x^3}{x^3}$ must equal 1. We can conclude from this that any number raised to the power 0 equals 1.

Key point

Any number raised to the power 0 equals 1, that is $a^0 = 1$.

WORKED EXAMPLE

6.7 Evaluate (a) 14^0, (b) 0.5^0.

Solution (a) Any number to the power 0 equals 1 and so $14^0 = 1$.
(b) Similarly, $0.5^0 = 1$.

The third law

$$(a^m)^n = a^{mn}$$

If a number is raised to a power, and the result is itself raised to a power, then the two powers are multiplied together.

Key point The third law: $(a^m)^n = a^{mn}$.

WORKED EXAMPLES

6.8 Simplify $(3^2)^4$.

Solution The third law states that the two powers are multiplied:

$$(3^2)^4 = 3^{2\times 4} = 3^8$$

6.9 Simplify $(x^4)^3$.

Solution Using the third law:

$$(x^4)^3 = x^{4\times 3} = x^{12}$$

6.10 Remove the brackets from the expression $(2a^2)^3$.

Solution $(2a^2)^3$ means $(2a^2) \times (2a^2) \times (2a^2)$. We can write this as

$$2 \times 2 \times 2 \times a^2 \times a^2 \times a^2$$

or simply $8a^6$. We could obtain the same result by noting that both terms in the brackets, that is the 2 and the a^2, must be raised to the power 3, that is

$$(2a^2)^3 = 2^3(a^2)^3 = 8a^6$$

The result of the previous example can be generalised to any term of the form $(a^m b^n)^k$. To simplify such an expression we make use of the formula $(a^m b^n)^k = a^{mk} b^{nk}$.

Key point $(a^m b^n)^k = a^{mk} b^{nk}$

WORKED EXAMPLE

6.11 Remove the brackets from the expression $(x^2y^3)^4$.

Solution Using the previous result we find

$$(x^2y^3)^4 = x^8y^{12}$$

We often need to use several laws of indices in one example.

WORKED EXAMPLES

6.12 Simplify $\frac{(x^3)^4}{x^2}$.

Solution

$$(x^3)^4 = x^{12} \qquad \text{using the third law of indices}$$

So

$$\frac{(x^3)^4}{x^2} = \frac{x^{12}}{x^2} = x^{10} \qquad \text{using the second law}$$

6.13 Simplify $(t^4)^2(t^2)^3$.

Solution

$$(t^4)^2 = t^8, \qquad (t^2)^3 = t^6 \qquad \text{using the third law}$$

So

$$(t^4)^2(t^2)^3 = t^8t^6 = t^{14} \qquad \text{using the first law}$$

Self-assessment questions 6.1

1. State the three laws of indices.
2. Explain what is meant by a^0.
3. Explain what is meant by x^1.

Exercise 6.1

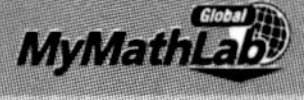

1. Simplify
   (a) $5^7 \times 5^{13}$ (b) $9^8 \times 9^5$
   (c) $11^2 \times 11^3 \times 11^4$
2. Simplify
   (a) $\frac{15^3}{15^2}$ (b) $\frac{4^{18}}{4^9}$ (c) $\frac{5^{20}}{5^{19}}$
3. Simplify
   (a) a^7a^3 (b) a^4a^5 (c) $b^{11}b^{10}b$
4. Simplify
   (a) $x^7 \times x^8$ (b) $y^4 \times y^8 \times y^9$
5. Explain why the laws of indices cannot be used to simplify $19^8 \times 17^8$.

6. Simplify
   (a) $(7^3)^2$ (b) $(4^2)^8$ (c) $(7^9)^2$

7. Simplify $\dfrac{1}{(5^3)^8}$.

VIDEO

8. Simplify
   (a) $(x^2y^3)(x^3y^2)$ (b) $(a^2bc^2)(b^2ca)$

9. Remove the brackets from
   (a) $(x^2y^4)^5$ (b) $(9x^3)^2$ (c) $(-3x)^3$
   (d) $(-x^2y^3)^4$

10. Simplify

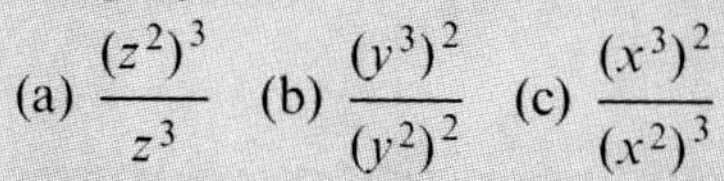

(a) $\dfrac{(z^2)^3}{z^3}$ (b) $\dfrac{(y^3)^2}{(y^2)^2}$ (c) $\dfrac{(x^3)^2}{(x^2)^3}$

6.2 Negative powers

Sometimes a number is raised to a negative power. This is interpreted as follows:

$$a^{-m} = \frac{1}{a^m}$$

This can also be rearranged and expressed in the form

$$a^m = \frac{1}{a^{-m}}$$

Key point

$$a^{-m} = \frac{1}{a^m}, \qquad a^m = \frac{1}{a^{-m}}$$

For example,

$$3^{-2} \text{ means } \frac{1}{3^2}, \text{ that is } \frac{1}{9}$$

Similarly,

$$\text{the number } \frac{1}{5^{-2}} \text{ can be written } 5^2, \text{ or simply } 25$$

To see the justification for this, note that because any number raised to the power 0 equals 1 we can write

$$\frac{1}{a^m} = \frac{a^0}{a^m}$$

Using the second law of indices to simplify the right-hand side we obtain $\frac{a^0}{a^m} = a^{0-m} = a^{-m}$ so that $\frac{1}{a^m}$ is the same as a^{-m}.

WORKED EXAMPLES

6.14 Evaluate

(a) 2^{-5} (b) $\dfrac{1}{3^{-4}}$

Solution (a) $2^{-5} = \dfrac{1}{2^5} = \dfrac{1}{32}$ (b) $\dfrac{1}{3^{-4}} = 3^4$ or simply 81

6.15 Evaluate

(a) 10^{-1} (b) 10^{-2}

Solution (a) 10^{-1} means $\frac{1}{10^1}$, or simply $\frac{1}{10}$. It is important to recognise that 10^{-1} is therefore the same as 0.1.

(b) 10^{-2} means $\frac{1}{10^2}$ or $\frac{1}{100}$. So 10^{-2} is therefore the same as 0.01.

6.16 Rewrite each of the following expressions using only positive powers:

(a) 7^{-3} (b) x^{-5}

Solution (a) 7^{-3} means the same as $\frac{1}{7^3}$. The expression has now been written using a positive power.

(b) $x^{-5} = \frac{1}{x^5}$.

6.17 Rewrite each of the following expressions using only positive powers:

(a) $\dfrac{1}{x^{-9}}$ (b) $\dfrac{1}{a^{-4}}$

Solution (a) $\dfrac{1}{x^{-9}} = x^9$ (b) $\dfrac{1}{a^{-4}} = a^4$

6.18 Rewrite each of the following using only negative powers:

(a) 6^8 (b) x^5 (c) z^a

Solution (a) $6^8 = \dfrac{1}{6^{-8}}$ (b) $x^5 = \dfrac{1}{x^{-5}}$ (c) $z^a = \dfrac{1}{z^{-a}}$

6.19 Simplify

(a) $x^{-2}x^7$ (b) $\dfrac{x^{-3}}{x^{-5}}$

Solution (a) To simplify $x^{-2}x^7$ we can use the first law of indices to write it as $x^{-2+7} = x^5$.

(b) To simplify $\frac{x^{-3}}{x^{-5}}$ we can use the second law of indices to write it as $x^{-3-(-5)} = x^{-3+5} = x^2$.

6.20 Simplify

(a) $(x^{-3})^5$ (b) $\dfrac{1}{(x^{-2})^2}$

Solution

(a) To simplify $(x^{-3})^5$ we can use the third law of indices and write it as $x^{-3\times 5} = x^{-15}$. The answer could also be written as $\frac{1}{x^{15}}$.

(b) Note that $(x^{-2})^2 = x^{-4}$ using the third law. So $\dfrac{1}{(x^{-2})^2} = \dfrac{1}{x^{-4}}$. This could also be written as x^4.

Self-assessment question 6.2

1. Explain how the negative power in a^{-m} is interpreted.

Exercise 6.2

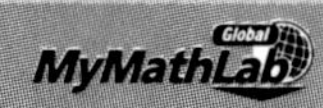

1. Without using a calculator express each of the following as a proper fraction:
(a) 2^{-2} (b) 2^{-3} (c) 3^{-2} (d) 3^{-3}
(e) 5^{-2} (f) 4^{-2} (g) 9^{-1} (h) 11^{-2}
(i) 7^{-1}

2. Express each of the following as decimal fractions:
(a) 10^{-1} (b) 10^{-2} (c) 10^{-6} (d) $\frac{1}{10^2}$
(e) $\frac{1}{10^3}$ (f) $\frac{1}{10^4}$

3. Write each of the following using only a positive power:
(a) x^{-4} (b) $\frac{1}{x^{-5}}$ (c) x^{-7} (d) y^{-2}
(e) $\frac{1}{y^{-1}}$ (f) y^{-1} (g) y^{-2} (h) z^{-1}
(i) $\frac{1}{z^{-1}}$

VIDEO

4. Simplify the following using the laws of indices and write your results using only positive powers:
(a) $x^{-2}x^{-1}$ (b) $x^{-3}x^{-2}$ (c) x^3x^{-4}
(d) $x^{-4}x^9$ (e) $\frac{x^{-2}}{x^{11}}$ (f) $(x^{-4})^2$
(g) $(x^{-3})^3$ (h) $(x^2)^{-2}$

5. Simplify
(a) $a^{13}a^{-2}$ (b) $x^{-9}x^{-7}$ (c) $x^{-21}x^2x$
(d) $(4^{-3})^2$

6. Evaluate
(a) 10^{-3} (b) 10^{-4} (c) 10^{-5}

7. Evaluate $4^{-8}/4^{-6}$ and $3^{-5}/3^{-8}$ without using a calculator.

6.3 Square roots, cube roots and fractional powers

Square roots

Consider the relationship between the numbers 5 and 25. We know that $5^2 = 25$ and so 25 is the square of 5. Equivalently we say that 5 is a **square root** of 25. The symbol $\sqrt[2]{\ }$, or simply $\sqrt{\ }$, is used to denote a square root and we write

$$5 = \sqrt{25}$$

We can picture this as follows:

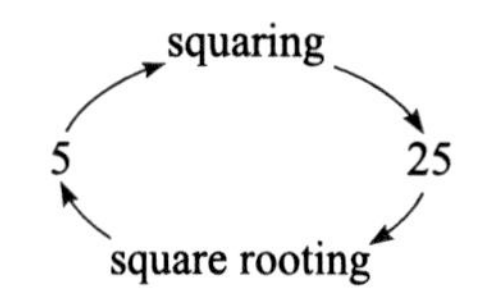

From this we see that taking the square root can be thought of as reversing the process of squaring.

We also note that

$$(-5) \times (-5) = (-5)^2$$
$$= 25$$

and so -5 is also a square root of 25. Hence we can write

$$-5 = \sqrt{25}$$

We can write both results together by using the 'plus or minus' sign $\pm$. We write

$$\sqrt{25} = \pm 5$$

In general, a **square root** of a number is a number that when squared gives the original number. Note that there are two square roots of any positive number but negative numbers possess no square roots.

Most calculators enable you to find square roots although only the positive value is normally given. Look for a $\sqrt{\ }$ or 'sqrt' button on your calculator.

WORKED EXAMPLE

6.21

(a) Use your calculator to find $\sqrt{79}$ correct to 4 decimal places.

(b) Check your answers are correct by squaring them.

Solution

(a) Using the $\sqrt{\ }$ button on the calculator you should verify that

$$\sqrt{79} = 8.8882 \text{ (to 4 decimal places)}$$

The second square root is -8.8882. Thus we can write

$$\sqrt{79} = \pm 8.8882$$

(b) Squaring either of the numbers ± 8.8882 we recover the original number, 79.

Cube roots

The **cube root** of a number is a number that when cubed gives the original number. The symbol for a cube root is $\sqrt[3]{\ }$. So, for example, since $2^3 = 8$ we can write $\sqrt[3]{8} = 2$.

We can picture this as follows:

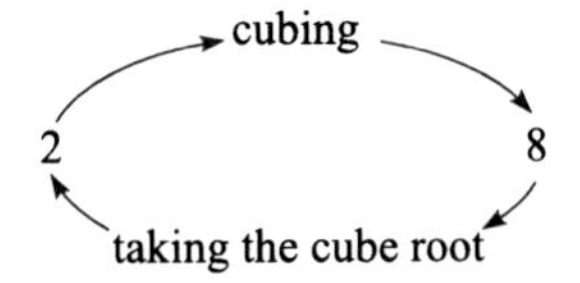

We can think of taking the cube root as reversing the process of cubing. As another example we note that $(-2)^3 = -8$ and hence $\sqrt[3]{-8} = -2$. All numbers, both positive and negative, possess a single cube root.

Your calculator may enable you to find a cube root. Look for a button marked $\sqrt[3]{}$. If so, check that you can use it correctly by verifying that

$$\sqrt[3]{46} = 3.5830$$

Fourth, fifth and other roots are defined in a similar way. For example, since

$$8^5 = 32768$$

we can write

$$\sqrt[5]{32768} = 8$$

Fractional powers

Sometimes fractional powers are used. The following example helps us to interpret a fractional power.

WORKED EXAMPLE

6.22 Simplify

(a) $x^{\frac{1}{2}}x^{\frac{1}{2}}$ (b) $x^{\frac{1}{3}}x^{\frac{1}{3}}x^{\frac{1}{3}}$

Use your results to interpret the fractional powers $\frac{1}{2}$ and $\frac{1}{3}$.

Solution (a) Using the first law we can write

$$x^{\frac{1}{2}}x^{\frac{1}{2}} = x^{\frac{1}{2}+\frac{1}{2}} = x^1 = x$$

(b) Similarly,

$$x^{\frac{1}{3}}x^{\frac{1}{3}}x^{\frac{1}{3}} = x^{\frac{1}{3}+\frac{1}{3}+\frac{1}{3}} = x^1 = x$$

From (a) we see that

$$(x^{\frac{1}{2}})^2 = x$$

So when $x^{\frac{1}{2}}$ is squared, the result is x. Thus $x^{\frac{1}{2}}$ is simply the square root of x, that is

$$x^{\frac{1}{2}} = \sqrt{x}$$

Similarly, from (b)

$$(x^{\frac{1}{3}})^3 = x$$

and so $x^{\frac{1}{3}}$ is the cube root of x, that is

$$x^{\frac{1}{3}} = \sqrt[3]{x}$$

Key point $x^{\frac{1}{2}} = \sqrt{x}, \quad x^{\frac{1}{3}} = \sqrt[3]{x}$

More generally we have the following result:

Key point $x^{\frac{1}{n}} = \sqrt[n]{x}$

Your scientific calculator will probably be able to find fractional powers. The button may be marked $x^{1/y}$ or $\sqrt[y]{x}$. Check that you can use it correctly by working through the following examples.

WORKED EXAMPLES

6.23 Evaluate to 3 decimal places, using a calculator:

(a) $3^{\frac{1}{4}}$ (b) $15^{1/5}$

Solution Use your calculator to obtain the following solutions:

(a) 1.316 (b) 1.719

Note in part (a) that although the calculator gives just a single fourth root, there is another, -1.316.

6.24 Evaluate $(81)^{1/2}$.

Solution $(81)^{1/2} = \sqrt{81} = \pm 9$.

6.25 Explain what is meant by the number $27^{1/3}$.

Solution $27^{1/3}$ can be written $\sqrt[3]{27}$, that is the cube root of 27. The cube root of 27 is 3, since $3 \times 3 \times 3 = 27$, and so $27^{1/3} = 3$. Note also that since $27 = 3^3$ we can write

$$\begin{aligned}(27)^{1/3} &= (3^3)^{1/3} = 3^{(3 \times 1/3)} \quad \text{using the third law}\\ &= 3^1 \quad = 3\end{aligned}$$

The following worked example shows how we deal with negative fractional powers.

6.26 Explain what is meant by the number $(81)^{-1/2}$.

Solution Recall from our work on negative powers that $a^{-m} = 1/a^m$. Therefore we can write $(81)^{-1/2}$ as $1/(81)^{1/2}$. Now $81^{1/2} = \sqrt{81} = \pm 9$ and so

$$(81)^{-1/2} = \frac{1}{\pm 9} = \pm \frac{1}{9}$$

6.27 Write each of the following using a single index:

(a) $(5^2)^{\frac{1}{3}}$ (b) $(5^{-2})^{\frac{1}{3}}$

Solution (a) Using the third law of indices we find

$$(5^2)^{\frac{1}{3}} = 5^{2 \times \frac{1}{3}} = 5^{\frac{2}{3}}$$

Note that $(5^2)^{\frac{1}{3}}$ is the cube root of 5^2, that is $\sqrt[3]{25}$ or 2.9240.

(b) Using the third law of indices we find

$$(5^{-2})^{\frac{1}{3}} = 5^{-2 \times \frac{1}{3}} = 5^{-\frac{2}{3}}$$

Note that there is a variety of equivalent ways in which this can be expressed, for example $\sqrt[3]{\frac{1}{5^2}}$ or $\sqrt[3]{\frac{1}{25}}$, or as $\frac{1}{5^{2/3}}$.

6.28 Write each of the following using a single index:

(a) $\sqrt{x^3}$ (b) $(\sqrt{x})^3$

Solution (a) Because the square root of a number can be expressed as that number raised to the power $\frac{1}{2}$ we can write

$$\begin{aligned}\sqrt{x^3} &= (x^3)^{\frac{1}{2}} \\ &= x^{3 \times \frac{1}{2}} \quad \text{using the third law} \\ &= x^{\frac{3}{2}}\end{aligned}$$

(b) $$\begin{aligned}(\sqrt{x})^3 &= (x^{\frac{1}{2}})^3 \\ &= x^{\frac{3}{2}} \quad \text{using the third law}\end{aligned}$$

Note from this example that $\sqrt{x^3} = (\sqrt{x})^3$.

Note that by generalising the results of the two previous worked examples we have the following:

Key point

$$a^{\frac{m}{n}} = \sqrt[n]{a^m} = (\sqrt[n]{a})^m$$

Self-assessment questions 6.3

1. Explain the meaning of the fractional powers $x^{1/2}$ and $x^{1/3}$.
2. What are the square roots of 100? Explain why the number -100 does not have any square roots.

Exercise 6.3 MyMathLab Global

1. Evaluate
   (a) $64^{1/3}$ (b) $144^{1/2}$ (c) $16^{-1/4}$
   (d) $25^{-1/2}$ (e) $\dfrac{1}{32^{-1/5}}$
2. Simplify and then evaluate
   (a) $(3^{-1/2})^4$ (b) $(8^{1/3})^{-1}$
3. Write each of the following using a single index:
   (a) $\sqrt{8}$ (b) $\sqrt[3]{12}$ (c) $\sqrt[4]{16}$ (d) $\sqrt{13^3}$
   (e) $\sqrt[3]{4^7}$
4. Write each of the following using a single index:
   (a) $\sqrt{x}$ (b) $\sqrt[3]{y}$ (c) $\sqrt[2]{x^5}$ (d) $\sqrt[3]{5^7}$

6.4 Multiplication and division by powers of 10

To multiply and divide decimal fractions by powers of 10 is particularly simple. For example, to multiply 256.875 by 10 the decimal point is moved one place to the right, that is

$$256.875 \times 10 = 2568.75$$

To multiply by 100 the decimal point is moved two places to the right. So

$$256.875 \times 100 = 25687.5$$

To divide a number by 10, the decimal point is moved one place to the left. This is equivalent to multiplying by 10^{-1}. To divide by 100, the decimal point is moved two places to the left. This is equivalent to multiplying by 10^{-2}.

In general, to multiply a number by 10^n, the decimal point is moved n places to the right if n is a positive integer, and n places to the left if n is a negative integer. If necessary, additional zeros are inserted to make up the required number of digits. Consider the following example.

WORKED EXAMPLE

6.29 Without the use of a calculator, write down

(a) 75.45×10^3 (b) 0.056×10^{-2} (c) 96.3×10^{-3} (d) 0.00743×10^5

Solution

(a) The decimal point is moved three places to the right:
$75.45 \times 10^3 = 75450$. It has been necessary to include an additional zero to make up the required number of digits.

(b) The decimal point is moved two places to the left:
$0.056 \times 10^{-2} = 0.00056$.

(c) $96.3 \times 10^{-3} = 0.0963$.

(d) $0.00743 \times 10^5 = 743$.

Exercise 6.4

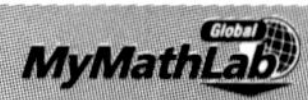

1. Without the use of a calculator write down:
   (a) 7.43×10^2 (b) 7.43×10^4 (c) 0.007×10^4 (d) 0.07×10^{-2}
2. Write each of the following as a multiple of 10^2:
   (a) 300 (b) 356 (c) 32 (d) 0.57

6.5 Scientific notation

It is often necessary to use very large numbers such as 65000000000 or very small numbers such as 0.000000001. **Scientific notation** can be used to express such numbers in a more concise form, which avoids writing very lengthy strings of numbers. Each number is written in the form

$$a \times 10^n$$

where a is usually a number between 1 and 10. We also make use of the fact that

$$10 = 10^1, \qquad 100 = 10^2, \qquad 1000 = 10^3 \text{ and so on}$$

and also that

$$10^{-1} = \frac{1}{10} = 0.1, \qquad 10^{-2} = \frac{1}{100} = 0.01 \text{ and so on}$$

Then, for example,

the number 4000 can be written $4 \times 1000 = 4 \times 10^3$

Similarly

the number 68000 can be written $6.8 \times 10000 = 6.8 \times 10^4$

and

the number 0.09 can be written $9 \times 0.01 = 9 \times 10^{-2}$

Note that all three numbers have been written in the form $a \times 10^n$ where a lies between 1 and 10.

WORKED EXAMPLES

6.30 Express the following numbers in scientific notation:

(a) 54 (b) −276 (c) 0.3

Solution

(a) 54 can be written as 5.4×10, so in scientific notation we have 5.4×10^1.

(b) Negative numbers cause no problem: $-276 = -2.76 \times 10^2$.

(c) We can write 0.3 as 3×0.1 or 3×10^{-1}.

6.31 Write out fully the following numbers:

(a) 2.7×10^{-1} (b) 9.6×10^5 (c) -8.2×10^2

Solution

(a) $2.7 \times 10^{-1} = 0.27$.

(b) $9.6 \times 10^5 = 9.6 \times 100000 = 960000$.

(c) $-8.2 \times 10^2 = -8.2 \times 100 = -820$.

6.32 Simplify the expression $(3 \times 10^2) \times (5 \times 10^3)$.

Solution

The order in which the numbers are written down does not matter, and so we can write

$$(3 \times 10^2) \times (5 \times 10^3) = 3 \times 5 \times 10^2 \times 10^3 = 15 \times 10^5$$

Noting that $15 = 1.5 \times 10$ we can express the final answer in scientific notation:

$$15 \times 10^5 = 1.5 \times 10 \times 10^5 = 1.5 \times 10^6$$

Hence

$$(3 \times 10^2) \times (5 \times 10^3) = 1.5 \times 10^6$$

Self-assessment question 6.5

1. What is the purpose of using scientific notation?

Exercise 6.5

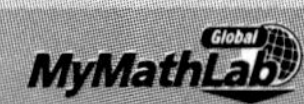

1. Express each of the following numbers in scientific notation:
   (a) 45 (b) 45000 (c) −450 (d) 90000000 (e) 0.15 (f) 0.00036 (g) 3.5
   (h) −13.2 (i) 1000000 (j) 0.0975 (k) 45.34
2. Write out fully the following numbers:
   (a) 3.75×10^2 (b) 3.97×10^1 (c) 1.875×10^{-1} (d) -8.75×10^{-3}
3. Simplify each of the following expressions, writing your final answer in scientific notation:
   (a) $(4 \times 10^3) \times (6 \times 10^4)$ (b) $(9.6 \times 10^4) \times (8.3 \times 10^3)$ (c) $(1.2 \times 10^{-3}) \times (8.7 \times 10^{-2})$
   (d) $\dfrac{9.37 \times 10^4}{6.14 \times 10^5}$ (e) $\dfrac{4.96 \times 10^{-2}}{9.37 \times 10^{-5}}$

Test and assignment exercises 6

1. Simplify

   (a) $\dfrac{z^5}{z^{-5}}$ (b) z^0 (c) $\dfrac{z^8 z^6}{z^{14}}$

2. Evaluate

   (a) $0.25^{1/2}$ (b) $(4096)^{1/3}$ (c) $(2601)^{1/2}$ (d) $16^{-1/2}$

3. Simplify $\dfrac{x^8 x^{-3}}{x^{-5} x^2}$.

4. Find the value of $(1/7)^0$.

5. Remove the brackets from

   (a) $(abc^2)^2$ (b) $(xy^2z^3)^2$ (c) $(8x^2)^{-3}$

6. Express each of the following numbers in scientific notation:

   (a) 5792 (b) 98.4 (c) 0.001 (d) −66.667

Simplifying algebraic expressions

7

Objectives: This chapter:

- describes a number of ways in which complicated algebraic expressions can be simplified

7.1 Addition and subtraction of like terms

Like terms are multiples of the same quantity. For example, $3y$, $72y$ and $0.5y$ are all multiples of y and so are like terms. Similarly, $5x^2$, $-3x^2$ and $\frac{1}{2}x^2$ are all multiples of x^2 and so are like terms. xy, $17xy$ and $-91xy$ are all multiples of xy and are therefore like terms. Like terms can be collected together and added or subtracted in order to simplify them.

WORKED EXAMPLES

7.1 Simplify $3x + 7x - 2x$.

Solution All three terms are multiples of x and so are like terms. Therefore $3x + 7x - 2x = 8x$.

7.2 Simplify $3x + 2y$.

Solution $3x$ and $2y$ are not like terms. One is a multiple of x and the other is a multiple of y. The expression $3x + 2y$ cannot be simplified.

7.3 Simplify $x + 7x + x^2$.

Solution The like terms are x and $7x$. These can be simplified to $8x$. Then $x + 7x + x^2 = 8x + x^2$. Note that $8x$ and x^2 are not like terms and so this expression cannot be simplified further.

7.4 Simplify $ab + a^2 - 7b^2 + 9ab + 8b^2$.

Solution The terms ab and $9ab$ are like terms. Similarly the terms $-7b^2$ and $8b^2$ are like terms. These can be collected together and then added or subtracted as appropriate. Thus

$$ab + a^2 - 7b^2 + 9ab + 8b^2 = ab + 9ab + a^2 - 7b^2 + 8b^2$$
$$= 10ab + a^2 + b^2$$

Exercise 7.1

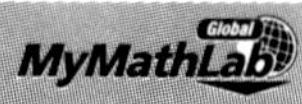

1. Simplify, if possible,
   (a) $5p - 10p + 11q + 8q$ (b) $-7r - 13s + 2r + z$ (c) $181z + 13r - 2$
   (d) $x^2 + 3y^2 - 2y + 7x^2$ (e) $4x^2 - 3x + 2x + 9$
2. Simplify
   (a) $5y + 8p - 17y + 9q$ (b) $7x^2 - 11x^3 + 14x^2 + y^3$ (c) $4xy + 3xy + y^2$
   (d) $xy + yx$ (e) $xy - yx$

7.2 Multiplying algebraic expressions and removing brackets

Recall that when multiplying two numbers together the order in which we write them is irrelevant. For example, both 5×4 and 4×5 equal 20.

When multiplying three or more numbers together the order in which we carry out the multiplication is also irrelevant. By this we mean, for example, that when asked to multiply $3 \times 4 \times 5$ we can think of this as either $(3 \times 4) \times 5$ or as $3 \times (4 \times 5)$. Check for yourself that the result is the same, 60, either way.

It is also important to appreciate that $3 \times 4 \times 5$ could have been written as $(3)(4)(5)$.

It is essential that you grasp these simple facts about numbers in order to understand the algebra that follows. This is because identical rules are applied. Rules for determining the sign of the answer when multiplying positive and negative algebraic expressions are also the same as those used for multiplying numbers.

Key point

When multiplying

positive × positive = positive
positive × negative = negative
negative × positive = negative
negative × negative = positive

We introduce the processes involved in removing brackets using some simple examples.

WORKED EXAMPLES

7.5 Simplify $3(4x)$.

Solution Just as with numbers $3(4x)$ could be written as $3 \times (4 \times x)$, and then as $(3 \times 4) \times x$, which evaluates to $12x$.

So $3(4x) = 12x$.

7.6 Simplify $5(3y)$.

Solution $5(3y) = 5 \times 3 \times y = 15y$.

7.7 Simplify $(5a)(3a)$.

Solution Here we can write $(5a)(3a) = (5 \times a) \times (3 \times a)$. Neither the order in which we carry out the multiplications nor the order in which we write down the terms matters, and so we can write this as

$$(5a)(3a) = (5 \times 3)(a \times a)$$

As we have shown, it is usual to write numbers at the beginning of an expression. This simplifies to $15 \times a^2$, that is $15a^2$. Hence

$$(5a)(3a) = 15a^2$$

7.8 Simplify $4x^2 \times 7x^5$.

Solution Recall that, when multiplying, the order in which we write down the terms does not matter. Therefore we can write

$$4x^2 \times 7x^5 = 4 \times 7 \times x^2 \times x^5$$

which equals $28x^{2+5} = 28x^7$.

7.9 Simplify $7(2b^2)$.

Solution $7(2b^2) = 7 \times (2 \times b^2) = (7 \times 2) \times b^2 = 14b^2$.

7.10 Simplify $(a) \times (-b)$

Solution Here we have the product of a positive and a negative quantity. The result will be negative. We write

$$(a) \times (-b) = -ab$$

7.11 Explain the distinction between ab^2 and $(ab)^2$.

Solution ab^2 means $a \times b \times b$ whereas $(ab)^2$ means $(ab) \times (ab)$ which equals $a \times b \times a \times b$. The latter could also be written as a^2b^2.

7.12 Simplify (a) $(6z)(8z)$, (b) $(6z) + (8z)$, noting the distinction between the two results.

Solution (a) $(6z)(8z) = 48z^2$.

(b) $(6z) + (8z)$ is the addition of like terms. This simplifies to $14z$.

7.13 Simplify (a) $(6x)(-2x)$, (b) $(-3y^2)(-2y)$.

Solution (a) $(6x)(-2x)$ means $(6x) \times (-2x)$, which equals $-12x^2$.

(b) $(-3y^2)(-2y) = (-3y^2) \times (-2y) = 6y^3$.

Self-assessment questions 7.2

1. Two negative expressions are multiplied together. State the sign of the resulting product.
2. Three negative expressions are multiplied together. State the sign of the resulting product.

Exercise 7.2

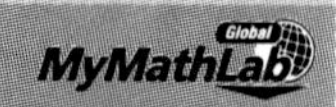

1. Simplify each of the following:
   (a) $(4)(3)(7)$ (b) $(7)(4)(3)$ (c) $(3)(4)(7)$
2. Simplify
   (a) $5 \times (4 \times 2)$ (b) $(5 \times 4) \times 2$
3. Simplify each of the following:
   (a) $7(2z)$ (b) $15(2y)$ (c) $(2)(3)x$
   (d) $9(3a)$ (e) $(11)(5a)$ (f) $2(3x)$
4. Simplify each of the following:
   (a) $5(4x^2)$ (b) $3(2y^3)$ (c) $11(2u^2)$
   (d) $(2 \times 4) \times u^2$ (e) $(13)(2z^2)$
5. Simplify
   (a) $(7x)(3x)$ (b) $3a(7a)$ (c) $14a(a)$
6. Simplify
   (a) $5y(3y)$ (b) $5y + 3y$
   Explain why the two results are not the same.
7. Simplify the following:
   (a) $(abc)(a^2bc)$ (b) $x^2y(xy)$
   (c) $(xy^2)(xy^2)$
8. Explain the distinction, if any, between $(xy^2)(xy^2)$ and xy^2xy^2.
9. Explain the distinction, if any, between $(xy^2)(xy^2)$ and $(xy^2) + (xy^2)$.
   In both cases simplify the expressions.
10. Simplify
    (a) $(3z)(-7z)$ (b) $3z - 7z$
11. Simplify
    (a) $(-x)(3x)$ (b) $-x + 3x$
12. Simplify
    (a) $(-2x)(-x)$ (b) $-2x - x$

7.3 Removing brackets from $a(b + c)$, $a(b - c)$ and $(a + b)(c + d)$

Recall from your study of arithmetic that the expression $(5 - 4) + 7$ is different from $5 - (4 + 7)$ because of the position of the brackets. In order to simplify an expression it is often necessary to remove brackets.

Removing brackets from expressions of the form $a(b + c)$ and $a(b - c)$

In an expression such as $a(b + c)$, it is intended that the a multiplies all the bracketed terms:

Key point

$a(b + c) = ab + ac$ Similarly: $a(b - c) = ab - ac$

WORKED EXAMPLES

7.14 Remove the brackets from

(a) $6(x + 5)$ (b) $8(2x - 4)$

Solution (a) In the expression $6(x + 5)$ it is intended that the 6 multiplies both terms in the brackets. Therefore

$$6(x + 5) = 6x + 30$$

(b) In the expression $8(2x - 4)$ the 8 multiplies both terms in the brackets so that

$$8(2x - 4) = 16x - 32$$

7.15 Remove the brackets from the expression $7(5x + 3y)$.

Solution The 7 multiplies both the terms in the bracket. Therefore

$$7(5x + 3y) = 7(5x) + 7(3y) = 35x + 21y$$

7.16 Remove the brackets from $-(x + y)$.

Solution The expression $-(x + y)$ actually means $-1(x + y)$. It is intended that the -1 multiplies both terms in the brackets, therefore

$$-(x + y) = -1(x + y) = (-1) \times x + (-1) \times y = -x - y$$

7.17 Remove the brackets from the expression

$$(x + y)z$$

Solution Note that the order in which we write down the terms to be multiplied does not matter, so that we can write $(x + y)z$ as $z(x + y)$. Then

$$z(x + y) = zx + zy$$

Alternatively note that $(x + y)z = xz + yz$, which is an equivalent form of the answer.

7.18 Remove the brackets from the expressions

VIDEO

(a) $5(x - 2y)$ (b) $(x + 3)(-1)$

Solution (a) $5(x - 2y) = 5x - 5(2y) = 5x - 10y$.

(b) $(x + 3)(-1) = (-1)(x + 3) = -1x - 3 = -x - 3$.

7.19 Simplify $x + 8(x - y)$.

Solution An expression such as this is simplified by first removing the brackets and then collecting together like terms. Removing the brackets we find

$$x + 8(x - y) = x + 8x - 8y$$

Collecting like terms we obtain $9x - 8y$.

7.20 Remove the brackets from

(a) $\frac{1}{2}(x + 2)$ (b) $\frac{1}{2}(x - 2)$ (c) $-\frac{1}{3}(a + b)$

Solution

(a) In the expression $\frac{1}{2}(x + 2)$ it is intended that the $\frac{1}{2}$ multiplies both the terms in the brackets. So

$$\frac{1}{2}(x + 2) = \frac{1}{2}x + \frac{1}{2}(2) = \frac{1}{2}x + 1$$

(b) Similarly,

$$\frac{1}{2}(x - 2) = \frac{1}{2}x - \frac{1}{2}(2) = \frac{1}{2}x - 1$$

(c) In the expression $-\frac{1}{3}(a + b)$ the term $-\frac{1}{3}$ multiplies both terms in the brackets. So

$$-\frac{1}{3}(a + b) = -\frac{1}{3}a - \frac{1}{3}b$$

Removing brackets from expressions of the form $(a + b)(c + d)$

In the expression $(a + b)(c + d)$ it is intended that the quantity $(a + b)$ multiplies both the c and the d in the second brackets. Therefore

$$(a + b)(c + d) = (a + b)c + (a + b)d$$

Each of these two terms can be expanded further to give

$$(a + b)c = ac + bc \quad \text{and} \quad (a + b)d = ad + bd$$

Therefore

Key point

$$(a + b)(c + d) = ac + bc + ad + bd$$

WORKED EXAMPLES

7.21 Remove the brackets from $(3 + x)(2 + y)$.

Solution $(3+x)(2+y) = (3+x)(2) + (3+x)y$

$$= 6 + 2x + 3y + xy$$

7.22 Remove the brackets from $(x+6)(x-3)$.

Solution $(x+6)(x-3) = (x+6)x + (x+6)(-3)$

$$= x^2 + 6x - 3x - 18$$

$$= x^2 + 3x - 18$$

7.23 Remove the brackets from

(a) $(1-x)(2-x)$ (b) $(-x-2)(2x-1)$

Solution (a) $(1-x)(2-x) = (1-x)2 + (1-x)(-x)$

$$= 2 - 2x - x + x^2$$

$$= 2 - 3x + x^2$$

(b) $(-x-2)(2x-1) = (-x-2)(2x) + (-x-2)(-1)$

$$= -2x^2 - 4x + x + 2$$

$$= -2x^2 - 3x + 2$$

7.24 Remove the brackets from the expression $3(x+1)(x-1)$.

VIDEO

Solution First consider the expression $(x+1)(x-1)$:

$(x+1)(x-1) = (x+1)x + (x+1)(-1)$

$$= x^2 + x - x - 1$$

$$= x^2 - 1$$

Then $3(x+1)(x-1) = 3(x^2-1) = 3x^2 - 3$.

Exercise 7.3

MyMathLab

1. Remove the brackets from
   (a) $4(x+1)$ (b) $-4(x+1)$
   (c) $4(x-1)$ (d) $-4(x-1)$

2. Remove the brackets from the following expressions:
   (a) $5(x-y)$ (b) $19(x+3y)$
   (c) $8(a+b)$ (d) $(5+x)y$
   (e) $12(x+4)$ (f) $17(x-9)$
   (g) $-(a-2b)$ (h) $\frac{1}{2}(2x+1)$
   (i) $-3m(-2+4m+3n)$

3. Remove the brackets and simplify the following:
   (a) $18-13(x+2)$ (b) $x(x+y)$

4. Remove the brackets and simplify the following expressions:
   (a) $(x+1)(x+6)$ (b) $(x+4)(x+5)$
   (c) $(x-2)(x+3)$ (d) $(x+6)(x-1)$
   (e) $(x+y)(m+n)$ (f) $(4+y)(3+x)$
   (g) $(5-x)(5+x)$
   (h) $(17x+2)(3x-5)$

5. Remove the brackets and simplify the following expressions:
   (a) $(x+3)(x-7)$ (b) $(2x-1)(3x+7)$
   (c) $(4x+1)(4x-1)$
   (d) $(x+3)(x-3)$ (e) $(2-x)(3+2x)$
6. Remove the brackets and simplify the following expressions:
   (a) $\frac{1}{2}(x+2y)+\frac{7}{2}(4x-y)$
   (b) $\frac{3}{4}(x-1)+\frac{1}{4}(2x+8)$
7. Remove the brackets from
   (a) $-(x-y)$ (b) $-(a+2b)$
   (c) $-\frac{1}{2}(3p+q)$
8. Remove the brackets from $(x+1)(x+2)$. Use your result to remove the brackets from $(x+1)(x+2)(x+3)$.

Test and assignment exercises 7

1. Simplify
   (a) $7x^2+4x^2+9x-8x$ (b) $y+7-18y+1$ (c) $a^2+b^2+a^3-3b^2$
2. Simplify
   (a) $(3a^2b)\times(-a^3b^2c)$ (b) $\frac{x^3}{-x^2}$
3. Remove the brackets from
   (a) $(a+3b)(7a-2b)$ (b) $x^2(x+2y)$ (c) $x(x+y)(x-y)$
4. Remove the brackets from
   (a) $(7x+2)(3x-1)$ (b) $(1-x)(x+3)$ (c) $(5+x)x$ (d) $(8x+4)(7x-2)$
5. Remove the brackets and simplify
   (a) $3x(x+2)-7x^2$ (b) $-(2a+3b)(a+b)$ (c) $4(x+7)+13(x-2)$
   (d) $5(2a+5)-3(5a-2)$ (e) $\frac{1}{2}(a+4b)+\frac{3}{2}a$

8 Factorisation

Objectives: This chapter:

- explains what is meant by the 'factors' of an algebraic expression
- shows how an algebraic expression can be factorised
- shows how to factorise quadratic expressions

8.1 Factors and common factors

Recall from Chapter 1 that a number is **factorised** when it is written as a product. For example, 15 may be factorised into 3×5. We say that 3 and 5 are **factors** of 15. The number 16 can be written as 8×2 or 4×4, or even as 16×1, and so the factorisation may not be unique.

Algebraic expressions can also be factorised. Consider the expression $5x + 20y$. Both $5x$ and $20y$ have the number 5 common to both terms. We say that 5 is a **common factor**. Any common factors can be written outside a bracketed term. Thus $5x + 20y = 5(x + 4y)$. Removal of the brackets will result in the original expression and can always be used to check your answer. We see that factorisation can be thought of as reversing the process of removing brackets. Similarly, if we consider the expression $x^2 + 2x$, we note that both terms contain the factor x, and so $x^2 + 2x$ can be written as $x(x + 2)$. Hence x and $x + 2$ are both factors of $x^2 + 2x$.

WORKED EXAMPLES

8.1 Factorise $3x + 12$.

Solution The number 12 can be factorised as 3×4 so that 3 is a common factor of $3x$ and 12. We can write $3x + 12 = 3x + 3(4)$. Any common factors are written in front of the brackets and the contents of the brackets are adjusted accordingly. So

$$3x + 3(4) = 3(x + 4)$$

Note again that this answer can be checked by removing the brackets.

8.2 List the ways in which $15x^2$ can be written as a product of its factors.

Solution $15x^2$ can be written in many different ways. Some of these are $15x^2 \times 1$, $15x \times x$, $15 \times x^2$, $5x \times 3x$, $5 \times 3x^2$ and $3 \times 5x^2$.

8.3 Factorise $8x^2 - 12x$.

Solution We can write $8x^2 - 12x = (4x)(2x) - (4x)3$ so that both terms contain the factor $4x$. This is placed at the front of the brackets to give

$$8x^2 - 12x = 4x(2x - 3)$$

8.4 What factors are common to the terms $5x^2$ and $15x^3$? Factorise $5x^2 + 15x^3$.

Solution Both terms contain a factor of 5. Because x^3 can be written as $x^2 \times x$, both $5x^2$ and $15x^3$ contain a factor x^2. Therefore

$$5x^2 + 15x^3 = 5x^2 + (5x^2)(3x) = 5x^2(1 + 3x)$$

8.5 Factorise $6x + 3x^2 + 9xy$.

Solution By careful inspection of all of the terms we see that $3x$ is a factor of each term. Hence

$$6x + 3x^2 + 9xy = 3x(2 + x + 3y)$$

Hence the factors of $6x + 3x^2 + 9xy$ are $3x$ and $2 + x + 3y$.

Self-assessment question 8.1

1. Explain what is meant by 'factorising an expression'.

Exercise 8.1

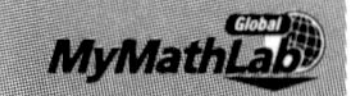

1. Remove the brackets from
   (a) $9(x + 3)$ (b) $-5(x - 2)$
   (c) $\frac{1}{2}(x + 1)$
   (d) $-(a - 3b)$ (e) $\frac{1}{2(x + y)}$
   (f) $\frac{x}{y(x - y)}$

2. List all the factors of each of (a) $4x^2$, (b) $6x^3$.

3. Factorise
   (a) $3x + 18$ (b) $3y - 9$ (c) $-3y - 9$
   (d) $-3 - 9y$ (e) $20 + 5t$ (f) $20 - 5t$
   (g) $-5t - 20$ (h) $3x + 12$ (i) $17t + 34$
   (j) $-36 + 4t$

4. Factorise
   (a) $x^4 + 2x$ (b) $x^4 - 2x$ (c) $3x^4 - 2x$
   (d) $3x^4 + 2x$ (e) $3x^4 + 2x^2$
   (f) $3x^4 + 2x^3$ (g) $17z - z^2$
   (h) $-xy + 3x$ (i) $-xy + 3y$
   (j) $x + 2xy + 3xyz$

5. Factorise
   (a) $10x + 20y$ (b) $12a + 3b$
   (c) $4x - 6xy$ (d) $7a + 14$
   (e) $10m - 15$ (f) $\frac{1}{5a + 35b}$
   (g) $\frac{1}{5a^2 + 35ab}$

6. Factorise
   (a) $15x^2 + 3x$ (b) $4x^2 - 3x$
   (c) $4x^2 - 8x$ (d) $15 - 3x^2$
   (e) $10x^3 + 5x^2 + 15x^2y$
   (f) $6a^2b - 12ab^2$
   (g) $16abc - 8ab^2 + 24bc$

8.2 Factorising quadratic expressions

Expressions of the form $ax^2 + bx + c$, where a, b and c are numbers, are called **quadratic expressions**. The numbers b or c may equal zero but a must not be zero. The number a is called the **coefficient** of x^2, b is the coefficient of x, and c is called the **constant term**.

We see that

$$2x^2 + 3x - 1, \quad x^2 + 3x + 2, \quad x^2 + 7 \quad \text{and} \quad 2x^2 - x$$

are all quadratic expressions.

Key point

An expression of the form $ax^2 + bx + c$, where a, b and c are numbers, is called a quadratic expression. The coefficient of x^2 is a, the coefficient of x is b, and the constant term is c.

To factorise such an expression means to express it as a product of two terms. For example, removing the brackets from $(x + 6)(x - 3)$ gives $x^2 + 3x - 18$ (see Worked Example 7.22). Reversing the process, $x^2 + 3x - 18$ can be factorised to $(x + 6)(x - 3)$. Not all quadratic expressions can be factorised in this way. We shall now explore how such factorisation is attempted.

Quadratic expressions where the coefficient of x^2 is 1

Consider the expression $(x + m)(x + n)$. Removing the brackets we find

$$\begin{aligned}(x + m)(x + n) &= (x + m)x + (x + m)n \\ &= x^2 + mx + nx + mn \\ &= x^2 + (m + n)x + mn\end{aligned}$$

Note that the coefficient of the x term is the sum $m+n$ and the constant term is the product mn. Using this information several quadratic expressions can be factorised by careful inspection. For example, suppose we wish to factorise x^2+5x+6. We know that $x^2+(m+n)x+mn$ can be factorised to $(x+m)(x+n)$. We seek values of m and n so that

$$x^2+5x+6=x^2+(m+n)x+mn$$

Comparing the coefficients of x on both sides we require

$$5=m+n$$

Comparing the constant terms on both sides we require

$$6=mn$$

By inspection we see that $m=3$ and $n=2$ have this property and so

$$x^2+5x+6=(x+3)(x+2)$$

Note that the answer can be easily checked by removing the brackets again.

WORKED EXAMPLES

8.6 Factorise the quadratic expression $x^2+8x+12$.

VIDEO

Solution The factorisation of $x^2+8x+12$ will be of the form $(x+m)(x+n)$. This means that mn must equal 12 and $m+n$ must equal 8. The two numbers must therefore be 2 and 6. So

$$x^2+8x+12=(x+2)(x+6)$$

Note again that the answer can be checked by removing the brackets.

8.7 Factorise $x^2+10x+25$.

Solution We try to factorise in the form $(x+m)(x+n)$. We require $m+n$ to equal 10 and mn to equal 25. If $m=5$ and $n=5$ this requirement is met. Therefore $x^2+10x+25=(x+5)(x+5)$. It is usual practice to write this as $(x+5)^2$.

8.8 Factorise x^2-121.

Solution In this example the x term is missing. We still attempt to factorise as $(x+m)(x+n)$. We require $m+n$ to equal 0 and mn to equal -121. Some thought shows that if $m=11$ and $n=-11$ this requirement is met. Therefore $x^2-121=(x+11)(x-11)$.

8.9 Factorise x^2-5x+6.

Solution We try to factorise in the form $(x+m)(x+n)$. We require $m+n$ to equal -5 and mn to equal 6. By inspection we see that if $m=-3$ and $n=-2$ this requirement is met. Therefore $x^2-5x+6=(x-3)(x-2)$.

Quadratic expressions where the coefficient of x^2 is not 1

These expressions are a little harder to factorise. All possible factors of the first and last terms must be found, and various combinations of these should be attempted until the required answer is found. This involves trial and error along with educated guesswork and practice.

WORKED EXAMPLES

8.10 Factorise, if possible, the expression $2x^2 + 11x + 12$.

Solution The factors of the first term, $2x^2$, are $2x$ and x. The factors of the last term, 12, are

$$12, 1 \quad -12, -1 \quad 6, 2 \quad -6, -2 \quad \text{and} \quad 4, 3 \quad -4, -3$$

We can try each of these combinations in turn to find which gives us a coefficient of x of 11. For example, removing the brackets from

$$(2x + 12)(x + 1)$$

gives

$$\begin{aligned}(2x + 12)(x + 1) &= (2x + 12)x + (2x + 12)(1)\\ &= 2x^2 + 12x + 2x + 12\\ &= 2x^2 + 14x + 12\end{aligned}$$

which has an incorrect middle term. By trying further combinations it turns out that the only one producing a middle term of $11x$ is $(2x + 3)(x + 4)$ because

$$\begin{aligned}(2x + 3)(x + 4) &= (2x + 3)(x) + (2x + 3)(4)\\ &= 2x^2 + 3x + 8x + 12\\ &= 2x^2 + 11x + 12\end{aligned}$$

so that $(2x + 3)(x + 4)$ is the correct factorisation.

8.11 Factorise, if possible, $4x^2 + 6x + 2$.

Solution Before we try to factorise this quadratic expression notice that there is a factor of 2 in each term so that we can write it as $2(2x^2 + 3x + 1)$. Now consider the quadratic expression $2x^2 + 3x + 1$. The factors of the first term, $2x^2$, are $2x$ and x. The factors of the last term, 1, are simply 1 and 1, or -1 and -1. We can try these combinations in turn to find which gives us a middle term of $3x$. Removing the brackets from $(2x + 1)(x + 1)$ gives $2x^2 + 3x + 1$, which has the correct middle term. Finally, we can write

$$4x^2 + 6x + 2 = 2(2x^2 + 3x + 1) = 2(2x + 1)(x + 1)$$

8.12 Factorise $6x^2 + 7x - 3$.

VIDEO

Solution The first term may be factorised as $6x \times x$ and also as $3x \times 2x$. The factors of the last term are

$$3, -1 \quad \text{and} \quad -3, 1$$

We need to try each combination in turn to find which gives us a coefficient of x of 7. For example, removing the brackets from

$$(6x + 3)(x - 1)$$

gives

$$\begin{aligned}(6x + 3)(x - 1) &= (6x + 3)x + (6x + 3)(-1)\\ &= 6x^2 + 3x - 6x - 3\\ &= 6x^2 - 3x - 3\end{aligned}$$

which has an incorrect middle term. By trying further combinations it turns out that the only one producing a middle term of $7x$ is $(3x - 1)(2x + 3)$ because

$$\begin{aligned}(3x - 1)(2x + 3) &= (3x - 1)2x + (3x - 1)(3)\\ &= 6x^2 - 2x + 9x - 3\\ &= 6x^2 + 7x - 3\end{aligned}$$

The correct factorisation is therefore $(3x - 1)(2x + 3)$.

Until you have sufficient experience at factorising quadratic expressions you must be prepared to go through the process of trying all possible combinations until the correct answer is found.

Self-assessment question 8.2

1. Not all quadratic expressions can be factorised. Try to find an example of one such expression.

Exercise 8.2

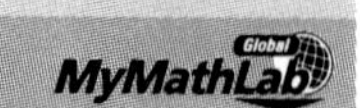

1. Factorise the following quadratic expressions:
(a) $x^2 + 3x + 2$ (b) $x^2 + 13x + 42$
(c) $x^2 + 2x - 15$ (d) $x^2 + 9x - 10$
(e) $x^2 - 11x + 24$ (f) $x^2 - 100$
(g) $x^2 + 4x + 4$ (h) $x^2 - 36$
(i) $x^2 - 25$ (j) $x^2 + 10x + 9$
(k) $x^2 + 8x - 9$ (l) $x^2 - 8x - 9$
(m) $x^2 - 10x + 9$ (n) $x^2 - 5x$

2. Factorise the following quadratic expressions:
(a) $2x^2 - 5x - 3$ (b) $3x^2 - 5x - 2$
(c) $10x^2 + 11x + 3$ (d) $2x^2 + 12x + 16$
(e) $2x^2 + 5x + 3$ (f) $3s^2 + 5s + 2$
(g) $3z^2 + 17z + 10$ (h) $9x^2 - 36$
(i) $4x^2 - 25$

3. (a) By removing the brackets show that

$$(x+y)(x-y) = x^2 - y^2$$

This result is known as the **difference of two squares**.

(b) Using the result in part (a) write down the factorisation of
(i) $16x^2 - 1$ (ii) $16x^2 - 9$
(iii) $25t^2 - 16r^2$

4. Factorise the following quadratic expressions:
(a) $x^2 + 3x - 10$ (b) $2x^2 - 3x - 20$
(c) $9x^2 - 1$ (d) $10x^2 + 14x - 12$
(e) $x^2 + 15x + 26$ (f) $-x^2 - 2x + 3$

5. Factorise
(a) $100 - 49x^2$ (b) $36x^2 - 25y^2$
(c) $\frac{1}{4} - 9v^2$ (d) $\frac{x^2}{y^2} - 4$

Test and assignment exercises 8

1. Factorise the following expressions:
(a) $7x + 49$ (b) $121x + 22y$ (c) $a^2 + ab$ (d) $ab + b^2$ (e) $ab^2 + ba^2$

2. Factorise the following quadratic expressions:
(a) $3x^2 + x - 2$ (b) $x^2 - 144$ (c) $s^2 - 5s + 6$ (d) $2y^2 - y - 15$

3. Factorise the following:
(a) $1 - x^2$ (b) $x^2 - 1$ (c) $9 - x^2$ (d) $x^2 - 81$ (e) $25 - y^2$

4. Factorise the denominators of the following expressions:
(a) $\dfrac{1}{x^2 + 6x}$ (b) $\dfrac{3}{s^2 + 3s + 2}$ (c) $\dfrac{3}{s^2 + s - 2}$ (d) $\dfrac{5}{x^2 + 11x + 28}$ (e) $\dfrac{x}{2x^2 - 17x - 9}$

9 Algebraic fractions

Objectives: This chapter:

- explains how to simplify algebraic fractions by cancelling common factors
- explains how algebraic fractions can be multiplied and divided
- explains how algebraic fractions can be added and subtracted
- explains how to express a fraction as the sum of its partial fractions

9.1 Introduction

Just as one whole number divided by another is a numerical fraction, so one algebraic expression divided by another is called an **algebraic fraction**.

$$\frac{x}{y} \qquad \frac{x^2+y}{x} \qquad \frac{3x+2}{7}$$

are all examples of algebraic fractions. The top line is known as the **numerator** of the fraction, and the bottom line is the **denominator**.

Rules for determining the sign of the answer when dividing positive and negative algebraic expressions are the same as those used for dividing numbers.

Key point

When dividing

$$\frac{\text{positive}}{\text{positive}} = \text{positive} \qquad \frac{\text{negative}}{\text{positive}} = \text{negative}$$

$$\frac{\text{positive}}{\text{negative}} = \text{negative} \qquad \frac{\text{negative}}{\text{negative}} = \text{positive}$$

Using these rules we see that an algebraic expression can often be written in different but equivalent forms. For example, note that

$$\frac{x}{-y} \text{ can be written as } -\frac{x}{y}$$

and that

$$\frac{-x}{y} \text{ can be written as } -\frac{x}{y}$$

and also that

$$\frac{-x}{-y} \text{ can be written as } \frac{x}{y}$$

9.2 Cancelling common factors

Cancellation of common factors was described in detail in §2.2.

Consider the numerical fraction $\frac{3}{12}$. To simplify this we factorise both the numerator and the denominator. Any factors which appear in both the numerator and the denominator are called **common factors**. These can be cancelled. For example,

$$\frac{3}{12} = \frac{1 \times 3}{4 \times 3} = \frac{1 \times \cancel{3}}{4 \times \cancel{3}} = \frac{1}{4}$$

The same process is applied when dealing with algebraic fractions.

WORKED EXAMPLES

9.1 For each pair of expressions, state which factors are common to both.

(a) $3xy$ and $6xz$ (b) xy and $5y^2$ (c) $3(x+2)$ and $(x+2)^2$
(d) $3(x-1)$ and $(x-1)(x+4)$

Solution

(a) The expression $6xz$ can be written $(3)(2)xz$. We see that factors common to both this and $3xy$ are 3 and x.

(b) The expression $5y^2$ can be written $5(y)(y)$. We see that the only factor common to both this and xy is y.

(c) $(x+2)^2$ can be written $(x+2)(x+2)$. Thus $(x+2)$ is a factor common to both $(x+2)^2$ and $3(x+2)$.

(d) $3(x-1)$ and $(x-1)(x+4)$ have a common factor of $(x-1)$.

9.2 Simplify

$$\frac{18x^2}{6x}$$

Solution First note that 18 can be factorised as 6×3. So there are factors of 6 and x in both the numerator and the denominator. Then common factors can be cancelled. That is,

$$\frac{18x^2}{6x} = \frac{(6)(3)x^2}{6x} = \frac{3x}{1} = 3x$$

Key point

When simplifying an algebraic fraction only factors common to both the numerator and denominator can be cancelled.

A fraction is expressed in its simplest form by factorising the numerator and denominator and cancelling any common factors.

WORKED EXAMPLES

9.3 Simplify

$$\frac{5}{25 + 15x}$$

Solution First of all note that the denominator can be factorised as $5(5 + 3x)$. There is therefore a factor of 5 in both the numerator and denominator. So 5 is a common factor. This can be cancelled. That is,

$$\frac{5}{25 + 15x} = \frac{1 \times 5}{5(5 + 3x)} = \frac{1 \times \not{5}}{\not{5}(5 + 3x)} = \frac{1}{5 + 3x}$$

It is very important to note that the number 5 that has been cancelled is a common factor. It is incorrect to try to cancel terms that are not common factors.

9.4 Simplify

VIDEO

$$\frac{5x}{25x + 10y}$$

Solution Factorising the denominator we can write

$$\frac{5x}{25x + 10y} = \frac{5x}{5(5x + 2y)}$$

We see that there is a common factor of 5 in both numerator and denominator that can be cancelled. Thus

$$\frac{5x}{25x+10y}=\frac{\cancel{5}x}{\cancel{5}(5x+2y)}=\frac{x}{5x+2y}$$

Note that no further cancellation is possible. x is not a common factor because it is not a factor of the denominator.

9.5 Simplify

$$\frac{4x}{3x^2+x}$$

Solution Note that the denominator factorises to $x(3x+1)$. Once both numerator and denominator have been factorised, any common factors are cancelled. So

$$\frac{4x}{3x^2+x}=\frac{4x}{x(3x+1)}=\frac{4\cancel{x}}{\cancel{x}(3x+1)}=\frac{4}{3x+1}$$

Note that the factor x is common to both numerator and denominator and so has been cancelled.

9.6 Simplify

$$\frac{x}{x^2+2x}$$

Solution Note that the denominator factorises to $x(x+2)$. Also note that the numerator can be written as $1 \times x$. So

$$\frac{x}{x^2+2x}=\frac{1\times x}{x(x+2)}=\frac{1\times\cancel{x}}{\cancel{x}(x+2)}=\frac{1}{x+2}$$

9.7 Simplify

(a) $\dfrac{2(x-1)}{(x+3)(x-1)}$ (b) $\dfrac{x-4}{(x-4)^2}$

Solution

(a) There is a factor of $(x-1)$ common to both the numerator and denominator. This is cancelled to give

$$\frac{2(x-1)}{(x+3)(x-1)}=\frac{2}{x+3}$$

(b) There is a factor of $x-4$ in both numerator and denominator. This is cancelled as follows:

$$\frac{x-4}{(x-4)^2}=\frac{1(x-4)}{(x-4)(x-4)}=\frac{1}{x-4}$$

9.8 Simplify

VIDEO

$$\frac{x+2}{x^2+3x+2}$$

Solution The denominator is factorised and then any common factors are cancelled:

$$\frac{x+2}{x^2+3x+2} = \frac{1(x+2)}{(x+2)(x+1)} = \frac{1}{x+1}$$

9.9 Simplify

(a) $\dfrac{3x+xy}{x^2+5x}$ (b) $\dfrac{x^2-1}{x^2+3x+2}$

Solution The numerator and denominator are both factorised and any common factors are cancelled:

(a) $\dfrac{3x+xy}{x^2+5x} = \dfrac{x(3+y)}{x^2+5x} = \dfrac{\not{x}(3+y)}{\not{x}(x+5)} = \dfrac{3+y}{5+x}$

(b) $\dfrac{x^2-1}{x^2+3x+2} = \dfrac{(x+1)(x-1)}{(x+1)(x+2)} = \dfrac{x-1}{x+2}$

Self-assessment questions 9.2

1. Explain why no cancellation is possible in the expression $\dfrac{3x}{3x+y}$.
2. Explain why no cancellation is possible in the expression $\dfrac{x+1}{x+3}$.
3. Explain why it is possible to perform a cancellation in the expression $\dfrac{x+1}{2x+2}$, and perform it.

Exercise 9.2

1. Simplify

(a) $\dfrac{9x}{3y}$ (b) $\dfrac{9x}{x^2}$ (c) $\dfrac{9xy}{3x}$ (d) $\dfrac{9xy}{3y}$

(e) $\dfrac{9xy}{xy}$ (f) $\dfrac{9xy}{3xy}$

2. Simplify

(a) $\dfrac{15x}{3y}$ (b) $\dfrac{15x}{5y}$ (c) $\dfrac{15xy}{x}$ (d) $\dfrac{15xy}{xy}$

(e) $\dfrac{x^5}{-x^3}$ (f) $\dfrac{-y^3}{y^7}$ (g) $\dfrac{-y}{-y^2}$ (h) $\dfrac{-y^{-3}}{-y^4}$

3. Simplify the following algebraic fractions:

(a) $\dfrac{4}{12+8x}$ (b) $\dfrac{5+10x}{5}$ (c) $\dfrac{2}{4+14x}$

(d) $\dfrac{2x}{4+14x}$ (e) $\dfrac{2x}{2+14x}$ (f) $\dfrac{7}{49x+7y}$

(g) $\dfrac{7y}{49x+7y}$ (h) $\dfrac{7x}{49x+7y}$

4. Simplify

(a) $\dfrac{15x+3}{3}$ (b) $\dfrac{15x+3}{3x+6y}$ (c) $\dfrac{12}{4x+8}$

(d) $\dfrac{12x}{4xy+8x}$ (e) $\dfrac{13x}{x^2+5x}$

(f) $\dfrac{17y}{9y^2+4y}$

5. Simplify the following:

(a) $\dfrac{5}{15+10x}$ (b) $\dfrac{2x}{x^2+7x}$

(c) $\dfrac{2x+8}{x^2+2x-8}$ (d) $\dfrac{7ab}{a^2b^2+9ab}$

(e) $\dfrac{xy}{xy+x}$

6. Simplify

(a) $\dfrac{x-4}{(x-4)(x-2)}$ (b) $\dfrac{2x-4}{x^2+x-6}$

(c) $\dfrac{3x}{3x^2+6x}$ (d) $\dfrac{x^2+2x+1}{x^2-2x-3}$

(e) $\dfrac{2(x-3)}{(x-3)^2}$ (f) $\dfrac{x-3}{(x-3)^2}$

(g) $\dfrac{x-3}{2(x-3)^2}$ (h) $\dfrac{4(x-3)}{2(x-3)^2}$

(i) $\dfrac{x+4}{2(x+4)^2}$ (j) $\dfrac{x+4}{2(x+4)}$

(k) $\dfrac{2(x+4)}{(x+4)}$ (l) $\dfrac{(x+4)(x-3)}{x-3}$

(m) $\dfrac{x+4}{(x-3)(x+4)}$ (n) $\dfrac{x+3}{x^2+7x+12}$

(o) $\dfrac{x+4}{2x+8}$ (p) $\dfrac{x+4}{2x+9}$

9.3 Multiplication and division of algebraic fractions

To multiply two algebraic fractions together we multiply their numerators together and multiply their denominators together:

Key point

$$\frac{a}{b}\times\frac{c}{d}=\frac{a\times c}{b\times d}$$

Any common factors in the result should be cancelled.

WORKED EXAMPLES

9.10 Simplify

$$\frac{4}{5}\times\frac{x}{y}$$

Solution We multiply the numerators together and multiply the denominators together. That is,

$$\frac{4}{5} \times \frac{x}{y} = \frac{4x}{5y}$$

9.11 Simplify

$$\frac{4}{x} \times \frac{3y}{16}$$

Solution The numerators are multiplied together and the denominators are multiplied together. Therefore

$$\frac{4}{x} \times \frac{3y}{16} = \frac{4 \times 3y}{16x}$$

Because $16x = 4 \times 4x$, the common factor 4 can be cancelled. So

$$\frac{4 \times 3y}{16x} = \frac{\not{4} \times 3y}{\not{4} \times 4x} = \frac{3y}{4x}$$

9.12 Simplify

(a) $\frac{1}{2} \times x$ (b) $\frac{1}{2} \times (a + b)$

Solution (a) Writing x as $\frac{x}{1}$ we can state

$$\frac{1}{2} \times x = \frac{1}{2} \times \frac{x}{1} = \frac{1 \times x}{2 \times 1} = \frac{x}{2}$$

(b) Writing $a + b$ as $\frac{a+b}{1}$ we can state

$$\frac{1}{2} \times (a + b) = \frac{1}{2} \times \frac{(a + b)}{1} = \frac{1 \times (a + b)}{2 \times 1} = \frac{a + b}{2}$$

9.13 Simplify

$$\frac{4x^2}{y} \times \frac{3x^3}{yz}$$

Solution We multiply the numerators together and multiply the denominators together:

$$\frac{4x^2}{y} \times \frac{3x^3}{yz} = \frac{4x^2 \times 3x^3}{y \times yz} = \frac{12x^5}{y^2 z}$$

9.14 Simplify

$$5 \times \left(\frac{x-3}{25}\right)$$

Solution This means

$$\frac{5}{1} \times \frac{x-3}{25}$$

which equals

$$\frac{5 \times (x-3)}{1 \times 25}$$

A common factor of 5 can be cancelled from the numerator and denominator to give

$$\frac{(x-3)}{5}$$

9.15 Simplify

$$-\frac{1}{5} \times \frac{3x-4}{8}$$

Solution We can write

$$-\frac{1}{5} \times \frac{3x-4}{8} = -\frac{1 \times (3x-4)}{5 \times 8} = -\frac{3x-4}{40}$$

Note that the answer can also be expressed as $\frac{(4-3x)}{40}$ because

$$-\frac{3x-4}{40} = \frac{-1}{1} \times \frac{3x-4}{40} = \frac{-3x+4}{40} = \frac{4-3x}{40}$$

You should be aware from the last worked example that a solution can often be expressed in a number of equivalent ways.

WORKED EXAMPLES

9.16 Simplify

$$\frac{a}{a+b} \times \frac{b}{5a^2}$$

Solution

$$\frac{a}{a+b} \times \frac{b}{5a^2} = \frac{ab}{5a^2(a+b)}$$

Cancelling the common factor of a in numerator and denominator gives

$$\frac{b}{5a(a+b)}$$

9.17 Simplify

VIDEO

$$\frac{x^2+4x+3}{2x+8} \times \frac{x+4}{x+1}$$

Solution Before multiplying the two fractions together we should try to factorise if possible so that common factors can be identified. By factorising, we can write the given expressions as

$$\frac{(x+1)(x+3)}{2(x+4)} \times \frac{x+4}{x+1} = \frac{(x+1)(x+3)(x+4)}{2(x+4)(x+1)}$$

Cancelling common factors this simplifies to just

$$\frac{x+3}{2}$$

Division is performed by inverting the second fraction and multiplying:

Key point

$$\frac{a}{b} \div \frac{c}{d} = \frac{a}{b} \times \frac{d}{c}$$

WORKED EXAMPLES

9.18 Simplify

$$\frac{10a}{b} \div \frac{a^2}{3b}$$

Solution The second fraction is inverted and then multiplied by the first. That is,

$$\frac{10a}{b} \div \frac{a^2}{3b} = \frac{10a}{b} \times \frac{3b}{a^2} = \frac{30ab}{a^2b} = \frac{30}{a}$$

9.19 Simplify

$$\frac{x^2y^3}{z} \div \frac{y}{x}$$

Solution

$$\frac{x^2y^3}{z} \div \frac{y}{x} = \frac{x^2y^3}{z} \times \frac{x}{y} = \frac{x^3y^3}{zy}$$

Any common factors in the result can be cancelled. So

$$\frac{x^3y^3}{zy} = \frac{x^3y^2}{z}$$

Self-assessment question 9.3

1. The technique of multiplying and dividing algebraic fractions is identical to that used for numbers. True or false?

Exercise 9.3

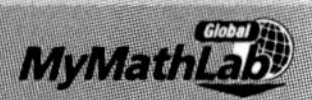

1. Simplify
   (a) $\frac{1}{2} \times \frac{y}{3}$ (b) $\frac{1}{3} \times \frac{z}{2}$ (c) $\frac{2}{5}$ of $\frac{1}{y}$
   (d) $\frac{2}{5}$ of $\frac{1}{x}$ (e) $\frac{3}{4}$ of $\frac{x}{y}$ (f) $\frac{3}{5} \times \frac{x^2}{y}$
   (g) $\frac{x}{y} \times \frac{3}{5}$ (h) $\frac{7}{8} \times \frac{x}{2y}$ (i) $\frac{1}{2} \times \frac{1}{2x}$
   (j) $\frac{1}{2} \times \frac{x}{2}$ (k) $\frac{1}{2} \times \frac{2}{x}$ (l) $\frac{1}{3} \times \frac{x}{3}$
   (m) $\frac{1}{3} \times \frac{3}{x}$ (n) $\frac{1}{3} \times \frac{1}{3x}$ (o) $\frac{1}{3} \times \frac{3x}{2}$

2. Simplify
   (a) $\frac{1}{2} \div \frac{x}{2}$ (b) $\frac{1}{2} \div \frac{2}{x}$ (c) $\frac{2}{x} \div \frac{2}{x}$
   (d) $\frac{x}{2} \div \frac{1}{2}$ (e) $\frac{2}{x} \div 2$ (f) $\frac{2}{x} \div \frac{1}{2}$
   (g) $\frac{3}{x} \div \frac{1}{2}$

3. Simplify the following:
   (a) $\frac{5}{4} \times \frac{a}{25}$ (b) $\frac{5}{4} \times \frac{a}{b}$ (c) $\frac{8a}{b^2} \times \frac{b}{16a^2}$
   (d) $\frac{9x}{3y} \times \frac{2x}{y^2}$ (e) $\frac{3}{5a} \times \frac{b}{a}$ (f) $\frac{1}{4} \times \frac{x}{y}$
   (g) $\frac{1}{3} \times \frac{x}{x+y}$ (h) $\frac{x-3}{x+4} \times \frac{1}{3x-9}$

4. Simplify the following:
   (a) $\frac{3}{x} \times \frac{xy}{z^3}$ (b) $\frac{(3+x)}{x} \div \frac{y}{x}$ (c) $\frac{4}{3} \div \frac{16}{x}$
   (d) $\frac{a}{bc^2} \times \frac{b^2c}{a}$

5. Simplify
   (a) $\frac{x+2}{(x+5)(x+4)} \times \frac{x+5}{x+2}$
   (b) $\frac{x-2}{4} \div \frac{x}{16}$ (c) $\frac{12ab}{5ef} \div \frac{4ab^2}{f}$
   (d) $\frac{x+3y}{2x} \div \frac{y}{4x^2}$ (e) $\frac{3}{x} \times \frac{3}{y} \times \frac{1}{z}$

6. Simplify

$$\frac{1}{x+1} \times \frac{2x+2}{x+3}$$

7. Simplify

$$\frac{x+1}{x+2} \times \frac{x^2+6x+8}{x^2+4x+3}$$

9.4 Addition and subtraction of algebraic fractions

For revision of adding and subtracting fractions see §2.3.

The method is the same as that for adding or subtracting numerical fractions. Note that it is not correct simply to add or subtract the numerator and denominator. The lowest common denominator must first be found. This is the simplest expression that contains all original denominators as its factors. Each fraction is then written with this common denominator. The fractions can then be added or subtracted by adding or subtracting just the numerators, and dividing the result by the common denominator.

WORKED EXAMPLES

9.20 Add the fractions $\frac{3}{4}$ and $\frac{1}{x}$.

Solution We must find $\frac{3}{4}+\frac{1}{x}$. To do this we must first rewrite the fractions to ensure they have a common denominator. The common denominator is the simplest expression that has the given denominators as its factors. The simplest such expression is $4x$. We write

$$\frac{3}{4} \text{ as } \frac{3x}{4x} \quad \text{and} \quad \frac{1}{x} \text{ as } \frac{4}{4x}$$

Then

$$\frac{3}{4}+\frac{1}{x}=\frac{3x}{4x}+\frac{4}{4x}$$

$$=\frac{3x+4}{4x}$$

No further simplification is possible.

9.21 Simplify

$$\frac{3}{x}+\frac{4}{x^2}$$

Solution The expression $\frac{3}{x}$ is rewritten as $\frac{3x}{x^2}$, which makes the denominators of both terms x^2 but leaves the value of the expression unaltered. Note that both the original denominators, x and x^2, are factors of the new denominator. We call this denominator the **lowest common denominator**. The fractions are then added by adding just the numerators. That is,

$$\frac{3x}{x^2}+\frac{4}{x^2}=\frac{3x+4}{x^2}$$

9.22 Express $\frac{5}{a} - \frac{4}{b}$ as a single fraction.

Solution Both fractions are rewritten to have the same denominator. The simplest expression containing both a and b as its factors is ab. Therefore ab is the lowest common denominator. Then

$$\frac{5}{a} - \frac{4}{b} = \frac{5b}{ab} - \frac{4a}{ab} = \frac{5b - 4a}{ab}$$

9.23 Write $\frac{4}{x+y} - \frac{3}{y}$ as a single fraction.

Solution The simplest expression that contains both denominators as its factors is $(x+y)y$. We must rewrite each term so that it has this denominator:

$$\frac{4}{x+y} = \frac{4}{x+y} \times \frac{y}{y} = \frac{4y}{(x+y)y}$$

Similarly,

$$\frac{3}{y} = \frac{3}{y} \times \frac{x+y}{x+y} = \frac{3(x+y)}{(x+y)y}$$

The fractions are then subtracted by subtracting just the numerators:

$$\frac{4}{x+y} - \frac{3}{y} = \frac{4y}{(x+y)y} - \frac{3(x+y)}{(x+y)y} = \frac{4y - 3(x+y)}{(x+y)y}$$

which simplifies to

$$\frac{y - 3x}{(x+y)y}$$

9.24 Express as a single fraction

$$\frac{2}{x+3} + \frac{5}{x-1}$$

Solution The simplest expression having both $x + 3$ and $x - 1$ as its factors is

$$(x+3)(x-1)$$

This is the lowest common denominator. Each term is rewritten so that it has this denominator. Thus

$$\frac{2}{x+3} = \frac{2(x-1)}{(x+3)(x-1)} \quad \text{and} \quad \frac{5}{x-1} = \frac{5(x+3)}{(x+3)(x-1)}$$

Then

$$\frac{2}{x+3}+\frac{5}{x-1}=\frac{2(x-1)}{(x+3)(x-1)}+\frac{5(x+3)}{(x+3)(x-1)}$$
$$=\frac{2(x-1)+5(x+3)}{(x+3)(x-1)}$$

which simplifies to

$$\frac{7x+13}{(x+3)(x-1)}$$

9.25 Express as a single fraction

$$\frac{1}{x-4}+\frac{1}{(x-4)^2}$$

Solution The simplest expression having $x-4$ and $(x-4)^2$ as its factors is $(x-4)^2$. Both fractions are rewritten with this denominator:

$$\frac{1}{x-4}+\frac{1}{(x-4)^2}=\frac{(x-4)}{(x-4)^2}+\frac{1}{(x-4)^2}$$
$$=\frac{x-4+1}{(x-4)^2}$$
$$=\frac{x-3}{(x-4)^2}$$

Self-assessment question 9.4

1. Explain what is meant by the 'lowest common denominator' and how it is found.

Exercise 9.4

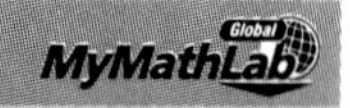

1. Express each of the following as a single fraction:

   (a) $\frac{z}{2}+\frac{z}{3}$ (b) $\frac{x}{3}+\frac{x}{4}$ (c) $\frac{y}{5}+\frac{y}{25}$

2. Express each of the following as a single fraction:

   (a) $\frac{1}{2}+\frac{1}{x}$ (b) $\frac{1}{2}+x$ (c) $\frac{1}{3}+y$ (d) $\frac{1}{3}+\frac{1}{y}$ (e) $8+\frac{1}{y}$

3. Express each of the following as a single fraction:

   (a) $\frac{5}{x}-\frac{1}{2}$ (b) $\frac{5}{x}+2$ (c) $\frac{3}{x}-\frac{1}{3}$ (d) $\frac{x}{3}-\frac{1}{2}$ (e) $\frac{3}{x}+\frac{1}{3}$

4. Express each of the following as a single fraction:
(a) $\frac{3}{x}+\frac{4}{y}$ (b) $\frac{3}{x^2}+\frac{4y}{x}$ (c) $\frac{4ab}{x}+\frac{3ab}{2y}$
(d) $\frac{4xy}{a}+\frac{3xy}{2b}$ (e) $\frac{3}{x}-\frac{6}{2x}$ (f) $\frac{3x}{2y}-\frac{7y}{4x}$
(g) $\frac{3}{x+y}-\frac{2}{y}$ (h) $\frac{1}{a+b}-\frac{1}{a-b}$
(i) $2x+\frac{1}{2x}$ (j) $2x-\frac{1}{2x}$

5. Express each of the following as a single fraction:
(a) $\frac{x}{y}+\frac{3x^2}{z}$ (b) $\frac{4}{a}+\frac{5}{b}$
(c) $\frac{6x}{y}-\frac{2y}{x}$ (d) $3x-\frac{3x+1}{4}$
(e) $\frac{5a}{12}+\frac{9a}{18}$ (f) $\frac{x-3}{4}+\frac{3}{5}$

6. Express each of the following as a single fraction:
(a) $\frac{1}{x+1}+\frac{1}{x+2}$ (b) $\frac{1}{x-1}+\frac{2}{x+3}$
(c) $\frac{3}{x+5}+\frac{1}{x+4}$ (d) $\frac{1}{x-2}+\frac{3}{x-4}$
(e) $\frac{3}{2x+1}+\frac{1}{x+1}$ (f) $\frac{3}{1-2x}+\frac{1}{x}$
(g) $\frac{3}{x+1}+\frac{4}{(x+1)^2}$
(h) $\frac{1}{x-1}+\frac{1}{(x-1)^2}$

9.5 Partial fractions

We have seen how to add and/or subtract algebraic fractions to yield a single fraction. Section 9.4 illustrates the process with some examples.

Sometimes we wish to use the reverse of this process. That is, starting with a single fraction we wish to express it as the sum of two or more simpler fractions. Each of these simpler fractions is known as a **partial fraction** because it is part of the original fraction.

Let us refer to Worked Example 9.24. From this example we can see that

$$\frac{2}{x+3}+\frac{5}{x-1}$$

can be expressed as the single fraction, $\frac{7x+13}{(x+3)(x-1)}$. Worked Example 9.26 illustrates the process of starting with $\frac{7x+13}{(x+3)(x-1)}$ and finding its partial fractions, $\frac{2}{x+3}$ and $\frac{5}{x-1}$.

WORKED EXAMPLES

9.26 Express $\frac{7x+13}{(x+3)(x-1)}$ as its partial fractions.

Solution The denominator has two factors: $x + 3$ and $x - 1$. It is these factors that determine the form of the partial fractions. Each factor in the denominator produces a partial fraction – this is an important point.

The factor $x + 3$ produces a partial fraction of the form $\frac{A}{x+3}$ where A is a constant. Similarly the factor $x - 1$ produces a partial fraction of the form $\frac{B}{x-1}$ where B is a constant. Hence we have

$$\frac{7x + 13}{(x + 3)(x - 1)} = \frac{A}{x + 3} + \frac{B}{x - 1} \tag{9.1}$$

We now need to find the values of A and B. By multiplying both sides of (9.1) by $(x + 3)(x - 1)$ we have

$$\frac{7x + 3}{(x + 3)(x - 1)} \times (x + 3)(x - 1) = \frac{A}{x + 3} \times (x + 3)(x - 1) + \frac{B}{x - 1} \times (x + 3)(x - 1) \tag{9.2}$$

By cancelling the common factors in each term, (9.2) simplifies to

$$7x + 13 = A(x - 1) + B(x + 3) \tag{9.3}$$

Note that (9.3) is true for *all* values of x. To find the values of A and B we can substitute into (9.3) any value of x we choose. We choose values of x that are most helpful and convenient. Let us choose x to be -3. When $x = -3$ is substituted into (9.3) we obtain

$$7(-3) + 13 = A(-3 - 1) + B(-3 + 3)$$

from which

$$-8 = A(-4)$$
$$A = 2$$

As you can see, the value of $x = -3$ was chosen in order to simplify (9.3) in such a way that the value of A could then be found.

Now we return to (9.3) and let $x = 1$. Then (9.3) simplifies to

$$7(1) + 13 = A(1 - 1) + B(1 + 3)$$

from which $B = 5$. Clearly we chose $x = 1$ in order to simplify (9.3) by eliminating the A term, thus allowing B to be found. Putting $A = 2$, $B = 5$ into (9.1) we have the partial fractions

$$\frac{7x + 13}{(x + 3)(x - 1)} = \frac{2}{x + 3} + \frac{5}{x - 1}$$

9.27 Express $\frac{5x+28}{x^2+7x+10}$ as partial fractions.

VIDEO

Solution The factors of the denominator must first be found, that is $x^2 + 7x + 10$ must be factorised:

$$x^2 + 7x + 10 = (x + 2)(x + 5)$$

As the denominator has two factors, then there are two partial fractions. The factor $x + 2$ leads to a partial fraction $\frac{A}{x+2}$ and the factor $x + 5$ leads to a partial fraction $\frac{B}{x+5}$, where A and B are constants whose values have yet to be found.

So

$$\frac{5x + 28}{x^2 + 7x + 10} = \frac{5x + 28}{(x + 2)(x + 5)} = \frac{A}{x + 2} + \frac{B}{x + 5} \qquad (9.4)$$

To find the values of A and B we multiply both sides of (9.4) by $(x + 2)(x + 5)$. After cancelling common factors in each term we have

$$5x + 28 = A(x + 5) + B(x + 2) \qquad (9.5)$$

We now select convenient values of x to simplify (9.5) so that A and B can be found. We see that choosing $x = -2$ will simplify (9.5) so that A can be determined. With $x = -2$, (9.5) becomes

$$5(-2) + 28 = A(-2 + 5) + B(-2 + 2)$$
$$18 = 3A$$
$$A = 6$$

Returning to (9.5), we let $x = -5$ so that B can be found:

$$5(-5) + 28 = A(-5 + 5) + B(-5 + 2)$$
$$3 = -3B$$
$$B = -1$$

Putting the values of A and B into (9.4) yields the partial fractions

$$\frac{5x + 28}{x^2 + 7x + 10} = \frac{6}{x + 2} - \frac{1}{x + 5}$$

If a denominator has a repeated factor then a slight variation of the previous method is employed.

WORKED EXAMPLE

9.28 Find the partial fractions of $\frac{6x-5}{4x^2-4x+1}$.

Solution As in the previous example the denominator must first be factorised:

$$4x^2 - 4x + 1 = (2x - 1)(2x - 1) = (2x - 1)^2$$

We note that although there are two factors, they are identical (that is, a repeated factor). As there are two factors, then two partial fractions are generated. The partial fractions in such a case are of the form

$$\frac{A}{2x-1}+\frac{B}{(2x-1)^2}$$

where A and B are constants. Note that the denominators $(2x-1)$ and $(2x-1)^2$ are used. So

$$\frac{6x-5}{4x^2-4x+1}=\frac{6x-5}{(2x-1)^2}=\frac{A}{2x-1}+\frac{B}{(2x-1)^2} \qquad (9.6)$$

We need to find the values of the constants A and B. Multiplying both sides of (9.6) by $(2x-1)^2$ and then cancelling any common factors yields:

$$6x-5=A(2x-1)+B \qquad (9.7)$$

The left-hand side and right-hand side of (9.7) are equal for all values of x. We choose convenient values of x to help us find the values of A and B. By substituting $x=\frac{1}{2}$ into (9.7) it simplifies to

$$-2=A(0)+B$$

and so $B=-2$. The value of $x=\frac{1}{2}$ was chosen as the factor $(2x-1)$ then evaluated to 0 and the A term on the right-hand side became 0, allowing B to be found directly. We now need to find the value of A. It is impossible to eliminate the B term from (9.7) by any choice of an x value. Hence we simply choose any value of x that simplifies (9.7) considerably. Let us for example substitute $x=0$ into (9.7) to obtain

$$-5=A(-1)+B$$
$$A=B+5$$

As we have already found the value of B to be -2, this is then substituted into the above equation to yield $A=3$.

Substituting the values of A and B into (9.6) produces the partial fractions

$$\frac{6x-5}{4x^2-4x+1}=\frac{3}{2x-1}-\frac{2}{(2x-1)^2}$$

From Worked Examples 9.26, 9.27 and 9.28 we note the following:

Key point

When calculating partial fractions:

- the denominator must be factorised
- each factor of the denominator generates a partial fraction
- for repeated factors, the partial fractions are generated by the factor and the square of the factor

There are more complicated examples of partial fractions that are not dealt with in this chapter. For example, some fractions have factors in the denominator that are quadratics that will not factorise. For a more in-depth study of partial fractions see *Mathematics for Engineers: A Modern Interactive Approach*, 3rd edition, by A. Croft and R. Davison (2008, Pearson Education).

Exercise 9.5

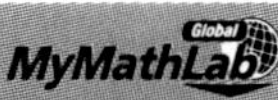

1. Find the partial fractions of the following:

   (a) $\dfrac{7x+18}{(x+2)(x+3)}$ (b) $\dfrac{2x-7}{x^2+5x+4}$ (c) $\dfrac{-9}{2x^2+15x+18}$ (d) $\dfrac{5x-11}{x^2-5x+4}$ (e) $\dfrac{3x+11}{2x^2+3x-2}$

2. Find the partial fractions of

   $$\frac{x+21}{2(2x+3)(3x-2)}$$

3. Find the partial fractions of

   (a) $\dfrac{x-35}{x^2-25}$ (b) $\dfrac{x-4}{x^2-6x+9}$ (c) $\dfrac{5x+4}{-x^2-x+2}$ (d) $\dfrac{12x-5}{9x^2-6x+1}$

Test and assignment exercises 9

1. Simplify

   (a) $\dfrac{5a}{4a+3ab}$ (b) $\dfrac{5a}{30a+15b}$ (c) $\dfrac{5ab}{30a+15b}$ (d) $\dfrac{5ab}{ab+7ab}$ (e) $\dfrac{y}{13y+y^2}$ (f) $\dfrac{13y+y^2}{y}$

2. Simplify the following:

   (a) $\dfrac{5a}{7}\times\dfrac{14b}{2}$ (b) $\dfrac{3}{x}+\dfrac{7}{3x}$ (c) $t-\dfrac{4-t}{2}$ (d) $4(x+3)-\dfrac{(4x-5)}{3}$

   (e) $\dfrac{7}{x}+\dfrac{3}{2x}+\dfrac{5}{3x}$ (f) $x+\dfrac{3x}{y}$ (g) $xy+\dfrac{1}{xy}$ (h) $\dfrac{1}{x+y}+\dfrac{2}{x-y}$

3. Simplify the following:

   (a) $\dfrac{y}{9}+\dfrac{2y}{7}$ (b) $\dfrac{3}{x}-\dfrac{5}{3x}+\dfrac{4}{5x}$ (c) $\dfrac{3x}{2y}+\dfrac{5y}{6x}$ (d) $m+\dfrac{m+n}{2}$

   (e) $m-\dfrac{m+n}{2}$ (f) $m-\dfrac{m-n}{2}$ (g) $\dfrac{3s-5}{10}-\dfrac{2s-3}{15}$

4. Simplify

   $$\frac{x^2-x}{x-1}$$

5. Find the partial fractions of

(a) $\dfrac{4x+13}{(x+2)(x+7)}$ (b) $\dfrac{x+4}{(x+1)(x+2)}$ (c) $\dfrac{x-14}{(x+4)(x-5)}$ (d) $\dfrac{-x}{(3x+1)(2x+1)}$

(e) $\dfrac{6x-13}{(2x+3)(3x-1)}$

6. Calculate the partial fractions of

(a) $\dfrac{8x+19}{x^2+5x+6}$ (b) $\dfrac{3x+11}{x^2+9x+20}$ (c) $\dfrac{3x+7}{x^2+4x+4}$ (d) $\dfrac{x-6}{x^2-6x+9}$ (e) $\dfrac{8x-7}{4x^2-4x+1}$

10 Transposing formulae

Objectives: This chapter:

- explains how formulae can be rearranged or transposed

10.1 Rearranging a formula

In the formula for the area of a circle, $A = \pi r^2$, we say that A is the **subject** of the formula. The subject appears by itself on one side, usually the left, of the formula, and nowhere else. If we are asked to **transpose** the formula for r, then we must rearrange the formula so that r becomes the subject. The rules for transposing formulae are quite simple. Essentially whatever you do to one side you must also do to the whole of the other side.

Key point

To transpose a formula you may:

- add the same quantity to both sides
- subtract the same quantity from both sides
- multiply or divide both sides by the same quantity
- perform operations on both sides, such as 'square both sides', 'square root both sides' etc.

WORKED EXAMPLES

10.1 The circumference C of a circle is given by the formula $C = 2\pi r$. Transpose this to make r the subject.

Solution The intention is to obtain r by itself on the left-hand side. Starting with $C = 2\pi r$ we must try to isolate r. Dividing both sides by 2π we find

$$C = 2\pi r$$

$$\frac{C}{2\pi} = \frac{2\pi r}{2\pi}$$

$$\frac{C}{2\pi} = r \quad \text{by cancelling the common factor } 2\pi \text{ on the right}$$

Finally we can write $r = \frac{C}{2\pi}$ and the formula has been transposed for r.

10.2 Transpose the formula $y = 3(x + 7)$ for x.

Solution We must try to obtain x on its own on the left-hand side. This can be done in stages. Dividing both sides by 3 we find

$$\frac{y}{3} = \frac{3(x+7)}{3} = x + 7 \quad \text{by cancelling the common factor 3}$$

Subtracting 7 from both sides then gives

$$\frac{y}{3} - 7 = x + 7 - 7$$

so that

$$\frac{y}{3} - 7 = x$$

Equivalently we can write $x = \frac{y}{3} - 7$, and the formula has been transposed for x.

Alternatively, again starting from $y = 3(x + 7)$ we could proceed as follows. Removing the brackets we obtain $y = 3x + 21$. Then, subtracting 21 from both sides,

$$y - 21 = 3x + 21 - 21 = 3x$$

Dividing both sides by 3 will give x on its own:

$$\frac{y-21}{3} = x \quad \text{so that} \quad x = \frac{y-21}{3}$$

Noting that

$$\frac{y-21}{3} = \frac{y}{3} - \frac{21}{3} = \frac{y}{3} - 7$$

we see that this answer is equivalent to the expression obtained previously.

10.3 Transpose the formula $y - z = 3(x + 2)$ for x.

Solution Dividing both sides by 3 we find

$$\frac{y - z}{3} = x + 2$$

Subtracting 2 from both sides gives

$$\frac{y - z}{3} - 2 = x + 2 - 2$$

$$= x$$

so that

$$x = \frac{y - z}{3} - 2$$

10.4 Transpose $x + xy = 7$, (a) for x, (b) for y.

Solution (a) In the expression $x + xy$, note that x is a common factor. We can then write the given formula as $x(1 + y) = 7$. Dividing both sides by $1 + y$ will isolate x. That is,

$$\frac{x(1 + y)}{1 + y} = \frac{7}{1 + y}$$

Cancelling the common factor $1 + y$ on the left-hand side gives

$$x = \frac{7}{1 + y}$$

and the formula has been transposed for x.

(b) To transpose $x + xy = 7$ for y we first subtract x from both sides to give $xy = 7 - x$. Finally, dividing both sides by x gives

$$\frac{xy}{x} = \frac{7 - x}{x}$$

Cancelling the common factor of x on the left-hand side gives $y = \frac{7 - x}{x}$, which is the required transposition.

10.5 Transpose the formula $I = \frac{V}{R}$ for R.

Solution Starting with $I = \frac{V}{R}$ we first multiply both sides by R. This gives

$$IR = \frac{V}{R} \times R = V$$

Then, dividing both sides by I gives

$$\frac{IR}{I} = \frac{V}{I}$$

so that $R = \frac{V}{I}$.

10.6 Make x the subject of the formula $y = \dfrac{4}{x-7}$.

Solution We must try to obtain x on its own on the left-hand side. It is often useful to multiply both sides of the formula by the same quantity in order to remove fractions. Multiplying both sides by $(x - 7)$ we find

$$(x-7)y = (x-7) \times \frac{4}{x-7} = 4$$

Removing the brackets on the left-hand side we find

$$xy - 7y = 4 \qquad \text{so that} \qquad xy = 7y + 4$$

Finally, dividing both sides by y we obtain

$$\frac{xy}{y} = \frac{7y+4}{y}$$

That is,

$$x = \frac{7y+4}{y}$$

10.7 Transpose the formula

VIDEO

$$T = 2\pi\sqrt{\frac{l}{g}}$$

to make l the subject.

Solution We must attempt to isolate l. We can do this in stages as follows. Dividing both sides by 2π we find

$$\frac{T}{2\pi} = \frac{2\pi\sqrt{\frac{l}{g}}}{2\pi} = \sqrt{\frac{l}{g}} \qquad \text{by cancelling the common factor } 2\pi$$

Recall from §6.3 that $(\sqrt{a})^2 = a$.

The square root sign over the l/g can be removed by squaring both sides. Recall that if a term containing a square root is squared then the square root will disappear:

$$\left(\frac{T}{2\pi}\right)^2 = \frac{l}{g}$$

Then, multiplying both sides by g we find

$$g\left(\frac{T}{2\pi}\right)^2 = l$$

Equivalently we have

$$l = g\left(\frac{T}{2\pi}\right)^2$$

and the formula has been transposed for l.

Self-assessment questions 10.1

1. Explain what is meant by the subject of a formula.
2. In what circumstances might you want to transpose a formula?
3. Explain what processes are allowed when transposing a formula.

Exercise 10.1

MyMathLab

1. Transpose each of the following formulae to make x the subject:

   (a) $y = 3x$ (b) $y = \frac{1}{x}$ (c) $y = 7x - 5$
   (d) $y = \frac{1}{2}x - 7$ (e) $y = \frac{1}{2x}$
   (f) $y = \frac{1}{2x + 1}$ (g) $y = \frac{1}{2x} + 1$
   (h) $y = 18x - 21$ (i) $y = 19 - 8x$

2. Transpose $y = mx + c$, (a) for m, (b) for x, (c) for c.

3. Transpose the following formulae for x:

   (a) $y = 13(x - 2)$ (b) $y = x\left(1 + \frac{1}{x}\right)$
   (c) $y = a + t(x - 3)$

4. Transpose each of the following formulae to make the given variable the subject:

   (a) $y = 7x + 11$, for x (b) $V = IR$, for I
   (c) $V = \frac{4}{3}\pi r^3$, for r (d) $F = ma$, for m

5. Make n the subject of the formula $l = a + (n - 1)d$.

6. If $m = n + t\sqrt{x}$, find an expression for x.

7. Make x the subject of the following formulae:
(a) $y = 1 - x^2$ (b) $y = \frac{1}{1 - x^2}$
(c) $y = \frac{1 - x^2}{1 + x^2}$

Test and assignment exercises 10

1. Make x the subject of the following formulae:
(a) $y = 13x - 18$ (b) $y = \frac{13}{x + 18}$ (c) $y = \frac{13}{x} + 18$

2. Transpose the following formulae:
(a) $y = \frac{7 + x}{14}$ for x (b) $E = \frac{1}{2} mv^2$ for v (c) $8x - 13y = 12$ for y (d) $y = \frac{x^2}{2g}$ for x

3. Make x the subject of the formula $v = k/\sqrt{x}$.

4. Transpose the formula $V = \pi r^2 h$ for h.

5. Make r the subject of the formula $y = H + Cr$.

6. Transpose the formula $s = ut + \frac{1}{2} at^2$ for a.

7. Transpose the formula $v^2 = u^2 + 2as$ for u.

8. Transpose the formula $k = pv^3$ for v.

11 Solving equations

Objectives: This chapter:

- explains what is meant by an equation and its solution
- shows how to solve linear, simultaneous and quadratic equations

An **equation** states that two quantities are equal, and will always contain an **unknown quantity** that we wish to find. For example, in the equation $5x + 10 = 20$ the unknown quantity is x. To **solve** an equation means to find all values of the unknown quantity that can be substituted into the equation so that the left side equals the right side. Each such value is called a **solution** or alternatively a **root** of the equation. In the example above the solution is $x = 2$ because when $x = 2$ is substituted both the left side and the right side equal 20. The value $x = 2$ is said to **satisfy** the equation.

11.1 Solving linear equations

A **linear** equation is one of the form $ax + b = 0$ where a and b are numbers and the unknown quantity is x. The number a is called the **coefficient** of x. The number b is called the **constant term**. For example, $3x + 7 = 0$ is a linear equation. The coefficient of x is 3 and the constant term is 7. Similarly, $-2x + 17.5 = 0$ is a linear equation. The coefficient of x is -2 and the constant term is 17.5. Note that the unknown quantity occurs only to the first power, that is as x, and not as x^2, x^3, $x^{1/2}$ etc. Linear equations may appear in other forms that may seem to be different but are nevertheless equivalent. Thus

$$4x + 13 = -7, \qquad 3 - 14x = 0 \qquad \text{and} \qquad 3x + 7 = 2x - 4$$

are all linear equations that could be written in the form $ax + b = 0$ if necessary.

An equation such as $3x^2 + 2 = 0$ is not linear because the unknown quantity occurs to the power 2. Linear equations are solved by trying to obtain the unknown quantity on its own on the left-hand side; that is, by making the unknown quantity the subject of the equation. This is done using the rules given for transposing formulae in Chapter 10. Consider the following examples.

WORKED EXAMPLE

11.1 Solve the equation $x + 10 = 0$.

Solution We make x the subject by subtracting 10 from both sides to give $x = -10$. Therefore $x = -10$ is the solution. It can be easily checked by substituting into the original equation:

$$(-10) + 10 = 0 \qquad \text{as required}$$

Note from the last worked example that the solution should be checked by substitution to ensure it satisfies the given equation. If it does not then a mistake has been made.

WORKED EXAMPLES

11.2 Solve the equation $4x + 8 = 0$.

Solution In order to find the unknown quantity x we attempt to make it the subject of the equation. Subtracting 8 from both sides we find

$$4x + 8 - 8 = 0 - 8 = -8$$

That is,

$$4x = -8$$

Then dividing both sides by 4 gives

$$\frac{4x}{4} = \frac{-8}{4}$$

so that $x = -2$. The solution of the equation $4x + 8 = 0$ is $x = -2$.

11.3 Solve the equation $5x + 17 = 4x - 3$.

Solution First we collect together all terms involving x . This is done by subtracting $4x$ from both sides to remove this term from the right. This gives

$$5x - 4x + 17 = -3$$

That is, $x + 17 = -3$. To make x the subject of this equation we now subtract 17 from both sides to give $x = -3 - 17 = -20$. The solution of the equation $5x + 17 = 4x - 3$ is $x = -20$. Note that the answer can be easily checked by substituting $x = -20$ into the original equation and verifying that the left side equals the right side.

11.4 Solve the equation $\dfrac{x-3}{4} = 1$.

Solution We attempt to obtain x on its own. First note that if we multiply both sides of the equation by 4 this will remove the 4 in the denominator. That is,

$$4 \times \left(\frac{x-3}{4}\right) = 4 \times 1$$

so that

$$x - 3 = 4$$

Finally, adding 3 to both sides gives $x = 7$.

Self-assessment questions 11.1

1. Explain what is meant by a root of an equation.
2. Explain what is meant by a linear equation.
3. State the rules that can be used to solve a linear equation.
4. You may think that a formula and an equation look very similar. Try to explain the distinction between a formula and an equation.

Exercise 11.1

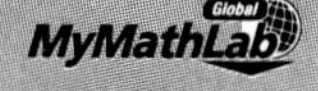

1. Verify that the given values of x satisfy the given equations:
   (a) $x = 7$ satisfies $3x + 4 = 25$
   (b) $x = -5$ satisfies $2x - 11 = -21$
   (c) $x = -4$ satisfies $-x - 8 = -4$
   (d) $x = \frac{1}{2}$ satisfies $8x + 4 = 8$
   (e) $x = -\frac{1}{3}$ satisfies $27x + 8 = -1$
   (f) $x = 4$ satisfies $3x + 2 = 7x - 14$
2. Solve the following linear equations:
   (a) $3x = 9$ (b) $\dfrac{x}{3} = 9$ (c) $3t + 6 = 0$
   (d) $3x - 13 = 2x + 9$ (e) $3x + 17 = 21$
   (f) $4x - 20 = 3x + 16$
   (g) $5 - 2x = 2 + 3x$
   (h) $\dfrac{x+3}{2} = 3$ (i) $\dfrac{3x+2}{2} + 3x = 1$
3. Solve the following equations:
   (a) $5(x + 2) = 13$
   (b) $3(x - 7) = 2(x + 1)$
   (c) $5(1 - 2x) = 2(4 - 2x)$
4. Solve the following equations:
   (a) $3t + 7 = 4t - 2$
   (b) $3v = 17 - 4v$
   (c) $3s + 2 = 14(s - 1)$

5. Solve the following linear equations:
   (a) $5t + 7 = 22$ (b) $7 - 4t = -13$
   (c) $7 - 4t = 27$ (d) $5 = 14 - 3t$
   (e) $4x + 13 = -x + 25$
   (f) $\frac{x+3}{2} = \frac{x-3}{4}$ (g) $\frac{1}{3}x + 6 = \frac{1}{2}x + 2$
   (h) $\frac{1}{5}x + 7 = \frac{1}{3}x + 5$
   (i) $\frac{2x+4}{5} = \frac{x-3}{2}$ (j) $\frac{x-7}{8} = \frac{3x+1}{5}$
6. The following equations may not appear to be linear at first sight but they can all be rewritten in the standard form of a linear equation. Find the solution of each equation.
   (a) $\frac{1}{x} = 5$ (b) $\frac{1}{x} = \frac{5}{2}$ (c) $\frac{1}{x+1} = \frac{5}{2}$
   (d) $\frac{1}{x-1} = \frac{5}{2}$ (e) $\frac{1}{x} = \frac{1}{2x+1}$
   (f) $\frac{3}{x} = \frac{1}{2x+1}$ (g) $\frac{1}{x+1} = \frac{1}{3x+2}$
   (h) $\frac{3}{x+1} = \frac{2}{4x+1}$

11.2 Solving simultaneous equations

Sometimes equations contain more than one unknown quantity. When this happens there are usually two or more equations. For example, in the two equations

$$x + 2y = 14 \qquad 3x + y = 17$$

the unknowns are x and y. Such equations are called **simultaneous equations** and to solve them we must find values of x and y that satisfy both equations at the same time. If we substitute $x = 4$ and $y = 5$ into either of the two equations above we see that the equation is satisfied. We shall demonstrate how simultaneous equations can be solved by removing, or **eliminating**, one of the unknowns.

WORKED EXAMPLES

11.5 Solve the simultaneous equations

$$x + 3y = 14 \tag{11.1}$$

$$2x - 3y = -8 \tag{11.2}$$

Solution Note that if these two equations are added the unknown y is removed or eliminated:

$$\begin{array}{r} x + 3y = 14 \\ 2x - 3y = -8 \\ \hline 3x \quad = 6 \\ \hline \end{array} \; +$$

so that $x = 2$. To find y we substitute $x = 2$ into either equation. Substituting into Equation 11.1 gives

$$2 + 3y = 14$$

Solving this linear equation will give y. We have

$$2 + 3y = 14$$

$$3y = 14 - 2 = 12$$

$$y = \frac{12}{3} = 4$$

Therefore the solution of the simultaneous equations is $x = 2$ and $y = 4$. Note that these solutions should be checked by substituting back into both given equations to check that the left-hand side equals the right-hand side.

11.6

VIDEO

Solve the simultaneous equations

$$5x + 4y = 7 \quad (11.3)$$

$$3x - y = 11 \quad (11.4)$$

Solution

Note that, in this example, if we multiply the first equation by 3 and the second by 5 we shall have the same coefficient of x in both equations. This gives

$$15x + 12y = 21 \quad (11.5)$$

$$15x - 5y = 55 \quad (11.6)$$

We can now eliminate x by subtracting Equation 11.6 from Equation 11.5, giving

$$\begin{array}{rl} 15x + 12y = & 21 \\ 15x - 5y = & 55 \\ \hline 17y = & -34 \\ \hline \end{array} \quad -$$

from which $y = -2$. In order to find x we substitute our solution for y into either of the given equations. Substituting into Equation 11.3 gives

$$5x + 4(-2) = 7 \qquad \text{so that} \qquad 5x = 15 \qquad \text{or} \qquad x = 3$$

The solution of the simultaneous equations is therefore $x = 3$, $y = -2$.

Exercise 11.2

MyMathLab

1. Verify that the given values of x and y satisfy the given simultaneous equations.
   (a) $x = 7$, $y = 1$ satisfy $2x - 3y = 11$, $3x + y = 22$
   (b) $x = -7$, $y = 2$ satisfy $2x + y = -12$, $x - 5y = -17$
   (c) $x = -1$, $y = -1$ satisfy $7x - y = -6$, $x - y = 0$

2. Solve the following pairs of simultaneous equations:
   (a) $3x + y = 1, \quad 2x - y = 2$
   (b) $4x + 5y = 21, \quad 3x + 5y = 17$
   (c) $2x - y = 17, \quad x + 3y = 12$
   (d) $-2x + y = -21, \quad x + 3y = -14$
   (e) $-x + y = -10, \quad 3x + 7y = 20$
   (f) $4x - 2y = 2, \quad 3x - y = 4$
3. Solve the following simultaneous equations:
   (a) $5x + y = 36, 3x - y = 20$
   (b) $x - 3y = -13, 4x + 2y = -24$
   (c) $3x + y = 30, -5x + 3y = -50$
   (d) $3x - y = -5, -7x + 3y = 15$
   (e) $11x + 13y = -24, x + y = -2$

11.3 Solving quadratic equations

A **quadratic equation** is an equation of the form $ax^2 + bx + c = 0$ where a, b and c are numbers and x is the unknown quantity we wish to find. The number a is the **coefficient** of x^2, b is the coefficient of x, and c is the **constant term**. Sometimes b or c may be zero, although a can never be zero. For example,

$$x^2 + 7x + 2 = 0 \qquad 3x^2 - 2 = 0 \qquad -2x^2 + 3x = 0 \qquad 8x^2 = 0$$

are all quadratic equations.

Key point

A quadratic equation has the form $ax^2 + bx + c = 0$ where a, b and c are numbers, and x represents the unknown we wish to find.

WORKED EXAMPLES

11.7 State the coefficient of x^2 and the coefficient of x in the following quadratic equations:

(a) $4x^2 + 3x - 2 = 0$ (b) $x^2 - 23x + 17 = 0$ (c) $-x^2 + 19 = 0$

Solution

(a) In the equation $4x^2 + 3x - 2 = 0$ the coefficient of x^2 is 4 and the coefficient of x is 3.

(b) In the equation $x^2 - 23x + 17 = 0$ the coefficient of x^2 is 1 and the coefficient of x is -23.

(c) In the equation $-x^2 + 19 = 0$ the coefficient of x^2 is -1 and the coefficient of x is 0, since there is no term involving just x.

11.8 Verify that both $x = -7$ and $x = 5$ satisfy the quadratic equation $x^2 + 2x - 35 = 0$.

Solution We substitute $x = -7$ into the left-hand side of the equation. This yields

$$(-7)^2 + 2(-7) - 35$$

That is,

$$49 - 14 - 35$$

This simplifies to zero, and so the left-hand side equals the right-hand side of the given equation. Therefore $x = -7$ is a solution.

Similarly, if $x = 5$ we find

$$(5^2) + 2(5) - 35 = 25 + 10 - 35$$

which also simplifies to zero. We conclude that $x = 5$ is also a solution.

Solution by factorisation

If the quadratic expression on the left-hand side of the equation can be factorised, solutions can be found using the method in the following examples.

WORKED EXAMPLES

11.9 Solve the quadratic equation $x^2 + 3x - 10 = 0$.

Solution The left-hand side of the equation can be factorised to give

$$x^2 + 3x - 10 = (x + 5)(x - 2) = 0$$

Whenever the product of two quantities equals zero, then one or both of these quantities must be zero. It follows that either $x + 5 = 0$ or $x - 2 = 0$ from which $x = -5$ and $x = 2$ are the required solutions.

11.10 Solve the quadratic equation $6x^2 + 5x - 4 = 0$.

VIDEO

Solution The left-hand side of the equation can be factorised to give

$$6x^2 + 5x - 4 = (2x - 1)(3x + 4) = 0$$

It follows that either $2x - 1 = 0$ or $3x + 4 = 0$ from which $x = \frac{1}{2}$ and $x = -\frac{4}{3}$ are the required solutions.

11.11 Solve the quadratic equation $x^2 - 8x = 0$.

Solution Factorising the left-hand side gives

$$x^2 - 8x = x(x - 8) = 0$$

from which either $x = 0$ or $x - 8 = 0$. The solutions are therefore $x = 0$ and $x = 8$.

11.12 Solve the quadratic equation $x^2 - 36 = 0$.

Solution The left-hand side factorises to give

$$x^2 - 36 = (x + 6)(x - 6) = 0$$

so that $x + 6 = 0$ or $x - 6 = 0$. The solutions are therefore $x = -6$ and $x = 6$.

11.13 Solve the equation $x^2 = 81$.

Solution Writing this in the standard form of a quadratic equation we obtain $x^2 - 81 = 0$. The left-hand side can be factorised to give

$$x^2 - 81 = (x - 9)(x + 9) = 0$$

from which $x - 9 = 0$ or $x + 9 = 0$. The solutions are then $x = 9$ and $x = -9$.

Solution of quadratic equations using the formula

When it is difficult or impossible to factorise the quadratic expression $ax^2 + bx + c$, solutions of a quadratic equation can be sought using the following formula:

Key point

If $ax^2 + bx + c = 0$, then $x = \dfrac{-b \pm \sqrt{b^2 - 4ac}}{2a}$

This formula gives possibly two solutions: one solution is obtained by taking the positive square root and the second solution by taking the negative square root.

WORKED EXAMPLES

11.14 Solve the equation $x^2 + 9x + 20 = 0$ using the formula.

Solution Comparing the given equation with the standard form $ax^2 + bx + c = 0$ we see that $a = 1$, $b = 9$ and $c = 20$. These values are substituted into the formula:

$$x = \frac{-b \pm \sqrt{b^2 - 4ac}}{2a}$$

$$= \frac{-9 \pm \sqrt{81 - 4(1)(20)}}{(2)(1)}$$

$$= \frac{-9 \pm \sqrt{81 - 80}}{2}$$

$$= \frac{-9 \pm \sqrt{1}}{2}$$

$$= \frac{-9 \pm 1}{2}$$

$$= \begin{cases} -4 & \text{by taking the positive square root} \\ -5 & \text{by taking the negative square root} \end{cases}$$

The two solutions are therefore $x = -4$ and $x = -5$.

11.15 Solve the equation $2x^2 - 3x - 7 = 0$ using the formula.

VIDEO

Solution In this example $a = 2$, $b = -3$ and $c = -7$. Care should be taken with the negative signs. Substituting these into the formula we find

$$x = \frac{-(-3) \pm \sqrt{(-3)^2 - 4(2)(-7)}}{2(2)}$$

$$= \frac{3 \pm \sqrt{9 + 56}}{4}$$

$$= \frac{3 \pm \sqrt{65}}{4}$$

$$= \frac{3 \pm 8.062}{4}$$

$$= \begin{cases} 2.766 & \text{by taking the positive square root} \\ -1.266 & \text{by taking the negative square root} \end{cases}$$

The two solutions are therefore $x = 2.766$ and $x = -1.266$.

If the values of a, b and c are such that $b^2 - 4ac$ is positive, the formula will produce two solutions known as **distinct real roots** of the equation. If $b^2 - 4ac = 0$ there will be a single root known as a **repeated root**. Some books refer to the equation having **equal roots**. If the equation is such that $b^2 - 4ac$ is negative, the formula requires us to find the square root of a negative number. In ordinary arithmetic this is impossible and we say that in such a case the quadratic equation does not possess any real roots.

However, it is still useful to be able to write down formally expressions for the roots. To do this mathematicians have developed a number system called **complex numbers**. When it is necessary to find the square root of a negative number, such as -4, we proceed as follows. We write

$$-4 = 4 \times -1$$

so that

$$\sqrt{-4} = \sqrt{4} \times \sqrt{-1} = 2 \times \sqrt{-1}$$

Now $\sqrt{-1}$ is not a real number, but we denote it by the symbol i and refer to it as an **imaginary number**. Using the symbol i enables us to write down $\sqrt{-4} = 2\mathrm{i}$. In a similar way we can write down the square root of any negative number. For example,

$$\sqrt{-9} = 3\mathrm{i}, \quad \sqrt{-16} = 4\mathrm{i} \quad \text{and} \quad \sqrt{-7} = \sqrt{7}\,\mathrm{i}$$

Then, a complex number is one which can have both a **real part** and an **imaginary part**. For example, the complex number $6 + 2\mathrm{i}$ has real part 6 and imaginary part 2. The imaginary part is always the number which multiplies the symbol i. We shall see how this helps us to solve the quadratic equation $x^2 - 6x + 10 = 0$ for which $b^2 - 4ac$ is negative in Example 11.17.

The quantity $b^2 - 4ac$ is called the **discriminant**, because it allows us to distinguish between the three possible cases. In summary:

Key point

Given

$$ax^2 + bx + c = 0$$

then

$b^2 - 4ac > 0$ two distinct real roots

$b^2 - 4ac = 0$ repeated (equal) root

$b^2 - 4ac < 0$ no real roots – there are two distinct complex roots

WORKED EXAMPLES

11.16 Use the formula to solve the equation $4x^2 + 4x + 1 = 0$.

Solution In this example $a = 4$, $b = 4$ and $c = 1$. Applying the formula gives

$$x = \frac{-4 \pm \sqrt{4^2 - 4(4)(1)}}{(2)(4)}$$

$$= \frac{-4 \pm \sqrt{16 - 16}}{8}$$

$$= \frac{-4 \pm 0}{8} = -\frac{1}{2}$$

There is a single, repeated root $x = -\frac{1}{2}$. Note that $b^2 - 4ac = 0$.

11.17 Use the formula to solve the quadratic equation $x^2 - 6x + 10 = 0$.

Solution We use the formula with $a = 1$, $b = -6$ and $c = 10$. We find

$$x = \frac{6 \pm \sqrt{(-6)^2 - 4(1)(10)}}{2}$$

$$= \frac{6 \pm \sqrt{36 - 40}}{2}$$

$$= \frac{6 \pm \sqrt{-4}}{2}$$

We are now faced with finding the square root of -4. We use the imaginary number i as explained on page 118. Writing -4 as 4×-1 we have $\sqrt{-4} = \sqrt{4} \times \sqrt{-1} = 2\mathrm{i}$. Then the solution of the equation becomes

$$x = \frac{6 \pm 2\mathrm{i}}{2} = 3 \pm \mathrm{i}$$

There are thus two roots, $x = 3 + \mathrm{i}$ and $x = 3 - \mathrm{i}$. These are complex roots with real part 3 and imaginary part ± 1.

For a fuller introduction to complex numbers refer to the companion text *Mathematics for Engineers*, 3rd edition (2008), by the same authors.

Self-assessment questions 11.3

1. Under what conditions will a quadratic equation possess distinct real roots?
2. Under what conditions will a quadratic equation possess a repeated root?
3. Under what conditions will a quadratic equation possess complex roots? What is it necessary to do in order to write down the roots in this case?

Exercise 11.3

MyMathLab Global

1. Solve the following quadratic equations by factorisation:

(a) $x^2 + x - 2 = 0$ (b) $x^2 - 8x + 15 = 0$ (c) $4x^2 + 6x + 2 = 0$
(d) $x^2 - 6x + 9 = 0$ (e) $x^2 - 81 = 0$ (f) $x^2 + 4x + 3 = 0$
(g) $x^2 + 2x - 3 = 0$ (h) $x^2 + 3x - 4 = 0$ (i) $x^2 + 6x + 5 = 0$
(j) $x^2 - 12x + 35 = 0$ (k) $x^2 + 12x + 35 = 0$ (l) $2x^2 + x - 3 = 0$
(m) $2x^2 - x - 6 = 0$ (n) $2x^2 - 7x - 15 = 0$ (o) $3x^2 - 2x - 1 = 0$
(p) $9x^2 - 12x - 5 = 0$ (q) $7x^2 + x = 0$ (r) $4x^2 + 12x + 9 = 0$

2. Solve the following quadratic equations using the formula:
   (a) $3x^2 - 6x - 5 = 0$ (b) $x^2 + 3x - 77 = 0$ (c) $2x^2 - 9x + 2 = 0$
   (d) $x^2 + 3x - 4 = 0$ (e) $3x^2 - 3x - 4 = 0$ (f) $4x^2 + x - 1 = 0$
   (g) $x^2 - 7x - 3 = 0$ (h) $x^2 + 7x - 3 = 0$ (i) $11x^2 + x + 1 = 0$
   (j) $2x^2 - 3x - 7 = 0$

3. Solve the following quadratic equations:
   (a) $6x^2 + 13x + 6 = 0$ (b) $3t^2 + 13t + 12 = 0$ (c) $t^2 - 7t + 3 = 0$

Test and assignment exercises 11

1. Solve the following equations:
   (a) $13t - 7 = 2t + 5$ (b) $-3t + 9 = 13 - 7t$ (c) $-5t = 0$

2. Solve the following equations:
   (a) $4x = 16$ (b) $\frac{x}{12} = 9$ (c) $4x - 13 = 3$ (d) $4x - 14 = 2x + 8$
   (e) $3x - 17 = 4$ (f) $7 - x = 9 + 3x$ (g) $4(2 - x) = 8$

3. Solve the following equations:
   (a) $2y = 8$ (b) $5u = 14u + 3$ (c) $5 = 4I$ (d) $13i + 7 = 2i - 9$
   (e) $\frac{1}{4}x + 9 = 3 - \frac{1}{2}x$ (f) $\frac{4x + 2}{2} + 8x = 0$

4. Solve the following equations:
   (a) $3x^2 - 27 = 0$ (b) $2x^2 + 18x + 14 = 0$ (c) $x^2 = 16$
   (d) $x^2 - x - 72 = 0$ (e) $2x^2 - 3x - 44 = 0$ (f) $x^2 - 4x - 21 = 0$

5. The solutions of the equation $x^2 + 4x = 0$ are identical to the solutions of $3x^2 + 12x = 0$. True or false?

6. Solve the following simultaneous equations:
   (a) $x - 2y = -11$, $7x + y = -32$ (b) $2x - y = 2$, $x + 3y = 29$
   (c) $x + y = 19$, $-x + y = 1$

Functions

16

Objectives: This chapter:

- explains what is meant by a function
- describes the notation used to write functions
- explains the terms 'independent variable' and 'dependent variable'
- explains what is meant by a composite function
- explains what is meant by the inverse of a function

16.1 Definition of a function

A **function** is a rule that receives an input and produces an output. It is shown schematically in Figure 16.1. For example, the rule may be 'add 2 to the input'. If 6 is the input, then $6 + 2 = 8$ will be the output. If -5 is the input, then $-5 + 2 = -3$ will be the output. In general, if x is the input then $x + 2$ will be the output. Figure 16.2 illustrates this function schematically.

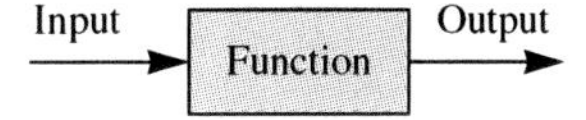

Figure 16.1
A function produces an output from an input

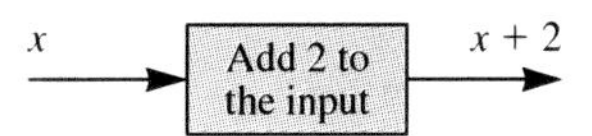

Figure 16.2
The function adds 2 to the input

For a rule to be a function then it is crucial that only *one* output is produced for any given input.

Key point A function is a rule that produces a *single* output for any given input.

The input to a function can usually take many values and so is called a **variable**. The output, too, varies depending upon the value of the input, and so is also a variable. The input is referred to as the **independent variable** because we are free to choose its value. The output is called the **dependent variable** because its value depends upon the value of the input.

16.2 Notation used for functions

We usually denote the input, the output and the function by letters or symbols. Commonly we use x to represent the input, y the output and f the function, although other letters will be used as well.

Consider again the example from §16.1. We let f be the function 'add 2 to the input' and we let x be the input. In mathematical notation we write

$$f: x \rightarrow x+2$$

This means that the function f takes an input x and produces an output $x+2$.

An alternative, but commonly used, notation is

$$f(x) = x+2$$

The quantity $f(x)$ does not mean f times x but rather indicates that the function f acts on the quantity in the brackets. Because we also call the output y we can write $y = f(x) = x+2$, or simply $y = x+2$.

We could represent the same function using different letters. If h represents the function and t the input then we can write

$$h(t) = t+2$$

WORKED EXAMPLES

16.1 A function multiplies the input by 4. Write down the function in mathematical notation.

Solution Let us call the function f and the input x. Then we have

$$f: x \rightarrow 4x \quad \text{or alternatively} \quad f(x) = 4x$$

If we call the output y, we can write $y = f(x) = 4x$, or simply $y = 4x$.

16.2 A function divides the input by 6 and then adds 3 to the result. Write the function in mathematical notation.

Solution Let us call the function z and the input t. Then we have

$$z(t) = \frac{t}{6} + 3$$

16.3 A function f is given by the rule $f: x \rightarrow 9$, or alternatively as $f(x) = 9$. Describe in words what this function does.

Solution Whatever the value of the input to this function, the output is always 9.

16.4 A function squares the input and then multiplies the result by 6. Write down the function using mathematical notation.

Solution Let us call the function f, the input x and the output y. Then

$$y = f(x) = 6x^2$$

16.5 Describe in words what the following functions do:

(a) $h(x) = \frac{1}{x}$ (b) $g(t) = t + t^2$

Solution (a) The function $h(x) = \frac{1}{x}$ divides 1 by the input.

(b) The function $g(t) = t + t^2$ adds the input to the square of the input.

Often we are given a function and need to calculate the output from a given input.

WORKED EXAMPLES

16.6 A function f is defined by $f(x) = 3x + 1$. Calculate the output when the input is (a) 4, (b) -1, (c) 0.

Solution The function f multiplies the input by 3 and then adds 1 to the result.

(a) When the input is 4, the output is $3 \times 4 + 1 = 12 + 1 = 13$. We write

$$f(x = 4) = 3(4) + 1 = 12 + 1 = 13$$

or more simply

$$f(4) = 13$$

Note that 4 has been substituted for x in the formula for f.

(b) We require the output when the input is -1, that is $f(-1)$:

$$f(x = -1) = f(-1) = 3(-1) + 1 = -3 + 1 = -2$$

The output is -2 when the input is -1.

(c) We require the output when the input is 0, that is $f(0)$:

$$f(x = 0) = f(0) = 3(0) + 1 = 0 + 1 = 1$$

16.7 A function g is defined by $g(t) = 2t^2 - 1$. Find

(a) $g(3)$ (b) $g(0.5)$ (c) $g(-2)$

Solution (a) We obtain $g(3)$ by substituting 3 for t:

$$g(3) = 2(3)^2 - 1 = 2(9) - 1 = 17$$

(b) $g(0.5) = 2(0.5)^2 - 1 = 0.5 - 1 = -0.5$.

(c) $g(-2) = 2(-2)^2 - 1 = 8 - 1 = 7$.

16.8 A function h is defined by $h(x) = \frac{x}{3} + 1$. Find

(a) $h(3)$ (b) $h(t)$ (c) $h(\alpha)$ (d) $h(2\alpha)$ (e) $h(2x)$

Solution (a) If $h(x) = \frac{x}{3} + 1$ then $h(3) = \frac{3}{3} + 1 = 1 + 1 = 2$.

(b) The function h divides the input by 3 and then adds 1. We require the output when the input is t, that is we require $h(t)$. Now

$$h(t) = \frac{t}{3} + 1$$

since the input t has been divided by 3 and then 1 has been added to the result. Note that $h(t)$ is obtained by substituting t in place of x in $h(x)$.

(c) We require the output when the input is α. This is obtained by substituting α in place of x. We find

$$h(\alpha) = \frac{\alpha}{3} + 1$$

(d) We require the output when the input is 2α. We substitute 2α in place of x. This gives

$$h(2\alpha) = \frac{2\alpha}{3} + 1$$

(e) We require the output when the input is $2x$. We substitute $2x$ in place of x. That is,

$$h(2x) = \frac{2x}{3} + 1$$

16.9 Given $f(x) = x^2 + x - 1$ write expressions for

(a) $f(\alpha)$ (b) $f(x+1)$ (c) $f(2t)$

Solution (a) Substituting α in place of x we obtain

$$f(\alpha) = \alpha^2 + \alpha - 1$$

(b) Substituting $x+1$ for x we obtain

$$f(x+1) = (x+1)^2 + (x+1) - 1$$
$$= x^2 + 2x + 1 + x + 1 - 1$$
$$= x^2 + 3x + 1$$

(c) Substituting $2t$ in place of x we obtain

$$f(2t) = (2t)^2 + 2t - 1 = 4t^2 + 2t - 1$$

$<$ is the symbol for less than.
$\leqslant$ is the symbol for less than or equal to.
$\geqslant$ is the symbol for greater than or equal to.

Sometimes a function uses different rules on different intervals. For example, we could define a function as

$$f(x) = \begin{cases} 3x & \text{when} \quad 0 \leqslant x \leqslant 4 \\ 2x+6 & \text{when} \quad 4 < x < 5 \\ 9 & \text{when} \quad x \geqslant 5 \end{cases}$$

Here the function is defined in three 'pieces'. The value of x determines which part of the definition is used to evaluate the function. The function is said to be a **piecewise** function.

WORKED EXAMPLE

16.10 A piecewise function is defined by

$$y(x) = \begin{cases} x^2+1 & \text{when} \quad -1 \leqslant x \leqslant 2 \\ 3x & \text{when} \quad 2 < x \leqslant 6 \\ 2x+1 & \text{when} \quad x > 6 \end{cases}$$

Evaluate
(a) $y(0)$ (b) $y(4)$ (c) $y(2)$ (d) $y(7)$

Solution

(a) We require the value of y when $x = 0$. Since 0 lies between -1 and 2 we use the first part of the definition, that is $y = x^2 + 1$. Hence

$$y(0) = 0^2 + 1 = 1$$

(b) We require y when $x = 4$. The second part of the definition must be used because x lies between 2 and 6. Therefore

$$y(4) = 3(4) = 12$$

(c) We require y when $x = 2$. The value $x = 2$ occurs in the first part of the definition. Therefore

$$y(2) = 2^2 + 1 = 5$$

(d) We require y when $x = 7$. The final part of the function must be used. Therefore

$$y(7) = 2(7) + 1 = 15$$

Self-assessment questions 16.2

1. Explain what is meant by a function.
2. Explain the meaning of the terms 'dependent variable' and 'independent variable'.
3. Given $f(x)$, is the statement '$f(1/x)$ means $1/f(x)$' true or false?
4. Give an example of a function $f(x)$ such that $f(2) = f(3)$, that is the outputs for the inputs 2 and 3 are identical.

Exercise 16.2

1. Describe in words each of the following functions:
   (a) $h(t) = 10t$ (b) $g(x) = -x + 2$
   (c) $h(t) = 3t^4$ (d) $f(x) = \dfrac{4}{x^2}$
   (e) $f(x) = 3x^2 - 2x + 9$ (f) $f(x) = 5$
   (g) $f(x) = 0$
2. Describe in words each of the following functions:
   (a) $f(t) = 3t^2 + 2t$ (b) $g(x) = 3x^2 + 2x$
   Comment upon your answers.
3. Write the following functions using mathematical notation:
   (a) The input is cubed and the result is divided by 12.
   (b) The input is added to 3 and the result is squared.
   (c) The input is squared and added to 4 times the input. Finally, 10 is subtracted from the result.
   (d) The input is squared and added to 5. Then the input is divided by this result.
   (e) The input is cubed and then 1 is subtracted from the result.
   (f) 1 is subtracted from the input and the result is squared.
   (g) Twice the input is subtracted from 7 and the result is divided by 4.
   (h) The output is always -13 whatever the value of the input.
4. Given the function $A(n) = n^2 - n + 1$ evaluate
   (a) $A(2)$ (b) $A(3)$ (c) $A(0)$
   (d) $A(-1)$
5. Given $y(x) = (2x - 1)^2$ evaluate
   (a) $y(1)$ (b) $y(-1)$ (c) $y(-3)$
   (d) $y(0.5)$ (e) $y(-0.5)$
6. The function f is given by $f(t) = 4t + 6$. Write expressions for
   (a) $f(t + 1)$ (b) $f(t + 2)$
   (c) $f(t + 1) - f(t)$ (d) $f(t + 2) - f(t)$
7. The function $f(x)$ is defined by $f(x) = 2x^2 - 3$. Write expressions for
   (a) $f(n)$ (b) $f(z)$ (c) $f(t)$ (d) $f(2t)$
   (e) $f\left(\dfrac{1}{z}\right)$ (f) $f\left(\dfrac{3}{n}\right)$ (g) $f(-x)$
   (h) $f(-4x)$ (i) $f(x + 1)$ (j) $f(2x - 1)$

8. Given the function $a(p) = p^2 + 3p + 1$ write an expression for $a(p + 1)$. Verify that $a(p + 1) - a(p) = 2p + 4$.

9. Sometimes the output from one function forms the input to another function. Suppose we have two functions: f given by $f(t) = 2t$, and h given by $h(t) = t + 1$. $f(h(t))$ means that t is input to h, and the output from h is input to f. Evaluate
(a) $f(3)$ (b) $h(2)$ (c) $f(h(2))$
(d) $h(f(3))$

10. The functions f and h are defined as in Question 9. Write down expressions for
(a) $f(h(t))$ (b) $h(f(t))$

11. A function is defined by

$$f(x) = \begin{cases} x & 0 \leqslant x < 1 \\ 2 & x = 1 \\ 1 & x > 1 \end{cases}$$

Evaluate
(a) $f(0.5)$ (b) $f(1.1)$ (c) $f(1)$

16.3 Composite functions

Sometimes we wish to apply two or more functions, one after the other. The output of one function becomes the input of the next function.

Suppose $f(x) = 2x$ and $g(x) = x + 3$. We note that the function $f(x)$ doubles the input while the function $g(x)$ adds 3 to the input. Now, we let the output of $g(x)$ become the input to $f(x)$. Figure 16.3 illustrates the position.

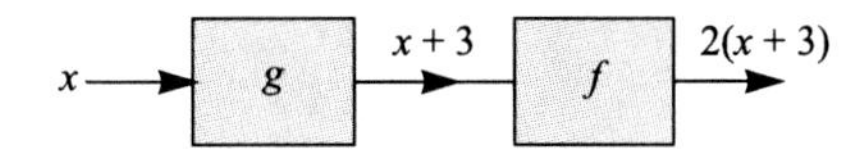

Figure 16.3
The output of g is the input of f

We have

$$g(x) = x + 3$$
$$f(x + 3) = 2(x + 3) = 2x + 6$$

Note that $f(x + 3)$ may be written as $f(g(x))$. Referring to Figure 16.3 we see that the initial input is x and that the final output is $2x + 6$. The functions $g(x)$ and $f(x)$ have been combined. We call $f(g(x))$ a **composite function**. It is composed of the individual functions $f(x)$ and $g(x)$. In this example we have

$$f(g(x)) = 2x + 6$$

WORKED EXAMPLES

16.11 Given $f(x) = 2x$ and $g(x) = x + 3$ find the composite function $g(f(x))$.

Solution The output of $f(x)$ becomes the input to $g(x)$. Figure 16.4 illustrates this.

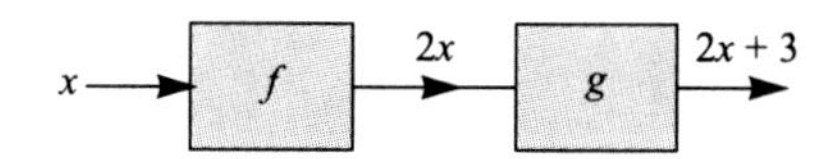

Figure 16.4
The composite function $g(f(x))$

We see that

$$g(f(x)) = g(2x)$$
$$= 2x + 3$$

Note that in general $f(g(x))$ and $g(f(x))$ are different functions.

16.12 Given $f(t) = t^2 + 1$, $g(t) = \frac{3}{t}$ and $h(t) = 2t$ determine each of the following composite functions:

(a) $f(g(t))$ (b) $g(h(t))$ (c) $f(h(t))$ (d) $f(g(h(t)))$ (e) $g(f(h(t)))$

Solution

(a) $f(g(t)) = f\left(\frac{3}{t}\right) = \left(\frac{3}{t}\right)^2 + 1 = \frac{9}{t^2} + 1$

(b) $g(h(t)) = g(2t) = \frac{3}{2t}$

(c) $f(h(t)) = f(2t) - (2t)^2 + 1 = 4t^2 + 1$

(d) $f(g(h(t))) = f\left(\frac{3}{2t}\right)$ using (b)

$$= \left(\frac{3}{2t}\right)^2 + 1$$

$$= \frac{9}{4t^2} + 1$$

(e) $g(f(h(t))) = g(4t^2 + 1)$ using (c)

$$= \frac{3}{4t^2 + 1}$$

Self-assessment questions 16.3

1. Explain the term 'composite function'.
2. Give examples of functions $f(x)$ and $g(x)$ such that $f(g(x))$ and $g(f(x))$ are equal.

Exercise 16.3

MyMathLab

1. Given $f(x) = 4x$ and $g(x) = 3x - 2$ find
   (a) $f(g(x))$ (b) $g(f(x))$

2. If $x(t) = t^3$ and $y(t) = 2t$ find
   (a) $y(x(t))$ (b) $x(y(t))$

3. Given $r(x) = \frac{1}{2x}$, $s(x) = 3x$ and $t(x) = x - 2$ find
   (a) $r(s(x))$ (b) $t(s(x))$ (c) $t(r(s(x)))$
   (d) $r(t(s(x)))$ (e) $r(s(t(x)))$

4. A function can be combined with itself. This is known as **self-composition**. Given $v(t) = 2t + 1$ find
   (a) $v(v(t))$ (b) $v(v(v(t)))$

5. Given $m(t) = (t+1)^3$, $n(t) = t^2 - 1$ and $p(t) = t^2$ find
   (a) $m(n(t))$ (b) $n(m(t))$ (c) $m(p(t))$
   (d) $p(m(t))$ (e) $n(p(t))$ (f) $p(n(t))$
   (g) $m(n(p(t)))$ (h) $p(p(t))$ (i) $n(n(t))$
   (j) $m(m(t))$

16.4 The inverse of a function

Note that the symbol f^{-1} does not mean $\frac{1}{f}$.

We have described a function f as a rule which receives an input, say x, and generates an output, say y. We now consider the reversal of that process, namely finding a function which receives y as input and generates x as the output. If such a function exists it is called the **inverse function** of f. Figure 16.5 illustrates this schematically. The inverse of $f(x)$ is denoted by $f^{-1}(x)$.

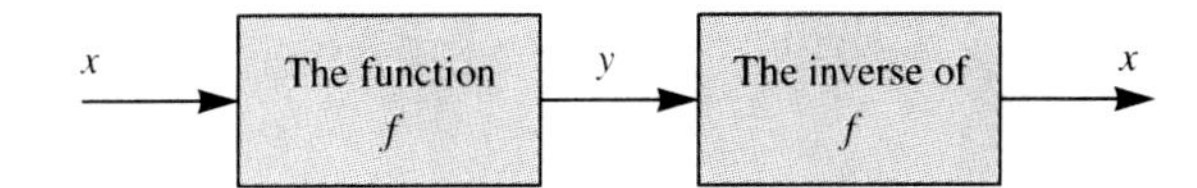

Figure 16.5
The inverse of f reverses the effect of f

WORKED EXAMPLES

16.13 The functions f and g are defined by

$$f(x) = 2x \qquad g(x) = \frac{x}{2}$$

(a) Verify that f is the inverse of g.

(b) Verify that g is the inverse of f.

Solution

(a) The function g receives an input of x and generates an output of $x/2$; that is, it halves the input. In order to reverse the process, the inverse of g should receive $x/2$ as input and generate x as output. Now consider the function $f(x) = 2x$. This function doubles the input. Hence

$$f\left(\frac{x}{2}\right) = 2\left(\frac{x}{2}\right) = x$$

The function f has received $x/2$ as input and generated x as output. Hence f is the inverse of g. This is shown schematically in Figure 16.6.

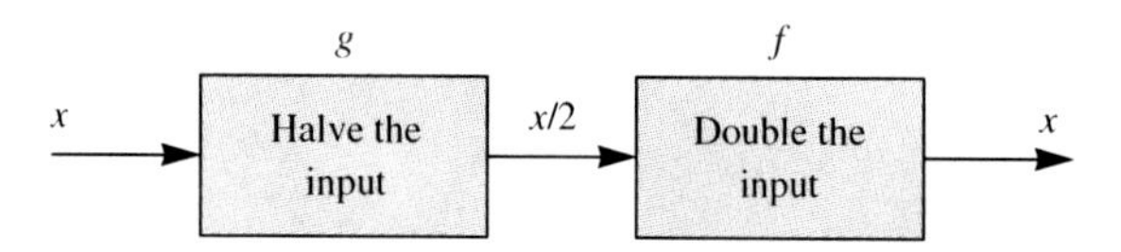

Figure 16.6
The function f is the inverse of g

(b) The function f receives x as input and generates $2x$ as output. In order to reverse the process, the inverse of f should receive $2x$ as input and generate x as output. Now $g(x) = x/2$, that is the input is halved, and so

$$g(2x) = \frac{2x}{2} = x$$

Hence g is the inverse of f. This is shown schematically in Figure 16.7.

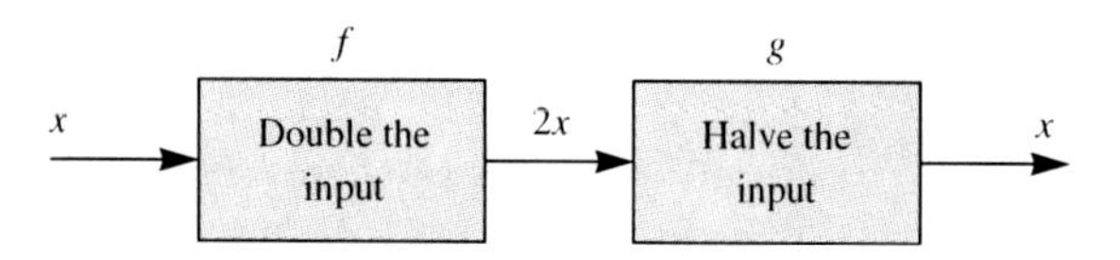

Figure 16.7
The function g is the inverse of f

16.14 Find the inverse of the function $f(x) = 3x - 4$.

Solution The function f multiplies the input by 3 and subtracts 4 from the result. To reverse the process, the inverse function, g say, must add 4 to the input and then divide the result by 3. Hence

$$g(x) = \frac{x+4}{3}$$

16.15 Find the inverse of $h(t) = -\frac{1}{2}t + 5$.

Solution The function h multiplies the input by $-\frac{1}{2}$ and then adds 5 to the result. Therefore the inverse function, g say, must subtract 5 from the input and then divide the result by $-\frac{1}{2}$. Hence

$$g(t) = \frac{(t-5)}{-1/2} = -2(t-5) = -2t + 10$$

There is an algebraic method of finding an inverse function that is often easier to apply. Suppose we wish to find the inverse of the function $f(x) = 6 - 2x$. We let

$$y = 6 - 2x$$

and then transpose this for x. This gives

$$x = \frac{6 - y}{2}$$

Finally, we interchange x and y to give $y = (6 - x)/2$. This is the required inverse function. To summarize these stages:

Key point

To find the inverse of $y = f(x)$,

- transpose the formula to make x the subject
- interchange x and y

The result is the required inverse function.

We shall meet some functions that do not have an inverse function. For example, consider the function $f(x) = x^2$. If 3 is the input, the output is 9. Now if -3 is the input, the output will also be 9 since $(-3)^2 = 9$. In order to reverse this process an inverse function would have to take an input of 9 and produce outputs of both 3 and -3. However, this contradicts the definition of a function, which states that a function must have only *one* output for a given input. We say that $f(x) = x^2$ does not have an inverse function.

Self-assessment questions 16.4

1. Explain what is meant by the inverse of a function.
2. Explain why the function $f(x) = 4x^4$ does not possess an inverse function.

Exercise 16.4

1. Find the inverse of each of the following functions:
(a) $f(x) = 3x$ (b) $f(x) = \frac{x}{4}$
(c) $f(x) = x + 1$ (d) $f(x) = x - 3$
(e) $f(x) = 3 - x$ (f) $f(x) = 2x + 6$
(g) $f(x) = 7 - 3x$ (h) $f(x) = \frac{1}{x}$
(i) $f(x) = \frac{3}{x}$ (j) $f(x) = -\frac{3}{4x}$
2. Find the inverse, $f^{-1}(x)$, when $f(x)$ is given by
(a) $6x$ (b) $6x + 1$ (c) $x + 6$
(d) $\frac{x}{6}$ (e) $\frac{6}{x}$

3. Find the inverse, $g^{-1}(t)$, when $g(t)$ is given by

   (a) $3t+1$ (b) $\dfrac{1}{3t+1}$

   (c) t^3 (d) $3t^3$

   (e) $3t^3+1$ (f) $\dfrac{3}{t^3+1}$

4. The functions $g(t)$ and $h(t)$ are defined by

   $$g(t) = 2t - 1, \; h(t) = 4t + 3$$

   Find
   (a) the inverse of $h(t)$, that is $h^{-1}(t)$
   (b) the inverse of $g(t)$, that is $g^{-1}(t)$
   (c) $g^{-1}(h^{-1}(t))$ (d) $h(g(t))$
   (e) the inverse of $h(g(t))$
   What observations do you make from (c) and (e)?

Test and assignment exercises 16

1. Given $r(t) = t^2 - t/2 + 4$ evaluate

   (a) $r(0)$ (b) $r(-1)$ (c) $r(2)$ (d) $r(3.6)$ (e) $r(-4.6)$

2. A function is defined as $h(t) = t^2 - 7$.

   (a) State the dependent variable. (b) State the independent variable.

3. Given the functions $a(x) = x^2 + 1$ and $b(x) = 2x + 1$ write expressions for

   (a) $a(\alpha)$ (b) $b(t)$ (c) $a(2x)$ (d) $b\left(\dfrac{x}{3}\right)$ (e) $a(x+1)$ (f) $b(x+h)$

   (g) $b(x-h)$ (h) $a(b(x))$ (i) $b(a(x))$

4. Find the inverse of each of the following functions:

   (a) $f(x) = \pi - x$ (b) $h(t) = \dfrac{t}{3} + 2$ (c) $r(n) = \dfrac{1}{n}$ (d) $r(n) = \dfrac{1}{n-1}$

   (e) $r(n) = \dfrac{2}{n-1}$ (f) $r(n) = \dfrac{a}{n-b}$ where a and b are constants

5. Given $A(n) = n^2 + n - 6$ find expressions for
   (a) $A(n+1)$ (b) $A(n-1)$
   (c) $2A(n+1) - A(n) + A(n-1)$

6. Find $h^{-1}(x)$ when $h(x)$ is given by

   (a) $\dfrac{x+1}{3}$ (b) $\dfrac{3}{x+1}$ (c) $\dfrac{x+1}{x}$ (d) $\dfrac{x}{x+1}$

7. Given $v(t) = 4t - 2$ find
   (a) $v^{-1}(t)$ (b) $v(v(t))$

8. Given $h(x) = 9x - 6$ and $g(x) = \dfrac{1}{3x}$ find

   (a) $h(g(x))$ (b) $g(h(x))$

9. Given $f(x) = (x + 1)^2$, $g(x) = 4x$ and $h(x) = x - 1$ find

   (a) $f(g(x))$ (b) $g(f(x))$ (c) $f(h(x))$ (d) $h(f(x))$ (e) $g(h(x))$
   (f) $h(g(x))$ (g) $f(g(h(x)))$ (h) $g(h(f(x)))$

10. Given $x(t) = t^3$, $y(t) = \dfrac{1}{t+1}$ and $z(t) = 3t - 1$ find

    (a) $y(x(t))$ (b) $y(z(t))$ (c) $x(y(z(t)))$ (d) $z(y(x(t)))$

11. Write each of the following functions using mathematical notation:

    (a) The input is multiplied by 7.
    (b) Five times the square of the input is subtracted from twice the cube of the input.
    (c) The output is 6.
    (d) The input is added to the reciprocal of the input.
    (e) The input is multiplied by 11 and then 6 is subtracted from this. Finally, 9 is divided by this result.

12. Find the inverse of

$$f(x) = \frac{x+1}{x-1}$$

17 Graphs of functions

Objectives: This chapter:

- shows how coordinates are plotted on the x–y plane
- introduces notation to denote intervals on the x axis
- introduces symbols to denote 'greater than' and 'less than'
- shows how to draw a graph of a function
- explains what is meant by the domain and range of a function
- explains how to use the graph of a function to solve equations
- explains how to solve simultaneous equations using graphs

In Chapter 16 we introduced functions and represented them by algebraic expressions. Another useful way of representing functions is pictorially, by means of **graphs**.

17.1 The x–y plane

We introduce horizontal and vertical **axes** as shown in Figure 17.1. These axes intersect at a point O called the **origin**. The horizontal axis is used to represent the independent variable, commonly x, and the vertical axis is used to represent the dependent variable, commonly y. The region shown is then referred to as the x–y plane.

A **scale** is drawn on both axes in such a way that at the origin $x = 0$ and $y = 0$. Positive x values lie to the right and negative x values to the left. Positive y values are above the origin, negative y values are below the origin. It is not essential that the scales on both axes are the same. Note that anywhere on the x axis the value of y must be zero. Anywhere on the y axis the value of x must be zero. Each point in the plane corresponds to a specific value of x and y. We usually call the x value the x **coordinate** and call the y

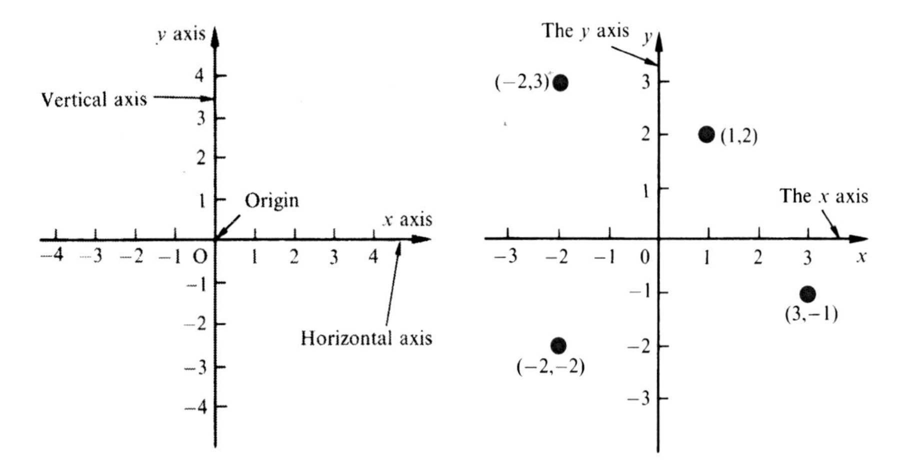

Figure 17.1
The x–y plane

Figure 17.2
Several points in the x–y plane

value the y **coordinate**. To refer to a specific point we give both its coordinates in brackets in the form (x, y), always giving the x coordinate first.

WORKED EXAMPLE

17.1 Draw the x–y plane and on it mark the points whose coordinates are $(1, 2)$, $(3, -1)$, $(-2, 3)$, $(-2, -2)$.

Solution The plane is drawn in Figure 17.2, and the given points are indicated. Note that the first coordinate in each bracket is the x coordinate.

Self-assessment questions 17.1

1. Explain the terms 'horizontal axis', 'vertical axis' and 'origin'.
2. When giving the coordinates of a point in the x–y plane, which coordinate is always given first?
3. State the coordinates of the origin.

Exercise 17.1

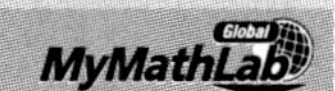

1. Plot and label the following points in the x–y plane:
   $(-2, 0), (2, 0), (0, -2), (0, 2), (0, 0)$.
2. A set of points is plotted and then these points are joined by a straight line that is parallel to the x axis. What can you deduce about the coordinates of these points?
3. A set of points is plotted and then these points are joined by a straight line which is parallel to the y axis. What can you deduce about the coordinates of these points?

17.2 Inequalities and intervals

We often need only part of the x axis when plotting graphs. For example, we may be interested only in that part of the x axis running from $x = 1$ to $x = 3$ or from $x = -2$ to $x = 7.5$. Such parts of the x axis are called **intervals**. In order to describe concisely intervals on the x axis, we introduce some new notation.

Greater than and less than

We use the symbol $>$ to mean 'greater than'. For example, $6 > 5$ and $-4 > -5$ are both true statements. If $x_1 > x_2$ then x_1 is to the right of x_2 on the x axis. The symbol $\geqslant$ means 'greater than or equal to', so, for example, $10 \geqslant 8$ and $8 \geqslant 8$ are both true.

The symbol $<$ means 'less than'. We may write $4 < 5$ for example. If $x_1 < x_2$ then x_1 is to the left of x_2 on the x axis. Finally, $\leqslant$ means 'less than or equal to'. Hence $6 \leqslant 9$ and $6 \leqslant 6$ are both true statements.

Intervals

$\mathbb{R}$ is called the set of real numbers.

We often need to represent intervals on the number line. To help us do this we introduce the set $\mathbb{R}$. $\mathbb{R}$ is the symbol we use to denote all numbers from minus infinity to plus infinity. All numbers, including integers, fractions and decimals, belong to $\mathbb{R}$. To show that a number, x, is in this set, we write $x \in \mathbb{R}$ where $\in$ means 'belongs to'.

There are three different kinds of interval:

(a) *The closed interval* An interval that includes its end-points is called a **closed interval**. All the numbers from 1 to 3, including both 1 and 3, comprise a closed interval, and this is denoted using square brackets, $[1, 3]$. Any number in this closed interval must be greater than or equal to 1, and also less than or equal to 3. Thus if x is any number in the interval, then $x \geqslant 1$ and also $x \leqslant 3$. We write this compactly as $1 \leqslant x \leqslant 3$. Finally we need to show explicitly that x is any number on the x axis, and not just an integer for example. So we write $x \in \mathbb{R}$. Hence the interval $[1, 3]$ can be expressed as

$$\{x : x \in \mathbb{R},\ 1 \leqslant x \leqslant 3\}$$

This means the set contains all the numbers x with $x \geqslant 1$ and $x \leqslant 3$.

(b) *The open interval* Any interval that does not include its end-points is called an **open interval**. For example, all the numbers from 1 to 3, but excluding 1 and 3, comprise an open interval. Such an interval is denoted using round brackets, $(1, 3)$. The interval may be written using set notation as

$$\{x : x \in \mathbb{R},\ 1 < x < 3\}$$

We say that x is **strictly greater** than 1, and **strictly less** than 3, so that the values of 1 and 3 are excluded from the interval.

(c) *The semi-open or semi-closed interval* An interval may be open at one end and closed at the other. Such an interval is called **semi-open** or, as some authors say, **semi-closed**. The interval $(1, 3]$ is a semi-open interval. The square bracket next to the 3 shows that 3 is included in the interval; the round bracket next to the 1 shows that 1 is not included in the interval. Using set notation we would write

$$\{x: x \in \mathbb{R},\ 1 < x \leqslant 3\}$$

When marking intervals on the x axis there is a notation to show whether or not the end-point is included. We use ● to show that an end-point is included (i.e. closed) whereas O is used to denote an end-point that is not included (i.e. open).

WORKED EXAMPLES

17.2 Describe the interval $[-3, 4]$ using set notation and illustrate it on the x axis.

Solution The interval $[-3, 4]$ is given in set notation by

$$\{x: x \in \mathbb{R},\ -3 \leqslant x \leqslant 4\}$$

Figure 17.3 illustrates the interval.

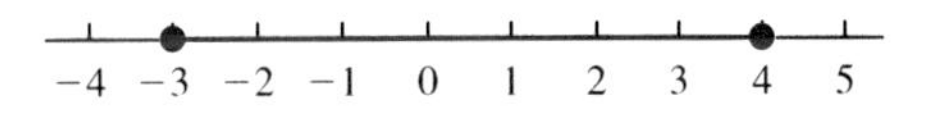

Figure 17.3
The interval $[-3, 4]$

17.3 Describe the interval $(1, 4)$ using set notation and illustrate it on the x axis.

Solution The interval $(1, 4)$ is expressed as

$$\{x: x \in \mathbb{R},\ 1 < x < 4\}$$

Figure 17.4 illustrates the interval on the x axis.

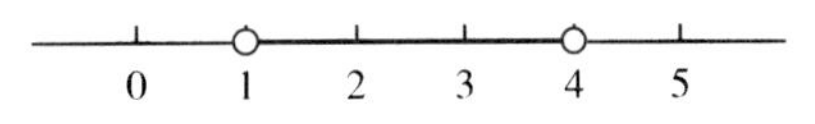

Figure 17.4
The interval $(1, 4)$

Self-assessment questions 17.2

1. Explain what is meant by (a) a closed interval, (b) an open interval, (c) a semi-closed interval.
2. Describe the graphical notation used to denote that an end-point is (a) included, (b) not included.

Exercise 17.2

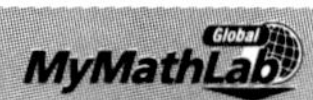

1. Describe the following intervals using set notation. Draw the intervals on the x axis.
   (a) $[2, 6]$ (b) $(6, 8]$ (c) $(-2, 0)$ (d) $[-3, -1.5)$
2. Which of the following are true?
   (a) $8 > 3$ (b) $-3 < 8$ (c) $-3 \leqslant -3$ (d) $0.5 \geqslant 0.25$ (e) $0 > 9$
   (f) $0 \leqslant 9$ (g) $-7 \geqslant 0$ (h) $-7 < 0$

17.3 Plotting the graph of a function

The method of plotting a graph is best illustrated by example.

WORKED EXAMPLE

17.4 Plot a graph of $y = 2x - 1$ for $-3 \leqslant x \leqslant 3$.

Solution We first calculate the value of y for several values of x. A table of x values and corresponding y values is drawn up as shown in Table 17.1.

Table 17.1
Values of x and y when $y = 2x - 1$

x	-3	-2	-1	0	1	2	3
y	-7	-5	-3	-1	1	3	5

The independent variable x varies from -3 to 3; the dependent variable y varies from -7 to 5. In order to accommodate all these values the scale on the x axis must vary from -3 to 3, and that on the y axis from -7 to 5. Each pair of values of x and y is represented by a unique point in the x–y plane, with coordinates (x, y). Each pair of values in the table is plotted as a point and then the points are joined to form the graph (Figure 17.5). By joining the points we see that, in this example, the graph is a straight line. We were asked to plot the graph for $-3 \leqslant x \leqslant 3$. Therefore, we have indicated each end-point of the graph by a ● to show that it is included. This follows the convention used for labelling closed intervals given in §17.2.

In the previous example, although we have used x and y as independent and dependent variables, clearly other letters could be used. The points to remember are that the axis for the independent variable is horizontal and

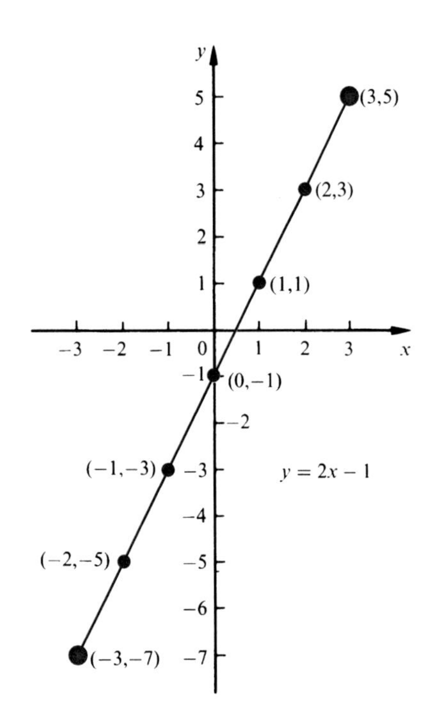

Figure 17.5
A graph of $y = 2x - 1$

the axis for the dependent variable is vertical. When coordinates are given the value of the independent variable is always given first.

WORKED EXAMPLE

17.5

(a) Plot a graph of $y = x^2 - 2x + 2$ for $-2 \leqslant x \leqslant 3$.

(b) Use your graph to determine which of the following points lie on the graph: (1, 2), (0, 2), and (1, 1).

Solution

(a) Table 17.2 gives values of x and the corresponding values of y. The independent variable is x and so the x axis is horizontal. The calculated points are plotted and joined. Figure 17.6 shows the graph. In this example the graph is not a straight line but rather a curve.

(b) The point (1, 2) has an x coordinate of 1 and a y coordinate of 2. The point is labelled by A on Figure 17.6. Points (0, 2) and (1, 1) are plotted and labelled by B and C. From the figure we can see that (0, 2) and (1, 1) lie on the graph.

Table 17.2
Values of x and y when $y = x^2 - 2x + 2$

x	−2	−1	0	1	2	3
y	10	5	2	1	2	5

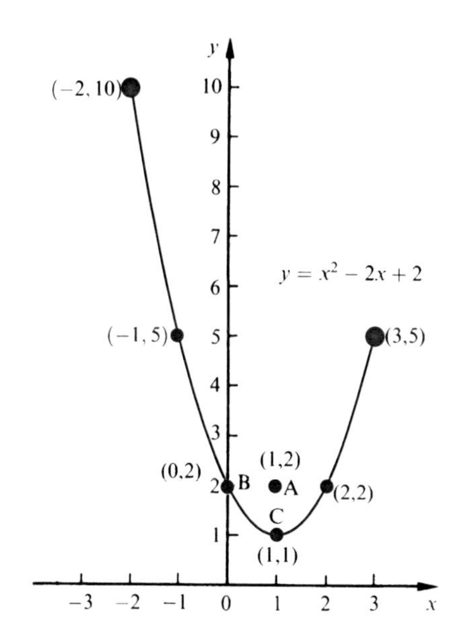

Figure 17.6
A graph of $y = x^2 - 2x + 2$

Self-assessment questions 17.3

1. Which variable is plotted vertically on a graph – the independent or dependent?
2. Suppose we wished to plot a graph of the function $g = 4t^2 + 3t - 2$. Upon which axis would the variable t be plotted?

Exercise 17.3

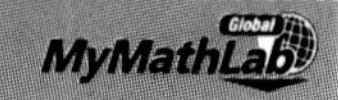

1. Plot graphs of the following functions:
   (a) $y = f(x) = 2x + 7$ for $-4 \leqslant x \leqslant 4$.
   (b) $y = f(x) = -2x$ for values of x between -3 and 3 inclusive.
   (c) $y = f(x) = x$ for $-5 \leqslant x \leqslant 5$.
   (d) $y = f(x) = -x$ for $-5 \leqslant x \leqslant 5$.
2. Plot a graph of the function $y = f(x) = 7$, for $-2 \leqslant x \leqslant 2$.
3. A function is given by $y = f(x) = x^3 + x$ for $-3 \leqslant x \leqslant 3$.
   (a) Plot a graph of the function.
   (b) Determine which of the following points lie on the curve: $(0, 1)$, $(1, 0)$, $(1, 2)$ and $(-1, -2)$.
4. Plot, on the same axes, graphs of the functions $f(x) = x + 2$ for $-3 \leqslant x \leqslant 3$, and $g(x) = -\frac{1}{2}x + \frac{1}{2}$ for $-3 \leqslant x \leqslant 3$. State the coordinates of the point where the graphs intersect.
5. Plot a graph of the function $f(x) = x^2 - x - 1$ for $-2 \leqslant x \leqslant 3$. State the coordinates of the points where the curve cuts (a) the horizontal axis, (b) the vertical axis.
6. (a) Plot a graph of the function $y = 1/x$ for $0 < x \leqslant 3$.
   (b) Deduce the graph of the function $y = -1/x$ for $0 < x \leqslant 3$ from your graph in part (a).

7. Plot a graph of the function defined by

$$r(t) = \begin{cases} t^2 & 0 \leqslant t \leqslant 3 \\ -2t + 15 & 3 < t \leqslant 7.5 \end{cases}$$

8. On the same axes plot a graph of $y = 2x$, $y = 2x + 1$ and $y = 2x + 3$, for $-3 \leqslant x \leqslant 3$. What observations can you make?

17.4 The domain and range of a function

The set of values that we allow the independent variable to take is called the **domain** of the function. If the domain is not actually specified in any particular example, it is taken to be the largest set possible. The set of values taken by the output is called the **range** of the function.

WORKED EXAMPLES

17.6 The function f is given by $y = f(x) = 2x$, for $1 \leqslant x \leqslant 3$.

(a) State the independent variable.

(b) State the dependent variable.

(c) State the domain of the function.

(d) Plot a graph of the function.

(e) State the range of the function.

Solution (a) The independent variable is x.

(b) The dependent variable is y.

(c) Since we are given $1 \leqslant x \leqslant 3$ the domain of the function is the interval $[1, 3]$, that is all values from 1 to 3 inclusive.

(d) The graph of $y = f(x) = 2x$ is shown in Figure 17.7.

Figure 17.7
Graph of $y = f(x) = 2x$

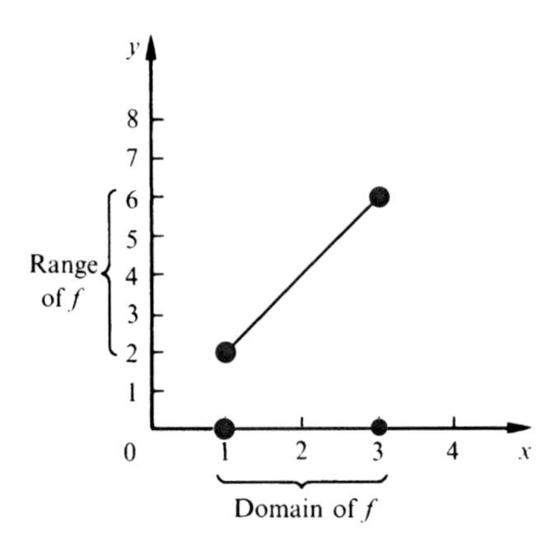

(e) The range is the set of values taken by the output, y. From the graph we see that as x varies from 1 to 3 then y varies from 2 to 6. Hence the range of the function is $[2, 6]$.

17.7 Consider the function $y = f(t) = 4t$.

(a) State which is the independent variable and which is the dependent variable.

(b) Plot a graph of the function and give its domain and range.

Solution

(a) The independent variable is t and the dependent variable is y.

(b) No domain is specified so it is taken to be the largest set possible. The domain is $\mathbb{R}$, the set of all (real) numbers. However, it would be impractical to draw a graph whose domain was the whole extent of the t axis and so a selected portion is shown in Figure 17.8. Similarly only a restricted portion of the range can be drawn, although in this example the range is also $\mathbb{R}$.

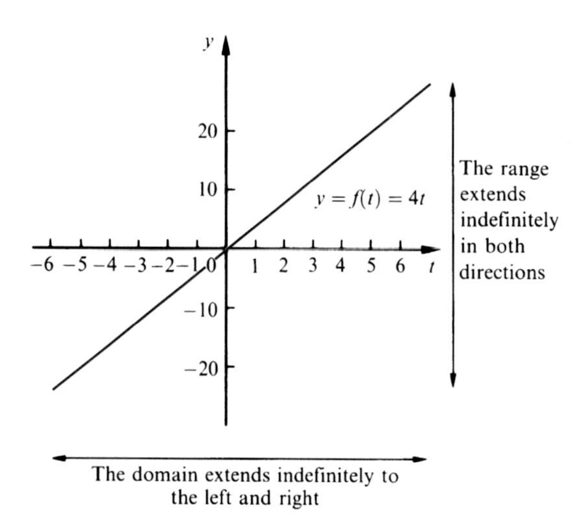

Figure 17.8
A graph of $f(t) = 4t$

17.8 Consider the function $y = f(x) = x^2 - 1$: (a) state the independent variable; (b) state the dependent variable; (c) state the domain; (d) plot a graph of f and determine the range.

Solution

(a) The independent variable is x.

(b) The dependent variable is y.

(c) No domain is specified so it is taken to be the largest set possible. The domain is $\mathbb{R}$.

(d) A table of values and graph of $f(x) = x^2 - 1$ are shown in Figure 17.9. From the graph we see that the smallest value of y is -1, which occurs when $x = 0$. Whether x increases or decreases from 0, the value of y increases. The range is thus all values greater than or equal to -1.

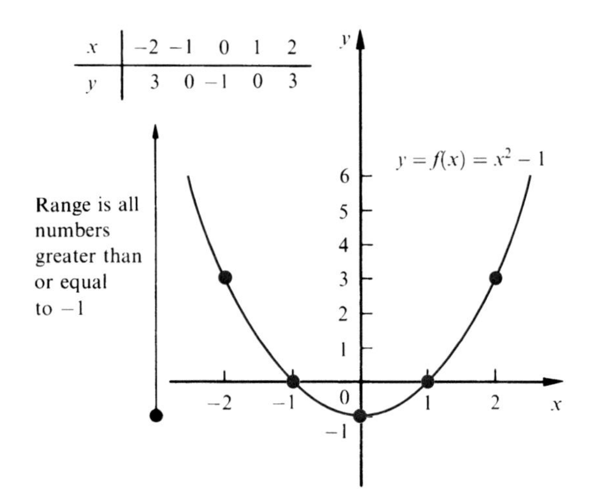

Figure 17.9
A graph of $f(x) = x^2 - 1$

We can write the range using the set notation introduced in §17.2 as

$$\text{range} = \{y : y \in \mathbb{R},\ y \geqslant -1\}$$

17.9 Given the function $y = 1/x$, state the domain and range.

Solution The domain is the largest set possible. All values of x are permissible except $x = 0$ since $\frac{1}{0}$ is not defined. It is never possible to divide by 0. Thus the domain comprises all real numbers except 0. A table of values and graph are shown in Figure 17.10. As x varies, y can take on any value except 0. In this example we see that the range is thus the same as the domain. We note that the graph is split at $x = 0$. For small, positive

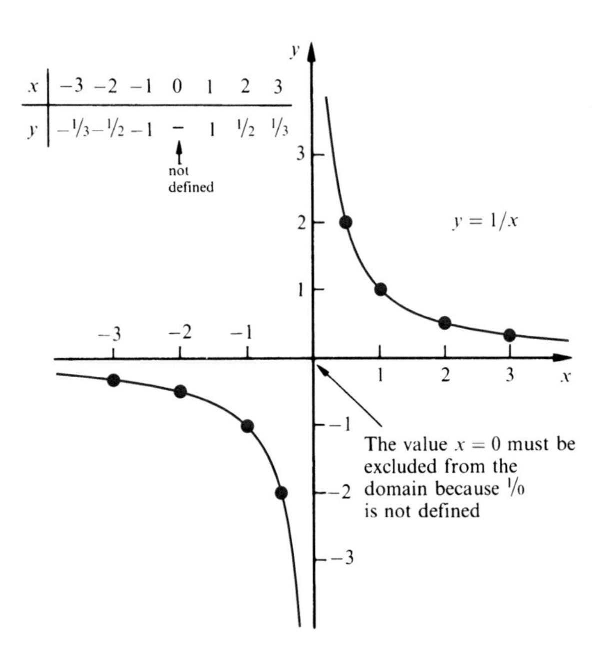

Figure 17.10
A graph of $y = 1/x$

values of x, y is large and positive. The graph approaches the y axis for very small, positive values of x. For small, negative values of x, y is large and negative. Again, the graph approaches the y axis for very small, negative values of x. We say that the y axis is an **asymptote**. In general, if the graph of a function approaches a straight line we call that line an asymptote.

WORKED EXAMPLE

17.10 Consider again the piecewise function in Worked Example 16.10:

$$y(x) = \begin{cases} x^2 + 1 & -1 \leqslant x \leqslant 2 \\ 3x & 2 < x \leqslant 6 \\ 2x + 1 & x > 6 \end{cases}$$

(a) Plot a graph of the function.

(b) State the domain of the function.

(c) State the range of the function.

Solution (a) Figure 17.11 shows a graph of $y(x)$.

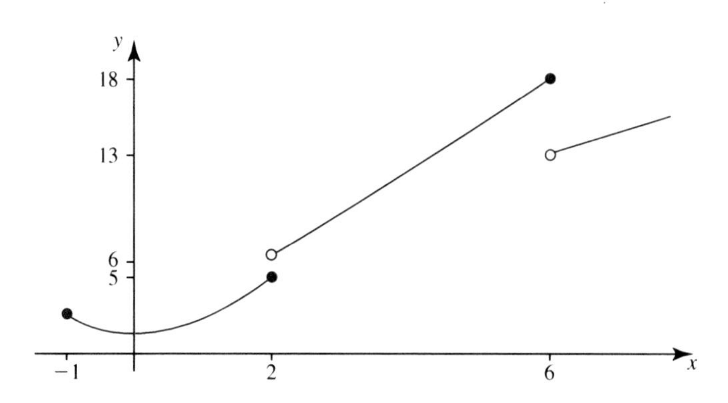

Figure 17.11
A graph of $y(x)$ for Worked Example 17.10

The different rules that make up the definition of y determine the three different pieces of the graph. Note the use of ● and ○ to denote the closed and open end-points.

(b) The domain is $[-1, \infty)$, that is

$$\text{domain} = \{x \in \mathbb{R}, x \geqslant -1\}$$

(c) The smallest value of y is $y = 1$, which occurs when $x = 0$. From the graph we see that the range is $[1, 5]$ and $(6, \infty)$, that is

$$\text{range} = \{y \in \mathbb{R}, 1 \leqslant y \leqslant 5 \text{ and } y > 6\}$$

Self-assessment questions 17.4

1. Explain the terms 'domain of a function' and 'range of a function'.
2. Explain why the value $x = 3$ must be excluded from any domain of the function $f(x) = 7/(x - 3)$.
3. Give an example of a function for which we must exclude the value $x = -2$ from the domain.

Exercise 17.4

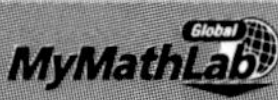

1. Plot graphs of the following functions. In each case state (i) the independent variable, (ii) the dependent variable, (iii) the domain, (iv) the range.
   (a) $y = x - 1,\quad 6 \leqslant x \leqslant 10$ (b) $h = t^2 + 3,\quad 4 \leqslant t < 5$ (c) $m = 3n - 2,\quad -1 < n < 1$
   (d) $y = 2x + 4$ (e) $X = y^3$ (f) $k = 4r^2 + 5,\quad 5 \leqslant r \leqslant 10$
2. A function is defined by

$$q(t) = \begin{cases} t^2 & 0 < t < 3 \\ 6 + t & 3 \leqslant t < 5 \\ -2t + 21 & 5 \leqslant t \leqslant 10.5 \end{cases}$$

   (a) Plot a graph of this function. (b) State the domain of the function.
   (c) State the range of the function. (d) Evaluate $q(1)$, $q(3)$, $q(5)$ and $q(7)$.

17.5 Solving equations using graphs

In Chapter 11 we used algebraic methods to solve equations. However, many equations cannot be solved exactly using such methods. In such cases it is often useful to use a graphical approach. The following examples illustrate the method of using graphs to solve equations.

WORKED EXAMPLES

17.11 Find graphically all solutions in the interval $[-3, 3]$ of the equation $x^2 + x - 3 = 0$.

Solution We draw a graph of $y(x) = x^2 + x - 3$. Table 17.3 gives x and y values and Figure 17.12 shows a graph of the function. We have plotted $y = x^2 + x - 3$ and wish to solve $0 = x^2 + x - 3$. Thus we read from the graph the coordinates of the points at which $y = 0$. Such points must be on the x axis. From the graph the points are (1.3, 0) and (−2.3, 0). So the solutions of

Table 17.3
Values of x and y when $y = x^2 + x - 3$

x	−3	−2	−1	0	1	2	3
y	3	−1	−3	−3	−1	3	9

Figure 17.12
A graph of $y = x^2 + x - 3$

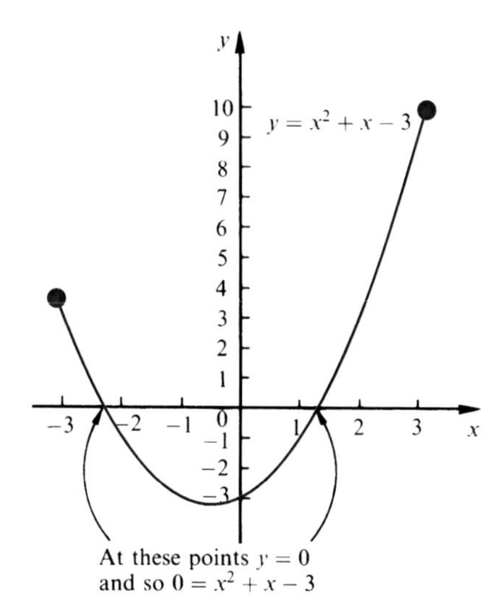

$x^2 + x - 3 = 0$ are $x = 1.3$ and $x = -2.3$. Note that these are approximate solutions, being dependent upon the accuracy of the graph and the x and y scales used. Increased accuracy can be achieved by drawing an enlargement of the graph around the values $x = 1.3$ and $x = -2.3$. So, for example, we could draw $y = x^2 + x - 3$ for $1.2 \leqslant x \leqslant 1.5$ and $y = x^2 + x - 3$ for $-2.4 \leqslant x \leqslant -2.2$.

17.12 Find solutions in the interval $[-2, 2]$ of the equation $x^3 = 2x + \frac{1}{2}$ using a graphical method.

Solution The problem of solving $x^3 = 2x + \frac{1}{2}$ is identical to that of solving $x^3 - 2x - \frac{1}{2} = 0$. So we plot a graph of $y = x^3 - 2x - \frac{1}{2}$ for $-2 \leqslant x \leqslant 2$ and then locate the points where the curve cuts the x axis; that is, where the y coordinate is 0. Table 17.4 gives x and y values and Figure 17.13 shows a graph of the function. We now consider points on the graph where the y coordinate is zero. These points are marked A, B and C. Their x coordinates are −1.27, −0.26 and 1.53. Hence the solutions of $x^3 = 2x + \frac{1}{2}$ are approximately $x = -1.27$, $x = -0.26$ and $x = 1.53$.

Table 17.4
Values of x and y when $y = x^3 - 2x - \frac{1}{2}$

x	−2	−1.5	−1	−0.5	0	0.5	1	1.5	2
y	−4.5	−0.875	0.5	0.375	−0.5	−1.375	−1.5	−0.125	3.5

Figure 17.13
A graph of
$y = x^3 - 2x - \frac{1}{2}$

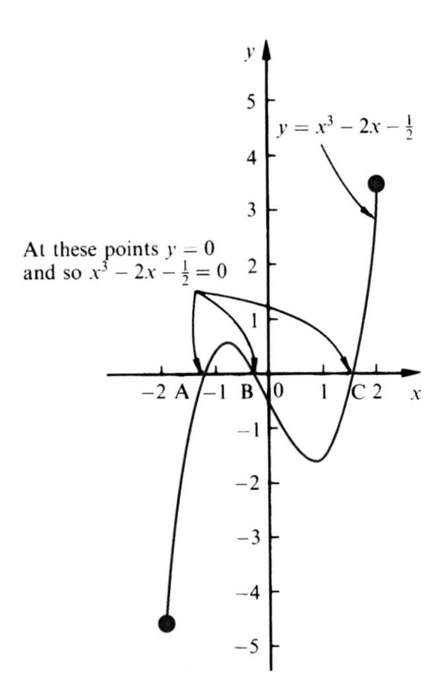

Self-assessment question 17.5

1. Explain how the graph of the function f can be used to solve the equation
   (a) $f(x) = 0$ (b) $f(x) = 1$

Exercise 17.5

MyMathLab

1. Plot a graph of
$$y = x^2 + \frac{3x}{2} - 2$$
for $-3 \leqslant x \leqslant 2$. Hence solve the equation
$$x^2 + \frac{3x}{2} - 2 = 0$$
2. Plot $y = x^2 - x - 1$ for $-3 \leqslant x \leqslant 3$. Hence solve $x^2 - x - 1 = 0$.
3. Plot $y = x^2 - 0.4x - 4$ for $-3 \leqslant x \leqslant 3$. Hence solve $x^2 - 0.4x - 4 = 0$.
4. (a) Plot $y = 3 + \frac{x}{2} - x^2$ for $-3 \leqslant x \leqslant 3$.
   (b) Use your graph to solve
$$x^2 - \frac{x}{2} - 3 = 0$$

17.6 Solving simultaneous equations graphically

In the previous section we saw how an approximate solution to an equation could be found by drawing an appropriate graph. We now extend that technique to find an approximate solution of simultaneous equations. Recall that simultaneous equations were introduced algebraically in Chapter 11.

Given two simultaneous equations in x and y, then a solution is a pair of x and y values that satisfy both equations. For example, $x = 2$, $y = -3$ is a solution of

$$3x - y = 9$$

$$x + 2y = -4$$

because both equations are satisfied by these particular values. Worked Example 17.13 illustrates the graphical method.

WORKED EXAMPLES

17.13 Solve graphically

$$3x - y = 9 \quad (17.1)$$

$$x + 2y = -4 \quad (17.2)$$

Solution Both equations are rearranged so that y is the subject. This gives

$$y = 3x - 9 \quad (17.3)$$

$$y = -\frac{x}{2} - 2 \quad (17.4)$$

Since Equations 17.3 and 17.4 are simple rearrangements of Equations 17.1 and 17.2 then they have the same solution. Equations 17.3 and 17.4 are drawn; Figure 17.14 illustrates this.

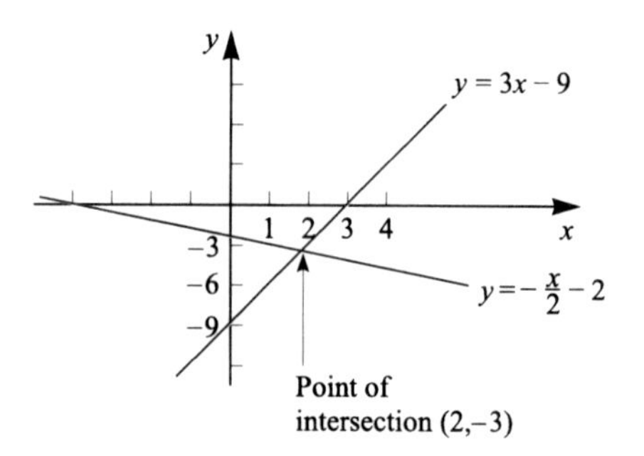

Figure 17.14 Points of intersection give the solution to simultaneous equations

We seek values of x and y that fit both equations. For a point to lie on both graphs, the point must be at the intersection of the graphs. In other words, solutions to simultaneous equations are given by the points of intersection. Reading from the graph, the point of intersection is at $x = 2$,

$y = -3$. Hence $x = 2$, $y = -3$ is a solution of the given simultaneous equations.

17.14 Solve graphically

$$4x - y = 0$$

$$3x + y = 7$$

Solution We write the equations with y as subject:

$$y = 4x$$

$$y = -3x + 7$$

These are now plotted as shown in Figure 17.15.

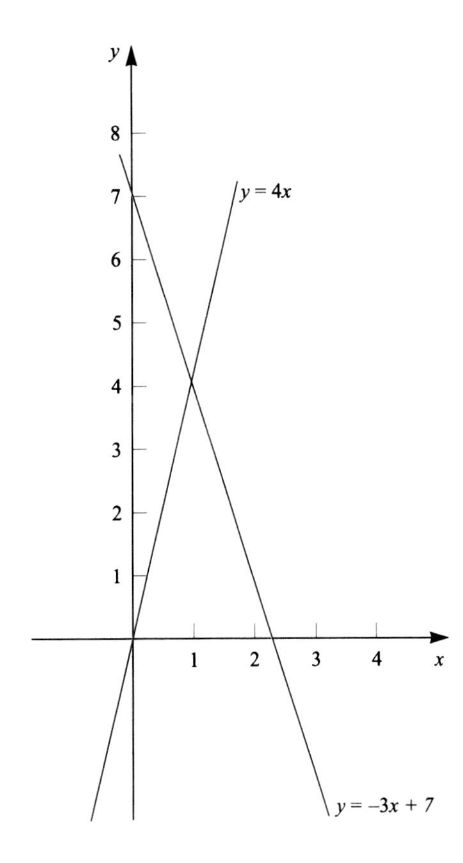

Figure 17.15
The graphs intersect at $x = 1$, $y = 4$

The point of intersection is $x = 1$, $y = 4$ and so this is the solution of the given simultaneous equations.

17.15 Solve graphically

$$x^2 - y = -1$$

$$2x - y = -3$$

Solution Writing the equations with y as subject gives

$$y = x^2 + 1$$

$$y = 2x + 3$$

These are plotted as shown in Figure 17.16.

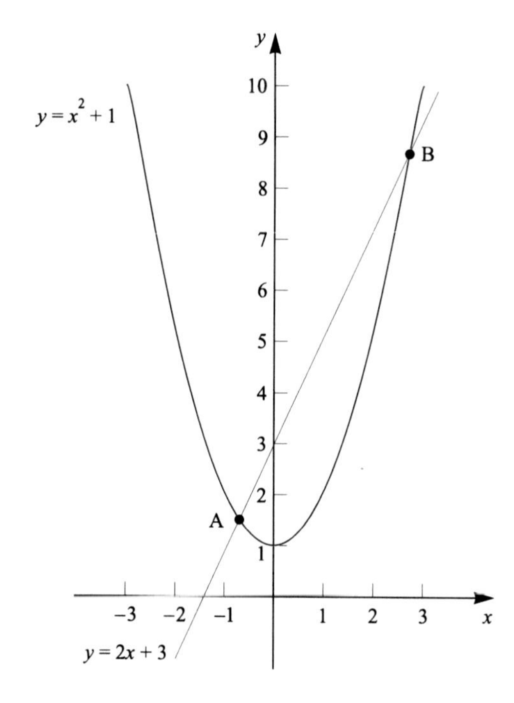

Figure 17.16
The graphs have two points of intersection and hence there are two solutions

There are two points of intersection, A and B. From the graph it is difficult to extract an accurate estimate. However, by using a graphics calculator or package, greater accuracy can be achieved. The coordinates of A are $x = -0.73$, $y = 1.54$; the coordinates of B are $x = 2.73$, $y = 8.46$. Hence there are two solutions to the given equations: $x = -0.73$, $y = 1.54$ and $x = 2.73$, $y = 8.46$.

17.16

(a) On the same axes draw graphs of $y = x^3$ and $y = 2x + \frac{1}{2}$ for $-2 \leqslant x \leqslant 2$.

(b) Note the x coordinates of the points where the two graphs intersect. By referring to Worked Example 17.12 what do you conclude? Can you explain your findings?

Solution

(a) Table 17.5 gives x values and the corresponding values of x^3 and $2x + \frac{1}{2}$. Figure 17.17 shows a graph of $y = x^3$ together with a graph of $y = 2x + \frac{1}{2}$.

(b) The two graphs intersect at A, B and C. The x coordinates of these points are -1.27, -0.26 and 1.53. We note from Worked Example 17.12 that these values are the solutions of $x^3 = 2x + \frac{1}{2}$. We

Table 17.5
Values of x and y when $y = x^3$ and $y = 2x + \frac{1}{2}$

x	-2	-1.5	-1	-0.5	0	0.5	1	1.5	2
x^3	-8	-3.375	-1	-0.125	0	0.125	1	3.375	8
$2x + \frac{1}{2}$	-3.5	-2.5	-1.5	-0.5	0.5	1.5	2.5	3.5	4.5

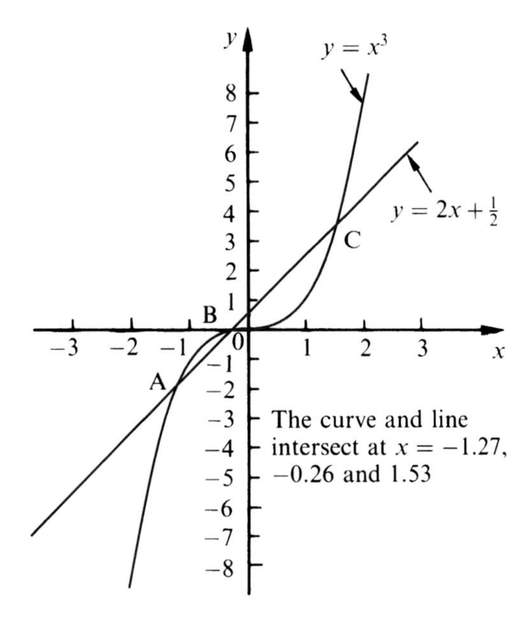

Figure 17.17
Graphs of $y = x^3$ and $y = 2x + \frac{1}{2}$

explain this as follows. Where $y = x^3$ and $y = 2x + \frac{1}{2}$ intersect, their y values are identical and so at these points

$$x^3 = 2x + \frac{1}{2}$$

By reading the x coordinates at these points of intersection we are finding those x values for which $x^3 = 2x + \frac{1}{2}$.

Self-assessment questions 17.6

1. Explain the significance of points of intersection when solving simultaneous equations graphically.
2. The graphs of a pair of simultaneous equations are drawn and they intersect at three points. How many solutions do the simultaneous equations have?
3. The graphs of a pair of simultaneous equations are drawn. The graphs do not intersect. What can you conclude, if anything, about the solution of the simultaneous equations?

Exercise 17.6

1. Solve the following pairs of simultaneous equations graphically. In each case use x values from -4 to 3.

   (a) $3x + 2y = 4$
   $x - y = 3$

   (b) $2x + y = -2$
   $-\frac{x}{2} + 2y = 5$

   (c) $2x + y = 4$
   $4x - 3y = -7$

   (d) $-x + 4y = 7$
   $2x - y = -7$

   (e) $x + 2y = 1$
   $\frac{x}{2} + 5y = -\frac{1}{2}$

2. Solve graphically the simultaneous equations

$$y = x^2$$
$$2x + y = 1$$

   Take x values from -3 to 2.

3. Solve graphically

$$x^2 + y = 3$$
$$4x - 3y = -3$$

   Take x values from -3 to 2.

4. Solve graphically

$$y = \frac{x^3}{2}$$
$$x + y = 3$$

5. Solve graphically, taking x values from -2 to 2:

$$x^3 - y = 0$$
$$-1.5x^2 + 5x - y = -2$$

Test and assignment exercises 17

1. Plot a graph of each of the following functions. In each case state the independent variable, the dependent variable, the domain and the range.

   (a) $y = f(x) = 3x - 7$
   (b) $y = f(x) = 3x - 7, \; -2 \leqslant x \leqslant 2$
   (c) $y = 4x^2, \; x \geqslant 0$
   (d) $y = \frac{1}{x^2}, \; x \neq 0$

2. A function is defined by

$$f(x) = \begin{cases} 2x + 4 & 0 \leqslant x \leqslant 4 \\ 6 & 4 < x < 6 \\ x & 6 \leqslant x \leqslant 8 \end{cases}$$

   (a) Plot a graph of this function, stating its domain and range.
   (b) Evaluate $f(0)$, $f(4)$ and $f(6)$.

3. Plot graphs of the functions $f(x) = x + 4$ and $g(x) = 2 - x$ for values of x between -4 and 4. Hence solve the simultaneous equations $y - x = 4$, $y + x = 2$.
4. Plot a graph of the function $f(x) = 3x^2 + 10x - 8$ for values of x between -5 and 3. Hence solve the equation $3x^2 + 10x - 8 = 0$.
5. Plot a graph of the function $f(t) = t^3 + 2t^2 - t - 2$ for $-3 \leqslant t \leqslant 3$. Hence solve the equation $t^3 + 2t^2 - t - 2 = 0$.
6. Plot a graph of $h = 2 + 2x - x^2$ for $-3 \leqslant x \leqslant 3$. Hence solve the equation $2 + 2x - x^2 = 0$.
7. (a) On the same axes plot graphs of $y = x^3$ and $y = 5x^2 - 3$ for $-2 \leqslant x \leqslant 2$.
   (b) Use your graphs to solve the equation $x^3 - 5x^2 + 3 = 0$.
8. (a) Plot a graph of $y = x^3 - 2x^2 - x + 1$ for $-2 \leqslant x \leqslant 3$.
   (b) Hence find solutions of $x^3 - 2x^2 - x + 1 = 0$.
   (c) Use your graph to solve $x^3 - 2x^2 - x + 2 = 0$.
9. Solve each of the following pairs of simultaneous equations graphically. In each case take x values from -3 to 3.

   (a) $2x - y = 0$
   $x + y = -3$

   (b) $3x + 4y = 6$
   $x - 2y = -8$

   (c) $3x + y = 8$
   $x - 3y = -1.5$

   (d) $2x - 3y = 4.7$
   $\frac{x}{2} + y = -0.4$

   (e) $x + 3y = 1.2$
   $-2x + 5y = 10.8$

10. Solve graphically

$$x^2 - y = -2$$
$$x^2 + 2x + y = 3$$

Take x values from -3 to 3.

11. Solve graphically

$$x^3 - y = 0$$
$$1.5x^2 + 6x - y = 9$$

Take x values from -3 to 3.

The straight line

18

Objectives: This chapter:

- describes some special properties of straight line graphs
- explains the equation $y = mx + c$
- explains the terms 'vertical intercept' and 'gradient'
- shows how the equation of a line can be calculated
- explains what is meant by a tangent to a curve
- explains what is meant by the gradient of a curve
- explains how the gradient of a curve can be estimated by drawing a tangent

18.1 Straight line graphs

In the previous chapter we explained how to draw graphs of functions. Several of the resulting graphs were straight lines. In this chapter we focus specifically on the straight line and some of its properties.

Any equation of the form $y = mx + c$, where m and c are constants, will have a straight line graph. For example,

$$y = 3x + 7 \qquad y = -2x + \frac{1}{2} \qquad y = 3x - 1.5$$

will all result in straight line graphs. It is important to note that the variable x only occurs to the power 1. The values of m and c may be zero, so that $y = -3x$ is a straight line for which the value of c is zero, and $y = 17$ is a straight line for which the value of m is zero.

Key point

Any straight line has an equation of the form $y = mx + c$ where m and c are constants.

WORKED EXAMPLES

18.1 Which of the following equations have straight line graphs? For those that do, identify the values of m and c.

(a) $y = 7x + 5$ (b) $y = \dfrac{13x - 5}{2}$ (c) $y = 3x^2 + x$ (d) $y = -19$

Solution (a) $y = 7x + 5$ is an equation of the form $y = mx + c$ where $m = 7$ and $c = 5$. This equation has a straight line graph.

(b) The equation

$$y = \frac{13x - 5}{2}$$

can be written as $y = \frac{13}{2}x - \frac{5}{2}$, which is in the form $y = mx + c$ with $m = \frac{13}{2}$ and $c = -\frac{5}{2}$. This has a straight line graph.

(c) $y = 3x^2 + x$ contains the term x^2. Such a term is not allowed in the equation of a straight line. The graph will not be a straight line.

(d) $y = -19$ is in the form of the equation of a straight line for which $m = 0$ and $c = -19$.

18.2 Plot each of the following graphs: $y = 2x + 3$, $y = 2x + 1$ and $y = 2x - 2$. Comment upon the resulting lines.

Solution The three graphs are shown in Figure 18.1. Note that all three graphs have the same steepness or **slope**. However, each one cuts the vertical

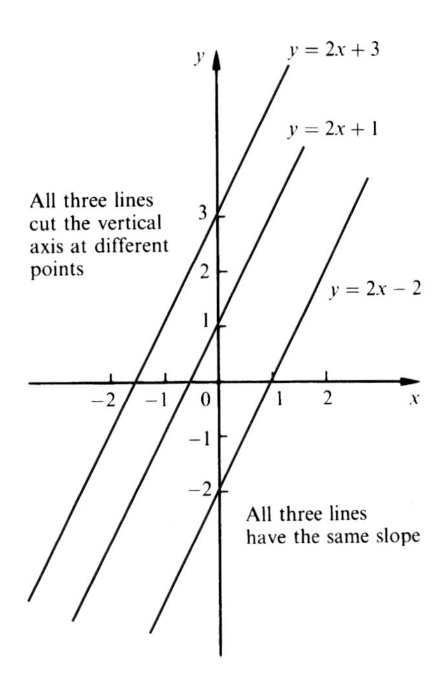

Figure 18.1
Graphs of $y = 2x + 3$, $y = 2x + 1$ and $y = 2x - 2$

axis at a different point. This point can be obtained directly from the equation $y = 2x + c$ by looking at the value of c. For example, $y = 2x + 1$ intersects the vertical axis at $y = 1$. The graph of $y = 2x + 3$ cuts the vertical axis at $y = 3$, and the graph of $y = 2x - 2$ cuts this axis at $y = -2$.

The point where a graph cuts the vertical axis is called the **vertical intercept**. The vertical intercept can be obtained from $y = mx + c$ by looking at the value of c.

Key point

In the equation $y = mx + c$ the value of c gives the y coordinate of the point where the line cuts the vertical axis.

WORKED EXAMPLE

18.3

On the same graph plot the following: $y = x + 2$, $y = 2x + 2$, $y = 3x + 2$. Comment upon the graphs.

Solution

The three straight lines are shown in Figure 18.2. Note that the steepness of the line is determined by the coefficient of x, with $y = 3x + 2$ being steeper than $y = 2x + 2$, and this in turn being steeper than $y = x + 2$.

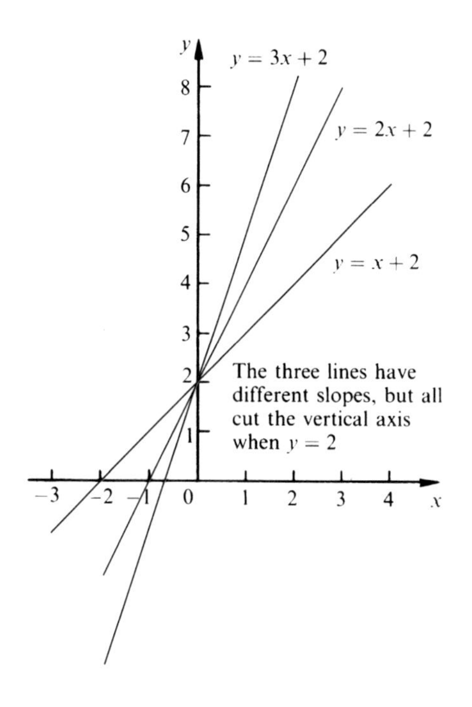

Figure 18.2
Graphs of $y = x + 2$, $y = 2x + 2$, and $y = 3x + 2$

The value of m in the equation $y = mx + c$ determines the steepness of the straight line. The larger the value of m, the steeper is the line. The value m is known as the slope or **gradient** of the straight line.

Key point

In the equation $y = mx + c$ the value m is known as the gradient and is a measure of the steepness of the line.

If m is positive, the line will rise as we move from left to right. If m is negative, the line will fall. If $m = 0$ the line will be horizontal. We say it has zero gradient. These points are summarised in Figure 18.3.

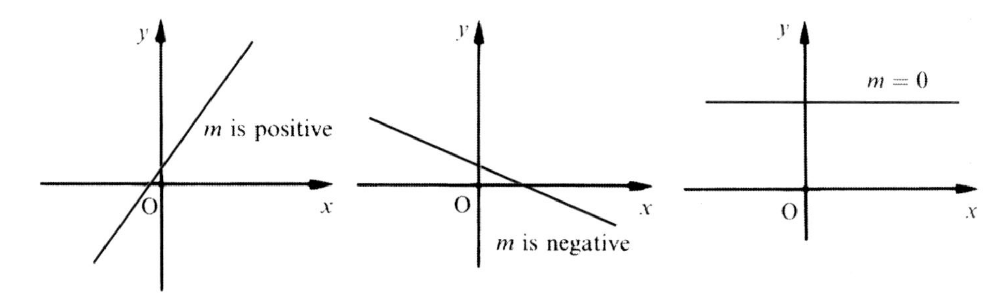

Figure 18.3
Straight line graphs with positive, negative and zero gradients

Self-assessment questions 18.1

1. Give the standard form of the equation of a straight line.
2. Explain the meaning of m and c in the equation $y = mx + c$.

Exercise 18.1

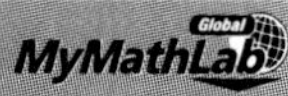

1. Identify, without plotting, which of the following functions will give straight line graphs:
   (a) $y = 3x + 9$ (b) $y = -9x + 2$ (c) $y = -6x$ (d) $y = x^2 + 3x$
   (e) $y = 17$ (f) $y = x^{-1} + 7$
2. Identify the gradient and vertical intercept of each of the following lines:
   (a) $y = 9x - 11$ (b) $y = 8x + 1.4$ (c) $y = \frac{1}{2}x - 11$ (d) $y = 17 - 2x$
   (e) $y = \frac{2x + 1}{3}$ (f) $y = \frac{4 - 2x}{5}$ (g) $y = 3(x - 1)$ (h) $y = 4$
3. Identify (i) the gradient and (ii) the vertical intercept of the following lines:
   (a) $y + x = 6$ (b) $y - 2x + 1 = 0$ (c) $2y - 4x + 3 = 0$ (d) $3x - 4y + 12 = 0$
   (e) $3x + \frac{y}{2} - 9 = 0$

18.2 Finding the equation of a straight line from its graph

If we are given the graph of a straight line it is often necessary to find its equation, $y = mx + c$. This amounts to finding the values of m and c. Finding the vertical intercept is straightforward because we can look directly for the point where the line cuts the y axis. The y coordinate of this point gives the value of c. The gradient m can be determined from knowledge of any two points on the line using the formula

Key point

$$\text{gradient} = \frac{\text{difference between the } y \text{ coordinates}}{\text{difference between the } x \text{ coordinates}}$$

WORKED EXAMPLES

18.4 A straight line graph is shown in Figure 18.4. Determine its equation.

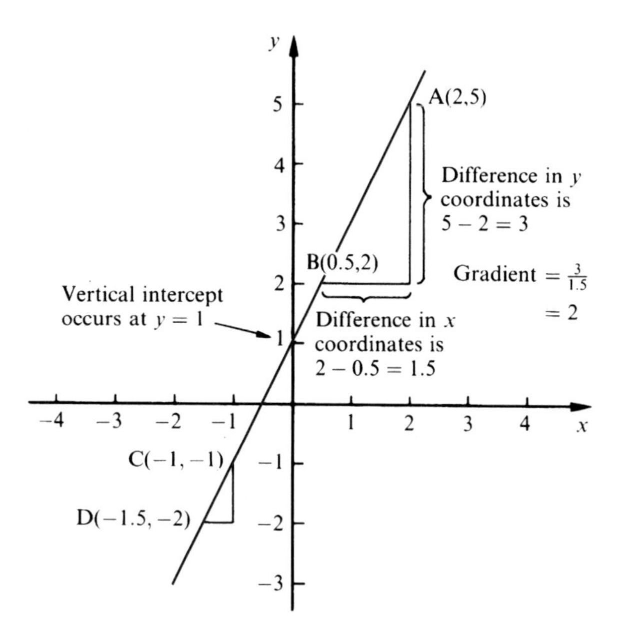

Figure 18.4 Graph for Worked Example 18.4

Solution We require the equation of the line in the form $y = mx + c$. From the graph it is easy to see that the vertical intercept occurs at $y = 1$. Therefore the value of c is 1. To find the gradient m we choose any two points on the line. We have chosen the point A with coordinates (2, 5) and the point B with coordinates (0.5, 2). The difference between their y coordinates is then

$5-2=3$. The difference between their x coordinates is $2-0.5=1.5$. Then

$$\text{gradient} = \frac{\text{difference between their } y \text{ coordinates}}{\text{difference between their } x \text{ coordinates}}$$

$$= \frac{3}{1.5} = 2$$

The gradient m is equal to 2. Note that as we move from left to right the line is rising and so the value of m is positive. The equation of the line is then $y = 2x + 1$. There is nothing special about the points A and B. Any two points are sufficient to find m. For example, using the points C with coordinates $(-1, -1)$ and D with coordinates $(-1.5, -2)$ we would find

$$\text{gradient} = \frac{\text{difference between their } y \text{ coordinates}}{\text{difference between their } x \text{ coordinates}}$$

$$= \frac{-1-(-2)}{-1-(-1.5)}$$

$$= \frac{1}{0.5} = 2$$

as before.

18.5 A straight line graph is shown in Figure 18.5. Find its equation.

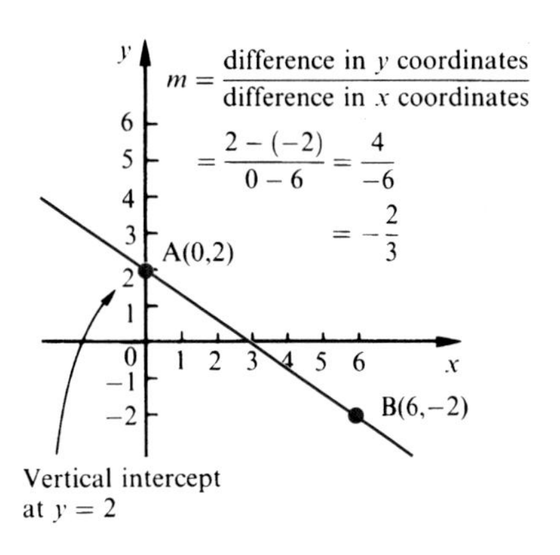

Figure 18.5 Graph for Worked Example 18.5

Solution We need to find the equation in the form $y = mx + c$. From the graph we see immediately that the value of c is 2. To find the gradient we have selected any two points, A(0, 2) and B(6, −2). The difference between their y coordinates is $2-(-2)=4$. The difference between their x coordinates is $0 - 6 = -6$. Then

$$\text{gradient} = \frac{\text{difference between their } y \text{ coordinates}}{\text{difference between their } x \text{ coordinates}}$$

$$= \frac{4}{-6}$$

$$= -\frac{2}{3}$$

The equation of the line is therefore $y = -\frac{2}{3}x + 2$. Note in particular that, because the line is sloping downwards as we move from left to right, the gradient is negative. Note also that the coordinates of A and B both satisfy the equation of the line. That is, for A(0, 2),

$$2 = -\frac{2}{3}(0) + 2$$

and for B(6, −2),

$$-2 = -\frac{2}{3}(6) + 2$$

The coordinates of any other point on the line must also satisfy the equation.

The point noted at the end of Worked Example 18.5 is important:

Key point

If the point (a, b) lies on the line $y = mx + c$ then this equation is satisfied by letting $x = a$ and $y = b$.

WORKED EXAMPLE

18.6 Find the equation of the line shown in Figure 18.6.

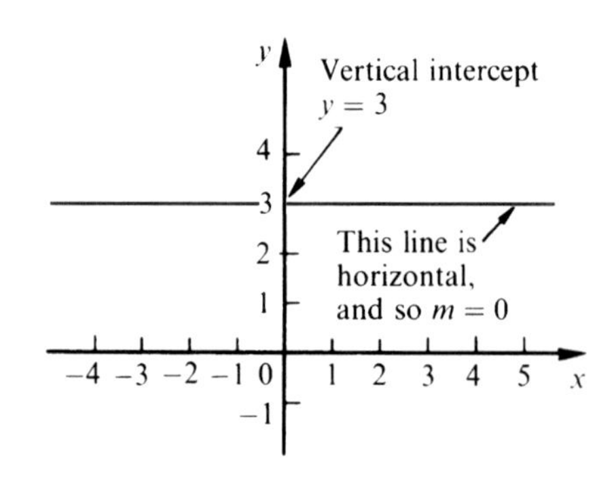

Figure 18.6 Graph for Worked Example 18.6

Solution We are required to express the equation in the form $y = mx + c$. From the graph we notice that the line is horizontal. This means that its gradient is 0, that is $m = 0$. Furthermore the line cuts the vertical axis at $y = 3$ and so the equation of the line is $y = 0x + 3$ or simply $y = 3$.

It is not necessary to sketch a graph in order to find the equation. Consider the following worked examples, which illustrate an algebraic method.

WORKED EXAMPLES

18.7 A straight line passes through A(7, 1) and B(−3, 2). Find its equation.

Solution The equation must be of the form $y = mx + c$. The gradient of the line can be found from

$$\begin{aligned}\text{gradient} = m &= \frac{\text{difference between their } y \text{ coordinates}}{\text{difference between their } x \text{ coordinates}}\\ &= \frac{1-2}{7-(-3)}\\ &= \frac{-1}{10}\\ &= -0.1\end{aligned}$$

Hence $y = -0.1x + c$. We can find c by noting that the line passes through (7, 1), that is the point where $x = 7$ and $y = 1$. Substituting these values into the equation $y = -0.1x + c$ gives

$$1 = -0.1(7) + c$$

so that $c = 1 + 0.7 = 1.7$. Therefore the equation of the line is $y = -0.1x + 1.7$.

18.8 Determine the equation of the line that passes through (4, −1) and has gradient −2.

Solution Let the equation of the line be $y = mx + c$. We are told that the gradient of the line is −2, that is $m = -2$, and so we have

$$y = -2x + c$$

The point (4, −1) lies on this line: hence when $x = 4$, $y = -1$. These values are substituted into the equation of the line:

$$\begin{aligned}-1 &= -2(4) + c\\ c &= 7\end{aligned}$$

The equation of the line is thus $y = -2x + 7$.

Self-assessment questions 18.2

1. State the formula for finding the gradient of a straight line when two points upon it are known. If the two points are (x_1, y_1) and (x_2, y_2) write down an expression for the gradient.
2. Explain how the value of c in the equation $y = mx + c$ can be found by inspecting the straight line graph.

Exercise 18.2

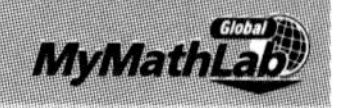

1. A straight line passes through the two points (1, 7) and (2, 9). Sketch a graph of the line and find its equation.
2. Find the equation of the line that passes through the two points (2, 2) and (3, 8).
3. Find the equation of the line that passes through (8, 2) and (−2, 2).
4. Find the equation of the straight line that has gradient 1 and passes through the origin.
5. Find the equation of the straight line that has gradient −1 and passes through the origin.
6. Find the equation of the straight line passing through (−1, 6) with gradient 2.
7. Which of the following points lie on the line $y = 4x - 3$?
   (a) (1, 2) (b) (2, 5) (c) (5, 17)
   (d) (−1, −7) (e) (0, 2)
8. Find the equation of the straight line passing through (−3, 7) with gradient −1.
9. Determine the equation of the line passing through (−1, −6) that is parallel to the line $y = 3x + 17$.
10. Find the equation of the line with vertical intercept −2 passing through (3, 10).

18.3 Gradients of tangents to curves

Figure 18.7 shows a graph of $y = x^2$. If you study the graph you will notice that as we move from left to right, at some points the y values are decreasing, whereas at others the y values are increasing. It is intuitively obvious that the slope of the curve changes from point to point. At some points, such as A, the curve appears quite steep and falling. At points such as B the curve appears quite steep and rising. Unlike a straight line, the slope of a curve is not fixed but changes as we move from one point to another. A useful way of measuring the slope at any point is to draw a **tangent** to the curve at that point. The tangent is a straight line that just touches the curve at the point of interest. In Figure 18.7 a tangent to the curve $y = x^2$ has been drawn at the point (2, 4). If we calculate the

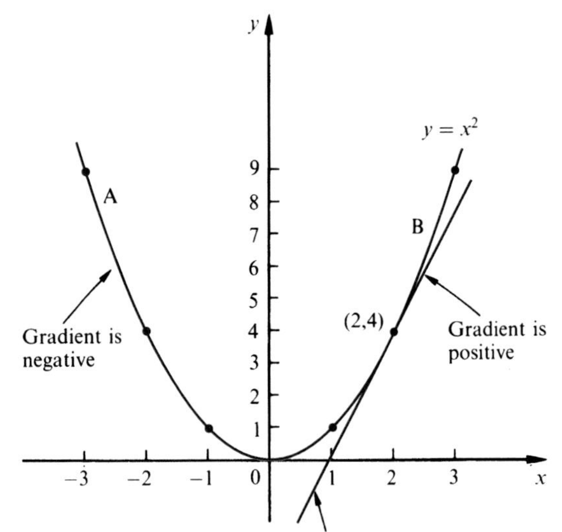

Figure 18.7
A graph of $y = x^2$

gradient of this tangent, this gives the gradient of the curve at the point (2, 4).

Key point

The gradient of a curve at any point is equal to the gradient of the tangent at that point.

WORKED EXAMPLE

18.9

(a) Plot a graph of $y = x^2 - x$ for values of x between -2 and 4.

(b) Draw in tangents at the points A$(-1, 2)$ and B$(3, 6)$.

(c) By calculating the gradients of these tangents find the gradient of the curve at A and at B.

Solution

(a) A table of values and the graph are shown in Figure 18.8.

(b) We now draw tangents at A and B. At present, the best we can do is estimate these by eye.

(c) We now calculate the gradient of the tangent at A. We select any two points on the tangent and calculate the difference between their y coordinates and the difference between their x coordinates. We have chosen the points $(-3, 8)$ and $(-2, 5)$. Referring to Figure 18.8 we see that

$$\text{gradient of tangent at A} = \frac{8-5}{-3-(-2)} = \frac{3}{-1}$$

$$= -3$$

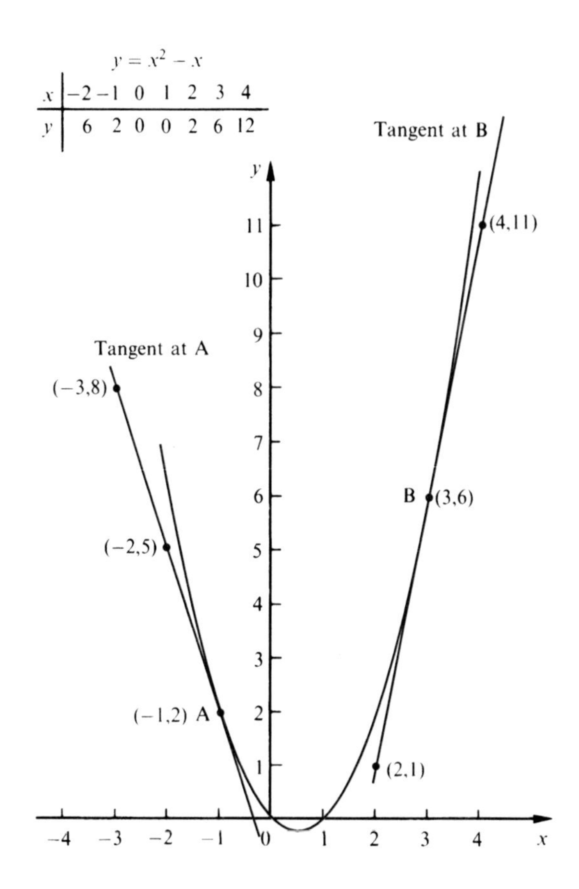

Figure 18.8
A graph of $y = x^2 - x$

Hence the gradient of the curve at A is −3. Similarly, to find the gradient of the tangent at B we have selected two points on this tangent, namely (4, 11) and (2, 1). We find

$$\text{gradient of tangent at B} = \frac{11 - 1}{4 - 2} = \frac{10}{2}$$

$$= 5$$

Hence the gradient of the tangent at B is 5. Thus the gradient of the curve at B(3, 6) is 5.

Clearly, the accuracy of our answer depends to a great extent upon how well we can draw and measure the gradient of the tangent.

WORKED EXAMPLE

18.10 (a) Sketch a graph of the curve $y = x^3$ for $-2 \leqslant x \leqslant 2$.

(b) Draw the tangent to the graph at the point where $x = 1$.

(c) Estimate the gradient of this tangent and find its equation.

Solution (a) A graph is shown in Figure 18.9.

Figure 18.9
Graph of $y = x^3$

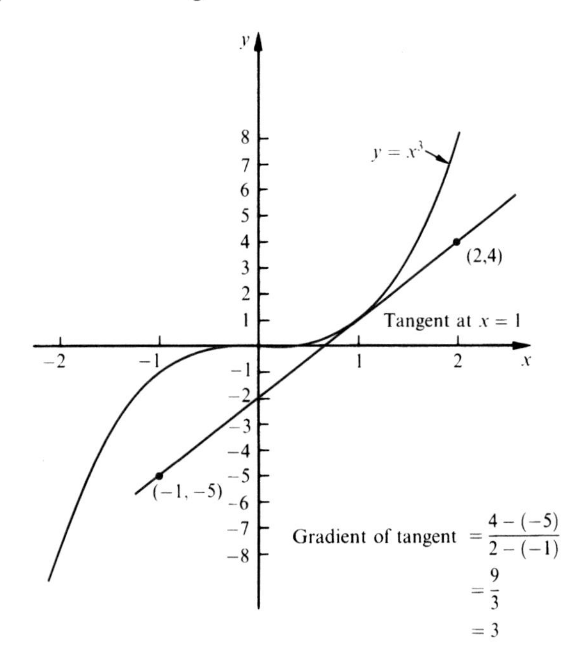

(b) The tangent has been drawn at $x = 1$.

(c) Let us write the equation of the tangent as $y = mx + c$. Two points on the tangent have been selected in order to estimate the gradient. These are $(2, 4)$ and $(-1, -5)$. From these we find

$$\text{gradient of tangent is approximately } \frac{4-(-5)}{2-(-1)} = \frac{9}{3} = 3$$

Therefore $m = 3$. The value of c is found by noting that the vertical intercept of the tangent is -2. The equation of the tangent is then $y = 3x - 2$.

Of course, this method will usually result in an approximation based upon how well we have drawn the graph and its tangent. A much more precise method for calculating gradients of curves is given in Chapter 33.

Self-assessment questions 18.3

1. Explain what is meant by the 'tangent' to a curve at a point.
2. Explain how a tangent is used to determine the gradient of a curve.

Exercise 18.3

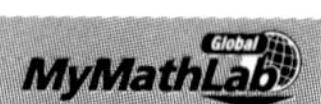

1. Draw the graph of $y = 2x^2 - 1$ for values of x between -3 and 3. By drawing a tangent, estimate the gradient of the curve at A(2, 7) and B(−1, 1).
2. Draw the graph of $y = -2x^2 + 2$ for values of x between -3 and 3. Draw the tangent at the point where $x = 1$ and calculate its equation.

Test and assignment exercises 18

1. Which of the following will have straight line graphs?

   (a) $y = 2x - 11$ (b) $y = 5x + 10$ (c) $y = x^2 - 1$ (d) $y = -3 + 3x$ (e) $y = \dfrac{2x + 3}{2}$

   For each straight line, identify the gradient and vertical intercept.
2. Find the equation of the straight line that passes through the points (1, 11) and (2, 18). Show that the line also passes through (−1, −3).
3. Find the equation of the line that has gradient −2 and passes through the point (1, 1).
4. Find the equation of the line that passes through (−1, 5) and (1, 5). Does the line also pass through (2, 6)?
5. Draw a graph of $y = -x^2 + 3x$ for values of x between -3 and 3. By drawing in tangents estimate the gradient of the curve at the points (−2, −10) and (1, 2).
6. Find the equations of the lines passing through the origin with gradients (a) −2, (b) −4, (c) 4.
7. Find the equation of the line passing through (4, 10) and parallel to $y = 6x - 3$.
8. Find the equation of the line with vertical intercept 3 and passing through (−1, 9).
9. Find where the line joining (−2, 4) and (3, 10) cuts (a) the x axis, (b) the y axis.
10. A line cuts the x axis at $x = -2$ and the y axis at $y = 3$. Determine the equation of the line.
11. Determine where the line $y = 4x - 1$ cuts

    (a) the y axis (b) the x axis (c) the line $y = 2$

The exponential function

19

Objectives: This chapter:

- shows how to simplify exponential expressions
- describes the form of the exponential function
- illustrates graphs of exponential functions
- lists the properties of the exponential function
- shows how to solve equations with exponential terms using a graphical technique

19.1 Exponential expressions

An expression of the form a^x is called an **exponential expression**. The number a is called the **base** and x is the **power** or **index**. For example, 4^x and 0.5^x are both exponential expressions. The first has base 4 and the second has base 0.5. An exponential expression can be evaluated using the power button on your calculator. For example, the exponential expression with base 4 and index 0.3 is

The letter e always denotes the constant 2.71828....

$$4^{0.3} = 1.516$$

One of the most commonly used values for the base in an exponential expression is 2.71828.... This number is denoted by the letter e.

Key point

The most common exponential expression is

$$e^x$$

where e is the exponential constant, 2.71828....

The expression e^x is found to occur in the modelling of many natural phenomena, for example population growth, spread of bacteria and radioactive decay. Scientific calculators are pre-programmed to evaluate e^x for any value of x. Usually the button is marked e^x or $\exp x$. Use the next worked example to check you can use this facility on your calculator.

WORKED EXAMPLE

19.1 Use a scientific calculator to evaluate

(a) e^2 (b) e^3 (c) $e^{1.3}$ (d) $e^{-1.3}$

Solution We use the e^x button giving:

(a) $e^2 = 7.3891$

(b) $e^3 = 20.0855$

(c) $e^{1.3} = 3.6693$

(d) $e^{-1.3} = 0.2725$

Exponential expressions can be simplified using the normal rules of algebra. The laws of indices apply to exponential expressions. We note

$$e^a e^b = e^{a+b}$$

$$\frac{e^a}{e^b} = e^{a-b}$$

$$e^0 = 1$$

$$(e^a)^b = e^{ab}$$

WORKED EXAMPLES

19.2 Simplify the following exponential expressions:

(a) $e^2 e^4$ (b) $\frac{e^4}{e^3}$ (c) $(e^2)^{2.5}$

Solution (a) $e^2 e^4 = e^{2+4} = e^6$

(b) $\frac{e^4}{e^3} = e^{4-3} = e^1 = e$

(c) $(e^2)^{2.5} = e^{2 \times 2.5} = e^5$

19.3 Simplify the following exponential expressions:

(a) $e^x e^{3x}$ (b) $\frac{e^{4t}}{e^{3t}}$ (c) $(e^t)^4$

Solution (a) $e^x e^{3x} = e^{x+3x} = e^{4x}$

(b) $\dfrac{e^{4t}}{e^{3t}} = e^{4t-3t} = e^t$

(c) $(e^t)^4 = e^{4t}$

19.4 Simplify

(a) $\sqrt{e^{6t}}$ (b) $e^{3y}(1+e^y) - e^{4y}$ (c) $(2e^t)^2(3e^{-t})$

Solution (a) $\sqrt{e^{6t}} = (e^{6t})^{\frac{1}{2}} = e^{6t/2} = e^{3t}$

(b) $$e^{3y}(1+e^y) - e^{4y} = e^{3y} + e^{3y}e^y - e^{4y}$$
$$= e^{3y} + e^{4y} - e^{4y}$$
$$= e^{3y}$$

$\sqrt{a} = a^{\frac{1}{2}}$; that is, a square root sign is equivalent to the power $\frac{1}{2}$.

(c) $$(2e^t)^2(3e^{-t}) = (2e^t)(2e^t)(3e^{-t})$$
$$= 2 \cdot 2 \cdot 3 \cdot e^t e^t e^{-t}$$
$$= 12e^{t+t-t}$$
$$= 12e^t$$

19.5 (a) Verify that

$$(e^t+1)^2 = e^{2t} + 2e^t + 1$$

(b) Hence simplify $\sqrt{e^{2t} + 2e^t + 1}$.

Solution (a) $$(e^t+1)^2 = (e^t+1)(e^t+1)$$
$$= e^t e^t + e^t + e^t + 1$$
$$= e^{2t} + 2e^t + 1$$

(b) $\sqrt{e^{2t} + 2e^t + 1} = \sqrt{(e^t+1)^2} = e^t + 1$

Exercise 19.1

MyMathLab

1. Use a scientific calculator to evaluate

   (a) $e^{2.3}$ (b) $e^{1.9}$ (c) $e^{-0.6}$

   (d) $\dfrac{1}{e^{-2}+1}$ (e) $\dfrac{3e^2}{e^2-10}$

   (f) $e^2\left(e + \dfrac{1}{e}\right) - e$ (g) $e^{1.5}e^{2.7}$

   (h) $\sqrt{e}\sqrt{4e}$

2. Simplify

   (a) $e^2.e^7$ (b) $\dfrac{e^7}{e^4}$ (c) $\dfrac{e^{-2}}{e^3}$

   (d) $\dfrac{e^3}{e^{-1}}$ (e) $\dfrac{(4e)^2}{(2e)^3}$

3. Simplify each of the following expressions as far as possible:

   (a) $e^x e^{-3x}$ (b) $e^{3x}e^{-x}$

   (c) $e^{-3x}e^{-x}$ (d) $(e^{-x})^2 e^{2x}$

   (e) $\dfrac{e^{3x}}{e^x}$ (f) $\dfrac{e^{-3x}}{e^{-x}}$

4. Simplify as far as possible

(a) $\dfrac{e^t}{e^3}$ (b) $\dfrac{2e^{3x}}{4e^x}$ (c) $\dfrac{(e^x)^3}{3e^x}$

(d) $e^x + 2e^{-x} + e^x$ (e) $\dfrac{e^x e^y e^z}{e^{x/2-y}}$

(f) $\dfrac{(e^{t/3})^6}{e^t + e^t}$ (g) $\dfrac{e^t}{e^{-t}}$ (h) $\dfrac{1+e^{-t}}{e^{-t}}$

(i) $(e^z e^{-z/2})^2$ (j) $\dfrac{e^{-3+t}}{2e^t}$

(k) $\left(\dfrac{e^{-1}}{e^{-x}}\right)^{-1}$

19.2 The exponential function and its graph

The **exponential function** has the form

$$y = e^x$$

The function $y = e^x$ is found to occur in the modelling of many natural phenomena. Table 19.1 gives values of x and e^x. Using Table 19.1, a graph of $y = e^x$ can be drawn. This is shown in Figure 19.1. From the table and graph we note the following important properties of $y = e^x$:

(a) The exponential function is never negative.

(b) When $x = 0$, the function value is 1.

(c) As x increases, then e^x increases. This is known as **exponential growth**.

Table 19.1

x	-3	-2	-1	0	1	2	3
e^x	0.0498	0.1353	0.3679	1	2.7183	7.3891	20.086

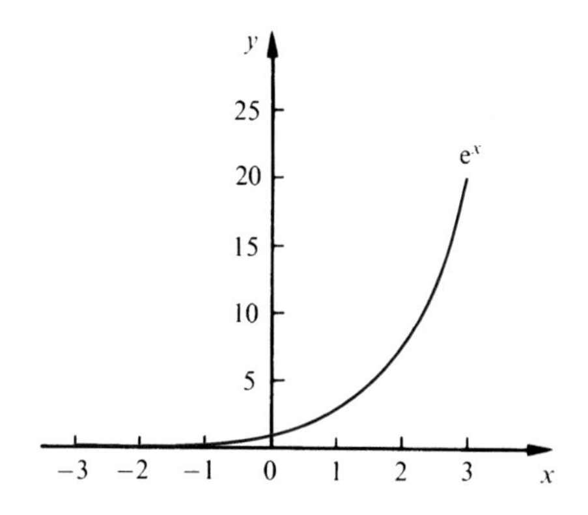

Figure 19.1
A graph of $y = e^x$

(d) By choosing x large enough, the value of e^x can be made larger than any given number. We say e^x increases without bound as x increases.

(e) As x becomes large and negative, e^x gets nearer and nearer to 0.

WORKED EXAMPLE

19.6 Sketch a graph of $y = 2e^x$ and $y = e^{2x}$ for $-3 \leqslant x \leqslant 3$.

Solution Table 19.2 gives values of $2e^x$ and e^{2x}, and the graphs are shown in Figure 19.2.

Table 19.2

x	-3	-2	-1	0	1	2	3
$2e^x$	0.0996	0.2707	0.7358	2	5.4366	14.7781	40.1711
e^{2x}	0.0025	0.0183	0.1353	1	7.3891	54.5982	403.429

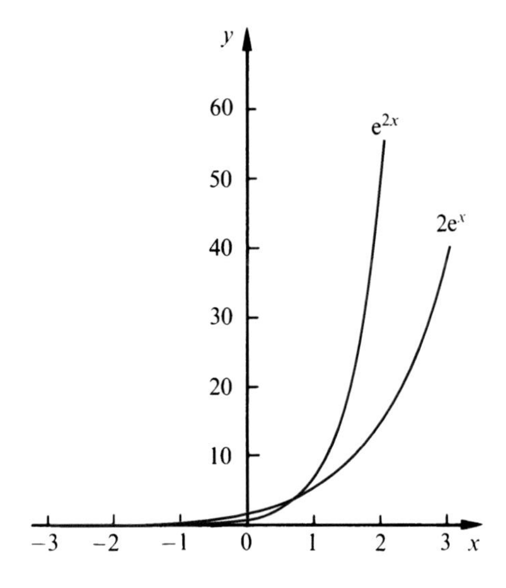

Figure 19.2 Graphs of $y = e^{2x}$ and $y = 2e^x$

Note that when $x > 0$, the graph of $y = e^{2x}$ rises more rapidly than that of $y = 2e^x$.

We now turn attention to an associated function: $y = e^{-x}$. Values are listed in Table 19.3 and the function is illustrated in Figure 19.3. From the table and the graph we note the following properties of $y = e^{-x}$:

(a) The function is never negative.

(b) When $x = 0$, the function has a value of 1.

(c) As x increases, then e^{-x} decreases, getting nearer to 0. This is known as **exponential decay**.

Table 19.3

x	−3	−2	−1	0	1	2	3
$-x$	3	2	1	0	−1	−2	−3
e^{-x}	20.086	7.3891	2.7183	1	0.3679	0.1353	0.0498

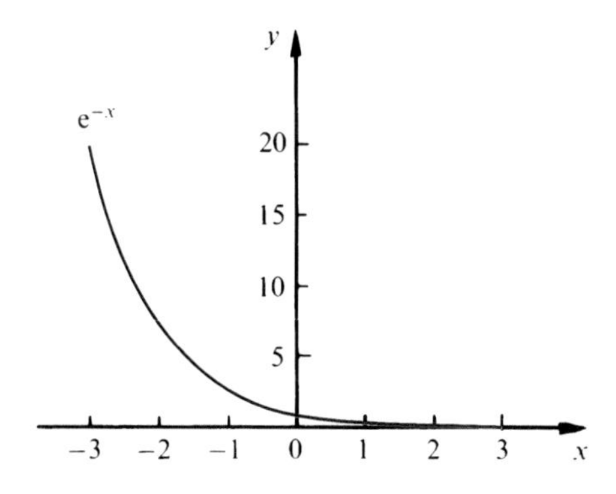

Figure 19.3
Graph of $y = e^{-x}$

Self-assessment questions 19.2

1. List properties that are common to $y = e^x$ and $y = e^{-x}$.
2. By choosing x large enough, then e^{-x} can be made to be negative. True or false?

Exercise 19.2

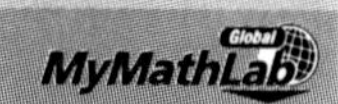

1. Draw up a table of values of $y = e^{3x}$ for x between −1 and 1 at intervals of 0.2. Sketch the graph of $y = e^{3x}$. Comment upon its shape and properties.
2. A species of animal has population $P(t)$ at time t given by

   $$P = 10 - 5e^{-t}$$

   (a) Sketch a graph of P against t for $0 \leqslant t \leqslant 5$.
   (b) What is the size of the population of the species as t becomes very large?
3. The concentration, $c(t)$, of a chemical in a reaction is modelled by the equation

   $$c(t) = 6 + 3e^{-2t}$$

   (a) Sketch a graph of c against t for $0 \leqslant t \leqslant 2$.
   (b) What value does the concentration approach as t becomes large?
4. The hyperbolic functions $\sinh x$ and $\cosh x$ are defined by

   $$\sinh x = \frac{e^x - e^{-x}}{2} \qquad \cosh x = \frac{e^x + e^{-x}}{2}$$

   (a) Show $e^x = \cosh x + \sinh x$.
   (b) Show $e^{-x} = \cosh x - \sinh x$.
5. Express $3 \sinh x + 7 \cosh x$ in terms of exponential functions.
6. Express $6e^x - 9e^{-x}$ in terms of $\sinh x$ and $\cosh x$.
7. Show $(\cosh x)^2 - (\sinh x)^2 = 1$.

19.3 Solving equations involving exponential terms using a graphical method

Many equations involving exponential terms can be solved using graphs. A graphical method will yield an approximate solution, which can then be refined by choosing a smaller interval for the domain. The following example illustrates the technique.

WORKED EXAMPLE

19.7

(a) Plot $y = e^{x/2}$ and $y = 2e^{-x}$ for $-1 \leqslant x \leqslant 1$.

(b) Hence solve the equation

$$e^{x/2} - 2e^{-x} = 0$$

Solution

(a) A table of values is drawn up for $e^{x/2}$ and $2e^{-x}$ so that the graphs can be drawn. Table 19.4 lists the values and Figure 19.4 shows the graphs of the functions.

Table 19.4

x	-1	-0.75	-0.5	-0.25	0	0.25	0.5	0.75	1
$e^{x/2}$	0.607	0.687	0.779	0.882	1	1.133	1.284	1.455	1.649
$2e^{-x}$	5.437	4.234	3.297	2.568	2	1.558	1.213	0.945	0.736

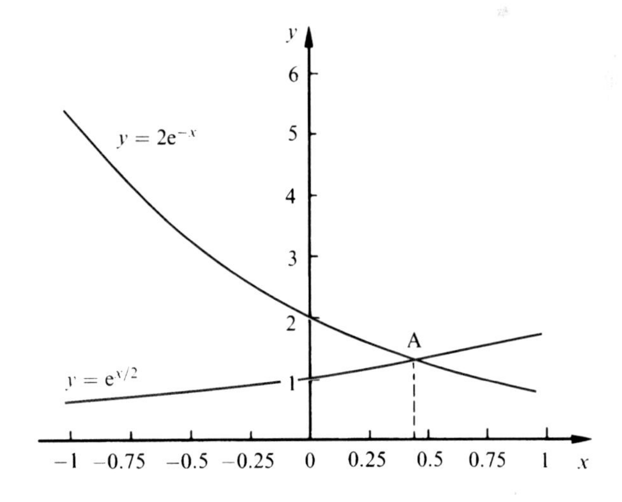

Figure 19.4 The graphs of $y = e^{x/2}$ and $y = 2e^{-x}$ intersect at A

(b) We note that the graphs intersect; the point of intersection is A. The equation

$$e^{x/2} - 2e^{-x} = 0$$

is equivalent to

$$e^{x/2} = 2e^{-x}$$

The graphs of $y = e^{x/2}$ and $y = 2e^{-x}$ intersect at A, where $x = 0.44$. Hence $x = 0.44$ is an approximate solution of $e^{x/2} - 2e^{-x} = 0$. The exact answer can be shown to be 0.46 (2 d.p.).

Exercise 19.3

MyMathLab

1. (a) Plot $y = 15 - x^2$ and $y = e^x$ for $0 \leqslant x \leqslant 3$.
   (b) Hence find an approximate solution to

   $$e^x + x^2 = 15$$

2. (a) Plot $y = 12x^2 - 1$ and $y = 3e^{-x}$ for $-1 \leqslant x \leqslant 1$.
   (b) Hence state two approximate solutions of $3e^{-x} = 12x^2 - 1$.

3. By drawing $y = e^x$ for $-3 \leqslant x \leqslant 3$ and additional appropriate lines solve the following equations:
   (a) $e^x + x = 0$
   (b) $e^x - 1.5 = 0$
   (c) $e^x - x - 5 = 0$
   (d) $\frac{e^x}{2} + \frac{x}{2} - 5 = 0$

4. (a) Draw $y = 2 + 6e^{-t}$ for $0 \leqslant t \leqslant 3$.
   (b) Use your graph to find an approximate solution to

   $$2 + 6e^{-t} = 5$$

   (c) Add the line $y = t + 4$ to your graph and hence find an approximate solution to

   $$6e^{-t} = t + 2$$

Test and assignment exercises 19

1. Use a calculator to evaluate

   (a) $e^{-3.1}$ (b) $e^{0.2}$ (c) $\frac{1}{e}$

2. (a) Sketch the graphs of

   $$y = e^{-x} \text{ and } y = \frac{x}{2}$$

   for $0 \leqslant x \leqslant 2$.
   (b) Hence solve the equation

   $$2e^{-x} - x = 0$$

3. Simplify where possible:

   (a) $e^x e^{2x} e^{-3x}$

   (b) $\frac{e^{2x}e^{3x}}{e^{4x}}$

   (c) $(e^{2x})^3 e^{-3x}$

   (d) $e^{2x} + e^{3x} - e^{5x}$

   (e) $\frac{e^{3x} + e^{5x}}{e^{2x}}$

4. Expand the brackets of

(a) $(e^x - 1)^2$ (b) $(e^{-x} - 1)^2$ (c) $e^{2x}(e^x + 1)$ (d) $2e^x(e^{-x} + 3e^x)$

5. Simplify

(a) $e^t e^{t/2}$ (b) $e^2 e^3 e$ (c) $e^{3x}(e^x - e^{2x}) - (e^{2x})^2$ (d) $e^{3+t}e^{5-2t}$

(e) $\dfrac{e^{2z+3}e^{5-z}}{e^{z+2}}$ (f) $\sqrt{e^{4z}}$ (g) $(e^{x/2}e^x)^2$ (h) $\sqrt{e^{2t} + 2e^{2t+1} + e^{2t+2}}$

(i) $\dfrac{e^t + e^{2t}}{e^{2t} + e^{3t}}$

6. Use a graphical method to solve

(a) $e^{2x} = 2x + 2$ (b) $e^{-x} = 2 - x^2$ (c) $e^x = 2 + \dfrac{x^3}{2}$

7. (a) Sketch the graphs of

$y = 2 + e^{-x}$ and $y = e^x$

for $0 \leqslant x \leqslant 2$.

(b) Use your graphs to solve

$e^x - e^{-x} = 2$

8. Remove the brackets from the following expressions:

(a) $(e^{2t})^4$ (b) $e^t(e^t - e^{-t})$ (c) $e^{2y}(e^{3y} + e^{-2y} + e^y)$ (d) $(1 + e^t)^2$

9. The height of mercury in an experiment, $H(t)$, varies according to

$H(t) = 10 + 3e^{-t} - 4e^{-2t}$

What value does H approach as t gets very large?

10. Given

$$X(t) = \frac{3 + 2e^{2t}}{4 + e^{2t}}$$

find the value that X approaches as t gets very large.

20 The logarithm function

Objectives: This chapter:

- explains the term 'base of a logarithm'
- shows how to calculate the logarithm of a number to any base
- states the laws of logarithms and uses them to simplify expressions
- shows how to solve exponential and logarithmic equations
- defines the logarithm function
- illustrates graphically the logarithm function

20.1 Introducing logarithms

Given an equation such as $125 = 5^3$, we call 5 the base and 3 the power or index. We can use **logarithms** to write the equation in another form. The logarithm form is

$$\log_5 125 = 3$$

This is read as 'logarithm to the base 5 of 125 is 3'. In general if

$$y = a^x$$

then

$$\log_a y = x$$

Key point

$y = a^x$ and $\log_a y = x$ are equivalent.

'The logarithm to the base a of y is x' is equivalent to saying that 'y is a to the power x'. The word logarithm is usually shortened to just 'log'.

WORKED EXAMPLE

20.1 Write down the logarithmic form of the following:

(a) $16 = 4^2$ (b) $8 = 2^3$ (c) $25 = 5^2$

Solution (a) $16 = 4^2$ may be written as

$$2 = \log_4 16$$

that is, 2 is the logarithm to the base 4 of 16.

(b) $8 = 2^3$ may be expressed as

$$3 = \log_2 8$$

which is read as '3 is the logarithm to the base 2 of 8'.

(c) $25 = 5^2$ may be expressed as

$$2 = \log_5 25$$

that is, 2 is the logarithm to the base 5 of 25.

Given an equation such as $16 = 4^2$ then 'taking logs' to base 4 will result in the logarithmic form $\log_4 16 = 2$.

WORKED EXAMPLE

20.2 Write the exponential form of the following:

(a) $\log_2 16 = 4$ (b) $\log_3 27 = 3$ (c) $\log_5 125 = 3$ (d) $\log_{10} 100 = 2$

Solution (a) Here the base is 2 and so we may write $16 = 2^4$.

(b) The base is 3 and so $27 = 3^3$.

(c) The base is 5 and so $125 = 5^3$.

(d) The base is 10 and so $100 = 10^2$.

Recall that e is the constant 2.71828..., which occurs frequently in natural phenomena.

Although the base of a logarithm can be any positive number other than 1, the commonly used bases are 10 and e. Logarithms to base 10 are often denoted by 'log' or '$\log_{10}$'; logarithms to base e are denoted by 'ln' or '$\log_e$' and referred to as **natural logarithms**. Most scientific calculators possess 'log' and 'ln' buttons, which are used to evaluate logarithms to base 10 and base e.

WORKED EXAMPLES

20.3 Use a scientific calculator to evaluate the following:

(a) $\log 71$ (b) $\ln 3.7$ (c) $\log 0.4615$ (d) $\ln 0.5$ (e) $\ln 1000$

Solution Using a scientific calculator we obtain

(a) $\log 71 = 1.8513$ (b) $\ln 3.7 = 1.3083$ (c) $\log 0.4615 = -0.3358$
(d) $\ln 0.5 = -0.6931$ (e) $\ln 1000 = 6.9078$

You should ensure that you know how to use your calculator to verify these results.

20.4 Given $10^{0.6990} = 5$ evaluate the following:

(a) $10^{1.6990}$ (b) $\log 5$ (c) $\log 500$

Solution

(a) $10^{1.6990} = 10^1.10^{0.6990} = 10(5) = 50$

(b) We are given $10^{0.6990} = 5$ and by taking logs to the base 10 we obtain $0.6990 = \log 5$.

(c) From $10^{0.6990} = 5$ we can see that

$$100(5) = 100(10^{0.6990})$$

and so

$$500 = 10^2.10^{0.6990} = 10^{2.6990}$$

Taking logs to the base 10 gives

$$\log 500 = 2.6990$$

Self-assessment question 20.1

1. The base of a logarithm is always a positive integer. True or false?

Exercise 20.1

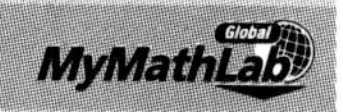

1. Use a scientific calculator to evaluate
   (a) $\log 150$ (b) $\ln 150$
   (c) $\log 0.316$ (d) $\ln 0.1$

2. Write down the logarithmic form of the following:
   (a) $3^8 = 6561$ (b) $6^5 = 7776$
   (c) $2^{10} = 1024$ (d) $10^5 = 100000$
   (e) $4^7 = 16384$
   (f) $\left(\frac{1}{2}\right)^5 = 0.03125$
   (g) $12^3 = 1728$ (h) $9^4 = 6561$

3. Write the exponential form of the following:
   (a) $\log_6 1296 = 4$ (b) $\log_{15} 225 = 2$
   (c) $\log_8 512 = 3$ (d) $\log_7 2401 = 4$
   (e) $\log_3 243 = 5$ (f) $\log_6 216 = 3$
   (g) $\log_{20} 8000 = 3$ (h) $\log_{16} 4096 = 3$
   (i) $\log_2 4096 = 12$

4. Given $10^{0.4771} = 3$, evaluate the following:
   (a) $\log 3$ (b) $\log 300$ (c) $\log 0.03$

5. Given $\log 7 = 0.8451$ evaluate the following:
   (a) $10^{0.8451}$ (b) $\log 700$ (c) $\log 0.07$

20.2 Calculating logarithms to any base

In §20.1 we showed how equations of the form $y = a^x$ could be written in logarithmic form as $\log_a y = x$. The number a, called the **base**, is always positive, that is $a > 0$. We also introduced the common bases 10 and e. Scientific calculators are programmed with logarithms to base 10 and to base e. Suppose that we wish to calculate logarithms to bases other than 10 or e. For example, in communication theory, logarithms to base 2 are commonly used. To calculate logarithms to base a we use one of the following two formulae:

Key point

$$\log_a X = \frac{\log_{10} X}{\log_{10} a} \qquad \log_a X = \frac{\ln X}{\ln a}$$

WORKED EXAMPLE

20.5 Evaluate (a) $\log_6 19$, (b) $\log_7 29$.

Solution (a) We use the formula

$$\log_a X = \frac{\log_{10} X}{\log_{10} a}$$

Comparing $\log_6 19$ with $\log_a X$ we see $X = 19$ and $a = 6$. So

$$\log_6 19 = \frac{\log_{10} 19}{\log_{10} 6} = \frac{1.2788}{0.7782} = 1.6433$$

We could have equally well used the formula

$$\log_a X = \frac{\ln X}{\ln a}$$

With this formula we obtain

$$\log_6 19 = \frac{\ln 19}{\ln 6} = \frac{2.9444}{1.7918} = 1.6433$$

(b) Comparing $\log_a X$ with $\log_7 29$ we see $X = 29$ and $a = 7$. Hence

$$\log_7 29 = \frac{\log_{10} 29}{\log_{10} 7} = \frac{1.4624}{0.8451} = 1.7304$$

Alternatively we use

$$\log_7 29 = \frac{\ln 29}{\ln 7} = \frac{3.3673}{1.9459} = 1.7304$$

By considering the formula

$$\log_a X = \frac{\log_{10} X}{\log_{10} a}$$

with $X = a$ we obtain

$$\log_a a = \frac{\log_{10} a}{\log_{10} a} = 1$$

Key point

$\log_a a = 1$

This same result could be derived by writing the logarithm form of $a = a^1$.

Exercise 20.2

MyMathLab Global

1. Evaluate the following:
(a) $\log_4 6$ (b) $\log_3 10$ (c) $\log_{20} 270$ (d) $\log_5 0.65$ (e) $\log_2 100$
(f) $\log_2 0.03$ (g) $\log_{100} 10$ (h) $\log_7 7$
2. Show that

$$2.3026 \log_{10} X = \ln X$$

3. Evaluate the following:
(a) $\log_3 7 + \log_4 7 + \log_5 7$ (b) $\log_8 4 + \log_8 0.25$ (c) $\log_{0.7} 2$ (d) $\log_2 0.7$

20.3 Laws of logarithms

Logarithms obey several laws, which we now examine. They are introduced via examples.

WORKED EXAMPLE

20.6 Evaluate (a) $\log 7$, (b) $\log 12$, (c) $\log 84$ and $\log 7 + \log 12$. Comment on your findings.

Solution

(a) $\log 7 = 0.8451$ (b) $\log 12 = 1.0792$
(c) $\log 84 = 1.9243$, and $\log 7 + \log 12 = 0.8451 + 1.0792 = 1.9243$

We note that $\log 7 + \log 12 = \log 84$.

Worked Example 20.6 illustrates the **first law** of logarithms, which states:

Key point

$$\log A + \log B = \log AB$$

This law holds true for any base. However, in any one calculation all bases must be the same.

WORKED EXAMPLES

20.7 Simplify to a single log term

(a) $\log 9 + \log x$
(b) $\log t + \log 4t$
(c) $\log 3x^2 + \log 2x$

Solution

(a) $\log 9 + \log x = \log 9x$
(b) $\log t + \log 4t = \log(t.4t) = \log 4t^2$
(c) $\log 3x^2 + \log 2x = \log(3x^2.2x) = \log 6x^3$

20.8 Simplify

(a) $\log 7 + \log 3 + \log 2$
(b) $\log 3x + \log x + \log 4x$

Solution

(a) We know $\log 7 + \log 3 = \log(7 \times 3) = \log 21$, and so

$$\begin{aligned} \log 7 + \log 3 + \log 2 &= \log 21 + \log 2 \\ &= \log(21 \times 2) = \log 42 \end{aligned}$$

(b) We have

$$\log 3x + \log x = \log(3x.x) = \log 3x^2$$

and so

$$\begin{aligned} \log 3x + \log x + \log 4x &= \log 3x^2 + \log 4x \\ &= \log(3x^2.4x) = \log 12x^3 \end{aligned}$$

We now consider an example that introduces the second law of logarithms.

WORKED EXAMPLE

20.9

(a) Evaluate log 12, log 4 and log 3.
(b) Compare the values of log 12 − log 4 and log 3.

Solution

(a) $\log 12 = 1.0792, \log 4 = 0.6021, \log 3 = 0.4771.$

(b) From part (a),

$$\log 12 - \log 4 = 1.0792 - 0.6021 = 0.4771$$

and also

$$\log 3 = 0.4771$$

We note that $\log 12 - \log 4 = \log 3$.

This example illustrates the **second law** of logarithms, which states:

Key point

$$\log A - \log B = \log\left(\frac{A}{B}\right)$$

WORKED EXAMPLES

20.10

Use the second law of logarithms to simplify the following to a single log term:

(a) $\log 20 - \log 10$ (b) $\log 500 - \log 75$ (c) $\log 4x^3 - \log 2x$
(d) $\log 5y^3 - \log y$

Solution

(a) Using the second law of logarithms we have

$$\log 20 - \log 10 = \log\left(\frac{20}{10}\right) = \log 2$$

(b) $$\log 500 - \log 75 = \log\left(\frac{500}{75}\right) = \log\left(\frac{20}{3}\right)$$

(c) $$\log 4x^3 - \log 2x = \log\left(\frac{4x^3}{2x}\right) = \log 2x^2$$

(d) $$\log 5y^3 - \log y = \log\left(\frac{5y^3}{y}\right) = \log 5y^2$$

20.11 Simplify

(a) $\log 20 + \log 3 - \log 6$
(b) $\log 18 - \log 24 + \log 2$

Solution (a) Using the first law of logarithms we see that

$$\log 20 + \log 3 = \log 60$$

and so

$$\log 20 + \log 3 - \log 6 = \log 60 - \log 6$$

Using the second law of logarithms we see that

$$\log 60 - \log 6 = \log\left(\frac{60}{6}\right) = \log 10$$

Hence

$$\log 20 + \log 3 - \log 6 = \log 10$$

(b)
$$\begin{aligned}\log 18 - \log 24 + \log 2 &= \log\left(\frac{18}{24}\right) + \log 2\\ &= \log\left(\frac{3}{4}\right) + \log 2\\ &= \log\left(\frac{3}{4} \times 2\right)\\ &= \log 1.5\end{aligned}$$

20.12 Simplify

(a) $\log 2 + \log 3x - \log 2x$
(b) $\log 5y^2 + \log 4y - \log 10y^2$

Solution (a)
$$\begin{aligned}\log 2 + \log 3x - \log 2x &= \log(2 \times 3x) - \log 2x\\ &= \log 6x - \log 2x\\ &= \log\left(\frac{6x}{2x}\right) = \log 3\end{aligned}$$

(b)
$$\begin{aligned}\log 5y^2 + \log 4y - \log 10y^2 &= \log(5y^2.4y) - \log 10y^2\\ &= \log 20y^3 - \log 10y^2\\ &= \log\left(\frac{20y^3}{10y^2}\right) = \log 2y\end{aligned}$$

We consider a special case of the second law. Consider $\log A - \log A$. This is clearly 0. However, using the second law we may write

$$\log A - \log A = \log\left(\frac{A}{A}\right) = \log 1$$

Thus

Key point

$\log 1 = 0$

In any base, the logarithm of 1 equals 0.

Finally we introduce the third law of logarithms.

WORKED EXAMPLE

20.13

(a) Evaluate log 16 and log 2.
(b) Compare log 16 and 4 log 2.

Solution

(a) $\log 16 = 1.204, \log 2 = 0.301$.
(b) $\log 16 = 1.204$, $4\log 2 = 1.204$. Hence we see that $4\log 2 = \log 16$.

Noting that $16 = 2^4$, Worked Example 20.13 suggests the **third law** of logarithms:

Key point

$n \log A = \log A^n$

This law applies if n is integer, fractional, positive or negative.

WORKED EXAMPLES

20.14

Write the following as a single logarithmic expression:

(a) $3\log 2$ (b) $2\log 3$ (c) $4\log 3$

Solution

(a) $3\log 2 = \log 2^3 = \log 8$
(b) $2\log 3 = \log 3^2 = \log 9$
(c) $4\log 3 = \log 3^4 = \log 81$

20.15

Write as a single log term

(a) $\frac{1}{2}\log 16$ (b) $-\log 4$ (c) $-2\log 2$ (d) $-\frac{1}{2}\log 0.5$

Solution

(a) $\frac{1}{2}\log 16 = \log 16^{\frac{1}{2}} = \log\sqrt{16} = \log 4$

(b) $-\log 4 = -1.\log 4 = \log 4^{-1} = \log\left(\frac{1}{4}\right) = \log 0.25$

(c) $-2\log 2 = \log 2^{-2} = \log\left(\frac{1}{2^2}\right) = \log\left(\frac{1}{4}\right) = \log 0.25$

(d) $-\frac{1}{2}\log 0.5 = -\frac{1}{2}\log\left(\frac{1}{2}\right) = \log\left(\frac{1}{2}\right)^{-\frac{1}{2}} = \log 2^{\frac{1}{2}} = \log\sqrt{2}$

20.16 Simplify

(a) $3\log x - \log x^2$
(b) $3\log t^3 - 4\log t^2$
(c) $\log Y - 3\log 2Y + 2\log 4Y$

Solution (a)
$$3\log x - \log x^2 = \log x^3 - \log x^2 = \log\left(\frac{x^3}{x^2}\right) = \log x$$

(b)
$$3\log t^3 - 4\log t^2 = \log(t^3)^3 - \log(t^2)^4 = \log t^9 - \log t^8 = \log\left(\frac{t^9}{t^8}\right) = \log t$$

(c)
$$\log Y - 3\log 2Y + 2\log 4Y = \log Y - \log(2Y)^3 + \log(4Y)^2 = \log Y - \log 8Y^3 + \log 16Y^2 = \log\left(\frac{Y.16Y^2}{8Y^3}\right) = \log 2$$

20.17 Simplify

(a) $2\log 3x - \frac{1}{2}\log 16x^2$

(b) $\frac{3}{2}\log 4x^2 - \log\left(\frac{1}{x}\right)$

(c) $2\log\left(\frac{2}{x^2}\right) - 3\log\left(\frac{2}{x}\right)$

Solution

(a)
$$2\log 3x - \frac{1}{2}\log 16x^2 = \log(3x)^2 - \log(16x^2)^{\frac{1}{2}}$$
$$= \log 9x^2 - \log 4x$$
$$= \log\left(\frac{9x^2}{4x}\right)$$
$$= \log\left(\frac{9x}{4}\right)$$

(b)
$$\frac{3}{2}\log 4x^2 - \log\left(\frac{1}{x}\right) = \log(4x^2)^{\frac{3}{2}} - \log(x^{-1})$$
$$= \log 8x^3 + \log x$$
$$= \log 8x^4$$

(c)
$$2\log\left(\frac{2}{x^2}\right) - 3\log\left(\frac{2}{x}\right) = \log\left(\frac{2}{x^2}\right)^2 - \log\left(\frac{2}{x}\right)^3$$
$$= \log\left(\frac{4}{x^4}\right) - \log\left(\frac{8}{x^3}\right)$$
$$= \log\left(\frac{4/x^4}{8/x^3}\right)$$
$$= \log\left(\frac{1}{2x}\right)$$

Self-assessment question 20.3

1. State the three laws of logarithms.

Exercise 20.3

MyMathLab Global

1. Write the following as a single log term using the laws of logarithms:
   (a) $\log 5 + \log 9$ (b) $\log 9 - \log 5$
   (c) $\log 5 - \log 9$ (d) $2\log 5 + \log 1$
   (e) $2\log 4 - 3\log 2$ (f) $\log 64 - 2\log 2$
   (g) $3\log 4 + 2\log 1 + \log 27 - 3\log 12$

2. Simplify as much as possible:
   (a) $\log 3 + \log x$
   (b) $\log 4 + \log 2x$
   (c) $\log 3X - \log 2X$
   (d) $\log T^3 - \log T$
   (e) $\log 5X + \log 2X$

3. Simplify
   (a) $3\log X - \log X^2$ (b) $\log y - 2\log\sqrt{y}$
   (c) $5\log x^2 + 3\log\dfrac{1}{x}$
   (d) $4\log X - 3\log X^2 + \log X^3$
   (e) $3\log y^{1.4} + 2\log y^{0.4} - \log y^{1.2}$
4. Simplify the following as much as possible by using the laws of logarithms:
   (a) $\log 4x - \log x$ (b) $\log t^3 + \log t^4$
   (c) $\log 2t - \log\left(\dfrac{t}{4}\right)$
   (d) $\log 2 + \log\left(\dfrac{3}{x}\right) - \log\left(\dfrac{x}{2}\right)$
   (e) $\log\left(\dfrac{t^2}{3}\right) + \log\left(\dfrac{6}{t}\right) - \log\left(\dfrac{1}{t}\right)$
   (f) $2\log y - \log y^2$
   (g) $3\log\left(\dfrac{1}{t}\right) + \log t^2$
   (h) $4\log\sqrt{x} + 2\log\left(\dfrac{1}{x}\right)$
   (i) $2\log x + 3\log t$ (j) $\log A - \dfrac{1}{2}\log 4A$
   (k) $\dfrac{\log 9x + \log 3x^2}{3}$
   (l) $\log xy + 2\log\left(\dfrac{x}{y}\right) + 3\log\left(\dfrac{y}{x}\right)$
   (m) $\log\left(\dfrac{A}{B}\right) - \log\left(\dfrac{B}{A}\right)$
   (n) $\log\left(\dfrac{2t}{3}\right) + \dfrac{1}{2}\log 9t - \log\left(\dfrac{1}{t}\right)$
5. Express as a single log term:
   $\log_{10} X + \ln X$
6. Simplify
   (a) $\log(9x-3) - \log(3x-1)$
   (b) $\log(x^2-1) - \log(x+1)$
   (c) $\log(x^2+3x) - \log(x+3)$

20.4 Solving equations with logarithms

This section illustrates the use of logarithms in solving certain types of equations. For reference we note from §20.1 the equivalence of

$$y = a^x \quad \text{and} \quad \log_a y = x$$

and from §20.3 the laws of logarithms:

$$\log A + \log B = \log AB$$

$$\log A - \log B = \log\left(\frac{A}{B}\right)$$

$$n\log A = \log A^n$$

WORKED EXAMPLES

20.18 Solve the following equations:

(a) $10^x = 59$ (b) $10^x = 0.37$ (c) $e^x = 100$ (d) $e^x = 0.5$

Solution

(a) $10^x = 59$
Taking logs to base 10 gives

$$x = \log 59 = 1.7709$$

(b) $10^x = 0.37$
Taking logs to base 10 gives

$$x = \log 0.37 = -0.4318$$

(c) $e^x = 100$
Taking logs to base e gives

$$x = \ln 100 = 4.6052$$

(d) $e^x = 0.5$
Taking logs to base e we have

$$x = \ln 0.5 = -0.6931$$

20.19 Solve the following equations:

(a) $\log x = 1.76$ (b) $\ln x = -0.5$ (c) $\log(3x) = 0.76$

(d) $\ln\left(\frac{x}{2}\right) = 2.6$ (e) $\log(2x - 4) = 1.1$ (f) $\ln(7 - 3x) = 1.75$

Solution

We note that if $\log_a X = n$, then $X = a^n$.

(a) $\log x = 1.76$ and so $x = 10^{1.76}$. Using the 'x^y' button of a scientific calculator we find

$$x = 10^{1.76} = 57.5440$$

(b) $\ln x = -0.5$

$$x = e^{-0.5} = 0.6065$$

(c) $\log 3x = 0.76$

$$3x = 10^{0.76}$$

$$x = \frac{10^{0.76}}{3} = 1.9181$$

(d) $\ln\left(\frac{x}{2}\right) = 2.6$

$$\frac{x}{2} = e^{2.6}$$

$$x = 2e^{2.6} = 26.9275$$

(e) $\log(2x - 4) = 1.1$

$$2x - 4 = 10^{1.1}$$

$$2x = 10^{1.1} + 4$$

$$x = \frac{10^{1.1} + 4}{2} = 8.2946$$

(f) $\ln(7 - 3x) = 1.75$

$$7 - 3x = e^{1.75}$$

$$3x = 7 - e^{1.75}$$

$$x = \frac{7 - e^{1.75}}{3} = 0.4151$$

20.20 Solve the following:

(a) $e^{3+x}.e^{x} = 1000$ (b) $\ln\left(\frac{x}{3} + 1\right) + \ln\left(\frac{1}{x}\right) = -1$

(c) $3e^{x-1} = 75$ (d) $\log(x + 2) + \log(x - 2) = 1.3$

Solution (a) Using the laws of logarithms we have

$$e^{3+x}.e^{x} = e^{3+2x}$$

Hence

$$e^{3+2x} = 1000$$

$$3 + 2x = \ln 1000$$

$$2x = \ln(1000) - 3$$

$$x = \frac{\ln(1000) - 3}{2} = 1.9539$$

(b) Using the laws of logarithms we may write

$$\ln\left(\frac{x}{3}+1\right)+\ln\left(\frac{1}{x}\right)=\ln\left(\left(\frac{x}{3}+1\right)\frac{1}{x}\right)=\ln\left(\frac{1}{3}+\frac{1}{x}\right)$$

So

$$\ln\left(\frac{1}{3}+\frac{1}{x}\right)=-1$$

$$\frac{1}{3}+\frac{1}{x}=e^{-1}$$

$$\frac{1}{x}=e^{-1}-\frac{1}{3}$$

$$\frac{1}{x}=0.0345$$

$$x=28.947$$

(c) $3e^{x-1}=75$

$$e^{x-1}=25$$

$$x-1=\ln 25$$

$$x=\ln 25+1=4.2189$$

(d) Using the first law we may write

$$\log(x+2)+\log(x-2)=\log(x+2)(x-2)=\log(x^2-4)$$

Hence

$$\log(x^2-4)=1.3$$

$$x^2-4=10^{1.3}$$

$$x^2=10^{1.3}+4$$

$$x=\sqrt{10^{1.3}+4}=4.8941$$

Exercise 20.4

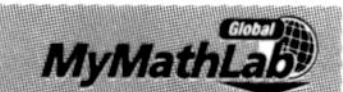

1. Solve the following equations, giving your answer to 4 d.p.:
   (a) $\log x = 1.6000$ (b) $10^x = 75$
   (c) $\ln x = 1.2350$ (d) $e^x = 36$

2. Solve each of the following equations, giving your answer to 4 d.p.:
   (a) $\log(3t) = 1.8$ (b) $10^{2t} = 150$
   (c) $\ln(4t) = 2.8$ (d) $e^{3t} = 90$

3. Solve the following equations, giving your answer to 4 d.p.:
(a) $\log x = 0.3940$ (b) $\ln x = 0.3940$
(c) $10^y = 5.5$ (d) $e^z = 500$
(e) $\log(3v) = 1.6512$ (f) $\ln\left(\frac{t}{6}\right) = 1$
(g) $10^{2r+1} = 25$ (h) $e^{(2t-1)/3} = 7.6700$
(i) $\log(4b^2) = 2.6987$
(j) $\log\left(\frac{6}{2+t}\right) = 1.5$
(k) $\ln(2r^3 + 1) = 3.0572$
(l) $\ln(\log t) = -0.3$ (m) $10^{t^2-1} = 180$
(n) $10^{3r^4} = 170000$
(o) $\log(10^t) = 1.6$ (p) $\ln(e^x) = 20000$

4. Solve the following equations:
(a) $e^{3x}.e^{2x} = 59$
(b) $10^{3t}.10^{4-t} = 27$
(c) $\log(5 - t) + \log(5 + t) = 1.2$
(d) $\log x + \ln x = 4$

20.5 Properties and graph of the logarithm function

Values of $\log x$ and $\ln x$ are given in Table 20.1, and graphs of the functions $y = \log x$ and $y = \ln x$ are illustrated in Figure 20.1. The following properties are noted from the graphs:

(a) As x increases, the values of $\log x$ and $\ln x$ increase.
(b) $\log 1 = \ln 1 = 0$.

Table 20.1

x	0.01	0.1	0.5	1	2	5	10	100
$\log x$	−2	−1	−0.30	0	0.30	0.70	1	2
$\ln x$	−4.61	−2.30	−0.69	0	0.69	1.61	2.30	4.61

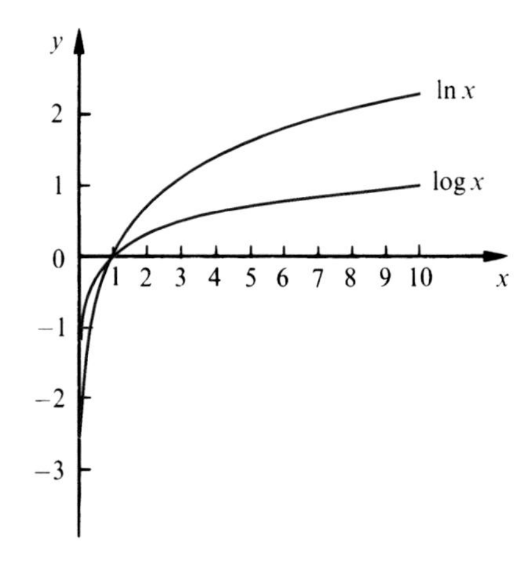

Figure 20.1
Graphs of $y = \log x$ and $y = \ln x$

(c) As x approaches 0 the values of $\log x$ and $\ln x$ increase negatively.
(d) When $x < 1$, the values of $\log x$ and $\ln x$ are negative.
(e) $\log x$ and $\ln x$ are not defined when $x \leqslant 0$.

Self-assessment questions 20.5

1. State two properties that are common to both $y = \log x$ and $y = \ln x$.
2. It is possible to find a value of x such that the value of $\log x$ exceeds 100000000. True or false?

Exercise 20.5

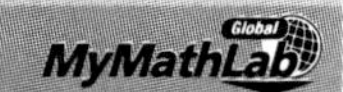

1. (a) Plot a graph of $y = \log x^2$ for $0 < x \leqslant 10$.
   (b) Plot a graph of $y = \log(1/x)$ for $0 < x \leqslant 10$.
2. (a) Plot a graph of $y = \log x$ for $0 < x \leqslant 3$.
   (b) Plot on the same axes $y = 1 - \frac{x}{3}$ for $0 \leqslant x \leqslant 3$.
   (c) Use your graphs to find an approximate solution to the equation

$$\log x = 1 - \frac{x}{3}$$

3. (a) Plot $y = \log x$ for $0 < x \leqslant 7$.
   (b) On the same axes, plot $y = 0.5 - 0.1x$.
   (c) Use your graphs to obtain an approximate solution to

$$0.5 - 0.1x = \log x$$

Test and assignment exercises 20

1. Use a scientific calculator to evaluate
   (a) $\log 107000$ (b) $\ln 0.0371$ (c) $\log 0.1$ (d) $\ln 150$
2. Evaluate
   (a) $\log_7 20$ (b) $\log_{16} 100$ (c) $\log_2 60$ (d) $\log_8 150$ (e) $\log_6 4$
3. Simplify the following expressions to a single log term:
   (a) $\log 7 + \log t$ (b) $\log t - \log x$ (c) $\log x + 2\log y$
   (d) $3\log t + 4\log r$ (e) $\frac{1}{2}\log 9v^4 - 3\log 1$ (f) $\log(a+b) + \log(a-b)$
   (g) $\frac{2}{3}\log x + \frac{1}{3}\log xy^3$

4. Solve the following logarithmic equations, giving your answer to 4 d.p.:
   (a) $\log 7x = 2.9$ (b) $\ln 2x = 1.5$ (c) $\ln x + \ln 2x = 3.6$
   (d) $\ln\left(\frac{x^2}{2}\right) - \ln x = 0.7$ (e) $\log(4t - 3) = 0.9$ (f) $\log 3t + 3\log t = 2$
   (g) $3\ln\left(\frac{2}{x}\right) - \frac{1}{2}\ln x = -1$

5. (a) Plot $y = \log x$ for $0 < x \leqslant 3$.
   (b) On the same axes plot $y = \sin x$, where x is measured in radians.
   (c) Hence find an approximate solution to $\log x = \sin x$.

6. Solve the following equations, giving your answer correct to 4 d.p.:
   (a) $10^{x+1} = 70$ (b) $e^{2x+3} = 500$ (c) $3(10^{2x}) = 750$ (d) $4(e^{-x+2}) = 1000$
   (e) $\frac{4}{6 + e^{3x}} = 0.1500$

7. By substituting $z = e^x$ solve the equation

   $e^{2x} - 9e^x + 14 = 0$

8. By substituting $z = 10^x$ solve the following equation, giving your answer correct to 4 d.p.:

   $10^{2x} - 9(10^x) + 20 = 0$

9. Solve the following equations:

   (a) $\log(x - 1) + \log(x + 1) = 2$ (b) $\log(10 + 10^x) = 2$ (c) $\frac{\log 2x}{\log x} = 2$

10. Solve

    (a) $8(10^{-x}) + 10^x = 6$ (b) $10^{3x} - 4(10^x) = 0$

Section Title 2

Statistics

1 Descriptive statistics

Contents

→

Contents
continued

Learning outcomes

By the end of this chapter you should be able to:

- recognise different types of data and use appropriate methods to summarise and analyse them;
- use graphical techniques to provide a visual summary of one or more data series;
- use numerical techniques (such as an average) to summarise data series;
- recognise the strengths and limitations of such methods;
- recognise the usefulness of data transformations to gain additional insight into a set of data.

Complete your diagnostic test for Chapter 1 now to create your personal study plan. Exercises with an icon ? are also available for practice in MathXL with additional supporting resources.

Introduction

The aim of descriptive statistical methods is simple: to present information in a clear, concise and accurate manner. The difficulty in analysing many phenomena, be they economic, social or otherwise, is that there is simply too much information for the mind to assimilate. The task of descriptive methods is therefore to summarise all this information and draw out the main features, without distorting the picture.

Consider, for example, the problem of presenting information about the wealth of British citizens (which follows later in this chapter). There are about 17 million adults for whom data are available: to present the data in raw form (i.e. the wealth holdings of each and every person) would be neither useful nor informative (it would take about 30 000 pages of a book, for example). It would be more useful to have much less information, but information that was still representative of the original data. In doing this, much of the original information would be deliberately lost; in fact, descriptive statistics might be described as the art of constructively throwing away much of the data!

There are many ways of summarising data and there are few hard and fast rules about how you should proceed. Newspapers and magazines often provide innovative (although not always successful) ways of presenting data. There are, however, a number of techniques that are tried and tested, and these are the subject of this chapter. These are successful because: (a) they tell us something useful about the underlying data; and (b) they are reasonably familiar to many people, so we can all talk in a common language. For example, the average tells us about the location of the data and is a familiar concept to most people. For example, my son talks of his day at school being 'average'.

The appropriate method of analysing the data will depend on a number of factors: the type of data under consideration; the sophistication of the audience; and the 'message' that it is intended to convey. One would use different methods to persuade academics of the validity of one's theory about inflation than one would use to persuade consumers that Brand X powder washes whiter than Brand Y. To illustrate the use of the various methods, three different topics are covered in this chapter. First we look at the relationship between educational attainment and employment prospects. Do higher qualifications improve your employment chances? The data come from people surveyed in 2004/5, so we have a sample of **cross-section** data giving a picture of the situation at one point in time. We look at the distribution of educational attainments amongst those surveyed, as well as the relationship to employment outcomes. In this example we simply count the numbers of people in different categories (e.g. the number of people with a degree qualification who are employed).

Second, we examine the distribution of wealth in the UK in 2003. The data are again cross-section, but this time we can use more sophisticated methods since wealth is measured on a **ratio scale**. Someone with £200 000 of wealth is twice as wealthy as someone with £100 000 for example, and there is a meaning to this ratio. In the case of education, one cannot say with any precision that one person is twice as educated as another (hence the perennial debate about educational standards). The educational categories may be ordered (so one person can be more educated than another, although even that may be ambiguous) but we cannot measure the 'distance' between them. We refer to this as education being measured on an **ordinal** scale. In contrast, there is not an obvious natural ordering to the three employment categories (employed, unemployed, inactive), so this is measured on a **nominal** scale.

Third, we look at national spending on investment over the period 1973 to 2005. This is **time series** data, as we have a number of observations on the variable measured at different points in time. Here it is important to take account of the time dimension of the data: things would look different if the observations were in the order 1973, 1983, 1977, . . . rather than in correct time order.

We also look at the relationship between two variables – investment and output – over that period of time and find appropriate methods of presenting it.

In all three cases we make use of both graphical and numerical methods of summarising the data. Although there are some differences between the methods used in the three cases these are not watertight compartments: the methods used in one case might also be suitable in another, perhaps with slight modification. Part of the skill of the statistician is to know which methods of analysis and presentation are best suited to each particular problem.

Summarising data using graphical techniques

Education and employment, or, after all this, will you get a job?

We begin by looking at a question which should be of interest to you: how does education affect your chances of getting a job? It is now clear that education improves one's life chances in various ways, one of the possible benefits being that it reduces the chances of being out of work. But by how much does it reduce those chances? We shall use a variety of graphical techniques to explore the question.

The raw data for this investigation come from the *Education and Training Statistics for the U.K.* 2006.[1] Some of these data are presented in Table 1.1 and show the numbers of people by employment status (either in work, unemployed, or inactive, i.e. not seeking work) and by educational qualification (higher education, A-levels, other qualification or no qualification). The table gives a **cross-tabulation** of employment status by educational qualification and is simply a count (the **frequency**) of the number of people falling into each of the 12 cells of the table. For example, there were 8 541 000 people in work who had experience of higher education. This is part of a total of just over 36 million people of working age. Note that the numbers in the table are in thousands, for the sake of clarity.

Table 1.1 Economic status and educational qualifications, 2006 *(numbers in 000s)*

	Higher education	A levels	Other qualification	No qualification	Total
In work	8541	5501	10 702	2260	27 004
Unemployed	232	247	758	309	1546
Inactive	1024	1418	3150	2284	7876
Total	9797	7166	14 610	4853	36 426

[1] This is now an internet-only publication, available at http://www.dcsf.gov.uk/rsgateway/DB/VOL/v000696/Vweb03-2006V1.pdf.

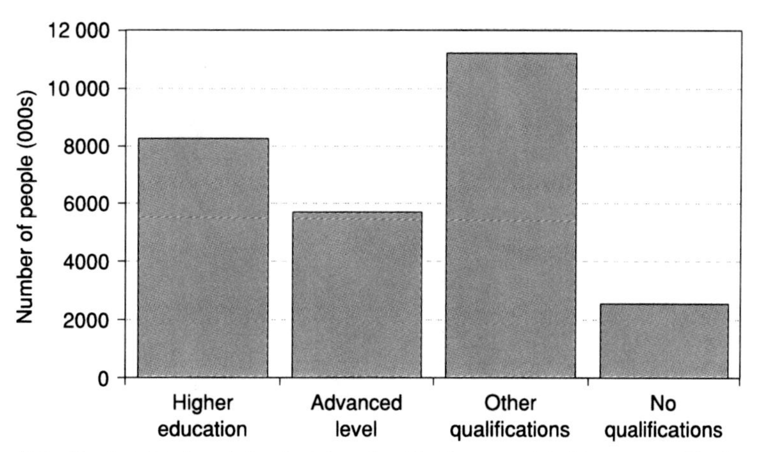

Figure 1.1 Educational qualifications of people in work in the UK, 2006

Note: The height of each bar is determined by the associated frequency. The first bar is 8541 units high, the second is 5501 units high and so on. The ordering of the bars could be reversed ('no qualifications' becoming the first category) without altering the message.

The bar chart

The first graphical technique we shall use is the **bar chart** and this is shown in Figure 1.1. This summarises the educational qualifications of those in work, i.e. the data in the first row of the table. The four educational categories are arranged along the horizontal (x) axis, while the frequencies are measured on the vertical (y) axis. The height of each bar represents the numbers in work for that category.

The biggest group is seen to be those with 'other qualifications', although this is now not much bigger than the 'higher education' category (the numbers entering higher education have been increasing substantially in the UK over time, although this is not evident in this chart, which uses cross-section data). The 'no qualifications' category is the smallest, although it does make up a substantial fraction of those in work.

It would be interesting to compare this distribution with those for the unemployed and inactive. This is done in Figure 1.2, which adds bars for these other two categories. This **multiple bar chart** shows that, as for the 'in work' category, among the inactive and unemployed, the largest group consists of those with 'other' qualifications (which are typically vocational qualifications). These findings simply reflect the fact that 'other qualifications' is the largest category. We can also begin to see whether more education increases your chance of having a job. For example, compare the height of the 'in work' bar to the 'inactive' bar. It is relatively much higher for those with higher education than for those with no qualifications. In other words, the likelihood of being inactive rather than employed is lower for graduates. However, we are having to make judgements about the relative heights of different bars simply by eye, and it is easy to make a mistake. It would be better if we could draw charts that would better highlight the differences. Figure 1.3 shows an alternative method of presentation: the **stacked bar chart**. In this case the bars are stacked one on top of another instead of being placed side by side. This is perhaps slightly better

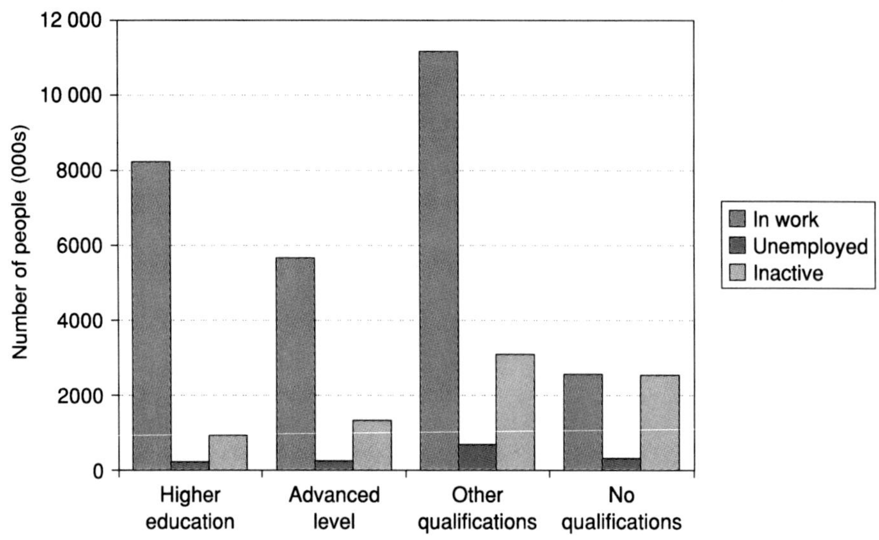

Figure 1.2 Educational qualifications by employment category

Note: The bars for the unemployed and inactive categories are constructed in the same way as for those in work: the height of the bar is determined by the frequency.

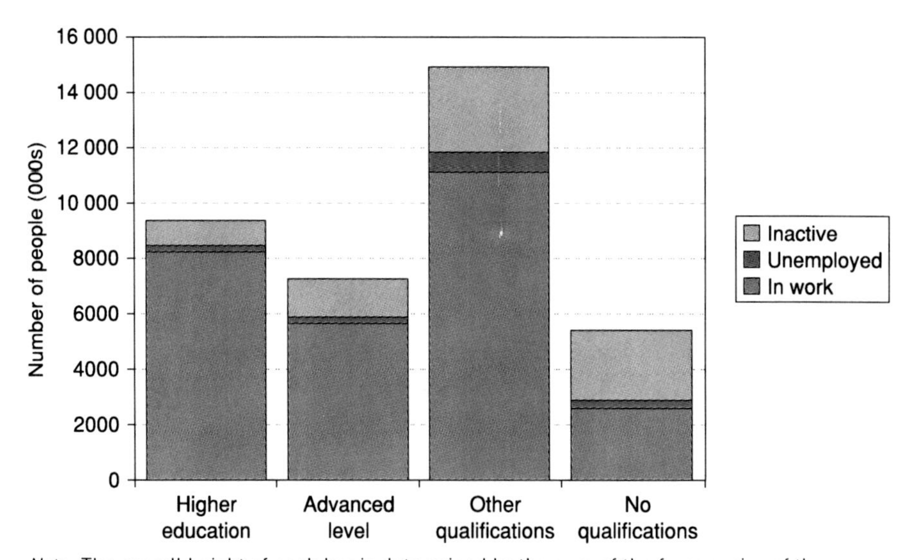

Figure 1.3 Stacked bar chart of educational qualifications and employment status

Note: The overall height of each bar is determined by the sum of the frequencies of the category, given in the final row of Table 1.1.

and the different overall sizes of the categories is clearly brought out. However, we are still having to make tricky visual judgements about proportions.

A clearer picture emerges if the data are **transformed** to (column) percentages, i.e. the columns are expressed as percentages of the column totals (e.g. the *proportion* of graduates are in work, rather than the number). This makes it easier directly to compare the different educational categories. These figures are shown in Table 1.2.

Having done this, it is easier to make a direct comparison of the different education categories (columns). This is shown in Figure 1.4, where all the bars

Table 1.2 Economic status and educational qualifications: column percentages

	Higher education	A levels	Other qualification	No qualification	All
In work	87%	77%	73%	47%	74%
Unemployed	2%	3%	5%	6%	4%
Inactive	10%	20%	22%	47%	22%
Totals	99%	100%	100%	100%	100%

Note: The column percentages are obtained by dividing each frequency by the column total. For example, 87% is 8541 divided by 9797; 77% is 5501 divided by 7166, and so on. Columns may not sum to 100% due to rounding.

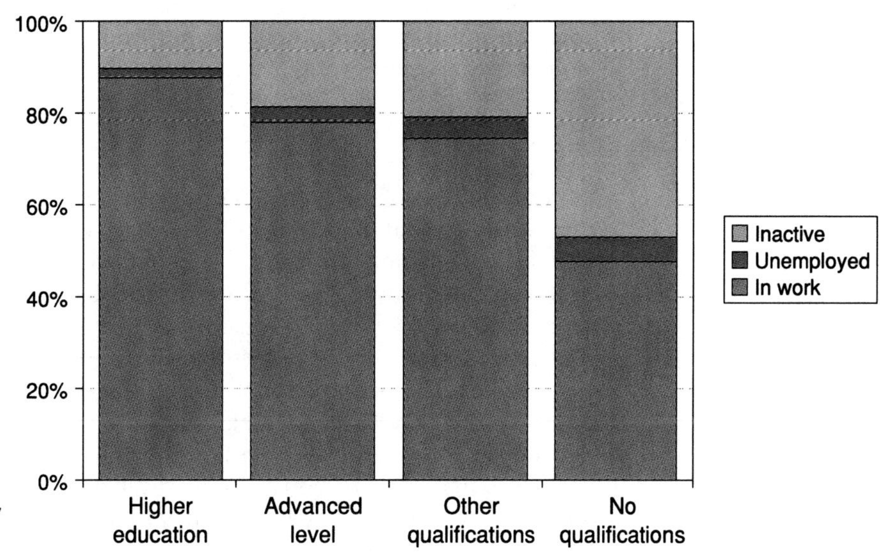

Figure 1.4 Percentages in each employment category, by educational qualification

are of the same height (representing 100%) and the components of each bar now show the *proportions* of people in each educational category either in work, unemployed or inactive.

It is now clear how economic status differs according to education and the result is quite dramatic. In particular:

- The probability of unemployment increases rapidly with lower educational attainment (this interprets proportions as probabilities, i.e. if 10% are out of work then the probability that a person picked at random is unemployed is 10%).
- The biggest difference is between the no qualifications category and the other three, which have relatively smaller differences between them. In particular, A-levels and other qualifications show a similar pattern.

Notice that we have looked at the data in different ways, drawing different charts for the purpose. You need to consider which type of chart of most suitable for the data you have and the questions you want to ask. There is no one graph that is ideal for all circumstances.

Can we safely conclude therefore that the probability of your being unemployed is significantly reduced by education? Could we go further and argue that the route to lower unemployment generally is through investment in education? The answer *may* be 'yes' to both questions, but we have not proved it. Two important considerations are as follows:

- Innate ability has been ignored. Those with higher ability are more likely to be employed *and* are more likely to receive more education. Ideally we would like to compare individuals of similar ability but with different amounts of education.
- Even if additional education does reduce a person's probability of becoming unemployed, this may be at the expense of someone else, who loses their job to the more educated individual. In other words, additional education does not reduce total unemployment but only shifts it around among the labour force. Of course it is still rational for individuals to invest in education if they do not take account of this externality.

The pie chart

Another useful way of presenting information graphically is the **pie chart**, which is particularly good at describing how a variable is distributed between different categories. For example, from Table 1.1 we have the distribution of people by educational qualification (the first row of the table). This can be shown in a pie chart as in Figure 1.5.

The area of each slice is proportional to the respective frequency and the pie chart is an alternative means of presentation to the bar chart shown in Figure 1.1. The percentages falling into each education category have been added around the chart, but this is not essential. For presentational purposes it is best not to have too many slices in the chart: beyond about six the chart tends to look crowded. It might be worth amalgamating less important categories to make a chart look clearer.

The chart reveals that 40% of those employed fall into the 'other qualification' category, and that just 8% have no qualifications. This may be

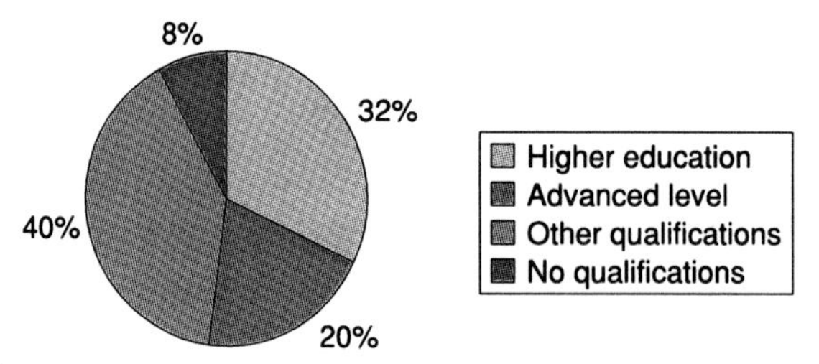

Note: If you have to draw a pie chart by hand, the angle of each slice can be calculated as follows:

$$angle = \frac{frequency}{total\ frequency} \times 360.$$

The angle of the first slice, for example, is

$$\frac{8541}{27\,004} \times 360 = 113.9°.$$

Figure 1.5 Educational qualifications of those in work

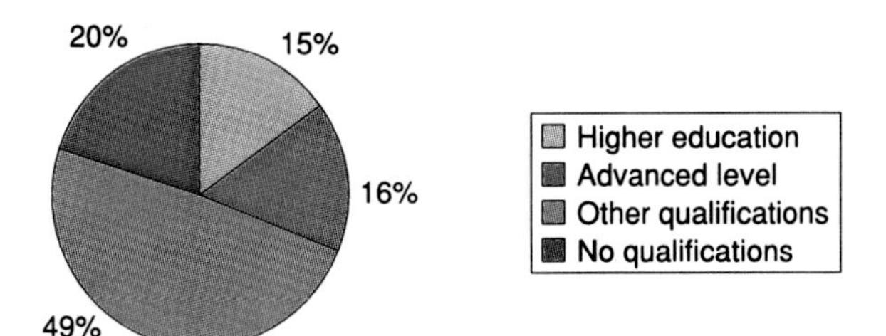

Figure 1.6 Educational qualifications of the unemployed

contrasted with Figure 1.6, which shows a similar chart for the unemployed (the second row of Table 1.1).

The 'other qualification' category is a little larger in this case, but the 'no qualification' group now accounts for 20% of the unemployed, a big increase. Further, the proportion with a degree approximately halves from 32% to 15%.

Producing charts using Microsoft Excel

Most of the charts in this book were produced using Excel's charting facility. Without wishing to dictate a precise style, you should aim for a similar, uncluttered look. Some tips you might find useful are:

- Make the grid lines dashed in a light grey colour (they are not actually part of the chart, hence should be discreet) or eliminate altogether.
- Get rid of the background fill (grey by default, alter to 'No fill'). It does not look great when printed.
- On the *x*-axis, make the labels horizontal or vertical, not slanted – it is then difficult to see which point they refer to. If they are slanted, double click on the *x*-axis then click the alignment tab.
- Colour charts look great on-screen but unclear if printed in black and white. Change the style type of the lines or markers (e.g. make some dashed) to distinguish them on paper.
- Both axes start at zero by default. If all your observations are large numbers this may result in the data points being crowded into one corner of the graph. Alter the scale on the axes to fix this: set the minimum value on the axis to be slightly less than the minimum observation.

Otherwise, Excel's default options will usually give a good result.

Exercise 1.1

The following table shows the total numbers (in millions) of tourists visiting each country and the numbers of English tourists visiting each country:

	France	Germany	Italy	Spain
All tourists	12.4	3.2	7.5	9.8
English tourists	2.7	0.2	1.0	3.6

(a) Draw a bar chart showing the total numbers visiting each country.

(b) Draw a stacked bar chart, which shows English and non-English tourists making up the total visitors to each country.

(c) Draw a pie chart showing the distribution of all tourists between the four destination countries.

(d) Do the same for English tourists and compare results.

Looking at cross-section data: wealth in the UK in 2003

Frequency tables and histograms

We now move on to examine data in a different form. The data on employment and education consisted simply of frequencies, where a characteristic (such as higher education) was either present or absent for a particular individual. We now look at the distribution of wealth – a variable that can be measured on a **ratio scale** so that a different value is associated with each individual. For example, one person might have £1000 of wealth, another might have £1 million. Different presentational techniques will be used to analyse this type of data. We use these techniques to investigate questions such as how much wealth does the average person have and whether wealth is evenly distributed or not.

The data are given in Table 1.3, which shows the distribution of wealth in the UK for the year 2003 (the latest available at the time of writing), available at http://www.hmrc.gov.uk/stats/personal_wealth/menu.htm. This is an example of a **frequency table**. Wealth is difficult to define and to measure; the data shown here refer to *marketable* wealth (i.e. items such as the right to a pension, which cannot be sold, are excluded) and are estimates for the population (of adults) as a whole based on taxation data.

Wealth is divided into 14 **class intervals**: £0 up to (but not including) £10 000; £10 000 up to £24 999, etc., and the number (or **frequency**) of

Table 1.3 The distribution of wealth, UK, 2003

Class interval (£)	Numbers (thousands)
0–9999	2448
10 000–24 999	1823
25 000–39 999	1375
40 000–49 999	480
50 000–59 999	665
60 000–79 999	1315
80 000–99 999	1640
100 000–149 999	2151
150 000–199 000	2215
200 000–299 000	1856
300 000–499 999	1057
500 000–999 999	439
1 000 000–1 999 999	122
2 000 000 or more	50
Total	17 636

Note: It would be impossible to show the wealth of all 18 million individuals, so it has been summarised in this **frequency table**.

individuals within each class interval is shown. Note that the widths of the intervals (the **class widths**) vary up the wealth scale: the first is £10 000, the second £15 000 (= 25 000 – 10 000); the third £15 000 also and so on. This will prove an important factor when it comes to graphical presentation of the data.

This table has been constructed from the original 17 636 000 observations on individuals' wealth, so it is already a summary of the original data (note that all the frequencies have been expressed in thousands in the table) and much of the original information is lost. The first decision to make if one had to draw up such a frequency table from the raw data is how many class intervals to have, and how wide they should be. It simplifies matters if they are all of the same width but in this case it is not feasible: if 10 000 were chosen as the **standard width** there would be many intervals between 500 000 and 1 000 000 (50 of them in fact), most of which would have a zero or very low frequency. If 100 000 were the standard width, there would be only a few intervals and the first (0–100 000) would contain 9746 observations (55% of all observations), so almost all the interesting detail would be lost. A compromise between these extremes has to be found.

A useful rule of thumb is that the number of class intervals should equal the square root of the total frequency, subject to a maximum of about 12 intervals. Thus, for example, a total of 25 observations should be allocated to five intervals; 100 observations should be grouped into 10 intervals; and 17 636 should be grouped into about 12 (14 are used here). The class widths should be equal in so far as this is feasible, but should increase when the frequencies become very small.

To present these data graphically one could draw a bar chart as in the case of education above, and this is presented in Figure 1.7. Before reading on, spend some time looking at it and ask yourself what is wrong with it.

The answer is that the figure gives a completely misleading picture of the data! (Incidentally, this is the picture that you will get using a spreadsheet computer program, as I have done here. All the standard packages appear to do this, so beware. One wonders how many decisions have been influenced by data presented in this incorrect manner.)

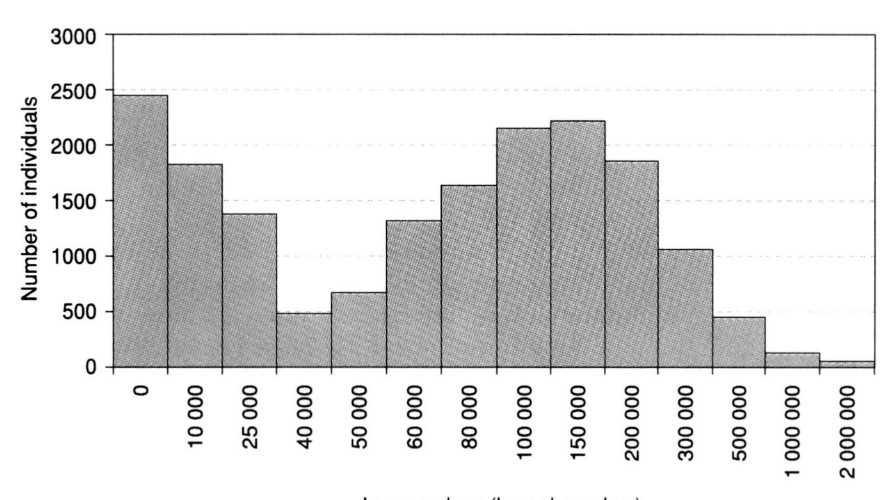

Figure 1.7
Bar chart of the distribution of wealth in the UK, 2003

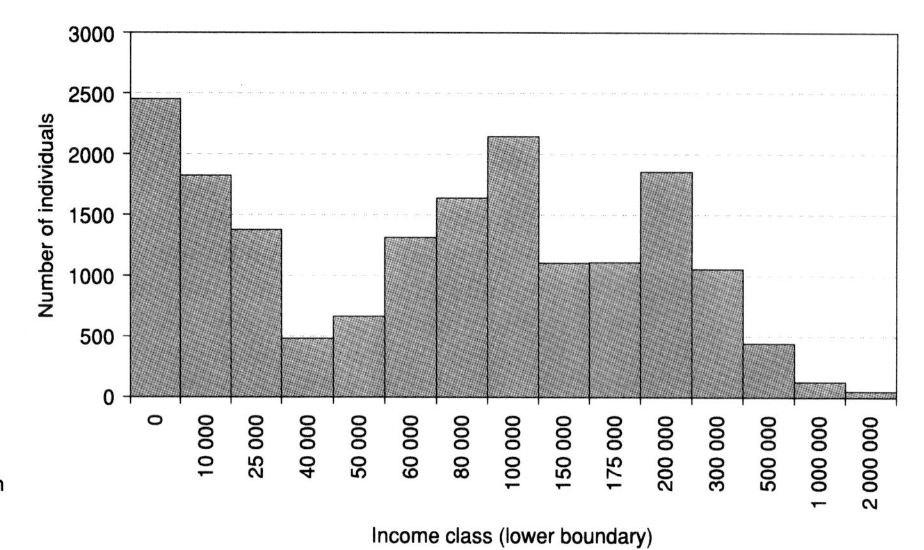

Figure 1.8
The wealth distribution with alternative class intervals

Why is the figure wrong? Consider the following argument. The diagram appears to show that there are few individuals around £40 000 to £60 000 (the frequency is at a low of 480 (thousand)) but many around £150 000. But this is just the result of the difference in the class width at these points (10 000 at £40 000 and 50 000 at £150 000). Suppose that we divide up the £150 000–£200 000 class into two: £150 000 to £175 000 and £175 000 to £200 000. We divide the frequency of 2215 equally between the two (this is an arbitrary decision but illustrates the point). The graph now looks like Figure 1.8.

Comparing Figures 1.7 and 1.8 reveals a difference: the hump around £150 000 has now disappeared, replaced by a small crater. But this is disturbing – it means that the shape of the distribution can be altered simply by altering the class widths. If so, how can we rely upon visual inspection of the distribution? What does the 'real' distribution look like? A better method would make the shape of the distribution independent of how the class intervals are arranged. This can be done by drawing a **histogram**.

The histogram

A histogram is similar to a bar chart except that it corrects for differences in class widths. If all the class widths are identical, then there is no difference between a bar chart and a histogram. The calculations required to produce the histogram are shown in Table 1.4.

The new column in the table shows the **frequency density**, which measures the frequency *per unit of class width*. Hence it allows a direct comparison of different class intervals, i.e. accounting for the difference in class widths.

The frequency density is defined as follows

$$\text{frequency density} = \frac{\text{frequency}}{\text{class width}} \tag{1.1}$$

Using this formula corrects the figures for differing class widths. Thus 0.2448 = 2448/10 000 is the first frequency density, 0.1215 = 1823/15 000 is the second,

Table 1.4 Calculation of frequency densities

Range	Number or frequency	Class width	Frequency density
0–	2448	10 000	0.2448
10 000–	1823	15 000	0.1215
25 000–	1375	15 000	0.0917
40 000–	480	10 000	0.0480
50 000–	665	10 000	0.0665
60 000–	1315	20 000	0.0658
80 000–	1640	20 000	0.0820
100 000–	2151	50 000	0.0430
150 000–	2215	50 000	0.0443
200 000–	3524	3 800 000	0.0009

Note: As an alternative to the frequency density, one could calculate the frequency per 'standard' class width, with the standard width chosen to be 10 000 (the narrowest class). The values in column 4 would then be 2448; 1215.3 (= 1823 ÷ 1.5); 916.7; etc. This would lead to the same shape of histogram as using the frequency density.

etc. Above £200 000 the class widths are very large and the frequencies small (too small to be visible on the histogram), so these classes have been combined.

The width of the final interval is unknown, so has to be estimated in order to calculate the frequency density. It is likely to be extremely wide since the wealthiest person may well have assets valued at several £m (or even £bn); the value we assume will affect the calculation of the frequency density and therefore of the shape of the histogram. Fortunately it is in the tail of the distribution and only affects a small number of observations. Here we assume (arbitrarily) a width of £3.8m to be a 'reasonable' figure, giving an upper class boundary of £4m.

The frequency density is then plotted on the vertical axis against wealth on the horizontal axis to give the histogram. One further point needs to be made: the scale on the wealth axis should be linear as far as possible, e.g. £50 000 should be twice as far from the origin as £25 000. However, it is difficult to fit all the values onto the horizontal axis without squeezing the graph excessively at lower levels of wealth, where most observations are located. Therefore the classes above £100 000 have been squeezed and the reader's attention is drawn to this. The result is shown in Figure 1.9.

The effect of taking frequency densities is to make the *area* of each block in the histogram represent the frequency, rather than the height, which now shows the density. This has the effect of giving an accurate picture of the shape of the distribution.

Having done all this, what does the histogram show?

- The histogram is heavily **skewed** to the right (i.e. the long tail is to the right).
- The **modal** class interval is £0–£10 000 (i.e. has the greatest density: no other £10 000 interval has more individuals in it).
- A little under half of all people (45.9% in fact) have less than £80 000 of marketable wealth.
- About 20% of people have more than £200 000 of wealth.[2]

[2] Due to the compressing of some class widths, it is difficult to see this accurately on the histogram. There are limitations to graphical presentation.

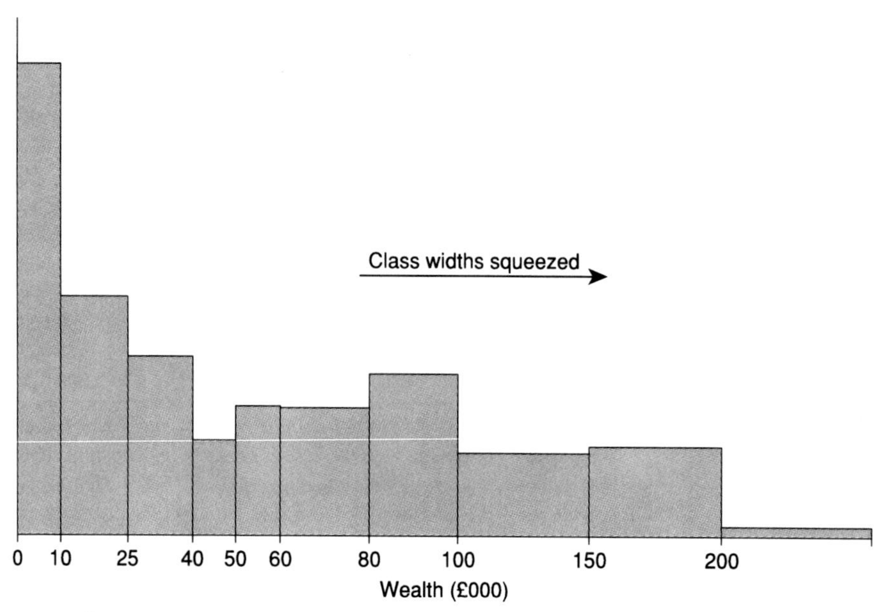

Figure 1.9 Histogram of the distribution of wealth in the UK, 2003

Note: A **frequency polygon** would be the result if, instead of drawing blocks for the histogram, lines were drawn connecting the centres of the top of each block. The diagram is better drawn with blocks, in general.

The figure shows quite a high degree of inequality in the wealth distribution. Whether this is acceptable or even desirable is a value judgement. It should be noted that part of the inequality is due to differences in age: younger people have not yet had enough time to acquire much wealth and therefore appear worse off, although in life-time terms this may not be the case. To obtain a better picture of the distribution of wealth would require some analysis of the acquisition of wealth over the life-cycle (or comparing individuals of a similar age). In fact, correcting for age differences does not make a big difference to the pattern of wealth distribution (on this point and on inequality in wealth in general, see Atkinson (1983), Chapters 7 and 8).

Relative frequency and cumulative frequency distributions

An alternative way of illustrating the wealth distribution uses the **relative** and **cumulative frequencies** of the data. The relative frequencies show the *proportion* of observations that fall into each class interval, so, for example, 2.72% of individuals have wealth holdings between £40 000 and £50 000 (480 000 out of 17 636 000 individuals). Relative frequencies are shown in the third column of Table 1.5, using the following formula[3]

$$\textit{Relative frequency} = \frac{\textit{frequency}}{\textit{sum of frequencies}} = \frac{f}{\Sigma f} \tag{1.2}$$

[3] If you are unfamiliar with the Σ notation then read Appendix 1A to this chapter before continuing.

Table 1.5 Calculation of relative and cumulative frequencies

Range	Frequency	Relative frequency (%)	Cumulative frequency
0–	2448	13.9	2448
10 000–	1823	10.3	4271
25 000–	1375	7.8	5646
40 000–	480	2.7	6126
50 000–	665	3.8	6791
60 000–	1315	7.5	8106
80 000–	1640	9.3	9746
100 000–	2151	12.2	11 897
150 000–	2215	12.6	14 112
200 000–	1856	10.5	15 968
300 000–	1057	6.0	17 025
500 000–	439	2.5	17 464
1 000 000–	122	0.7	17 586
2 000 000–	50	0.3	17 636
Total	17 636	100.00	

Note: Relative frequencies are calculated in the same way as the column percentages in Table 1.2. Thus for example, 13.9% is 2448 divided by 17 636. Cumulative frequencies are obtained by cumulating, or successively adding, the frequencies. For example, 4271 is 2448 + 1823, 5646 is 4271 + 1375, etc.

The AIDS epidemic

To show how descriptive statistics can be helpful in presenting information we show below the 'population pyramid' for Botswana (one of the countries most seriously affected by AIDS), projected for the year 2020. This is essentially two bar charts (one for men, one for women) laid on their sides, showing the frequencies in each age category (rather than wealth categories). The inner pyramid (in the darker colour) shows the projected population given the existence of AIDS; the outer pyramid assumes no deaths from AIDS.

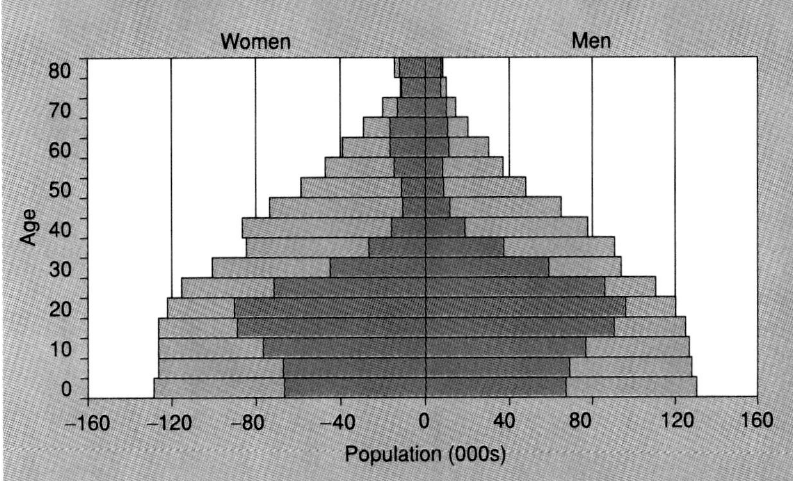

Original source of data: US Census Bureau, *World Population Profile 2000*. Graph adapted from the UNAIDS web site at http://www.unaids.org/epidemic_update/report/Epi_report.htm#thepopulation.

→

One can immediately see the huge effect of AIDS, especially on the 40–60 age group (currently aged 20–40), for both men and women. These people would normally be in the most productive phase of their lives but, with AIDS, the country will suffer enormously with many old and young people dependent on a small working population. The severity of the future problems is brought out vividly in this simple graphic, based on the bar chart.

The sum of the relative frequencies has to be 100% and this acts as a check on the calculations.

The cumulative frequencies, shown in the fourth column, are obtained by cumulating (successively adding) the frequencies. The cumulative frequencies show the total number of individuals with wealth *up to* a given amount; for example, about 10 million people have less than £100 000 of wealth.

Both relative and cumulative frequency distributions can be drawn, in a similar way to the histogram. In fact, the relative frequency distribution has exactly the same shape as the frequency distribution. This is shown in Figure 1.10. This time we have written the relative frequencies above the appropriate column, although this is not essential.

The cumulative frequency distribution is shown in Figure 1.11, where the blocks increase in height as wealth increases. The simplest way to draw this is to cumulate the frequency densities (shown in the final column of Table 1.4) and to use these values as the y-axis coordinates.

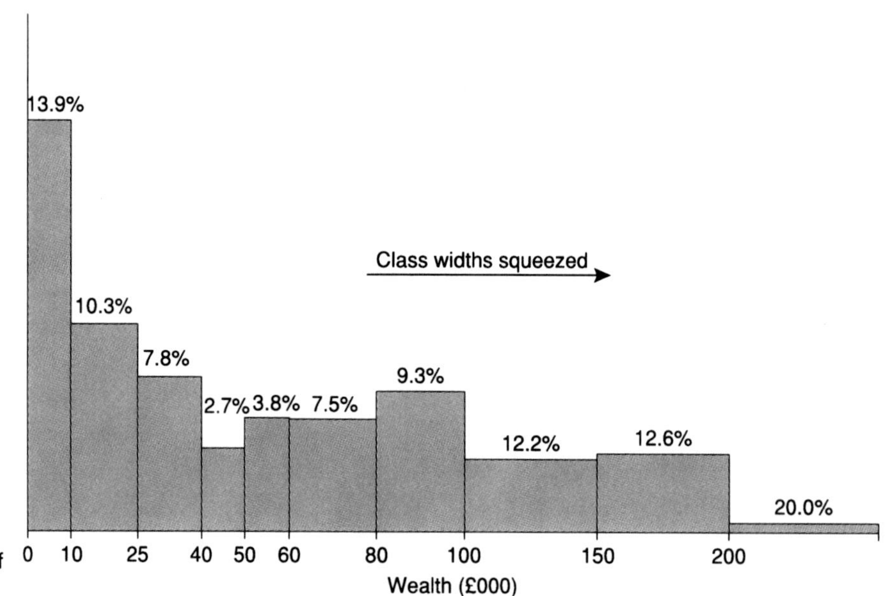

Figure 1.10
The relative density frequency distribution of wealth in the UK, 2003

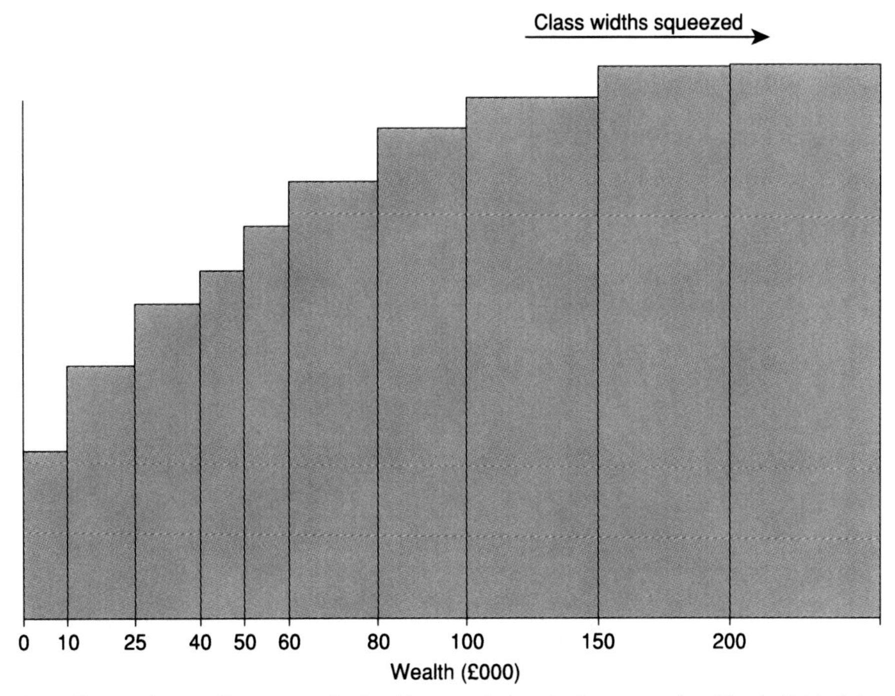

Figure 1.11
The cumulative frequency distribution of wealth in the UK, 2003

Note: The *y*-axis coordinates are obtained by cumulating the frequency densities in Table 1.4 above. For example, the first two *y* coordinates are 0.2448, 0.3663.

Worked example 1.1

There is a mass of detail in the sections above, so this worked example is intended to focus on the essential calculations required to produce the summary graphs. Simple artificial data are deliberately used to avoid the distraction of a lengthy interpretation of the results and their meaning. The data on the variable X and its frequencies f are shown in the following table, with the calculations required:

X	Frequency, f	Relative frequency	Cumulative frequency, F
10	6	0.17	6
11	8	0.23	14
12	15	0.43	29
13	5	0.14	34
14	1	0.03	35
Total	35	1.00	

Notes:
The X values are unique but could be considered the mid-point of a range, as earlier.
The relative frequencies are calculated as $0.17 = 6/35$, $0.23 = 8/35$, etc.
The cumulative frequencies are calculated as $14 = 6 + 8$, $29 = 6 + 8 + 15$, etc.
The symbol F usually denotes the cumulative frequency in statistical work.

The resulting bar chart and cumulative frequency distribution are:

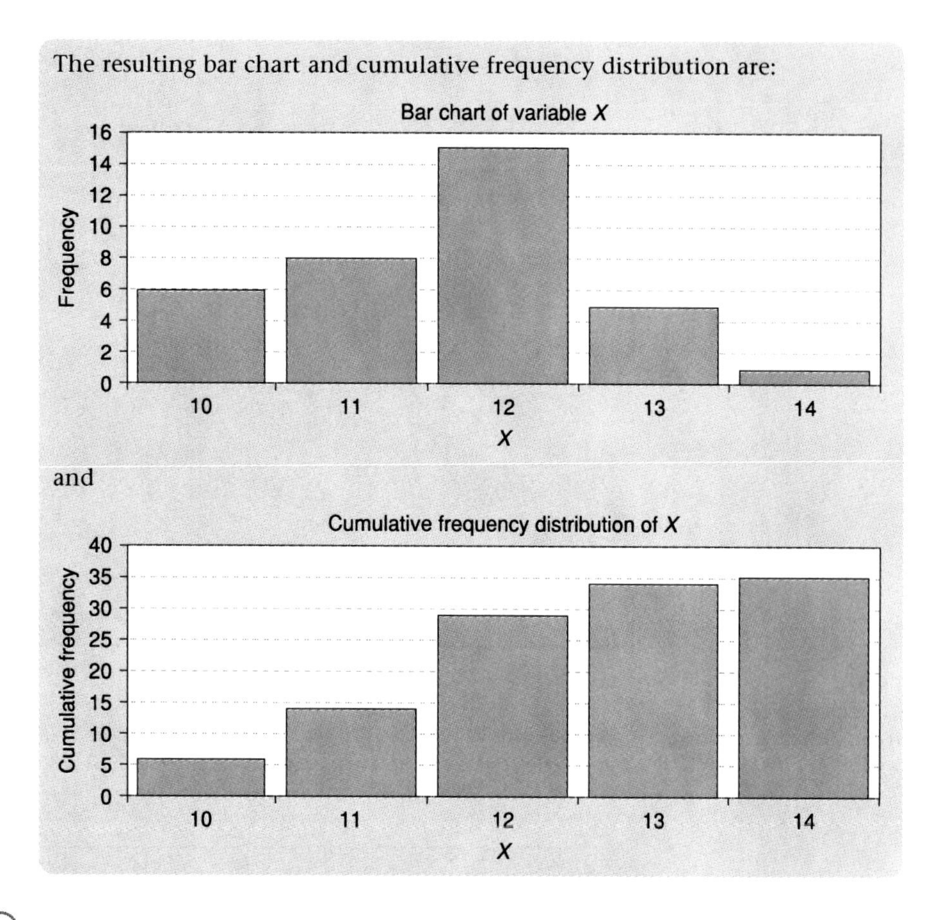

Exercise 1.2

Given the following data:

Range	Frequency
0–10	20
11–30	40
31–60	30
60–100	20

(a) Draw both a bar chart and a histogram of the data and compare them.

(b) Calculate cumulative frequencies and draw a cumulative frequency diagram.

Summarising data using numerical techniques

Graphical methods are an excellent means of obtaining a quick overview of the data, but they are not particularly precise, nor do they lend themselves to further analysis. For this we must turn to numerical measures such as the average. There are a number of different ways in which we may describe a distribution such as that for wealth. If we think of trying to describe the histogram, it is useful to have:

- A **measure of location** giving an idea of whether people own a lot of wealth or a little. An example is the average, which gives some idea of where the distribution is located along the x-axis. In fact, we will encounter three different measures of the 'average':
  - the mean;
  - the median;
  - the mode.
- A **measure of dispersion** showing how wealth is dispersed around (usually) the average, whether it is concentrated close to the average or is generally far away from it. An example here is the standard deviation.
- A **measure of skewness** showing how symmetric or not the distribution is, i.e. whether the left half of the distribution is a mirror image of the right half or not. This is obviously not the case for the wealth distribution.

We consider each type of measure in turn.

Measures of location: the mean

The **arithmetic mean**, commonly called the average, is the most familiar measure of location, and is obtained simply by adding all the observations and dividing by the number of observations. If we denote the wealth of the ith household by x_i (so that the index i runs from 1 to N, where N is the number of observations; as an example, x_3 would be the wealth of the third household) then the mean is given by the following formula

$$\mu = \frac{\sum_{i=1}^{i=N} x_i}{N} \tag{1.3}$$

where μ (the Greek letter mu, pronounced 'myu'[4]) denotes the mean and $\sum_{i=1}^{i=N} x_i$ (read 'sigma x i, from $i = 1$ to N', Σ being the Greek capital letter sigma) means the sum of the x values. We may simplify this to

$$\mu = \frac{\Sigma x}{N} \tag{1.4}$$

when it is obvious which x values are being summed (usually all the available observations). This latter form is more easily readable and we will generally use this.

Worked example 1.2

We will find the mean of the values 17, 25, 28, 20, 35. The total of these five numbers is 125, so we have $N = 5$ and $\Sigma x = 125$. Therefore the mean is

$$\mu = \frac{\Sigma x}{N} = \frac{125}{5} = 25$$

Formula (1.3) can only be used when all the individual x values are known. The frequency table for wealth does not show all 17 million observations, however,

[4] Well, mathematicians pronounce it like this, but modern Greeks do not. For them it is 'mi'.

but only the range of values for each class interval and the associated frequency. In this case of grouped data the following equivalent formula may be used

$$\mu = \frac{\sum_{i=1}^{i=C} f_i x_i}{\sum_{i=1}^{i=C} f_i} \tag{1.5}$$

or, more simply

$$\mu = \frac{\Sigma fx}{\Sigma f} \tag{1.6}$$

In this formula

- x denotes the **mid-point** of each class interval, since the individual x values are unknown. The mid-point is used as the representative x value for each class. In the first class interval, for example, we do not know precisely where each of the 2448 observations lies. Hence we *assume* they all lie at the mid-point, £5000. This will cause a slight inaccuracy – because the distribution is so skewed, there are more households below the mid-point than above it in every class interval except, perhaps, the first. We ignore this problem here, and it is less of a problem for most distributions which are less skewed than this one.
- The summation runs from 1 to C, the number of class intervals, or distinct x values. f times x gives the total wealth in each class interval. If we sum over the 14 class intervals we obtain the total wealth of all individuals.
- $\Sigma f_i = N$ gives the total number of observations, the sum of the individual frequencies. The calculation of the mean, μ, for the wealth data is shown in Table 1.6.

Table 1.6 The calculation of average wealth

Range	x	f	fx
0–	5.0	2448	12 240
10 000–	17.5	1823	31 902
25 000–	32.5	1375	44 687
40 000–	45.0	480	21 600
50 000–	55.0	665	36 575
60 000–	70.0	1315	92 050
80 000–	90.0	1640	147 600
100 000–	125.0	2151	268 875
150 000–	175.0	2215	387 625
200 000–	250.0	1856	464 000
300 000–	400.0	1057	422 800
500 000–	750.0	439	329 250
1 000 000–	1500.0	122	183 000
2 000 000–	3000.0	50	150 000
Total		17 636	2 592 205

Note: The fx column gives the product of the values in the f and x columns (so, for example, $5.0 \times 2448 = 12\ 240$, which is the total wealth held by those in the first class interval). The sum of the fx values gives total wealth.

From this we obtain

$$\mu = \frac{2\,592\,205}{17\,636} = 146.984$$

Note that the x values are expressed in £000, so we must remember that the mean will also be in £000; the average wealth holding is therefore £146 984. Note that the frequencies have also been divided by 1000, but this has no effect upon the calculation of the mean since f appears in both numerator and denominator of the formula for the mean.

The mean tells us that if the total wealth were divided up equally between all individuals, each would have £146 984. This value may seem surprising, since the histogram clearly shows most people have wealth below this point (approximately 65% of individuals are below the mean, in fact). The mean does not seem to be typical of the wealth that most people have. The reason the mean has such a high value is that there are some individuals whose wealth is way above the figure of £146 984 – up into the £millions, in fact. The mean is the 'balancing point' of the distribution – if the histogram were a physical model, it would balance on a fulcrum placed at 146 984. The few very high wealth levels exert a lot of leverage and counter-balance the more numerous individuals below the mean.

Worked example 1.3

Suppose we have 10 families with a single television in their homes, 12 families with two televisions each and 3 families with three. You can probably work out in your head that there are 43 televisions in total (10 + 24 + 9) owned by the 25 families (10 + 12 + 3). The average number of televisions per family is therefore 43/25 = 1.72.

Setting this out formally, we have (as for the wealth distribution, but simpler):

x	f	fx
1	10	10
2	12	24
3	3	9
Totals	25	43

This gives our resulting mean as 1.72. Note that our data are discrete values in this case and we have the actual values, not a broad class interval.

The mean as the expected value

We also refer to the mean as the **expected value** of x and write

$$E(x) = \mu = 146\,984 \tag{1.7}$$

$E(x)$ is read 'E of x' or 'the expected value of x'. The mean is the expected value in the sense that, if we selected a household at random from the population we would 'expect', its wealth to be £146 984. It is important to note that this is a *statistical* expectation, rather than the everyday use of the term. Most of the random individuals we encounter have wealth substantially below this value. Most people might therefore 'expect' a lower value because that is their everyday experience; but statisticians are different, they always expect the mean value.

The expected value notation is particularly useful in keeping track of the effects upon the mean of certain data transformations (e.g. dividing wealth by 1000 also divides the mean by 1000); Appendix 1B provides a detailed explanation. Use is also made of the E operator in inferential statistics, to describe the properties of estimators (see Chapter 4).

The sample mean and the population mean

Very often we have only a sample of data (as in the worked example above), and it is important to distinguish this case from the one where we have all the possible observations. For this reason, the sample mean is given by

$$\bar{x} = \frac{\sum x}{n} \text{ or } \bar{x} = \frac{\sum fx}{\sum f} \text{ for grouped data} \tag{1.8}$$

Note the distinctions between μ (the population mean) and $\bar{x}$ (the sample mean), and between N (the size of the population) and n (the sample size). Otherwise, the calculations are identical. It is a convention to use Greek letters, such as μ, to refer to the population and Roman letters, such as $\bar{x}$, to refer to a sample.

The weighted average

Sometimes observations have to be given different weightings in calculating the average, as the following example. Consider the problem of calculating the average spending per pupil by an education authority. Some figures for spending on primary (ages 5 to 11), secondary (11 to 16) and post-16 pupils are given in Table 1.7.

Clearly, significantly more is spent on secondary and post-16 pupils (a general pattern throughout England and most other countries) and the overall average should lie somewhere between 1750 and 3820. However, taking a simple average of these values would give the wrong answer, because there are different numbers of children in the three age ranges. The numbers and proportions of children in each age group are given in Table 1.8.

Table 1.7 Cost per pupil in different types of school (£ p.a.)

	Primary	Secondary	Post-16
Unit cost	1750	3100	3820

Table 1.8 Numbers and proportions of pupils in each age range

	Primary	Secondary	Post-16	Total
Numbers	8000	7000	3000	18 000
Proportion	44%	39%	17%	

As there are relatively more primary school children than secondary, and relatively fewer post-16 pupils, the primary unit cost should be given greatest weight in the averaging process and the post-16 unit cost the least. The **weighted average** is obtained by multiplying each unit cost figure by the proportion of children in each category and summing. The weighted average is therefore

$$0.44 \times 1750 + 0.39 \times 3100 + 0.17 \times 3820 = 2628 \quad (1.9)$$

The weighted average gives an answer closer to the primary unit cost than does the simple average of the three figures (2890 in this case), which would be misleading. The formula for the weighted average is

$$\bar{x}_w = \Sigma_i w_i x_i \quad (1.10)$$

where w represents the weights, *which must sum to one*, i.e.

$$\Sigma_i w_i = 1 \quad (1.11)$$

and x represents the unit cost figures.

Notice that what we have done is equivalent to multiplying each unit cost by its frequency (8000, etc.) and then dividing the sum by the grand total of 18 000. This is the same as the procedure we used for the wealth calculation. The difference with weights is that we first divide 8000 by 18 000 (and 7000 by 18 000, etc.) to obtain the weights, which must then sum to one, and use these weights in formula (1.10).

Calculating your degree result

If you are a university student your final degree result will probably be calculated as a weighted average of your marks on the individual courses. The weights may be based on the credits associated with each course or on some other factors. For example, in my university the average mark for a year is a weighted average of the marks on each course, the weights being the credit values of each course.

The grand mean G, on which classification is based, is then a weighted average of the averages for the different years, as follows

$$G = \frac{0 \times Year\ 1 + 40 \times Year\ 2 + 60 \times Year\ 3}{100}$$

i.e. the year 3 mark has a weight of 60%, year 2 is weighted 40% and the first year is not counted at all.

For students taking a year abroad the formula is slightly different

$$G = \frac{0 \times Year\ 1 + 40 \times Year\ 2 + 25 \times Yabroad + 60 \times Year\ 3}{125}$$

Note that, to accommodate the year abroad mark, the weights on years 2 and 3 are reduced (to 40/125 = 32% and 60/125 = 48% respectively).

The median

Returning to the study of wealth, the unrepresentative result for the mean suggests that we may prefer a measure of location which is not so strongly affected by outliers (extreme observations) and skewness.

The **median** is a measure of location which is more robust to such extreme values; it may be defined by the following procedure. Imagine everyone in a line from poorest to wealthiest. Go to the individual located halfway along the line. Ask what their wealth is. Their answer is the median. The median is clearly unaffected by extreme values, unlike the mean: if the wealth of the richest person were doubled (with no reduction in anyone else's wealth) there would be no effect upon the median. The calculation of the median is not so straightforward as for the mean, especially for grouped data. The following worked example shows how to calculate the median for ungrouped data.

Worked example 1.4 The median

Calculate the median of the following values: 45, 12, 33, 80, 77.

First we put them into ascending order: 12, 33, 45, 77, 80.

It is then easy to see that the middle value is 45. This is the median. Note that if the value of the largest observation changes to, say, 150, the value of the median is unchanged. This is not the case for the mean, which would change from 49.4 to 63.4.

If there is an even number of observations, then there is no middle observation. The solution is to take the average of the two middle observations. For example:

Find the median of 12, 33, 45, 63, 77, 80.

Note the new observation, 63, making six observations. The median value is halfway between the third and fourth observations, i.e. (45 + 63)/2 = 54.

For grouped data there are two stages to the calculation: first we must first identify the class interval which contains the median person, then we must calculate where in the interval that person lies.

(1) To find the appropriate class interval: since there are 17 636 000 observations, we need the wealth of the person who is 8 818 000 in rank order. The table of cumulative frequencies (see Table 1.5 above) is the most suitable for this. There are 8 106 000 individuals with wealth of less than £80 000 and 9 746 000 with wealth of less than £100 000. The middle person therefore falls into the £80 000–100 000 class. Furthermore, given that 8 818 000 falls roughly half way between 8 106 000 and 9 746 000 it follows that the median is close to the middle of the class interval. We now go on to make this statement more precise.

(2) To find the position in the class interval, we can now use formula (1.12)

$$median = x_L + (x_U - x_L)\left\{\frac{\frac{N+1}{2} - F}{f}\right\} \tag{1.12}$$

where

x_L = the lower limit of the class interval containing the median

x_U = the upper limit of this class interval

N = the number of observations (using $N + 1$ rather than N in the formula is only important when N is relatively small)

F = the cumulative frequency of the class intervals up to (but not including) the one containing the median

f = the frequency for the class interval containing the median.

For the wealth distribution we have

$$median = 80\,000 + (100\,000 - 80\,000)\left\{\frac{\dfrac{17\,636\,000}{2} - 8\,106\,000}{1\,640\,000}\right\} = £90\,829$$

This alternative measure of location gives a very different impression: it is less than two-thirds of the mean. Nevertheless, it is equally valid despite having a different meaning. It demonstrates that the person 'in the middle' has wealth of £90 829 and in this sense is typical of the UK population. Before going on to compare these measures further we examine a third: the mode.

Generalising the median – quantiles

The idea of the median as the middle of the distribution can be extended: **quartiles** divide the distribution into four equal parts, **quintiles** into five, **deciles** into 10, and finally **percentiles** divide the distribution into 100 equal parts. Generically they are known as **quantiles**. We shall illustrate the idea by examining deciles (quartiles are covered below).

The first decile occurs one-tenth of the way along the line of people ranked from poorest to wealthiest. This means we require the wealth of the person ranked 1 763 600 (= N/10) in the distribution. From the table of cumulative frequencies, this person lies in the first class interval. Adapting formula (1.12), we obtain

$$first\ decile = 0 + (10\,000 - 0) \times \left\{\frac{1\,763\,600 - 0}{2\,448\,000}\right\} = £7203$$

Thus we estimate that any household with less than £7203 of wealth falls into the bottom 10% of the wealth distribution. In a similar fashion, the ninth decile can be found by calculating the wealth of the household ranked 15 872 400 (= $N \times 9/10$) in the distribution.

The mode

The **mode** is defined as that level of wealth which occurs with the greatest frequency, in other words the value that occurs most often. It is most useful and easiest to calculate when one has all the data and there are relatively few distinct observations. This is the case in the simple example below.

Suppose we have the following data on sales of dresses by a shop, according to size

Size	Sales
8	7
10	25
12	36
14	11
16	3
18	1

The modal size is 12. There are more women buying dresses of this size than any other. This may be the most useful form of average as far as the shop is concerned. Although it needs to stock a range of sizes, it knows it needs to order more dresses in size 12 than in any other size. The mean would not be so helpful in this case (it is $\bar{x} = 11.7$) as it is not an actual dress size.

In the case of grouped data matters are more complicated. It is the modal class interval which is required, once the intervals have been corrected for width (otherwise a wider class interval is unfairly compared with a narrower one). For this, we can again make use of the frequency densities. From Table 1.4 it can be seen that it is the first interval, from £0 to £10 000, which has the highest frequency density. It is 'typical' of the distribution because it is the one which occurs most often (using the frequency densities, *not* frequencies). The wealth distribution is most concentrated at this level and more people are like this in terms of wealth than anything else. Once again it is notable how different it is from both the median and the mean.

The three measures of location give different messages because of the skewness of the distribution: if it were symmetric they would all give approximately the same answer. Here we have a rather extreme case of skewness, but it does serve to illustrate how the different measures of location compare. When the distribution is skewed to the right, as here, they will be in the order mode, median, mean; if skewed to the left the ordering is reversed. If the distribution has more than one peak then this rule for orderings may not apply.

Which of the measures is 'correct' or most useful? In this particular case the mean is not very useful: it is heavily influenced by extreme values. The median is therefore often used when discussing wealth (and income) distributions. Where inequality is even more pronounced, as in some less developed countries, then the mean is even less informative. The mode is also quite useful in telling us about a large section of the population, although it can be sensitive to how the class intervals are arranged. If the data were arranged such that there was a class interval of £5000 to £15 000, then this might well be the modal class, conveying a slightly different impression.

The three different measures of location are marked on the histogram in Figure 1.12. This brings out the substantial difference between the measures for a skewed distribution such as for wealth.

Exercise 1.3

(a) For the data in Exercise 2, calculate the mean, median and mode of the data.

(b) Mark these values on the histogram you drew for Exercise 2.

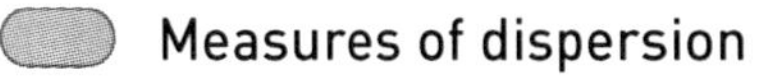

Measures of dispersion

Two different distributions (e.g. wealth in two different countries) might have the same mean yet look very different, as shown in Figure 1.13 (the distributions have been drawn using smooth curves rather than bars to improve clarity). In one country everyone might have a similar level of wealth (curve B). In another, although the average is the same there might be extremes of great wealth and poverty (curve A). A measure of dispersion is a number which allows us to distinguish between these two situations.

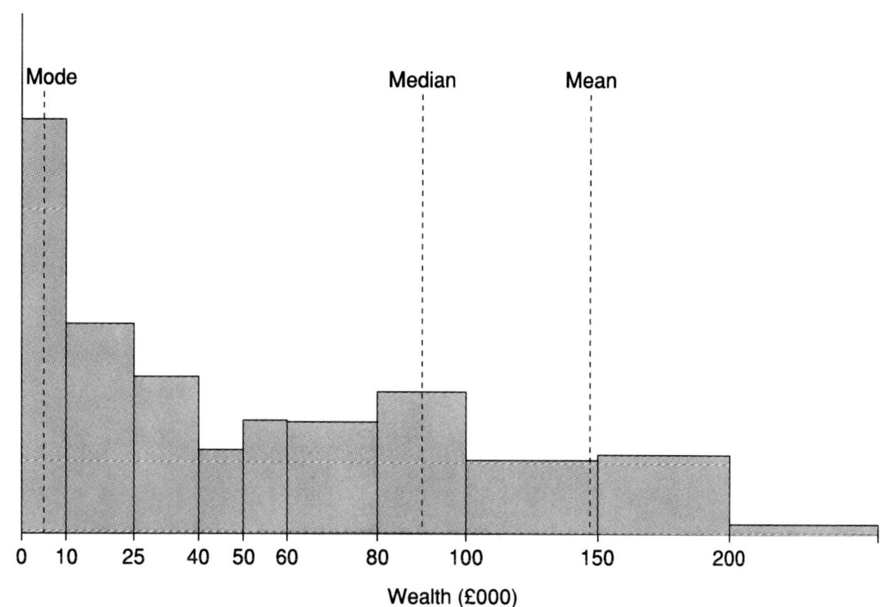

Figure 1.12
The histogram with mean, median and mode marked

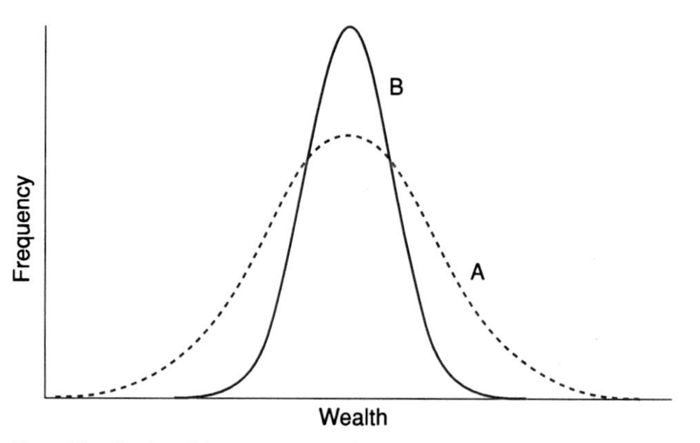

Figure 1.13
Two distributions with different degrees of dispersion

Note: Distribution A has a greater degree of dispersion than B, where everyone has a similar level of wealth.

The simplest measure of dispersion is the **range**, which is the difference between the smallest and largest observations. It is impossible to calculate accurately from the table of wealth holdings since the largest observation is not available. In any case, it is not a very useful figure since it relies on two extreme values and ignores the rest of the distribution. In simpler cases it might be more informative. For example, in an exam the marks may range from a low of 28% to a high of 74%. In this case the range is 74 − 28 = 46 and this tells us something useful.

An improvement is the **inter-quartile range** (IQR), which is the difference between the first and third quartiles. It therefore defines the limits of wealth of the middle half of the distribution and ignores the very extremes of the distribution. To calculate the first quartile (which we label Q_1) we have to go one-quarter of the way along the line of wealth holders (ranked from poorest to wealthiest) and ask the person in that position what their wealth is. Their answer is the first quartile. The calculation is as follows:

- one-quarter of 17 636 is 4409;
- the person ranked 4409 is in the £25 000–40 000 class;
- adapting formula (1.12)

$$Q_1 = 25\,000 + (40\,000 - 25\,000)\left\{\frac{4409 - 4271}{1375}\right\} = 26\,505.5 \tag{1.13}$$

The third quartile is calculated in similar fashion:

- three-quarters of 17 636 is 13 227;
- the person ranked 13 227 is in the £150 000–200 000 class;
- again using formula (1.12)

$$Q_3 = 150\,000 + (200\,000 - 150\,000)\left\{\frac{13\,227 - 11\,897}{2215}\right\} = 180\,022.6$$

and therefore the inter-quartile range is $Q_3 - Q_1 = 180\,022 - 26\,505 = 153\,517$. This might be reasonably rounded to £150 000 given the approximations in our calculation, and is a much more memorable figure.

This gives one summary measure of the dispersion of the distribution: the higher the value the more spread-out is the distribution. Two different wealth distributions might be compared according to their inter-quartile ranges therefore, with the country having the larger figure exhibiting greater inequality. Note that the figures would have to be expressed in a common unit of currency for this comparison to be valid.

Worked example 1.5 The range and inter-quartile range

Suppose 110 children take a test, with the following results:

Mark, X	Frequency, f	Cumulative frequency, F
13	5	5
14	13	18
15	29	47
16	33	80
17	17	97
18	8	105
19	4	109
20	1	110
Total	110	

The range is simply $20 - 13 = 7$. The inter-quartile range requires calculation of the quartiles. Q_1 is given by the value of the 27.5th observation (= 110/4), which is 15. Q_3 is the value of the 82.5th observation (= 110×0.75) which is 17. The IQR is therefore $17 - 15 = 2$ marks. Half the students achieve marks within this range.

Notice that a slight change in the data (three more students getting 16 rather than 17 marks) would alter the IQR to 1 mark (16–15). The result should be treated with some caution therefore. This is a common problem when there are few distinct values of the variable (eight in this example). It is often worth considering whether a few small changes to the data could alter the calculation considerably. In such a case, the original result might not be very robust.

The variance

A more useful measure of dispersion is the **variance**, which makes use of all of the information available, rather than trimming the extremes of the distribution. The variance is denoted by the symbol σ^2. σ is the Greek lower-case letter sigma, so σ^2 is read 'sigma squared'. It has a completely different meaning from Σ (capital sigma) used before. Its formula is

$$\sigma^2 = \frac{\Sigma(x - \mu)^2}{N} \tag{1.14}$$

In this formula, $x - \mu$ measures the distance from each observation to the mean. Squaring these makes all the deviations positive, whether above or below the mean. We then take the average of all the squared deviations from the mean. A more dispersed distribution (such as A in Figure 1.13) will tend to have larger deviations from the mean, and hence a larger variance. In comparing two distributions with similar means, therefore, we could examine their variances to see which of the two has the greater degree of dispersion. With grouped data the formula becomes

$$\sigma^2 = \frac{\Sigma f(x - \mu)^2}{\Sigma f} \tag{1.15}$$

The calculation of the variance is shown in Table 1.9 and from this we obtain

$$\sigma^2 = \frac{1\,001\,772\,261.83}{17\,636} = 56\,802.69$$

This calculated value is before translating back into the original units of measurement, as was done for the mean by multiplying by 1000. In the case of the variance, however, we must multiply by 1 000 000 which is the *square* of 1000. The variance is therefore 56 802 690 000. Multiplying by the square of 1000 is a consequence of using squared deviations in the variance formula (see Appendix 1B on E and V operators for more details of this).

One needs to be a little careful about the units of measurement therefore. If the mean is reported at 146.984 then it is appropriate to report the variance as 56 802.69. If the mean is reported as 146 984 then the variance should be reported as 56 802 690 000. Note that it is only the presentation that changes: the underlying facts are the same.

The standard deviation

In what units is the variance measured? As we have used a squaring procedure in the calculation, we end up with something like 'squared' £s, which is not very

Table 1.9 The calculation of the variance of wealth

Range	Mid-point x (£000)	Frequency, f	Deviation $(x - \mu)$	$(x - \mu)^2$	$f(x - \mu)^2$
0	5.0	2448	−142.0	20 159.38	49 350 158.77
10 000–	17.5	1823	−129.5	16 766.04	30 564 482.57
25 000–	32.5	1375	−114.5	13 106.52	18 021 469.99
40 000–	45.0	480	−102.0	10 400.68	4 992 326.62
50 000–	55.0	665	−92.0	8461.01	5 626 568.95
60 000–	70.0	1315	−77.0	5926.49	7 793 339.80
80 000–	90.0	1640	−57.0	3247.15	5 325 317.93
100 000–	125.0	2151	−22.0	483.28	1 039 544.38
150 000–	175.0	2215	28.0	784.91	1 738 579.16
200 000–	250.0	1856	103.0	10 612.35	19 696 526.45
300 000–	400.0	1057	253.0	64 017.23	67 666 217.05
500 000–	750.0	439	603.0	363 628.63	159 632 966.88
1 000 000–	1500.0	122	1353.0	1 830 653.04	223 339 670.45
2 000 000–	3000.0	50	2853.0	8 139 701.86	406 985 092.85
Total		17 636			1 001 772 261.83

convenient. Because of this, we define the square root of the variance to be the **standard deviation**, which is therefore back in £s. The standard deviation is therefore given by

$$\sigma = \sqrt{\frac{\Sigma(x - \mu)^2}{N}} \tag{1.16}$$

or, for grouped data

$$\sigma = \sqrt{\frac{\Sigma f(x - \mu)^2}{N}} \tag{1.17}$$

These are simply the square roots of equations (1.14) and (1.15). The standard deviation of wealth is therefore $\sqrt{56\,802.69} = 238.333$. This is in £000, so the standard deviation is actually £238 333 (note that this is the square root of 56 802 690 000, as it should be). On its own the standard deviation (and the variance) is not easy to interpret since it is not something we have an intuitive feel for, unlike the mean. It is more useful when used in a comparative setting. This will be illustrated later on.

The variance and standard deviation of a sample

As with the mean, a different symbol is used to distinguish a variance calculated from the population and one calculated from a sample. In addition, the sample variance is calculated using a slightly different formula from the one for the population variance. The sample variance is denoted by s^2 and its formula is given by equations (1.18) and (1.19) below

$$s^2 = \frac{\Sigma(x - \bar{x})^2}{n - 1} \tag{1.18}$$

and, for grouped data

$$s^2 = \frac{\Sigma f(x - \bar{x})^2}{n - 1} \qquad (1.19)$$

where n is the sample size. The reason $n - 1$ is used in the denominator rather than n (as one might expect) is the following. Our real interest is in the population variance, and the sample variance is an estimate of it. The former is measured by the dispersion around μ, and the sample variance should ideally be measured around μ also. However, μ is unknown, so $\bar{x}$ is used instead. But the variation of the sample observations around $\bar{x}$ tends to be smaller than that around μ. Using $n - 1$ rather than n in the formula compensates for this and the result is an **unbiased**[5] (i.e. correct on average) estimate of the population variance.

Using the correct formula is more important the smaller is the sample size, as the proportionate difference between $n - 1$ and n increases. For example, if $n = 10$, the adjustment amounts to 10% of the variance; when $n = 100$ the adjustment is only 1%.

The sample standard deviation is given by the square root of equation (1.18) or (1.19).

Worked example 1.6 The variance and standard deviation

We continue with the previous worked example, relating to students' marks. The variance and standard deviation can be calculated as:

X	f	fx	$x - \mu$	$(x - \mu)^2$	$f(x - \mu)^2$
13	5	65	−2.81	7.89	39.45
14	13	182	−1.81	3.27	42.55
15	29	435	−0.81	0.65	18.98
16	33	528	0.19	0.04	1.20
17	17	289	1.19	1.42	24.11
18	8	144	2.19	4.80	38.40
19	4	76	3.19	10.18	40.73
20	1	20	4.19	17.56	17.56
Totals	110	1739			222.99

The mean is calculated as 1739/110 = 15.81 and from this the deviations column $(x - \mu)$ is calculated (so −2.81 = 13 − 15.81, etc.).

The variance is calculated as $\Sigma f(x - \mu)^2/(n - 1) = 222.99/109 = 2.05$. The standard deviation is therefore 1.43, the square root of 2.05. (Calculations are shown to two decimal places but have been calculated using exact values.)

For distributions which are approximately symmetric and bell-shaped (i.e. the observations are clustered around the mean) there is an approximate relationship between the standard deviation and the inter-quartile range. This rule of thumb is that the IQR is 1.3 times the standard deviation. In this case, $1.3 \times 1.43 = 1.86$, close to the value calculated earlier, 2.

[5] The concept of *bias* is treated in more detail in Chapter 4.

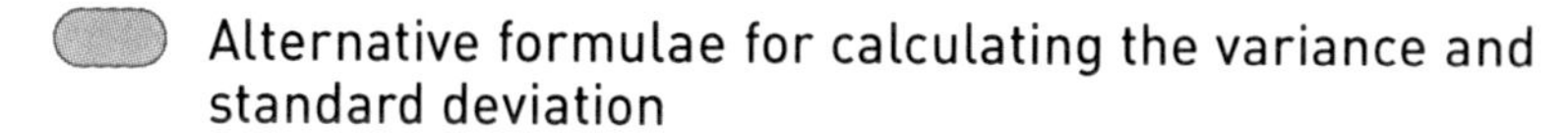

Alternative formulae for calculating the variance and standard deviation

The following formulae give the same answers as equations (1.14) to (1.17) but are simpler to calculate, either by hand or using a spreadsheet. For the population variance one can use

$$\sigma^2 = \frac{\Sigma x^2}{N} - \mu^2 \tag{1.20}$$

or, for grouped data

$$\sigma^2 = \frac{\Sigma fx^2}{\Sigma f} - \mu^2 \tag{1.21}$$

The calculation of the variance using equation (1.21) is shown in Figure 1.14.

Microsoft Excel - wealth data 2003.xls

H7 =E20/C20-H6^2

	A	B	C	D	E	F	G	H
1			WEALTH DATA 2003					
2								
3	Wealth	Mid-point	Frequency					
4	Range	x	f	fx	fx squared		Summary statistics	
5	0	5.0	2 448	12 240.0	61 200.00			
6	10 000	17.5	1 823	31 902.5	558 293.75		Mean	146 984
7	25 000	32.5	1 375	44 687.5	1452 343.75		Variance	56 802.691
8	40 000	45.0	480	21 600.0	972 000.00		Std devn	238.333
9	50 000	55.0	665	36 575.0	2011 625.00		Coef varn	1.621
10	60 000	70.0	1 315	92 050.0	6443 500.00			
11	80 000	90.0	1 640	147 600.0	13284 000.00			
12	100 000	125.0	2 151	268 875.0	33609 375.00			
13	150 000	175.0	2 215	387 625.0	67834 375.00			
14	200 000	250.0	1 856	464 000.0	116000 000.00			
15	300 000	400.0	1 057	422 800.0	169120 000.00			
16	500 000	750.0	439	329 250.0	246937 500.00			
17	1000 000	1 500.0	122	183 000.0	274500 000.00			
18	2000 000	3 000.0	50	150 000.0	450 000 000.00			
19								
20	Totals		17 636	2592 205.0	1382 784 212.50			
21								
22								
23								

Figure 1.14 Descriptive statistics calculated using *Excel*

The sample variance can be calculated using

$$s^2 = \frac{\Sigma x^2 - n\bar{x}^2}{n - 1} \tag{1.22}$$

or, for grouped data

$$s^2 = \frac{\Sigma fx^2 - n\bar{x}^2}{n - 1} \tag{1.23}$$

The standard deviation may of course be obtained as the square root of these formulae.

Using a calculator or computer for calculation

Electronic calculators and (particularly) computers have simplified the calculation of the mean, etc. Figure 1.14 shows how to set out the above calculations in a spreadsheet (*Microsoft Excel* in this case) including some of the appropriate cell formulae.

The variance in this case is calculated using the formula $\sigma^2 = \frac{\sum fx^2}{\sum f} - \mu^2$, which is the formula given in equation (1.21) above. Note that it gives the same result as that calculated in the text.

The following formulae are contained in the cells:

D5: = C5*B5	to calculate f times x
E5: = D5*B5	to calculate f times x^2
C20: = SUM(C5:C18)	to sum the frequencies
H6: = D20/C20	calculates $\sum fx/\sum f$
H7: = E20/C20 – H6^2	calculates $\sum fx^2/\sum f - \mu^2$
H8: = SQRT(H7)	calculates σ
H9: = H8/H6	calculates σ/μ

The coefficient of variation

The measures of dispersion examined so far are all measures of **absolute dispersion** and, in particular, their values depend upon the units in which the variable is measured. It is therefore difficult to compare the degrees of dispersion of two variables which are measured in different units. For example, one could not compare wealth in the UK with that in Germany if the former uses £s and the latter euros for measurement. Nor could one compare the wealth distribution in one country between two points in time because inflation alters the value of the currency over time. The solution is to use a measure of **relative dispersion**, which is independent of the units of measurement. One such measure is the **coefficient of variation**, defined as

$$\text{Coefficient of variation} = \frac{\sigma}{\mu} \tag{1.24}$$

i.e. the standard deviation divided by the mean. Whenever the units of measurement are changed, the effect upon the mean and the standard deviation is the same, hence the coefficient of variation is unchanged. For the wealth distribution its value is 238.333/146.984 = 1.621, i.e. the standard deviation is 162% of the mean. This may be compared directly with the coefficient of variation of a different wealth distribution to see which exhibits a greater relative degree of dispersion.

Independence of units of measurement

It is worth devoting a little attention to this idea that some summary measures are independent of the units of measurement and some are not, as it occurs quite often in statistics and is not often appreciated at first. A statistic that is independent of the units of measurement is one which is unchanged even when the units of measurement are changed. It is therefore more useful in general than a statistic which is not independent, since one can use it to make comparisons, or judgements, without worrying about how it was measured.

The mean is not independent of the units of measurement. If we are told the average income in the UK is 20 000, for example, we need to know whether it is measured in pounds sterling, euros or even dollars. The underlying level of income is the same, of course, but it is measured differently. By contrast, the rate of growth (described in detail shortly) is independent of the units of measurement. If we are told it is 3% per annum, it would be the same whether it were calculated in pounds, euros or dollars. If told that the rate of growth in the US is 2% per annum, we can immediately conclude that the UK is growing faster, no further information is needed.

Most measures we have encountered so far, such as the mean and variance, do depend on units of measurement. The coefficient of variation is one that does not. We now go on to describe another means of measuring dispersion that avoids the units of measurement problem.

The standard deviation of the logarithm

Another solution to the problem of different units of measurement is to use the logarithm[6] of wealth rather than the actual value. The reason why this works can best be illustrated by an example. Suppose that between 1997 and 2003 each individual's wealth doubled, so that $X_i^{2003} = 2X_i^{1997}$, where X_i^t indicates the wealth of individual *i* in year *t*. It follows that the standard deviation of wealth in 2003, X^{2003}, is therefore exactly twice that of 1997, X^{1997}. Taking logs, we have $\ln X_i^{2003} = \ln 2 + \ln X_i^{1997}$, so it follows that the distribution of $\ln X^{2003}$ is the same as that of $\ln X^{1997}$, except that it is shifted to the right by ln 2 units. The variances (and hence standard deviations) of the two logarithmic distributions must therefore be the same, indicating no change in the *relative* dispersion of the two wealth distributions.

The standard deviation of the logarithm of wealth is calculated from the data in Table 1.10. The variance turns out to be

Table 1.10 The calculation of the standard deviation of the logarithm of wealth

Range	Mid-point *x* (£000)	ln (*x*)	Frequency, *f*	*fx*	*fx*²
0–	5.0	1.609	2448	3939.9	6341.0
10 000–	17.5	2.862	1823	5217.8	14 934.4
25 000–	32.5	3.481	1375	4786.7	16 663.7
40 000–	45.0	3.807	480	1827.2	6955.5
50 000–	55.0	4.007	665	2664.9	10 679.0
60 000–	70.0	4.248	1315	5586.8	23 735.4
80 000–	90.0	4.500	1640	7379.7	33 207.2
100 000–	125.0	4.828	2151	10 385.7	50 145.4
150 000–	175.0	5.165	2215	11 440.0	59 085.2
200 000–	250.0	5.521	1856	10 247.8	56 583.0
300 000–	400.0	5.991	1057	6333.0	37 943.8
500 000–	750.0	6.620	439	2906.2	19 239.3
1 000 000–	1500.0	7.313	122	892.2	6524.9
2 000 000–	3000.0	8.006	50	400.3	3205.1
Totals			17 636	74 008.2	345 243.0

Note: Use the 'ln' key on your calculator or the = LN() function in a spreadsheet to obtain natural logarithms of the data. You should obtain ln 5 = 1.609, ln 17.5 = 2.862, etc.

[6] See Appendix 1C if you are unfamiliar with logarithms. Note that we use the natural logarithm here, but the effect would be the same using logs to base 10.

$$\sigma^2 = \frac{345\ 243.0}{17\ 636} - \left(\frac{74\ 008.2}{17\ 636}\right)^2 = 1.966$$

and the standard deviation $\sigma = 1.402$.

For comparison, the standard deviation of log income in 1979 (discussed in more detail later on) is 1.31, so there appears to have been a slight increase in relative dispersion over this time period.

Measuring deviations from the mean: z-scores

Imagine the following problem. A man and a woman are arguing over their career records. The man says he earns more than she does, so is more successful. The woman replies that women are discriminated against and that, relative to women, she is doing better than the man is, relative to other men. Can the argument be resolved?

Suppose the data are as follows: the average male salary is £19 500, the average female salary £16 800. The standard deviation of male salaries is £4750, for women it is £3800. The man's salary is £31 375 while the woman's is £26 800. The man is therefore £11 875 above the mean, the woman £10 000. However, women's salaries are less dispersed than men's, so the woman has done well to reach £26 800.

One way to resolve the problem is to calculate the **z-score**, which gives the salary in terms of the *number of standard deviations from the mean*. Thus for the man, the z-score is

$$z = \frac{X - \mu}{\sigma} = \frac{31\ 375 - 19\ 500}{4750} = 2.50 \tag{1.25}$$

Thus the man is 2.5 standard deviations above the male mean salary. For the woman the calculation is

$$z = \frac{26\ 800 - 16\ 800}{3800} = 2.632 \tag{1.26}$$

The woman is 2.632 standard deviations above her mean and therefore wins the argument – she is nearer the top of her distribution than is the man and so is more of an outlier. Actually, this probably will not end the argument, but is the best the statistician can do! The z-score is an important concept which will be used again later in the book when we cover hypothesis testing (Chapter 5).

Chebyshev's inequality

Use of the z-score leads on naturally to **Chebyshev's inequality**, which tells us about the proportion of observations that fall into the tails of any distribution, regardless of its shape. The theorem is expressed as follows

At least $(1 - 1/k^2)$ of the observations in any distribution
lie within k standard deviations of the mean (1.27)

If we take the female wage distribution given above, we can ask what proportion of women lie beyond 2.632 standard deviations from the mean (in both tails of the distribution). Setting $k = 2.632$, then $(1 - 1/k^2) = (1 - 1/2.632^2) = 0.8556$.

So at least 85% of women have salaries within ±2.632 standard deviations of the mean, i.e. between £6 800 (= 16 800 − 2.632 × 3800) and £26 800 (= 16 800 + 2.632 × 3800). 15% of women therefore lie outside this range.

Chebyshev's inequality is a very conservative rule since it applies to *any* distribution; if we know more about the shape of a particular distribution (for example, men's heights follow a Normal distribution – see Chapter 3) then we can make a more precise statement. In the case of the Normal distribution, over 99% of men are within 2.632 standard deviations of the average height, because there is a concentration of observations near the centre of the distribution.

We can also use Chebyshev's inequality to investigate the inter-quartile range. The formula (1.27) implies that 50% of observations lie within $\sqrt{2} = 1.41$ standard deviations of the mean, a more conservative value than our previous 1.3.

Exercise 1.4

(a) For the data in Exercise 2, calculate the inter-quartile range, the variance and the standard deviation.

(b) Calculate the coefficient of variation.

(c) Check if the relationship between the IQR and the standard deviation stated in the text is approximately true for this distribution.

(d) Approximately how much of the distribution lies within one standard deviation either side of the mean? How does this compare with the prediction from Chebyshev's inequality?

Measuring skewness

The **skewness** of a distribution is the third characteristic that was mentioned earlier, in addition to location and dispersion. The wealth distribution is heavily skewed to the right, or **positively** skewed; it has its long tail in the right-hand end of the distribution. A measure of skewness gives a numerical indication of how asymmetric is the distribution.

One measure of skewness, known as the **coefficient of skewness**, is

$$\frac{\Sigma f(x - \mu)^3}{N\sigma^3} \tag{1.28}$$

and it is based upon *cubed* deviations from the mean. The result of applying formula (1.28) is positive for a right-skewed distribution (such as wealth), zero for a symmetric one, and negative for a left-skewed one. Table 1.11 shows the calculation for the wealth data (some rows are omitted for brevity). From this we obtain

$$\frac{\Sigma f(x - \mu)^3}{N} = \frac{1\,563\,796\,357\,499}{17\,636} = 88\,670\,693.89$$

and dividing by σ^3 gives $\frac{88\,670\,693.89}{13\,537\,964} = 6.550$, which is positive, as expected.

The measure of skewness is much less useful in practical work than measures of location and dispersion, and even knowing the value of the coefficient does not always give much idea of the shape of the distribution: two quite different distributions can share the same coefficient. In descriptive work it is probably better to draw the histogram itself.

Table 1.11 Calculation of the skewness of the wealth data

Range	Mid-point x (£000)	Frequency f	Deviation $x-\mu$	$(x-\mu)^3$	$f(x-\mu)^3$
0	5.0	2448	−142.0	−2 862 304	−7 006 919 444
10 000	17.5	1823	−129.5	−2 170 929	−3 957 603 101
⋮		⋮		⋮	⋮
1 000 000	1500.0	122	1353.0	2 476 903 349	302 182 208 638
2 000 000	3000.0	50	2853.0	23 222 701 860	1 161 135 092 991
Totals		17 636	4457.2	25 927 167 232	1 563 796 357 499

Comparison of the 2003 and 1979 distributions of wealth

Some useful lessons may be learned by comparing the 2003 distribution with its counterpart from 1979. This covers the period of Conservative government starting with Mrs Thatcher in 1979 up until the first six years of Labour administration. This shows how useful the various summary statistics are when it comes to comparing two different distributions. The wealth data for 1979 are given in Problem 1.5 below, where you are asked to confirm the following calculations.

Average wealth in 1979 was £16 399, about one-ninth of its 2003 value. The average increased substantially therefore (at about 10% per annum, on average), but some of this was due to inflation rather than a real increase in the quantity of assets held. In fact, between 1979 and 2003 the retail price index rose from 52.0 to 181.3, i.e. it increased approximately three and a half times. Thus the nominal[7] increase (i.e. in cash terms, before any adjustment for rising prices) in wealth is made up of two parts: (i) an inflationary part which more than tripled measured wealth and (ii) a real part, consisting of a 2.5 fold increase (thus $3.5 \times 2.5 = 9$, approximately). Price indexes are covered in Chapter 10 where it is shown more formally how to divide a nominal increase into price and real (quantity) components. It is likely that the extent of the real increase in wealth is overstated here due to the use of the retail price index rather than an index of asset prices. A substantial part of the increase in asset values over the period is probably due to the very rapid rise in house prices (houses form a significant part of the wealth of many households).

The standard deviation is similarly affected by inflation. The 1979 value is 25 552 compared to 2003's 238 333, which is about nine times larger. The spread of the distribution appears to have increased therefore (even if we take account of the general price effect). Looking at the coefficient of variation, however, shows that it has increased from 1.56 to 1.62 which is a modest difference. The spread of the distribution *relative to its mean* has not changed by much. This is confirmed by calculating the standard deviation of the logarithm: for 1979 this gives a figure of 1.31, slightly smaller than the 2003 figure (of 1.40).

[7] This is a different meaning of the term 'nominal' from that used earlier to denote data measured on a nominal scale, i.e. data grouped into categories without an obvious ordering. Unfortunately, both meanings of the word are in common (statistical) usage, although it should be obvious from the context which use is meant.

The measure of skewness for the 1979 data comes out as 5.723, smaller that the 2003 figure (of 6.550). This suggests that the 1979 distribution is less skewed than is the 1994 one. Again, these two figures can be directly compared because they do not depend upon the units in which wealth is measured. However, the relatively small difference is difficult to interpret in terms of how the shape of the distribution has changed.

The box and whiskers diagram

Having calculated these various summary statistics we can now return to a useful graphical method of presentation. This is the **box and whiskers diagram** (sometimes called a **box plot**) which shows the median, quartiles and other aspects of a distribution on a single diagram. Figure 1.15 shows the box plot for the wealth data.

Wealth is measured on the vertical axis. The rectangular box stretches (vertically) from the first to third quartile and therefore encompasses the middle half of the distribution. The horizontal line through it is at the median and lies less than halfway up the box. This tells us that there is a degree of skewness even within the central half of the distribution, although it does not appear very severe. The two 'whiskers' extend above and below the box as far as the highest and lowest observations, *excluding outliers*. An outlier is defined to be any observation which is more than 1.5 times the inter-quartile range (which is the same as the height of the box) above or below the box. Earlier we found the IQR to be 153 517 and the upper quartile to be 180 022, so an (upper) outlier lies beyond

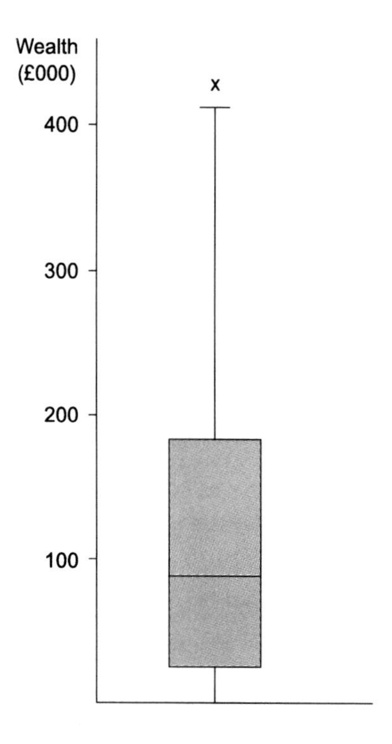

Figure 1.15
Box plot of the wealth distribution

180 022 + 1.5 × 153 517 = 410 298. There are no outliers below the box as wealth cannot fall below zero. The top whisker is thus substantially longer than the bottom one, and indicates the extent of dispersion towards the tails of the distribution. The crosses indicate the outliers and in reality extend far beyond those shown in the diagram.

A simple diagram thus reveals a lot of information about the distribution. Other boxes and whiskers could be placed alongside in the same diagram (perhaps representing other countries) making comparisons straightforward. Some statistical software packages, such as *SPSS* and *STATA*, can generate box plots from the original data, without the need for the user to calculate the median, etc. However, spreadsheet packages do not yet have this useful facility.

Time-series data: investment expenditures 1973–2005

The data on the wealth distribution give a snapshot of the situation at particular points in time, and comparisons can be made between the 1979 and 2003 snapshots. Often, however, we wish to focus on the time-path of a variable and therefore we use **time-series data**. The techniques of presentation and summarising are slightly different than for cross-section data. As an example, we use data on investment in the UK for the period 1973–2005. These data were taken from Statbase (http://www.statistics.gov.uk/statbase/) although you can find the data in *Economic Trends Annual Supplement*. Investment expenditure is important to the economy because it is one of the primary determinants of growth. Until recent years, the UK economy's growth record had been poor by international standards and lack of investment may have been a cause. The variable studied here is total gross (i.e. before depreciation is deducted) domestic fixed capital formation, measured in £m. The data are shown in Table 1.12.

It should be remembered that the data are in current prices so that the figures reflect price increases as well as changes in the volume of physical investment. The series in Table 1.12 thus shows the actual amount of cash that was

Table 1.12 UK investment, 1973–2005

Year	Investment	Year	Investment	Year	Investment
1973	15 227	1984	58 589	1995	118 031
1974	18 134	1985	64 400	1996	126 593
1975	21 856	1986	68 546	1997	133 620
1976	25 516	1987	78 996	1998	151 083
1977	28 201	1988	96 243	1999	156 344
1978	32 208	1989	111 324	2000	161 468
1979	38 211	1990	114 300	2001	165 472
1980	43 238	1991	105 179	2002	173 525
1981	43 331	1992	101 111	2003	178 751
1982	47 394	1993	101 153	2004	194 491
1983	51 490	1994	108 534	2005	205 843

Note: Time-series data consist of observations on one or more variables over several time periods. The observations can be daily, weekly, monthly, quarterly or, as here, annually.

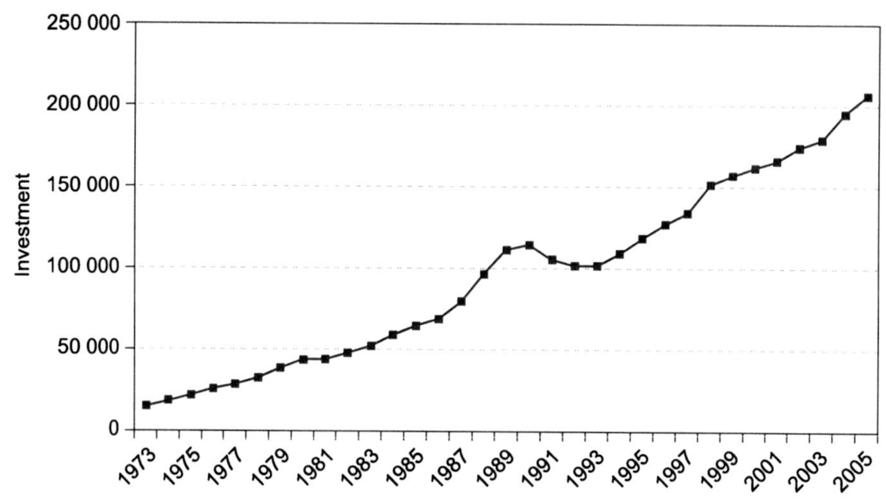

Figure 1.16 Time-series graph of investment in the UK, 1973–2005

Note: The *X*, *Y* coordinates are the values {year, investment}; the first data point has the coordinates {1973, 15 227}, for example.

spent each year on investment. The techniques used below for summarising the investment data could equally well be applied to a series showing the volume of investment.

First of all we can use graphical techniques to gain an insight into the characteristics of investment. Figure 1.16 shows a **time-series graph** of investment. The graph plots the time periods on the horizontal axis and the investment variable on the vertical.

Plotting the data in this way brings out clearly some key features of the series:

- The **trend** in investment is upwards, with only a few years in which there was either no increase or a decrease.
- There is a 'hump' in the data in the late 1980s/early 1990s, before the series returns to its trend. Something unusual must have happened around that time. If we want to know what factors determine investment (or the effect of investment upon other economic magnitudes) we should get some useful insights from this period of the data.
- The trend is slightly **non-linear** – it follows an increasingly steep curve over time. This is essentially because investment grows by a *percentage* or *proportionate* amount each year. As we shall see shortly, it grows by about 8.5% each year. Therefore, as the level of investment increases each year, so does the increase in the level, giving a non-linear graph.
- Successive values of the investment variable are similar in magnitude, i.e. the value in year t is similar to that in $t - 1$. Investment does not change from £40bn in one year to £10bn the next, then back to £50bn, for instance. In fact, the value in one year appears to be based on the value in the previous year, plus (in general) 8.5% or so. We refer to this phenomenon as **serial correlation** and it is one of the aspects of the data that we might wish to investigate. The *ordering* of the data matters, unlike the case with cross-section data where the ordering is usually irrelevant. In deciding how to model investment behaviour, we might focus on *changes* in investment from year to year.

Table 1.13 The change in investment

Year	Δ Investment	Year	Δ Investment	Year	Δ Investment
1973	2880	1984	7099	1995	9497
1974	2907	1985	5811	1996	8562
1975	3722	1986	4146	1997	7027
1976	3660	1987	10 450	1998	17 463
1977	2685	1988	17 247	1999	5261
1978	4007	1989	15 081	2000	5124
1979	6003	1990	2976	2001	4004
1980	5027	1991	−9121	2002	8053
1981	93	1992	−4068	2003	5226
1982	4063	1993	42	2004	15 740
1983	4096	1994	7381	2005	11 352

Note: The change in investment is obtained by taking the difference between successive observations. For example, 2907 is the difference between 18 134 and 15 227.

- The series seems 'smoother' in the earlier years (up to perhaps 1986) and exhibits greater volatility later on. In other words, there are greater fluctuations *around* the trend in the later years. We could express this more formally by saying that the variance of investment around its trend appears to change (increase) over time. This is known as **heteroscedasticity**; a constant variance is termed **homoscedasticity**.

We may gain further insight into how investment evolves over time by focusing on the *change* in investment from year to year. If we denote investment in year t by I_t then the change in investment, ΔI_t, is given by $I_t - I_{t-1}$. Table 1.13 shows the changes in investment each year and Figure 1.17 provides a time-series graph.

The series is made up of mainly positive values, indicating that investment increases over time. It also shows that the increase grows each year, with perhaps some greater volatility (of the increase) towards the end of the period. The graph also shows dramatically the change that occurred around 1990.

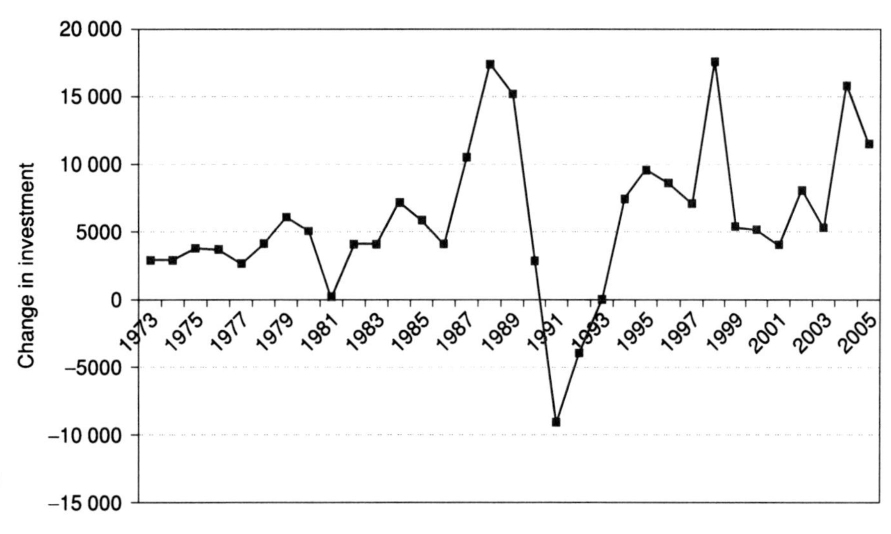

Figure 1.17
Time-series graph of the change in investment

Outliers

Graphing data also allows you to see **outliers** (unusual observations). Outliers might be due to an error in inputting the data (e.g. typing 97 instead of 970) or because something unusual happened (e.g. the investment figure for 1991). Either of these should be apparent from an appropriate graph. For example, the graph of the change in investment highlights the 1991 figure. In the case of a straightforward error you should obviously correct it. If you are satisfied that the outlier is not simply a typo, you might want to think about the possible reasons for its existence and whether it distorts the descriptive picture you are trying to paint.

Another useful way of examining the data is to look at the **logarithm** of investment. This transformation has the effect of straightening out the non-linear investment series. Table 1.14 shows the transformed values and Figure 1.18 graphs the series. In this case we use the natural (base e) logarithm.

Table 1.14 The logarithm of investment and the change in the logarithm

Year	ln Investment	Δ ln Investment	Year	ln Investment	Δ ln Investment	Year	ln Investment	Δ ln Investment
1973	9.631	0.210	1984	10.978	0.129	1995	11.679	0.084
1974	9.806	0.175	1985	11.073	0.095	1996	11.749	0.070
1975	9.992	0.187	1986	11.135	0.062	1997	11.803	0.054
1976	10.147	0.155	1987	11.277	0.142	1998	11.926	0.123
1977	10.247	0.100	1988	11.475	0.197	1999	11.960	0.034
1978	10.380	0.133	1989	11.620	0.146	2000	11.992	0.032
1979	10.551	0.171	1990	11.647	0.026	2001	12.017	0.024
1980	10.674	0.124	1991	11.563	−0.083	2002	12.064	0.048
1981	10.677	0.002	1992	11.524	−0.039	2003	12.094	0.030
1982	10.766	0.090	1993	11.524	0.000	2004	12.178	0.084
1983	10.849	0.083	1994	11.595	0.070	2005	12.235	0.057

Note: For 1973, 9.631 is the natural logarithm of 15 227 , i.e. ln 15 227 = 9.631.

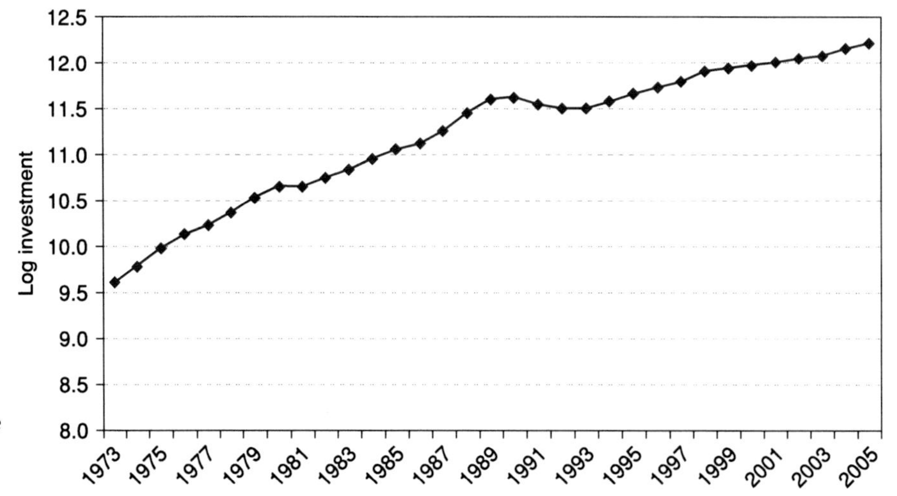

Figure 1.18 Time-series graph of the logarithm of investment expenditures

This new series is much smoother than the original one (as is usually the case when taking logs) and is helpful in showing the long-run trend, though it tends to mask some of the volatility of investment. The slope of the graph gives a close approximation to the average rate of growth of investment over the period (expressed as a decimal). This is calculated as follows

$$slope = \frac{change\ in\ (\ln)\ investment}{number\ of\ years} = \frac{12.235 - 9.631}{32} = 0.081 \tag{1.29}$$

i.e. 8.1% per annum. Note that although there are 33 observations, there are only 32 years of growth. A word of warning: you must use natural (base *e*) logarithms, not logarithms to the base 10, for this calculation to work. Remember also that the growth of the *volume* of investment will be less than 8.1% per annum, because part of it is due to price increases.

The logarithmic presentation is useful when comparing two different data series: when graphed in logs it is easy to see which is growing faster – just see which series has the steeper slope.

A corollary of equation (1.29) is that change in the natural logarithm of investment from one year to the next represents the *percentage* change in the data over that year. For example, the natural logarithm of investment in 1973 is 9.631, while in 1974 it is 9.806. The difference is 0.175, so the rate of growth is 17.5%. Remember that this is an approximation and the result of a quick and easy calculation. It is reasonably accurate up to a figure of about 20%.

Finally we can graph the difference of the logarithm, as we graphed the difference of the level. This is shown in Figure 1.19 (the calculations are in Table 1.14).

This is quite revealing. It shows the series fluctuating about the value of approximately 0.08 (the average calculated in equation (1.29) above), with a slight downwards trend. Furthermore, the series does not seem to show increasing volatility over time, as the others did. The graph therefore demonstrates that in *proportionate* terms there is no increasing volatility; the variance of the series around 0.08 does not change much over time (although 1991 still seems to be an 'unusual' observation).

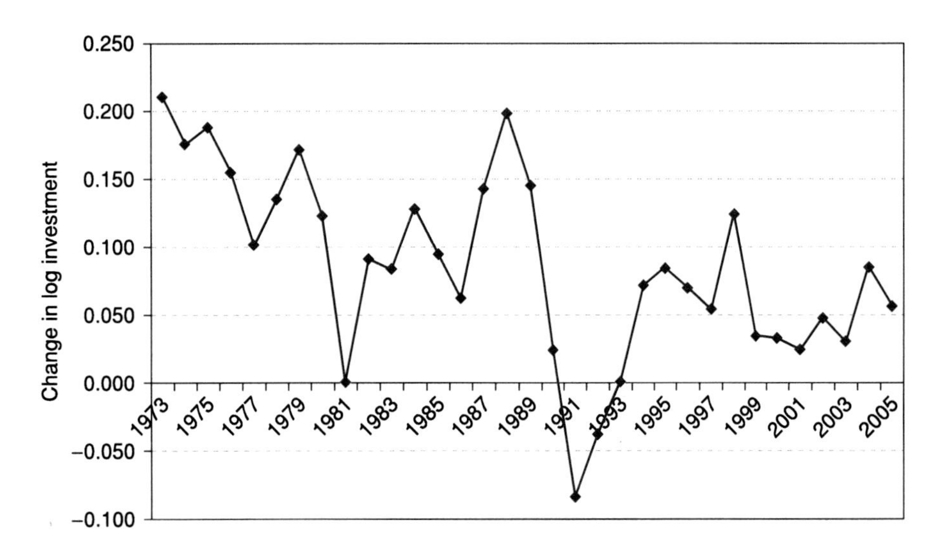

Figure 1.19
Time-series graph of the difference of the logarithmic series

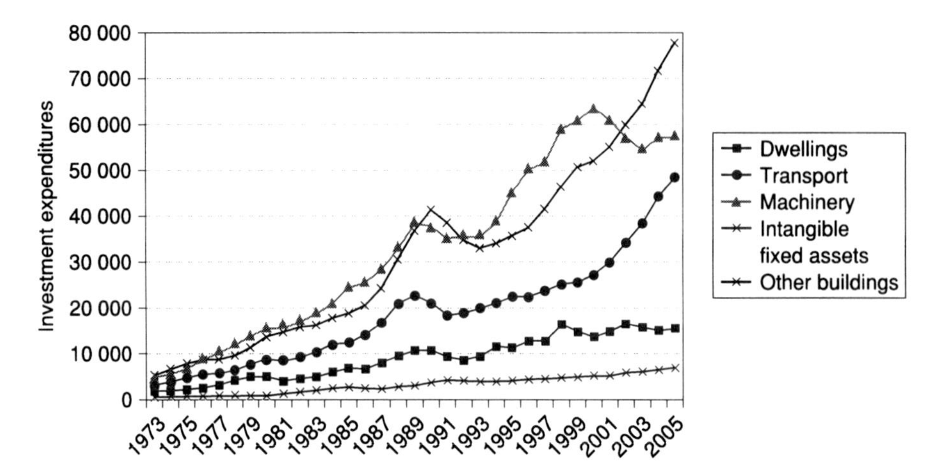

Figure 1.20
A multiple time-series graph of investment

Graphing multiple series

Investment is made up of different categories: the table in Problem 1.14 presents investment data under four different headings: dwellings; transport; machinery; intangible fixed assets; and other buildings. Together they make up total investment. It is often useful to show all of the series together on one graph. Figure 1.20 shows a **multiple time-series graph** of the investment data.

Construction of this type of graph is straightforward; it is just an extension of the technique for presenting a single series. The chart shows that all investment categories have increased over time in a fairly similar way, including the hump then fall around 1990. It is noticeable, however, that investment in machinery fell significantly around 2000 while other categories, particularly dwellings, continued to increase. It is difficult from the graph to tell which categories have increased most rapidly over time: the 1973 values are relatively small and hard to distinguish. In fact, it is the 'intangible fixed assets' category (the smallest one) that has increased fastest in proportionate terms. This is easier to observe with a few numerical calculations (covered later in this chapter) rather than trying to read a cramped graph.

One could also produce a multiple series graph of the logarithms of the variables and also of the change, as was done for the total investment series. Since the log transformation tends to squeeze the values (on the y-axis) closer together (compare Figures 1.16 and 1.18) it might be easier to see the relative rates of growth of the series using this method. This is left as an exercise for the reader.

Another complication arises when the series are of different orders of magnitude and it is difficult to make all the series visible on the chart. In this case you can chart some of the series against a second vertical scale, on the right-hand axis. An example is shown in Figure 1.21, plotting the (total) investment data with the interest rate, which has much smaller numerical values. If the same axis were used for both series, the interest rate would appear as a horizontal line coinciding with the x-axis. This would reveal no useful information to the viewer.

It would usually be inappropriate to use this technique on data such as the investment categories graphed in Figure 1.20. Those are directly comparable to each other and to magnify one of the series by plotting it on a separate axis risks

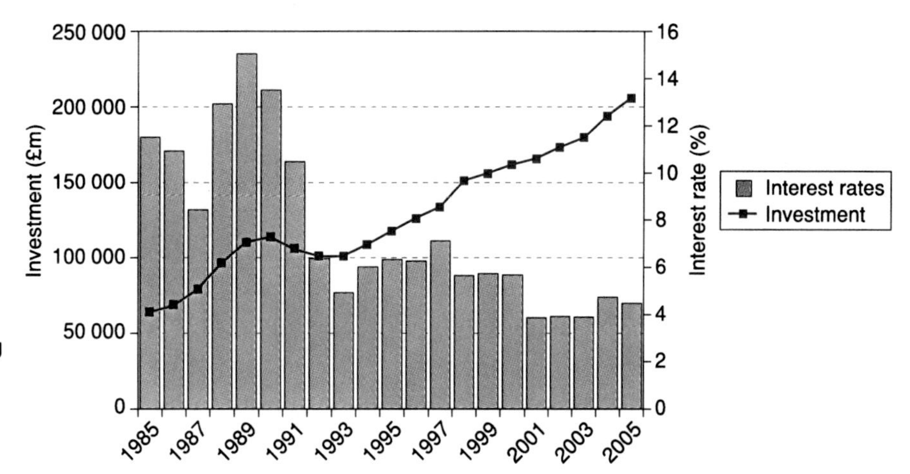

Figure 1.21
Time-series graph using two vertical scales: investment (LH scale) and the interest rate (RH scale), 1985–2005

distorting the message for the reader. However, investment and interest rates are measured in inherently different ways and one cannot directly compare their sizes, hence it is acceptable to use separate axes. The graph allows one to observe the *movements* of the series together and hence perhaps infer something about the relationship between them. The rising investment and falling interest rate possibly suggest an inverse relationship between them.

Overlapping the ranges of the data series

The graph below, taken from the *Treasury Briefing*, February 1994, provides a nice example of how to plot multiple time-series and compare them. The aim is to compare the recessions and recoveries of 1974–78, 1979–83 and 1990–93. Instead of plotting time on the horizontal axis, the number of quarters since the start of each recession is used, so that the series overlap. This makes it easy to see the depth of the last recession and the long time before recovery commenced. By contrast, the 1974–78 recession ended quite quickly and recovery was quite rapid.

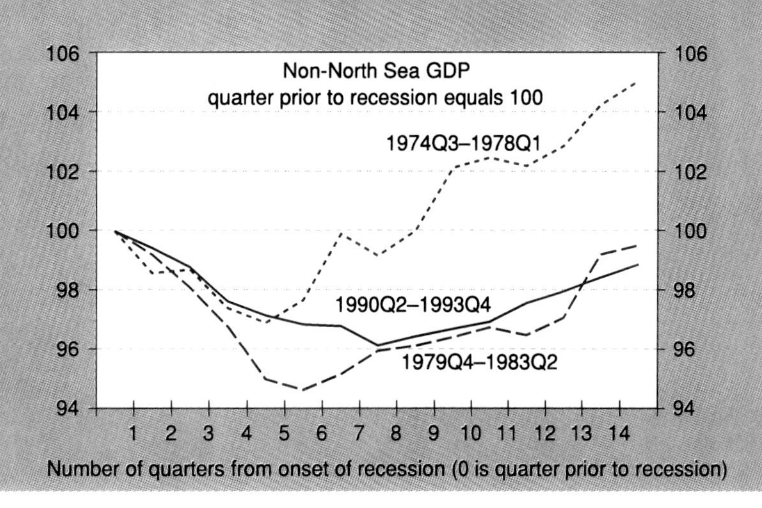

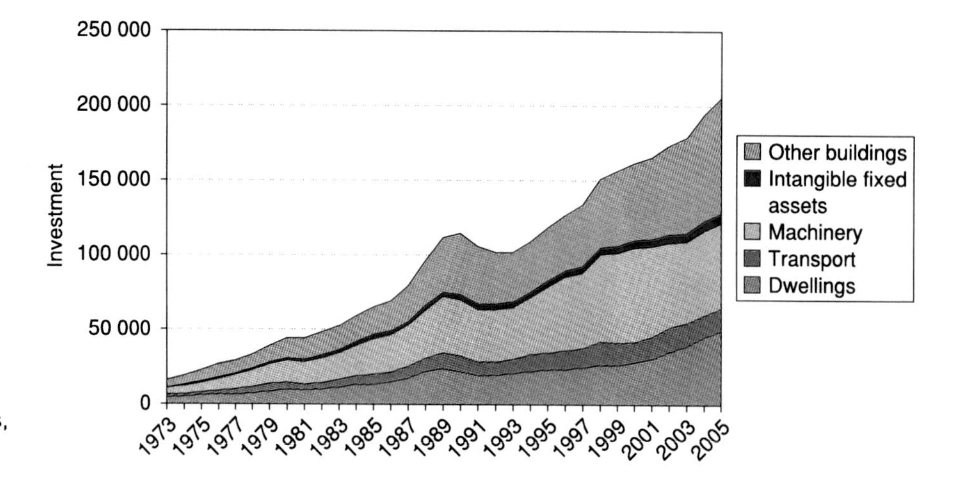

Figure 1.22
Area graph of investment categories, 1973–2005

The investment categories may also be illustrated by means of an **area graph**, which plots the four series stacked one on top of the other, as illustrated in Figure 1.22.

This shows, for example, the 'dwellings' and 'machinery' categories each take up about one quarter of total investment. This is easier to see from the area graph than from the multiple series graph in Figure 1.20.

'Chart junk'

With modern computer software it is easy to get carried away and produce a chart that actually hides more than it reveals. There is a great temptation to add some 3D effects, liven it up with a bit of colour, rotate and tilt the viewpoint, etc. This sort of stuff is generally known as 'chart junk'. As an example, look at Figure 1.23 which is an alternative to the area graph in Figure 1.22 above. It was fun to create, but it does not get the message across at all! Taste is of course personal, but moderation is usually an essential part of it.

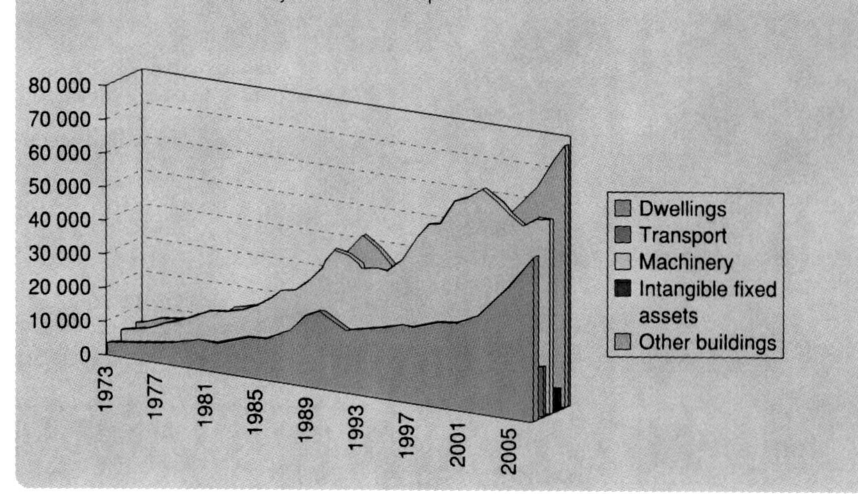

Figure 1.23
Over-the-top graph of investment

Exercise 1.5

Given the following data:

	1990	1991	1992	1993	1994	1995	1996	1997	1998	1999
Profit	50	60	25	−10	10	45	60	50	20	40
Sales	300	290	280	255	260	285	300	310	300	330

(a) Draw a multiple time series graph of the two variables. Label both axes appropriately and provide a title for the graph.

(b) Adjust the graph by using the right-hand axis to measure profits, the left-hand axis sales. What difference does this make?

Numerical summary statistics

The graphs have revealed quite a lot about the data already, but we can also calculate numerical descriptive statistics as we did for the cross-section data. First we consider the mean, then the variance and standard deviation.

The mean of a time series

We could calculate the mean of investment itself, but would this be helpful? Because the series is trended, it passes through the mean at some point between 1973 and 2005, but never returns to it. The mean of the series is actually £95.103bn, which is not very informative since it tells nothing about its value today, for instance. The problem is that the variable is trended, so that the mean is not typical of the series. The annual increase in investment is also trended, so is subject to the same criticism (see Figure 1.17).

It is better in this case to calculate the **average growth rate**, as this is more likely to be representative of the whole time period. It seems more reasonable to say that a series is growing at (for example) 8% per annum than that it is growing at 5000 per annum. The average growth rate was calculated in equation (1.29) as 8.1% per annum, by measuring the slope of the graph of the log investment series. That was stated to be an approximate answer. We can obtain an accurate value in the following way:

(1) Calculate the overall **growth factor** of the series, i.e. x_T/x_1 where x_T is the final observation and x_1 is the initial observation. This is $\frac{x_T}{x_1} = \frac{205\,843}{15\,227}$ = 13.518, i.e. investment expenditure is 13.5 times larger in 2005 than in 1973.

(2) Take the $T - 1$ root of the growth factor. Since $T = 33$ we calculate $\sqrt[32]{13.518}$ = 1.085. (This can be performed on a scientific calculator by raising 13.518 to the power 1/32, i.e. $13.518^{(1/32)} = 1.085$.)

(3) Subtract 1 from the result in the previous step, giving the growth rate as a decimal. In this case we have $1.085 - 1 = 0.085$.

Thus the average growth rate of investment is 8.5% per annum, rather than the 8.1% calculated earlier.

The power of compound growth

The *Economist* magazine provided some amusing and interesting examples of how a $1 investment can grow over time. They assumed that an investor (they named her Felicity Foresight, for reasons that become obvious) started with $1 in 1900 and had the foresight or luck to invest, each year, in the best performing asset of the year. Sometimes she invested in equities, some years in gold and so on. By the end of the century she had amassed $9.6 quintillion ($\$9.6 \times 10^{18}$, more than world gross domestic product (GDP), so highly unrealistic). This is equivalent to an average annual growth rate of 55%. In contrast, Henry Hindsight did the same, but invested in the *previous year's* best asset. This might be thought more realistic. Unfortunately, his $1 turned into only $783, a still respectable annual growth rate of 6.9%. This, however, is beaten by the strategy of investing in the previous year's *worst* performing asset (what goes down must come up . . .). This turned $1 into $1730, a return of 7.7%. Food for thought!

Source: The Economist, 12 February 2000, p. 111.

Note that we could also obtain the accurate answer from our earlier calculation as follows:

- the slope of the graph is 0.0814 (from equation (1.29) above, but to four decimal places for accuracy);
- calculate the anti-log (e^x) of this: $e^{0.0814} = 1.085$;
- subtract 1, giving a growth rate of $1.085 - 1 = 0.085 = 8.5\%$ (p.a.).

Note that, as the calculated growth rate is based only upon the initial and final observations, it could be unreliable if either of these two values is an outlier. With a sufficient span of time, however, this is unlikely to be a serious problem.

The geometric mean

In calculating the average growth rate of investment we have implicitly calculated the **geometric mean** of a series. If we have a series of n values, then their geometric mean is calculated as the nth root of the *product* of the values, i.e.

$$\textit{geometric mean} = \sqrt[n]{\prod_{i=1}^{n} x_i} \tag{1.30}$$

The x values in this case are the growth factors in each year, as in Table 1.15 (the values in intermediate years are omitted). The 'Π' symbol is similar to the use of Σ, but means 'multiply together' rather than 'add up'.

The product of the 32 growth factors is 13.518 (the same as is obtained by dividing the final observation by the initial one – why?) and the 32nd root of this is 1.085. This latter figure, 1.085, is the geometric mean of the growth factors and from it we can derive the growth rate of 8.5% p.a. by subtracting 1.

Whenever one is dealing with growth data (or any series that is based on a multiplicative process) one should use the geometric mean rather than the arithmetic mean to get the answer. However, using the arithmetic mean in this case generally gives only a small error, as is indicated below.

Table 1.15 Calculation of the geometric mean – annual growth factors

	Investment	Growth factors	
1973	15 227		
1974	18 134	1.191	(= 18 134/15 227)
1975	21 856	1.205	(= 21 856/18 134)
1976	25 516	1.167	Etc.
⋮	⋮	⋮	
2002	173 525	1.049	
2003	178 751	1.030	
2004	194 491	1.088	
2005	205 843	1.058	

Note: Each growth factor simply shows the ratio of that year's investment to the previous year's.

Another approximate way of obtaining the average growth rate

We have seen that when calculating rates of growth one should use the geometric mean, but if the growth rate is reasonably small then taking the arithmetic mean of the growth factors will give approximately the right answer. The arithmetic mean of the growth factors is

$$\frac{1.191 + 1.205 + \ldots + 1.088 + 1.058}{32} = 1.087$$

giving an estimate of the growth rate of 1.087 – 1 = 0.087 = 8.7% p.a. – close to the correct value. Note also that one could equivalently take the average of the annual growth rates (0.191, 0.205, etc.), giving 0.087, to obtain the same result. Use of the arithmetic mean is justified in this context if one needs only an approximation to the right answer and annual growth rates are reasonably small. It is usually quicker and easier to calculate the arithmetic rather than geometric mean, especially if one does not have a computer to hand.

By now you might be feeling a little overwhelmed by the various methods we have used, all to get an idea of the average – methods which give similar but not always identical answers. Let us summarise the findings:

(a) measuring the slope of the log graph: gives approximately the right answer;
(b) transforming the slope using the formula $e^b - 1$: gives the precise answer (b is the measured slope);
(c) calculating $\sqrt[T-1]{\frac{x_T}{x_1}} - 1$: gives the precise answer (as in (b));
(d) calculating the geometric mean of the growth factors: gives the precise answer;
(e) calculating the arithmetic mean of the growth factors: gives approximately the right answer (although not the same approximation as (a) above).

Remember also that the 'precise' answer could be slightly misleading if either initial or final value is an outlier.

Compound interest

The calculations we have performed relating to growth rates are analogous to computing **compound interest**. If we invest £100 at a rate of interest of 10% per annum, then the investment will grow at 10% p.a. (assuming all the interest is reinvested). Thus after one year the total will have grown to £100 × 1.1 (£110), after two years to £100 × 1.1^2 (£121) and after t years to £100 × 1.1^t. The general formula for the terminal value S_t of a sum S_0 invested for t years at a rate of interest r is

$$S_t = S_0(1 + r)^t \tag{1.31}$$

where r is expressed as a decimal. Rearranging (1.31) to make r the subject yields

$$r = \sqrt[t]{S_t / S_0} - 1 \tag{1.32}$$

which is precisely the formula for the average growth rate. To give a further example: suppose an investment fund turns an initial deposit of £8000 into £13 500 over 12 years. What is the average rate of return on the investment? Setting $S_0 = 8$, $S_t = 13.5$, $t = 12$ and using equation (1.32) we obtain

$$r = \sqrt[12]{13.5 / 8} - 1 = 0.045$$

or 4.5% per annum.

Formula (1.32) can also be used to calculate the **depreciation rate** and the amount of annual depreciation on a firm's assets. In this case, S_0 represents the initial value of the asset, S_t represents the final or scrap value, and the annual rate of depreciation (as a negative number) is given by r from equation (1.32).

The variance of a time series

How should we describe the variance of a time series? The variance of the investment data can be calculated, but it would be uninformative in the same way as the mean. As the series is trended, and this is likely to continue in the longer run, the variance is in principle equal to infinity. The calculated variance would be closely tied to the sample size: the larger it is, the larger the variance. Again it makes more sense to calculate the variance of the growth rate, which has little trend in the long run.

This variance can be calculated from the formula

$$s^2 = \frac{\Sigma(x - \bar{x})^2}{n - 1} = \frac{\Sigma x^2 - n\bar{x}^2}{n - 1} \tag{1.33}$$

where $\bar{x}$ is the average rate of growth. The calculation is set out in Table 1.16 using the right-hand formula in equation (1.33).

The variance is therefore

$$s^2 = \frac{0.3990 - 32 \times 0.087^2}{31} = 0.0051$$

and the standard deviation is 0.071, the square root of the variance. The coefficient of variation is

Table 1.16 Calculation of the variance of the growth rate

Year	Investment	Growth rate	
		x	x^2
1974	18 134	0.191	0.036
1975	21 856	0.205	0.042
1976	25 516	0.167	0.028
⋮	⋮	⋮	⋮
2002	173 525	0.049	0.002
2003	178 751	0.030	0.001
2004	194 491	0.088	0.008
2005	205 843	0.058	0.003
Totals		2.7856	0.3990

$$cv = \frac{0.071}{0.087} = 0.816$$

i.e. the standard deviation of the growth rate is about 80% of the mean.

Note three things about this calculation: first, we have used the arithmetic mean (using the geometric mean makes very little difference); second, we have used the formula for the sample variance since the period 1974–2005 constitutes a sample of all the possible data we could collect; and third, we could have equally used the growth factors for the calculation of the variance (why?).

Worked example 1.7

Given the following data

Year	1999	2000	2001	2002	2003
Price of a laptop PC	1100	900	800	750	700

we can work out the average rate of price growth per annum as follows. The overall growth factor is $\frac{700}{1100} = 0.6363$. The fact that this number is less than one simply reflects the fact that the price has fallen over time. It has fallen to 64% of its original value. To find the annual rate, we take the fourth root of 0.6363 (four years of growth). Hence we obtain $\sqrt[4]{0.6363} = 0.893$, i.e. each year the price falls to 89% of its value the previous year. This implies price is falling at $0.893 - 1 = -0.107$, or approximately an 11% fall each year.

We can see if the fall is more or less the same, by calculating each year's growth factor. These are:

Year	1999	2000	2001	2002	2003
Laptop price	1100	900	800	750	700
Growth factor	–	0.818	0.889	0.9375	0.933
Price fall	–	−19%	−11%	−6%	−7%

The price fall was larger in the earlier years, in percentage as well as absolute terms. Calculating the standard deviation of the values in the final row →

provides a measure of the variability from year to year. The variance is given by

$$s^2 = \frac{(19-11)^2 + (11-11)^2 + (6-11)^2 + (7-11)^2}{3} = 30.7$$

and the standard deviation is then 5.54%. (The calculations are shown rounded but the answer is accurate.)

Exercise 1.6

(a) Using the data in Exercise 1.5, calculate the average level of profit over the time period and the average growth rate of profit over the period. Which appears more useful?

(b) Calculate the variance of profit and compare it to the variance of sales.

Graphing bivariate data: the scatter diagram

The analysis of investment is an example of the use of **univariate methods**: only a single variable is involved. However, we often wish to examine the relationship between two (or sometimes more) variables and we have to use **bivariate** (or **multivariate**) **methods**. To illustrate the methods involved we shall examine the relationship between investment expenditures and gross domestic product (GDP). Economics tells us to expect a positive relationship between these variables, higher GDP is usually associated with higher investment. Table 1.17 provides data on GDP for the UK.

A **scatter diagram** (also called an ***XY* chart**) plots one variable (in this case investment) on the y axis, the other (GDP) on the x axis, and therefore shows the relationship between them. For example, one can see whether high values of one variable tend to be associated with high values of the other. Figure 1.24 shows the relationship for investment and GDP.

The chart shows a strong linear relationship between the two variables, apart from a curious dip in the middle. This reflects the sharp fall in investment after 1990, which is *not* matched by a fall in GDP (if it were, the *XY* chart would show

Table 1.17 GDP data

Year	GDP	Year	GDP	Year	GDP
1973	74 020	1984	324 633	1995	719 747
1974	83 793	1985	355 269	1996	765 152
1975	105 864	1986	381 782	1997	811 194
1976	125 203	1987	420 211	1998	860 796
1977	145 663	1988	469 035	1999	906 567
1978	167 905	1989	514 921	2000	953 227
1979	197 438	1990	558 160	2001	996 987
1980	230 800	1991	587 080	2002	1 048 767
1981	253 154	1992	611 974	2003	1 110 296
1982	277 198	1993	642 656	2004	1 176 527
1983	302 973	1994	680 978	2005	1 224 715

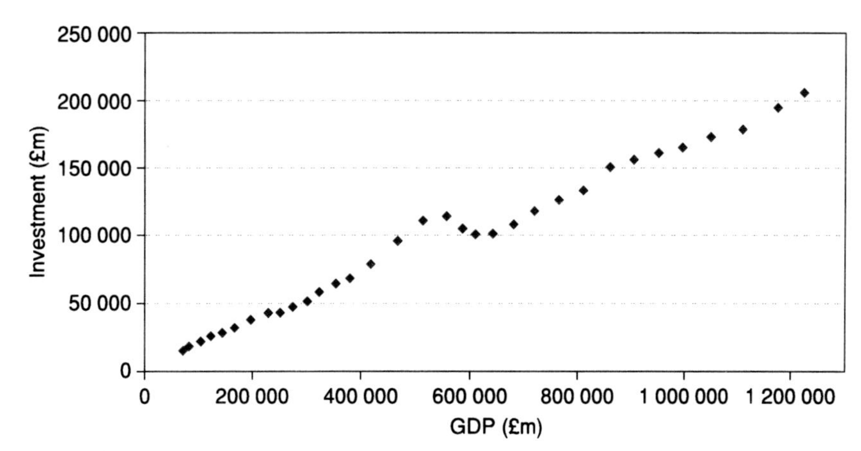

Figure 1.24
Scatter diagram of investment (vertical axis) against GDP (horizontal axis) (nominal values)

Note: The (*x*, *y*) coordinates of each point are given by the values of investment and GDP respectively. Thus the first (1973) data point is drawn 15 227 units above the horizontal axis and 74 020 units from the vertical one.

a linear relationship without the dip). It is important to recognise the difference between the time-series plot and the *XY* chart. Because of inflation later observations tend to be towards the top right of the *XY* chart (both investment and GDP are increasing over time) but this does not *have* to happen; if both variables fluctuated up and down, later observations could be at the bottom left (or centre, or anywhere). By contrast, in a time series plot, later observations are always further to the right.

Note that both variables are in nominal terms, i.e. they make no correction for inflation over the time period. This may be seen algebraically: investment expenditure is made up of the *volume* of investment (I) times its *price* (P_I). Similarly, nominal GDP is real GDP (Y) times its price (P_Y). Thus the scatter diagram actually charts $P_I \times I$ against $P_Y \times Y$. It is likely that the two prices follow a similar trend over time and that this dominates the movements in real investment and GDP. The chart then shows the relationship between a mixture of prices and quantities, when the more interesting relationship is between the *quantities* of investment and output.

Figure 1.25 shows the relationship between the quantities of investment and output, i.e. after the strongly trending price effects have been removed. It is not so straightforward as the nominal graph. There is now a 'knot' of points in the centre where perhaps both (real) investment and GDP fluctuated up and down. Overall it is clear that something 'interesting' happened around 1990 that merits additional investigation.

Chapter 10, on index numbers, explains in detail how to derive real variables from nominal ones, as we have done here, and generally describes how to correct for the effects of inflation on economic magnitudes.

Exercise 1.7

(a) Once again using the data from Exercise 1.5, draw an *XY* chart with profits on the vertical axis, sales on the horizontal axis. Choose the scale of the axes appropriately.

(b) (If using Excel to produce graphs) Right click on the graph, choose 'Add trendline' and choose a linear trend. This gives the 'line of best fit' (covered in detail in Chapter 7). What does this appear to show?

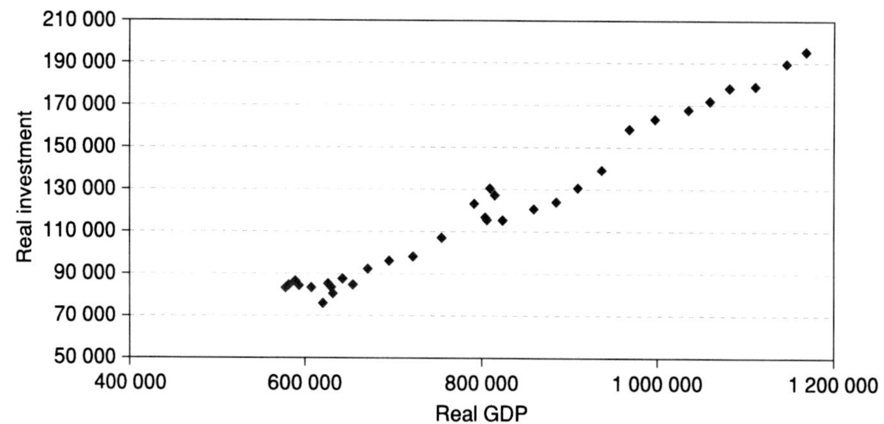

Figure 1.25
The relationship between real investment and real output

Data transformations

In analysing employment and investment data in the examples above we have often changed the variables in some way in order to bring out the important characteristics. In statistics one usually works with data that have been transformed in some way rather than using the original numbers. It is therefore worth summarising the main data transformations available, providing justifications for their use and exploring the implications of such adjustments to the original data. We briefly deal with the following transformations:

- rounding;
- grouping;
- dividing or multiplying by a constant;
- differencing;
- taking logarithms;
- taking the reciprocal;
- deflating.

Rounding

Rounding improves readability. Too much detail can confuse the message, so rounding the answer makes it more memorable. To give an example, the average wealth holding calculated earlier in this chapter is actually £146 983.726 (to three decimal places). It would be absurd to present it in this form, however. We do not know for certain that this figure is accurate (in fact, it almost certainly is not). There is a spurious degree of precision which might mislead the reader. How much should this be rounded for presentational purposes therefore? Remember that the figures have already been effectively rounded by allocation to classes of width 10 000 or more (all observations have been rounded to the mid-point of the interval). However, much of this rounding is offsetting, i.e. numbers rounded up offset those rounded down, so the mean is reasonably accurate. Rounding to £147 000 makes the figure much easier to remember, and is only a change of 0.01% (147 000/146 984 = 1.000 111), so is a reasonable compromise. In the text above, the answer was not rounded to such an extent since the purpose was to highlight the methods of calculation.

Inflation in Zimbabwe

'Zimbabwe's rate of inflation surged to 3731.9%, driven by higher energy and food costs, and amplified by a drop in its currency, official figures show.'

BBC news online, 17 May 2007.

Whether official or not, it is impossible that the rate of inflation is known with such accuracy (to one decimal place!), especially when prices are rising so fast. It would be more reasonable to report a figure of 3700% in this case. Sad to say, inflation rose even further in subsequent months.

Rounding is a 'trap door' function: you cannot obtain the original value from the transformed (rounded) value. Therefore, if you are going to need the original value in further calculations you should not round your answer. Furthermore, small rounding errors can cumulate, leading to a large error in the final answer. Therefore, you should *never* round an intermediate answer, only the final one. Even if you only round the intermediate answer by a small amount, the final answer could be grossly inaccurate. Try the following: calculate $60.29 \times 30.37 - 1831$ both before and after rounding the first two numbers to integers. In the first case you obtain 0.0073, in the second –31.

Grouping

When there is too much data to present easily, grouping solves the problem, although at the cost of hiding some of the information. The examples relating to education and unemployment and to wealth used grouped data. Using the raw data would have given us far too much information, so grouping is a first stage in data analysis. Grouping is another trap door transformation: once it is done you cannot recover the original information.

Dividing/multiplying by a constant

This transformation is carried out to make numbers more readable or to make calculation simpler by removing trailing zeros. The data on wealth were divided by 1000 to ease calculation; otherwise the fx^2 column would have contained extremely large values. Some summary statistics (e.g. the mean) will be affected by the transformation, but not all (e.g. the coefficient of variation). Try to remember which are affected! E and V operators (see Appendix 1B) can help. The transformation is easy to reverse.

Differencing

In time-series data there may be a trend, and it is better to describe the features of the data relative to the trend. The result may also be more economically meaningful, for example governments are often more concerned about the growth of output than about its level. Differencing is one way of eliminating the trend (see Chapter 11 for other methods of detrending data). Differencing was used for the investment data for both of these reasons. One of the implications of differencing is that information about the *level* of the variable is lost.

Taking logarithms

Taking logarithms is used to linearise a non-linear series, in particular one that is growing at a fairly constant rate. It is often easier to see the important features of such a series if the logarithm is graphed rather than the raw data. The logarithmic transformation is also useful in regression (see Chapter 9) because it yields estimates of **elasticities** (e.g. of demand). Taking the logarithm of the investment data linearised the series and tended to smooth it. The inverses of the logarithmic transformations are 10^x (for common logarithms) and e^x (for natural logarithms) so one can recover the original data.

Taking the reciprocal

The reciprocal of a variable might have a useful interpretation and provide a more intuitive explanation of a phenomenon. The reciprocal transformation will also turn a linear series into a non-linear one. The reciprocal of turnover in the labour market (i.e. the number leaving unemployment divided by the number unemployed) gives an idea of the duration of unemployment. If a half of those unemployed find work each year (turnover = 0.5) then the average duration of unemployment is 2 years (= 1/0.5). If a graph of turnover shows a linear decline over time, then the average duration of unemployment will be rising, at a faster and faster rate. Repeating the reciprocal transformation recovers the original data.

Deflating

Deflating turns a nominal series into a real one, i.e. one that reflects changes in quantities without the contamination of price changes. This is dealt with in more detail in Chapter 10. It is often more meaningful in economic terms to talk about a real variable than a nominal one. Consumers are more concerned about their real income than about their money income, for example.

Confusing real and nominal variables is dangerous! For example, someone's nominal (money) income may be rising yet their real income falling (if prices are rising faster than money income). It is important to know which series you are dealing with (this is a common failing among students new to statistics and economics). An income series that is growing at 2–3% per annum is probably a real series; one that is growing at 10% per annum or more is likely to be nominal.

Guidance to the student: how to measure your progress

Now you have reached the end of the chapter your work is not yet over! It is very unlikely that you have fully understood everything after one read through. What you should do now is:

- Check back over the learning outcomes at the start of the chapter. Do you feel you have achieved them? For example, can you list the various different data types you should be able to recognise (the first learning outcome)?
- Read the chapter summary below to help put things in context. You should recognise each topic and be aware of the main issues, techniques, etc., within them. There should be no surprises or gaps!
- Read the list of key terms. You should be able to give a brief and precise definition or description of each one. Do not worry if you cannot remember all the formulae (although you should try to memorise simple ones such as that for the mean).
- Try out the problems (most important!). Answers to odd-numbered problems are at the back of the book, so you can check your answers. There is more detail for some of the answers on the book's web site.

From all of this you should be able to work out whether you have really mastered the chapter. Do not be surprised if you have not – it will take more than one reading. Go back over those parts where you feel unsure of your knowledge. Use these same learning techniques for each chapter of the book.

Summary

- Descriptive statistics are useful for summarising large amounts of information, highlighting the main features but omitting the detail.
- Different techniques are suited to different types of data, e.g. bar charts for cross-section data and rates of growth for time series.
- Graphical methods, such as the bar chart, provide a picture of the data. These give an informal summary but they are unsuitable as a basis for further analysis.
- Important graphical techniques include the bar chart, frequency distribution, relative and cumulative frequency distributions, histogram and pie chart. For time-series data a time-series chart of the data is informative.
- Numerical techniques are more precise as summaries. Measures of location (such as the mean), of dispersion (the variance) and of skewness form the basis of these techniques.
- Important numerical summary statistics include the mean, median and mode; variance, standard deviation and coefficient of variation; coefficient of skewness.
- For bivariate data the scatter diagram (or *XY* graph) is a useful way of illustrating the data.
- Data are often transformed in some way before analysis, for example by taking logs. Transformations often make it easier to see key features of the data in graphs and sometimes make summary statistics easier to interpret. For example, with time-series data the average rate of growth may be more appropriate than the mean of the series.

Key terms and concepts

- bar chart
- box and whiskers plot
- coefficient of variation
- compound growth
- cross-section data
- cross-tabulation
- data transformation
- frequencies
- frequency table
- histogram
- mean
- median
- mode
- outliers
- pie chart
- quantiles
- relative and cumulative frequencies
- scatter diagram (*XY* chart)
- skewness
- standard deviation
- time-series data
- variance
- *z*-score

Reference

Atkinson, A. B. *The Economics of Inequality*, 1983, 2nd edn., Oxford University Press.

Problems

Some of the more challenging problems are indicated by highlighting the problem number in colour.

1.1 The following data show the education and employment status of women aged 20–29 (from the *General Household Survey*):

	Higher education	A levels	Other qualification	No qualification	Total
In work	209	182	577	92	1060
Unemployed	12	9	68	32	121
Inactive	17	34	235	136	422
Sample	238	225	880	260	1603

(a) Draw a bar chart of the numbers in work in each education category. Can this be easily compared with the similar diagram for in Figure 1.1?

(b) Draw a stacked bar chart using all the employment states, similar to Figure 1.3. Comment upon any similarities and differences from the diagram in the text.

(c) Convert the table into (column) percentages and produce a stacked bar chart similar to Figure 1.4. Comment upon any similarities and differences.

(d) Draw a pie chart showing the distribution of educational qualifications of those in work and compare it to Figure 1.5 in the text.

1.2 The data below show the median weekly earnings (in £s) of those in full-time employment in Great Britain in 1992, by category of education.

	Degree	Other higher education	A level	GCSE A–C	GCSE D–G	None
Males	433	310	277	242	226	220
Females	346	278	201	183	173	146

(a) In what fundamental way do the data in this table differ from those in Problem 1.1?

(b) Construct a bar chart showing male and female earnings by education category. What does it show?

(c) Why would it be inappropriate to construct a stacked bar chart of the data? How should one graphically present the combined data for males and females? What extra information is necessary for you to do this?

1.3 Using the data from Problem 1.1:

(a) Which education category has the highest proportion of women in work? What is the proportion?

(b) Which category of employment status has the highest proportion of women with a degree? What is the proportion?

1.4 Using the data from Problem 1.2:

(a) What is the premium, in terms of median earnings, of a degree over A levels? Does this differ between men and women?

(b) Would you expect *mean* earnings to show a similar picture? What differences, if any, might you expect?

1.5 The distribution of marketable wealth in 1979 in the UK is shown in the table below (taken from *Inland Revenue Statistics*, 1981, p. 105):

Range	Number 000s	Amount £m
0–	1606	148
1000–	2927	5985
3000–	2562	10 090
5000–	3483	25 464
10 000–	2876	35 656
15 000–	1916	33 134
20 000–	3425	104 829
50 000–	621	46 483
100 000–	170	25 763
200 000–	59	30 581

Draw a bar chart and histogram of the data (assume the final class interval has a width of 200 000). Comment on the differences between the two. Comment on any differences between this histogram and the one for 1994 given in the text.

1.6 The data below show the number of manufacturing plants in the UK in 1991/92 arranged according to employment:

Number of employees	Number of firms
1–	95 409
10–	15 961
20–	16 688
50–	7229
100–	4504
200–	2949
500–	790
1000–	332

Draw a bar chart and histogram of the data (assume the mid-point of the last class interval is 2000). What are the major features apparent in each and what are the differences?

1.7 Using the data from Problem 1.5:

(a) Calculate the mean, median and mode of the distribution. Why do they differ?

(b) Calculate the inter-quartile range, variance, standard deviation and coefficient of variation of the data.

(c) Calculate the skewness of the distribution.

(d) From what you have calculated, and the data in the chapter, can you draw any conclusions about the degree of inequality in wealth holdings, and how this has changed?

(c) What would be the effect upon the mean of assuming the final class width to be £10m? What would be the effects upon the median and mode?

1.8 Using the data from Problem 1.6:

(a) Calculate the mean, median and mode of the distribution. Why do they differ?

(b) Calculate the inter-quartile range, variance, standard deviation and coefficient of variation of the data.

(c) Calculate the coefficient of skewness of the distribution.

1.9 A motorist keeps a record of petrol purchases on a long journey, as follows:

Petrol station	1	2	3
Litres purchased	33	40	25
Price per litre	55.7	59.6	57.0

Calculate the average petrol price for the journey.

1.10 Demonstrate that the weighted average calculation given in equation (1.9) is equivalent to finding the total expenditure on education divided by the total number of pupils.

1.11 On a test taken by 100 students, the average mark is 65, with variance 144. Student A scores 83, student B scores 47.

(a) Calculate the z-scores for these two students.

(b) What is the maximum number of students with a score either better than A's or worse than B's?

(c) What is the maximum number of students with a score better than A's?

1.12 The average income of a group of people is £8000. 80% of the group have incomes within the range £6000–10 000. What is the minimum value of the standard deviation of the distribution?

1.13 The following data show car registrations in the UK during 1970–91 (source: *ETAS*, 1993, p. 57):

Year	Registrations	Year	Registrations	Year	Registrations
1970	91.4	1978	131.6	1986	156.9
1971	108.5	1979	142.1	1987	168.0
1972	177.6	1980	126.6	1988	184.2
1973	137.3	1981	124.5	1989	192.1
1974	102.8	1982	132.1	1990	167.1
1975	98.6	1983	150.5	1991	133.3
1976	106.5	1984	146.6	–	–
1977	109.4	1985	153.5	–	–

(a) Draw a time-series graph of car registrations. Comment upon the main features of the series.

(b) Draw time-series graphs of the change in registrations, the (natural) log of registrations, and the change in the ln. Comment upon the results.

1.14 The table below shows the different categories of investment, 1986–2005.

Year	Dwellings	Transport	Machinery	Intangible fixed assets	Other buildings
1986	14 140	6527	25 218	2184	20 477
1987	16 548	7872	28 225	2082	24 269
1988	21 097	9227	32 614	2592	30 713
1989	22 771	10 624	38 417	2823	36 689
1990	21 048	10 571	37 776	3571	41 334
1991	18 339	9051	35 094	4063	38 632
1992	18 826	8420	35 426	3782	34 657
1993	19 886	9315	35 316	3648	32 988
1994	21 155	11 395	38 426	3613	33 945
1995	22 448	11 036	45 012	3939	35 596
1996	22 516	12 519	50 102	4136	37 320
1997	23 928	12 580	51 465	4249	41 398
1998	25 222	16 113	58 915	4547	46 286
1999	25 700	14 683	60 670	4645	50 646
2000	27 394	13 577	63 535	4966	51 996
2001	29 806	14 656	60 929	5016	55 065
2002	34 499	16 314	57 152	5588	59 972
2003	38 462	15 592	54 441	5901	64 355
2004	44 299	14 939	57 053	6395	71 805
2005	48 534	15 351	57 295	6757	77 906

Use appropriate graphical techniques to analyse the properties of any one of the investment series. Comment upon the results.

1.15 Using the data from Problem 1.13:

(a) Calculate the average rate of growth of the series.

(b) Calculate the standard deviation around the average growth rate.

(c) Does the series appear to be more or less volatile than the investment figures used in the chapter? Suggest reasons.

1.16 Using the data from Problem 1.14:

(a) Calculate the average rate of growth of the series for dwellings.

(b) Calculate the standard deviation around the average growth rate.

(c) Does the series appear to be more or less volatile than the investment figures used in the chapter? Suggest reasons.

1.17 How would you *expect* the following time-series variables to look when graphed? (e.g. Trended? Linear trend? Trended up or down? Stationary? Homoscedastic? Autocorrelated? Cyclical? Anything else?)

(a) Nominal national income.

(b) Real national income.

(c) The nominal interest rate.

1.18 How would you expect the following time-series variables to look when graphed?

(a) The price level.

(b) The inflation rate.

(c) The £/$ exchange rate.

1.19 (a) A government bond is issued, promising to pay the bearer £1000 in five years' time. The prevailing market rate of interest is 7%. What price would you expect to pay now for the bond? What would its price be after two years? If, after two years, the market interest rate jumped to 10%, what would the price of the bond be?

(b) A bond is issued which promises to pay £200 per annum over the next five years. If the prevailing market interest rate is 7%, how much would you be prepared to pay for the bond? Why does the answer differ from the previous question? (Assume interest is paid at the end of each year.)

1.20 A firm purchases for £30 000 a machine that is expected to last for 10 years, after which it will be sold for its scrap value of £3000. Calculate the average rate of depreciation per annum, and calculate the written-down value of the machine after one, two and five years.

1.21 Depreciation of BMW and Mercedes cars is given in the following table:

Age	BMW 525i	Mercedes 200E
Current	22 275	21 900
1 year	18 600	19 700
2 years	15 200	16 625
3 years	12 600	13 950
4 years	9750	11 600
5 years	8300	10 300

(a) Calculate the average rate of depreciation of each type of car.

(b) Use the calculated depreciation rates to estimate the value of the car after 1, 2, etc., years of age. How does this match the actual values?

(c) Graph the values and estimated values for each car.

1.22 A bond is issued which promises to pay £400 per annum in perpetuity. How much is the bond worth now, if the interest rate is 5%? (Hint: the sum of an infinite series of the form

$$\frac{1}{1+r} + \frac{1}{(1+r)^2} + \frac{1}{(1+r)^3} + \ldots$$

is $1/r$, as long as $r > 0$.)

1.23 Demonstrate, using Σ notation, that $E(x + k) = E(x) + k$.

1.24 Demonstrate, using Σ notation, that $V(kx) = k^2V(x)$.

1.25 Criticise the following statistical reasoning. The average price of a dwelling is £54 150. The average mortgage advance is £32 760. So purchasers have to find £21 390, that is, about 40% of the purchase price. On any basis that is an enormous outlay which young couples, in particular, who are buying a house for the first time would find incredibly difficult, if not impossible, to raise.

1.26 Criticise the following statistical reasoning. Among arts graduates 10% fail to find employment. Among science graduates only 8% remain out of work. Therefore, science graduates are better than arts graduates. (Hint: imagine there are two types of job: popular and unpopular. Arts graduates tend to apply for the former, scientists for the latter.)

1.27 **Project 1**: Is it true that the Conservative government in the UK 1979–1997 lowered taxes, while the Labour government 1997–2007 raised them?

You should gather data that you think are appropriate to the task, summarise them as necessary and write a brief report of your findings. You might like to consider the following points:

- Should one consider tax revenue, or revenue as a proportion of gross national product (GNP)?
- Should one distinguish between tax rates and the tax base (i.e. what is taxed)?
- Has the balance between direct and indirect taxation changed?
- Have different sections of the population fared differently?

You might like to consider other points, and do the problem for a different country. Suitable data sources for the UK are: *Inland Revenue Statistics*, *UK National Accounts*, *Annual Abstract of Statistics* or *Financial Statistics*.

1.28 **Project 2**: Is the employment and unemployment experience of the UK economy worse than that of its competitors? Write a report on this topic in a similar manner to the project above. You might consider rates of unemployment in the UK and other countries; trends in unemployment in each of the countries; the growth in employment in each country; the structure of employment (e.g. full-time/part-time) and unemployment (e.g. long-term/short-term).

You might use data for a number of countries, or concentrate on two in more depth. Suitable data sources are: *OECD Main Economic Indicators*; *European Economy* (published by the European Commission); *Employment Gazette*.

Answers to exercises

Exercise 1.1

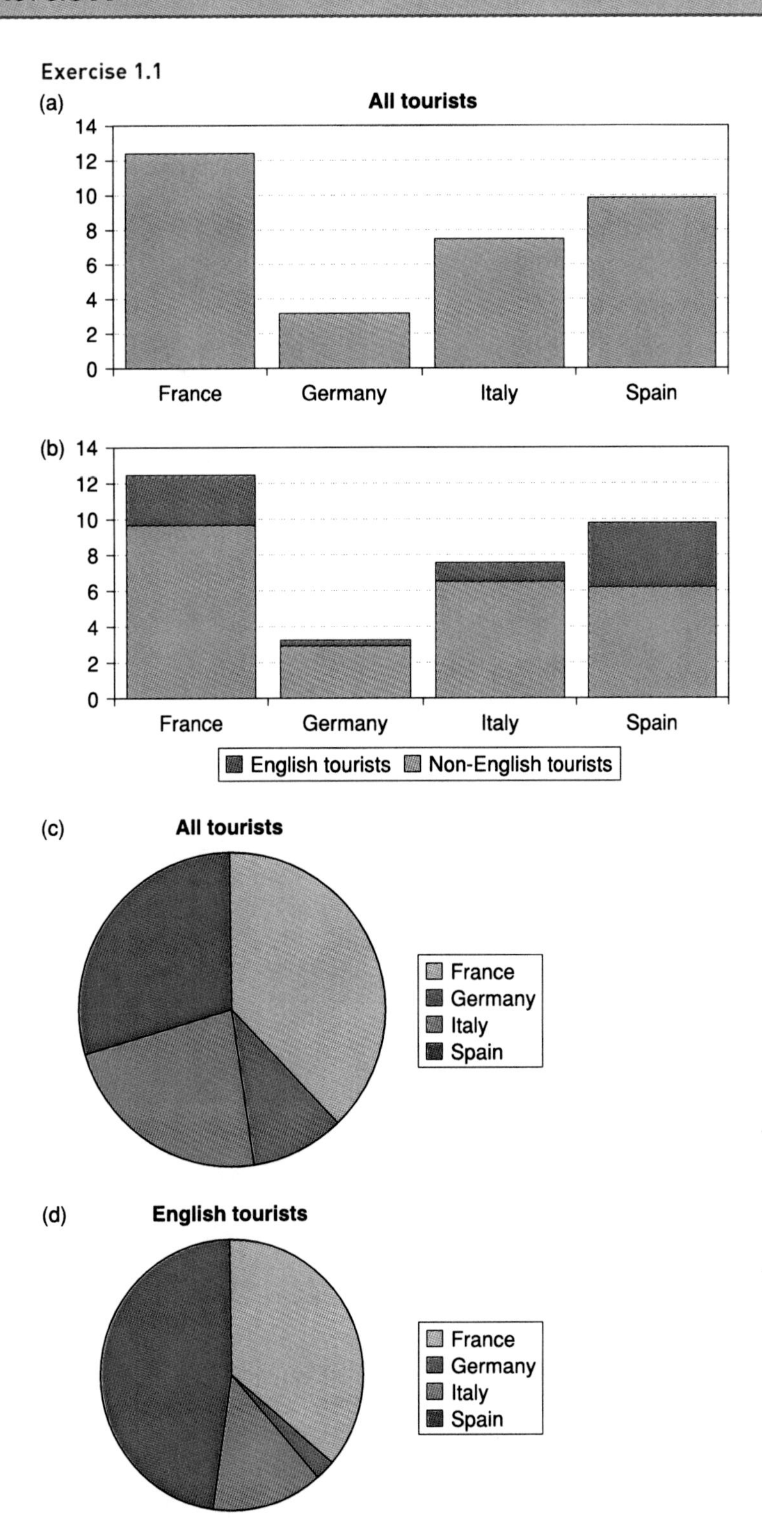

It is clear the English are more likely to visit Spain than are other nationalities.

Exercise 1.2

(a) Bar chart

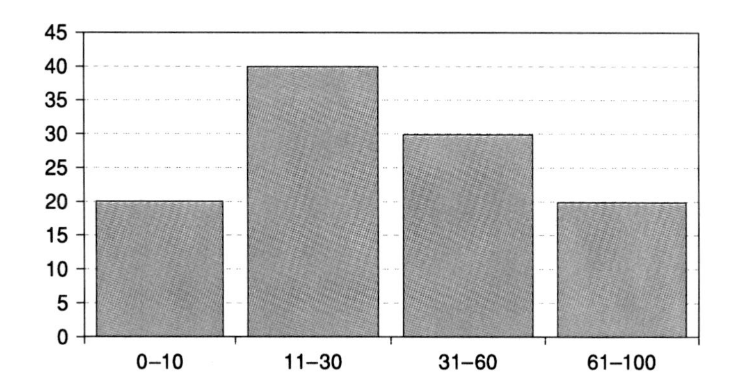

Histogram

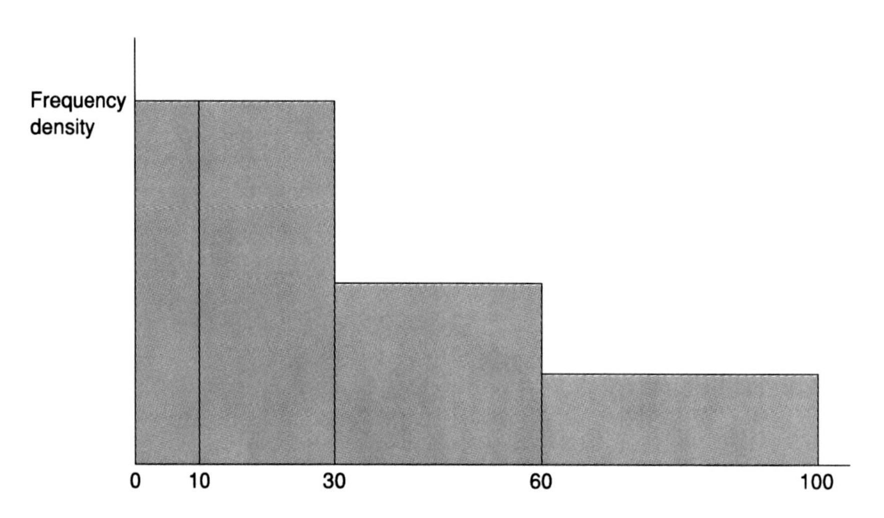

Exercise 1.3

(a)

	Midpoint, x	Frequency, f	fx
0–10	5	20	100
11–30	20	40	800
31–60	45	30	1350
60–100	80	20	1600
–	–	110	3850

Hence the mean = 3850/110 = 35.

The median is contained in the 11–30 group and is 35/40 of the way through the interval (20 + 35 moves us to observation 55). Hence the median is 11 + 35/40 × 19 = 27.625.

The mode is anywhere in the 0–30 range; the frequency density is the same throughout this range.

(b)

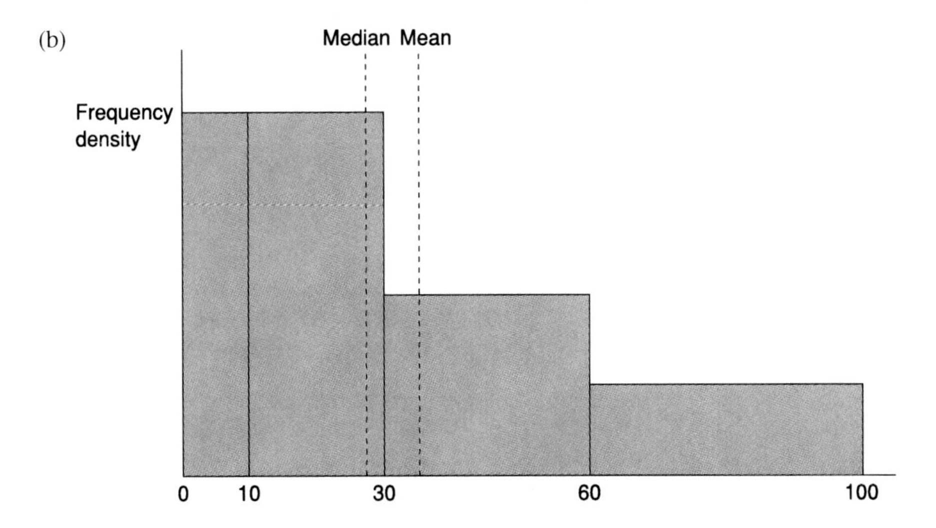

Exercise 1.4

(a) Q1 relates to observation 27.5 (= 110/4). This observation lies in the 11–30 range. There are 20 observations in the first class interval, so Q1 will relate to observation 7.5 in the second interval. Hence we need to go 7.5/40 of the way through the interval. This gives $11 + (7.5/40) \times 19 = 14.6$. Similarly, Q3 is 22.5/30 of the way through the third interval, yielding $Q3 = 31 + 22.5/30 \times 29 = 52.8$. The IQR is therefore 38, approximately. For the variance we obtain $\Sigma fx = 3850$ and $\Sigma fx^2 = 205\ 250$. The variance is therefore $\sigma^2 = 205\ 250/110 - 35^2 = 640.9$ and the standard deviation 25.3.

(b) $CV = 25.3/35 = 0.72$.

(c) $1.3 \times 25.3 = 32.9$, not far from the IQR value of 38.

(d) 1 standard deviation either side of the mean takes us from 9.7 up to 60.3. This contains all 70 observations in the second and third intervals plus perhaps one from the first interval. Thus we obtain approximately 71 observations within this range. Chebyshev's inequality does not help us here as it is not defined for $k \leq 1$.

Exercise 1.5

(a)

(b)

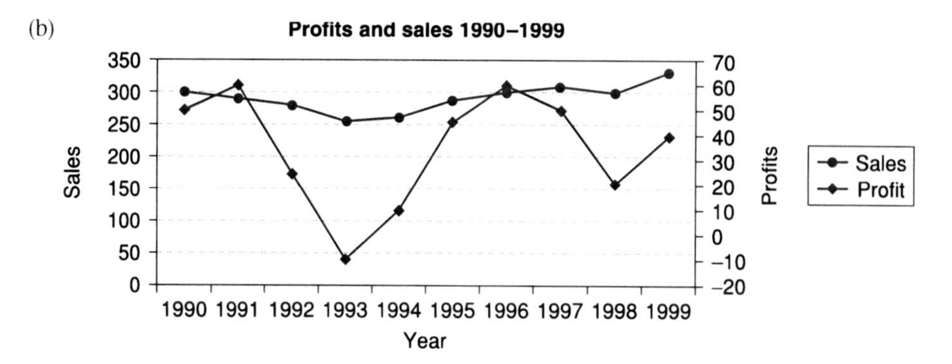

Using the second axis brings out the variability of profits relative to sales.

Exercise 1.6

(a) The average profit is 35. The average rate of growth is calculated by comparing the end values 50 and 40, over the 10-year period. The ratio is 0.8. Taking the ninth root of this (nine years of growth) gives $\sqrt[9]{0.8} = 0.926$ so the annual rate of growth is $0.976 - 1 = -2.4\%$.

(b) The variances are (using the sample variance formula): for profits, $\Sigma(x - \mu)^2 = 4800$ and dividing by 9 gives 533.3. For sales, the mean is 291 and $\Sigma(x - \mu)^2 = 4540$. The variance is therefore $4540/9 = 504.4$. This is similar in absolute size to the variance of profits, but relative to the mean it is much smaller.

Exercise 1.7

(a/b)

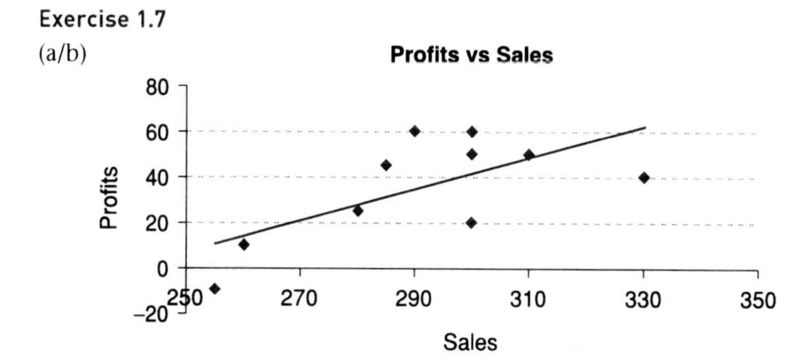

The trend line seems to show a positive relationship between the variables: higher profits are associated with higher sales.

Appendix 1A Σ notation

The Greek symbol Σ (capital sigma) means 'add up' and is a shorthand way of writing what would otherwise be long algebraic expressions. Instead of writing out each term in the series, we provide a template, or typical term of the series, with instructions about how many terms there are.

For example, given the following observations on x:

x_1	x_2	x_3	x_4	x_5
3	5	6	4	8

then

$$\sum_{i=1}^{5} x_i = x_1 + x_2 + x_3 + x_4 + x_5 = 3 + 5 + 6 + 4 + 8 = 26$$

The template is simply x in this case, representing a number to be added in the series. To expand the sigma expression, the subscript i is replaced by successive integers, beginning with the one below the Σ sign and ending with the one above it (1 to 5 in the example above). Hence the instruction is to add the terms x_1 to x_5. Similarly

$$\sum_{i=2}^{4} x_i = x_2 + x_3 + x_4 = 5 + 6 + 4 = 15$$

The instruction tells us to add up only the second, third and fourth terms of the series. When it is clear what range of values i takes (usually when we are to add all available values), the formula can be simplified to $\sum_i x_i$ or Σx_i or even Σx.

When frequencies are associated with each of the observations, as in the data below:

i	1	2	3	4	5
x_i	3	5	6	4	8
f_i	2	2	4	3	1

then

$$\sum_{i=1}^{i=5} f_i x_i = f_1 x_1 + \ldots + f_5 x_5 = 2 \times 3 + 2 \times 5 + \ldots + 1 \times 8 = 60$$

And also

$$\Sigma f_i = 2 + 2 + 4 + 3 + 1 = 12$$

Thus the sum of the 12 observations is 60 and the mean is

$$\frac{\Sigma fx}{\Sigma f} = \frac{60}{12} = 5$$

We are not limited just to adding the x values. For example, we might wish to square each observation before adding them together. This is expressed as

$$\Sigma x^2 = x_1^2 + x_2^2 + \ldots + x_5^2 = 150$$

Note that this is different from

$$(\Sigma x)^2 = (x_1 + x_2 + \ldots + x_5)^2 = 676$$

Part of the formula for the variance calls for the following calculation

$$\Sigma fx^2 = f_1x_1^2 + f_2x_2^2 + \ldots + f_5x_5^2 = 2 \times 3^2 + 2 \times 5^2 + \ldots + 1 \times 8^2 = 324$$

Using Σ notation we can see the effect of transforming x by dividing by 1000, as was done in calculating the average level of wealth. Instead of working with x we used kx, where $k = 1/1000$. In finding the mean we calculated

$$\frac{\Sigma kx}{N} = \frac{kx_1 + kx_2 + \ldots}{N} = \frac{k(x_1 + x_2 + \ldots)}{N} = k\frac{\Sigma x}{N} \quad (1.34)$$

So, to find the mean of the original variable x, we had to divide by k again, i.e. multiply by 1000. In general, whenever each observation in a sum is multiplied by a constant, the constant can be taken outside the summation operator, as in equation (1.34) above.

Problems on Σ notation

1A.1 Given the following data on x_i: {4, 6, 3, 2, 5}, evaluate

$$\Sigma x_i, \Sigma x_i^2, (\Sigma x_i)^2, \Sigma(x_i - 3), \Sigma x_i - 3, \sum_{i=2}^{4} x_i$$

1A.2 Given the following data on x_i: {8, 12, 6, 4, 10}, evaluate

$$\Sigma x_i, \Sigma x_i^2, (\Sigma x_i)^2, \Sigma(x_i - 3), \Sigma x_i - 3, \sum_{i=2}^{4} x_i$$

1A.3 Given the following frequencies, f_i, associated with the x values in Problem 1A.1: {5, 3, 3, 8, 5}, evaluate

$$\Sigma fx, \Sigma fx^2, \Sigma f(x - 3), \Sigma fx - 3$$

1A.4 Given the following frequencies, f_i, associated with the x values in Problem 1A.2: {10, 6, 6, 16, 10}, evaluate

$$\Sigma fx, \Sigma fx^2, \Sigma f(x - 3), \Sigma fx - 3$$

1A.5 Given the pairs of observations on x and y

x	4	3	6	8	12
y	3	9	1	4	3

evaluate Σxy, $\Sigma x(y - 3)$, $\Sigma(x + 2)(y - 1)$

1A.6 Given the pairs of observations on x and y

x	3	7	4	1	9
y	1	2	5	1	2

evaluate Σxy, $\Sigma x(y-2)$, $\Sigma(x-2)(y+1)$.

1A.7 Demonstrate that

$$\frac{\Sigma f(x-k)}{\Sigma f} = \frac{\Sigma fx}{\Sigma f} - k$$

where k is a constant.

1A.8 Demonstrate that

$$\frac{\Sigma f(x-\mu)^2}{\Sigma f} = \frac{\Sigma fx^2}{\Sigma f} - \mu^2$$

Appendix 1B E and V operators

These operators are an extremely useful form of notation that we shall make use of later in the book. It is quite easy to keep track of the effects of data transformations using them. There are a few simple rules for manipulating them that allow some problems to be solved quickly and elegantly.

E(x) is the mean of a distribution and V(x) is its variance. We showed above in equation (1.34) that multiplying each observation by a constant k multiplies the mean by k. Thus we have

$$\mathrm{E}(kx) = k\mathrm{E}(x) \tag{1.35}$$

Similarly, if a constant is added to every observation the effect is to add that constant to the mean (see Problem 1.23)

$$\mathrm{E}(x + a) = \mathrm{E}(x) + a \tag{1.36}$$

(Graphically, the whole distribution is shifted a units to the right and hence so is the mean.) Combining equations (1.35) and (1.36)

$$\mathrm{E}(kx + a) = k\mathrm{E}(x) + a \tag{1.37}$$

Similarly for the variance operator it can be shown that

$$\mathrm{V}(x + k) = \mathrm{V}(x) \tag{1.38}$$

Proof

$$\mathrm{V}(x+k) = \frac{\Sigma((x+k)-(\mu+k))^2}{N} = \frac{\Sigma((x-\mu)+(k-k))^2}{N} = \frac{\Sigma(x-\mu)^2}{N} = \mathrm{V}(x)$$

(A shift of the whole distribution leaves the variance unchanged.) Also

$$\mathrm{V}(kx) = k^2\mathrm{V}(x) \tag{1.39}$$

(See Problem 1.24 above.) This is why, when the wealth figures were divided by 1000, the variance became divided by 1000^2. Applying (1.38) and (1.39)

$$V(kx + a) = k^2V(x) \tag{1.40}$$

Finally we should note that V itself can be expressed in terms of E

$$V(x) = E(x - E(x))^2 \tag{1.41}$$

Appendix 1C Using logarithms

Logarithms are less often used now that cheap electronic calculators are available. Formerly logarithms were an indispensable aid to calculation. However, the logarithmic transformation is useful in other contexts in statistics and economics so its use is briefly set out here.

The logarithm (to the base 10) of a number x is defined as the power to which 10 must be raised to give x. For example, $10^2 = 100$, so the log of 100 is 2 and we write $\log_{10} 100 = 2$ or simply $\log 100 = 2$.

Similarly, the log of 1000 is 3 ($1000 = 10^3$), of 10 000 it is 4, etc. We are not restricted to integer (whole number) powers of 10, so for example $10^{2.5} = 316.227766$ (try this if you have a scientific calculator), so the log of 316.227766 is 2.5. Every number x can therefore be represented by its logarithm.

Multiplication of two numbers

We can use logarithms to multiply two numbers x and y, based on the property[x]

$$\log xy = \log x + \log y$$

For example, to multiply 316.227766 by 10

$$\begin{aligned}\log(316.227766 \times 10) &= \log 316.227766 + \log 10 \\ &= 2.5 + 1 \\ &= 3.5\end{aligned}$$

The *anti-log* of 3.5 is given by $10^{3.5} = 3162.27766$ which is the answer.

Taking the anti-log (i.e. 10 raised to a power) is the inverse of the log transformation. Schematically we have

$$x \rightarrow \text{take logarithms} \rightarrow a\ (= \log x) \rightarrow \text{raise 10 to the power } a \rightarrow x$$

Division

To divide one number by another we subtract the logs. For example, to divide 316.227766 by 100

$$\begin{aligned}\log(316.227766/100) &= \log 316.227766 - \log 100 \\ &= 2.5 - 2 \\ &= 0.5\end{aligned}$$

and $10^{0.5} = 3.16227766$.

[x] This is equivalent to saying $10^x \times 10^y = 10^{x+y}$.

Powers and roots

Logarithms simplify the process of raising a number to a power. To find the square of a number, multiply the logarithm by 2, e.g. to find 316.227766^2:

$$\log(316.227766^2) = 2 \log(316.227766) = 5$$

and $10^5 = 100\ 000$.

To find the square root of a number (equivalent to raising it to the power $\frac{1}{2}$) divide the log by 2. To find the *n*th root, divide the log by *n*. For example, in the text we have to find the 32nd root of 13.518

$$\frac{\log(13.518)}{32} = \frac{1.1309}{32} = 0.0353$$

and $10^{0.0353} = 1.085$.

Common and natural logarithms

Logarithms to the base 10 are known as common logarithms but one can use any number as the base. *Natural* logarithms are based on the number e (= 2.71828 . . .) and we write $\ln x$ instead of $\log x$ to distinguish them from common logarithms. So, for example

$$\ln 316.227766 = 5.756462732$$

since $e^{5.756462732} = 316.227766$.

Natural logarithms can be used in the same way as common logarithms and have the similar properties. Use the 'ln' key on your calculator just as you would the 'log' key, but remember that the inverse transformation is e^x rather than 10^x.

Problems on logarithms

1C.1 Find the common logarithms of: 0.15, 1.5, 15, 150, 1500, 83.7225, 9.15, –12.

1C.2 Find the log of the following values: 0.8, 8, 80, 4, 16, –37.

1C.3 Find the natural logarithms of: 0.15, 1.5, 15, 225, –4.

1C.4 Find the ln of the following values: 0.3, e, 3, 33, –1.

1C.5 Find the anti-log of the following values: –0.823909, 1.1, 2.1, 3.1, 12.

1C.6 Find the anti-log of the following values: –0.09691, 2.3, 3.3, 6.3.

1C.7 Find the anti-ln of the following values: 2.70805, 3.70805, 1, 10.

1C.8 Find the anti-ln of the following values: 3.496508, 14, 15, –1.

1C.9 Evaluate: $\sqrt[2]{10}$, $\sqrt[4]{3.7}$, $4^{1/4}$, 12^{-3}, $25^{-3/2}$.

1C.10 Evaluate: $\sqrt[3]{30}$, $\sqrt[5]{17}$, $8^{1/4}$, 15^0, 12^0, $3^{-1/3}$.

7 Correlation and regression

Contents

Learning outcomes

By the end of this chapter you should be able to:

- understand the principles underlying correlation and regression;
- calculate and interpret a correlation coefficient and relate it to an *XY* graph of the two variables;
- calculate the line of best fit (regression line) and interpret the result;
- recognise the statistical significance of the results, using confidence intervals and hypothesis tests;
- recognise the importance of the units in which the variables are measured and of transformations to the data;
- use computer software (*Excel*) to derive the regression line and interpret the computer output.

Complete your diagnostic test for Chapter 7 now to create your personal study plan. Exercises with an icon ? are also available for practice in MathXL with additional supporting resources.

Introduction

Correlation and regression are techniques for investigating the statistical relationship between two, or more, variables. In Chapter 1 we examined the relationship between investment and gross domestic product (GDP) using graphical methods (the *XY* chart). Although visually helpful, this did not provide any precise measurement of the strength of the relationship. In Chapter 6 the χ^2 test did provide a test of the significance of the association between two category-based variables, but this test cannot be applied to variables measured on a ratio scale. Correlation and regression fill in these gaps: the strength of the relationship between two (or more) ratio scale variables can be measured and the significance tested.

Correlation and regression are the techniques most often used by economists and forecasters. They can be used to answer such questions as

- Is there a link between the money supply and the price level?
- Do bigger firms produce at lower cost than smaller firms?
- Does instability in a country's export performance hinder its growth?

Each of these questions is about economics or business as much as about statistics. The statistical analysis is part of a wider investigation into the problem; it cannot provide a complete answer to the problem but, used sensibly, is a vital input. Correlation and regression techniques may be applied to time-series or cross-section data. The methods of analysis are similar in each case, although there are differences of approach and interpretation which are highlighted in this chapter and the next.

This chapter begins with the topic of correlation and simple (i.e. two variable) regression, using as an example the determinants of the birth rate in developing countries. In Chapter 8, multiple regression is examined, where a single dependent variable is explained by more than one explanatory variable. This is illustrated using time-series data pertaining to imports into the UK. This shows how a small research project can be undertaken, avoiding the many possible pitfalls along the way. Finally, a variety of useful additional techniques, tips and traps is set out, to help you understand and overcome a number of problems that can arise in regression analysis.

What determines the birth rate in developing countries?

This example follows the analysis in Michael Todaro's book, *Economic Development in the Third World* (3rd edn, pp. 197–200) where he tries to establish which of three variables (gross national product (GNP) per capita, the growth rate per capita or income inequality) is most important in determining a country's birth rate. (This analysis has been dropped from later editions of Todaro's book.) The analysis is instructive as an example of correlation and regression techniques in a number of ways. First, the question is an important one; it was discussed at the UN International Conference on Population and Development in Cairo in 1995. It is felt by many that reducing the birth rate is a vital factor in economic

Table 7.1 Todaro's data on birth rate, GNP, growth and inequality

Country	Birth rate	1981 GNP p.c.	GNP growth	Income ratio
Brazil	30	2200	5.1	9.5
Colombia	29	1380	3.2	6.8
Costa Rica	30	1430	3.0	4.6
India	35	260	1.4	3.1
Mexico	36	2250	3.8	5.0
Peru	36	1170	1.0	8.7
Philippines	34	790	2.8	3.8
Senegal	48	430	−0.3	6.4
South Korea	24	1700	6.9	2.7
Sri Lanka	27	300	2.5	2.3
Taiwan	21	1170	6.2	3.8
Thailand	30	770	4.6	3.3

Source: Adapted from Todaro, M. (1992).

development (birth rates in developed countries average around 12 per 1000 population, in developing countries around 30). Second, Todaro uses the statistical analysis to arrive at an unjustified conclusion (it is always best to learn from others' mistakes).

The data used by Todaro are shown in Table 7.1 using a sample of 12 developing countries. Two points need to be made initially. First, the sample only includes developing countries, so the results will not give an all-embracing explanation of the birth rate. Different factors might be relevant to developed countries, for example. Second, there is the important question of why these particular countries were chosen as the sample and others ignored. The choice of country was, in fact, limited by data availability, and one should ask whether countries with data available are likely to be representative of all countries. Data were, in fact, available for more than 12 countries, so Todaro was selective. You are asked to explore the implications of this in some of the problems at the end of the chapter.

The variables are defined as follows:

Birth rate: the number of births per 1000 population in 1981.
GNP per capita: 1981 gross national product p.c., in US dollars.
Growth rate: the growth rate of GNP p.c. per annum, 1961–1981.
Income ratio: the ratio of the income share of the richest 20% to that of the poorest 40%. A higher value of this ratio indicates greater inequality.

We leave aside the concerns about the sample until later and concentrate now on analysing the figures. The first thing it is useful to do is to graph the variables to see if anything useful is revealed. *XY* graphs are the most suitable in this case and they are shown in Figure 7.1. From these we see a reasonably tidy relationship between the birth rate and the growth rate, with a negative slope; there is a looser relationship with the income ratio, with a positive slope; and there is little discernible pattern (apart from a flat line) in the graph of birth rate against GNP. Todaro asserts that the best relationship is between the birth rate and income inequality. He rejects the growth rate as an important determinant of the birth rate because of the four countries at the top of the chart, which have

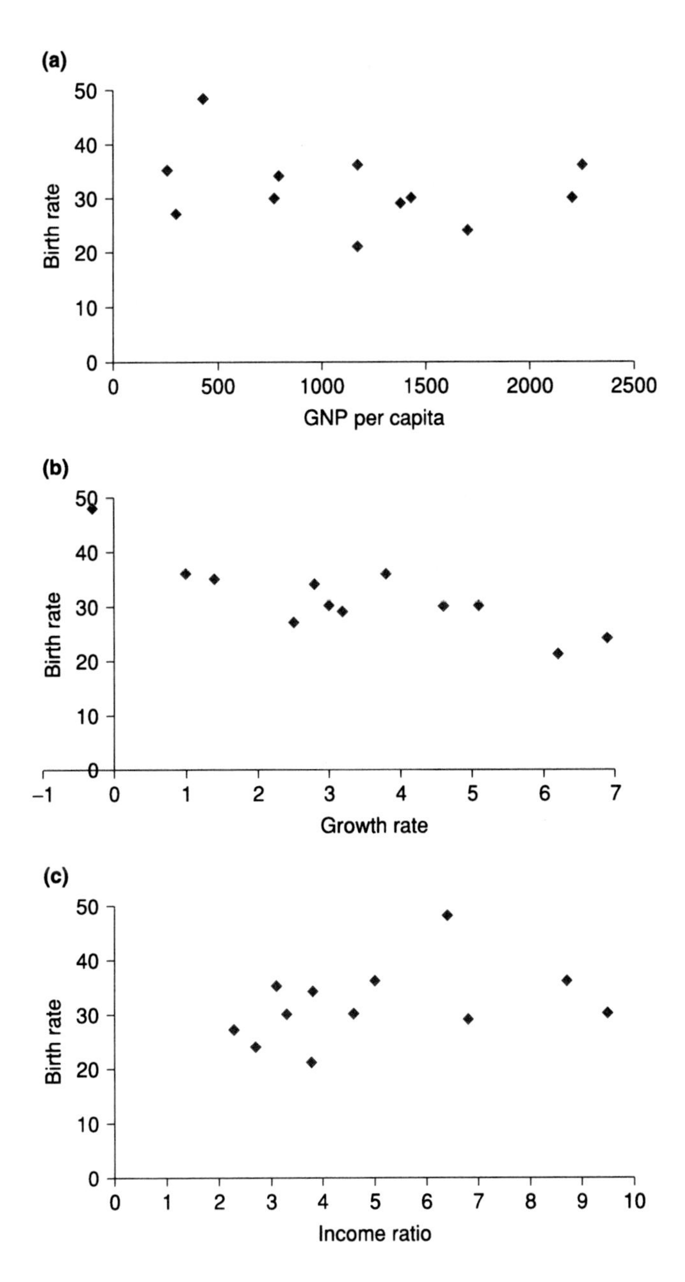

Figure 7.1
Graphs of the birth rate against (a) GNP, (b) growth and (c) income ratio

very different growth rates, yet similar birth rates. In the following sections we shall see whether Todaro's conclusions are justified.

Correlation

The relationships graphed in Figure 7.1 can first be summarised numerically by measuring the **correlation coefficient** between any pair of variables. We illustrate this by calculating the correlation coefficient between the birth rate (B) and growth (G), although we also present the results for the other cases. Just as the mean is a number that summarises information about a single variable, so the correlation coefficient is a number which summarises the relationship between two variables.

The different types of possible relationship between any two variables, X and Y, may be summarised as follows:

- High values of X tend to be associated with low values of Y and vice versa. This is termed **negative correlation**, and appears to be the case for B and G.
- High (low) values of X tend to be associated with high (low) values of Y. This is **positive correlation** and reflects (rather weakly) the relationship between B and the income ratio (IR).
- No relationship between X and Y exists. High (low) values of X are associated about equally with high and low values of Y. This is **zero**, or the absence of, **correlation**. There appears to be little correlation between the birth rate and per capita GNP.

It should be noted that positive correlation does not mean that high values of X are *always* associated with high values of Y, but usually they are. It is also the case that correlation only represents a *linear* relationship between the two variables. As a counter-example, consider the backwards-bending labour supply curve, as suggested by economic theory (higher wages initially encourage extra work effort, but above a certain point the benefit of higher wage rates is taken in the form of more leisure). The relationship is non-linear and the measured degree of correlation between wages and hours of work is likely to be low, even though the former obviously influences the latter.

The sample correlation coefficient, r, is a numerical statistic which distinguishes between the types of cases shown in Figure 7.1. It has the following properties:

- It always lies between –1 and +1. This makes it relatively easy to judge the strength of an association.
- A positive value of r indicates positive correlation, a higher value indicating a stronger correlation between X and Y (i.e. the observations lie closer to a straight line). $r = 1$ indicates perfect positive correlation and means that all the observations lie precisely on a straight line with positive slope, as Figure 7.2 illustrates.
- A negative value of r indicates negative correlation. Similar to the above, a larger negative value indicates stronger negative correlation and $r = -1$ signifies perfect negative correlation.

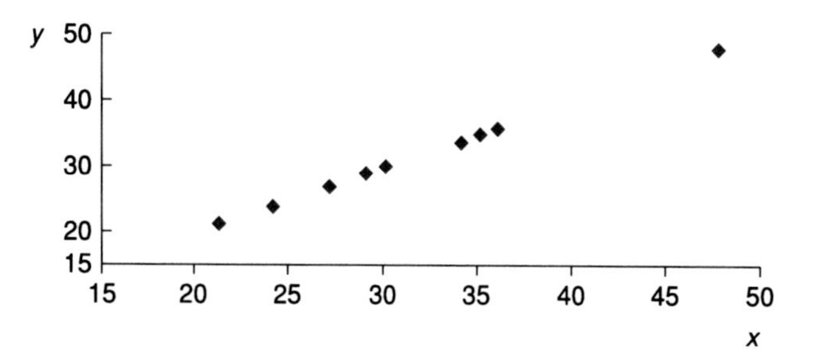

Figure 7.2
Perfect positive correlation

Table 7.2 Calculation of the correlation coefficient, *r*

Country	Birth rate Y	GNP growth X	Y^2	X^2	XY
Brazil	30	5.1	900	26.01	153.0
Colombia	29	3.2	841	10.24	92.8
Costa Rica	30	3.0	900	9.00	90.0
India	35	1.4	1225	1.96	49.0
Mexico	36	3.8	1296	14.44	136.8
Peru	36	1.0	1296	1.00	36.0
Philippines	34	2.8	1156	7.84	95.2
Senegal	48	−0.3	2304	0.09	−14.4
South Korea	24	6.9	576	47.61	165.6
Sri Lanka	27	2.5	729	6.25	67.5
Taiwan	21	6.2	441	38.44	130.2
Thailand	30	4.6	900	21.16	138.0
Totals	380	40.2	12 564	184.04	1139.7

Note: In addition to the X and Y variables in the first two columns, three other columns are needed, for X^2, Y^2 and XY values.

- A value of $r = 0$ (or close to it) indicates a lack of correlation between X and Y.
- The relationship is symmetric, i.e. the correlation between X and Y is the same as between Y and X. It does not matter which variable is labelled Y and which is labelled X.

The formula[1] for calculating the correlation coefficient is given in equation (7.1)

$$r = \frac{n\Sigma XY - \Sigma X\Sigma Y}{\sqrt{(n\Sigma X^2 - (\Sigma X)^2)(n\Sigma Y^2 - (\Sigma Y)^2)}} \tag{7.1}$$

The calculation of r for the relationship between birth rate (Y) and growth (X) is shown in Table 7.2 and equation (7.2). From the totals in Table 7.2 we calculate

$$r = \frac{12 \times 1139.7 - 40.2 \times 380}{\sqrt{(12 \times 184.04 - 40.2^2)(12 \times 12\,564 - 380^2)}} = -0.824 \tag{7.2}$$

This result indicates a fairly strong negative correlation between the birth rate and growth. Countries which have higher economic growth rates also tend to have lower birth rates. The result of calculating the correlation coefficient for the case of the birth rate and the income ratio is $r = 0.35$, which is positive as expected. Greater inequality (higher IR) is associated with a higher birth rate, though the degree of correlation is not particularly strong and less than the correlation with the growth rate. Between the birth rate and GNP per capita the value of r is only −0.26 indicating only a modest degree of correlation. All of this begins to cast doubt upon Todaro's interpretation of the data.

[1] The formula for r can be written in a variety of different ways. The one given here is the most convenient for calculation.

Exercise 7.1

(a) Perform the required calculations to confirm that the correlation between the birth rate and the income ratio is 0.35.

(b) In *Excel*, use the = CORREL() function to confirm your calculations in the previous two exercises. (For example, the function = CORREL(A1:A12,B1:B12) would calculate the correlation between a variable X in cells A1:A12 and Y in cells B1:B12.)

(c) Calculate the correlation coefficient between the birth rate and the growth rate again, but expressing the birth rate per 100 population and the growth rate as a decimal. (In other words, divide Y by 10 and X by 100.) Your calculation should confirm that changing the units of measurement leaves the correlation coefficient unchanged.

Are the results significant?

These results come from a (small) sample, one of many that could have been collected. Once again we can ask the question, what can we infer about the population (of all developing countries) from the sample? *Assuming* the sample was drawn at random (which may not be justified) we can use the principles of hypothesis testing introduced in Chapter 5. As usual, there are two possibilities.

(1) The truth is that there is no correlation (in the population) and that our sample exhibits such a large (absolute) value by chance.
(2) There really is a correlation between the birth rate and the growth rate and the sample correctly reflects this.

Denoting the true but unknown population correlation coefficient by ρ (the Greek letter 'rho') the possibilities can be expressed in terms of a hypothesis test

$$H_0: \rho = 0$$
$$H_1: \rho \neq 0$$

The test statistic in this case is not r itself but a transformation of it

$$t = \frac{r\sqrt{n-2}}{\sqrt{1-r^2}} \tag{7.3}$$

which has a t distribution with $n - 2$ degrees of freedom. The five steps of the test procedure are therefore:

(1) Write down the null and alternative hypotheses (shown above).
(2) Choose the significance level of the test: 5% by convention.
(3) Look up the critical value of the test for $n - 2 = 10$ degrees of freedom: $t^*_{10} = 2.228$ for a two-tail test.
(4) Calculate the test statistic using equation (7.3)

$$t = \frac{-0.824\sqrt{12-2}}{\sqrt{1-(-0.824)^2}} = -4.59$$

(5) Compare the test statistic with the critical value. In this case $t < -t^*_{10}$ so H_0 is rejected. There is a less than 5% chance of the sample evidence occurring if the null hypothesis were true, so the latter is rejected. There does appear to be a genuine association between the birth rate and the growth rate.

Performing similar calculations (see Exercise 7.2 below) for the income ratio and for GNP reveals that in both cases the null hypothesis cannot be rejected at the 5% significance level. These observed associations could well have arisen by chance.

Are significant results important?

Following the discussion in Chapter 5, we might ask if a certain value of the correlation coefficient is economically important as well as being significant. We saw earlier that 'significant' results need not be important. The difficulty in this case is that we have little intuitive understanding of the correlation coefficient. Is $\rho = 0.5$ important, for example? Would it make much difference if it were only 0.4?

Our understanding may be helped if we look at some graphs of variables with different correlation coefficients. Three are shown in Figure 7.3. Panel (a) of the figure graphs two variables with a correlation coefficient of 0.2. Visually there seems little association between the variables, yet the correlation coefficient is (just) significant: $t = 2.06$ ($n = 100$ and the Prob-value is 0.046). This is a significant result which does not impress much.

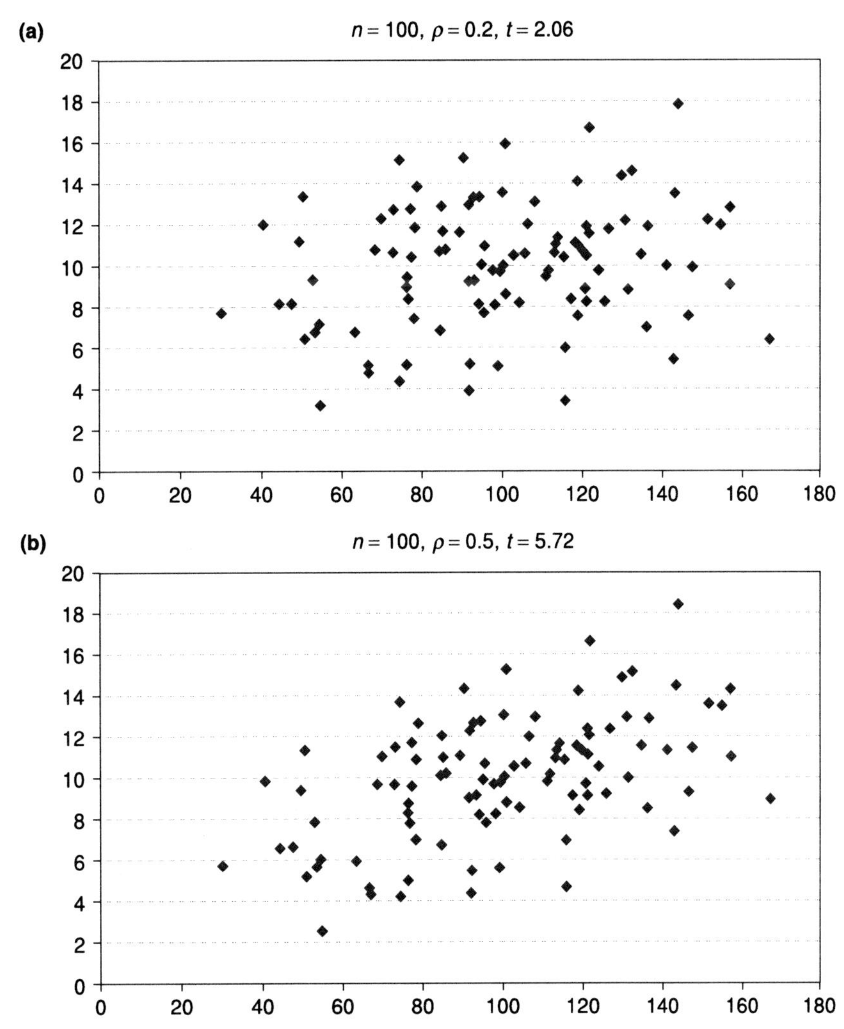

Figure 7.3
Variables with different correlations

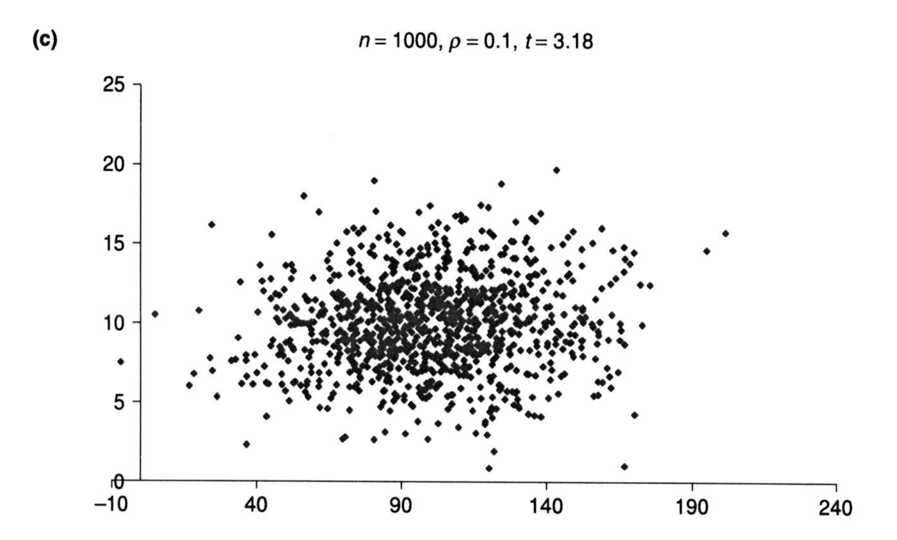

Figure 7.3 (cont'd)

In panel (b) the correlation coefficient is 0.5 and the association seems a little stronger visually, though there is still a substantial scatter of the observations around a straight line. Yet the t statistic in this case is 5.72, highly significant (Prob-value 0.000).

Finally, panel (c) shows an example where n = 1000. To the eye this looks much like a random scatter, with no discernable pattern. Yet the correlation coefficient is 0.1 and the t statistic is 3.18, again highly significant (Prob-value – 0.002).

The lessons from this seem fairly clear. What looks like a random scatter on a chart may in fact reveal a relationship between variables which is statistically significant, especially if there are a large number of observations. On the other hand, a high t-statistic and correlation coefficient can still mean there is a lot of variation in the data, revealed by the chart. Panel (b) suggests, for example, that we are unlikely to get a very reliable prediction of the value of y, even if we know the value of x.

Exercise 7.2

(a) Test the hypothesis that there is no association between the birth rate and the income ratio.

(b) Look up the Prob-value associated with the test statistic and confirm that it does not reject the null hypothesis.

Correlation and causality

It is important to test the significance of any result because almost every pair of variables will have a non-zero correlation coefficient, even if they are totally unconnected (the chance of the sample correlation coefficient being *exactly* zero is very, very small). Therefore it is important to distinguish between correlation coefficients which are significant and those which are not, using the t test just outlined. But even when the result is significant one should beware of the danger of 'spurious' correlation. Many variables which clearly cannot be related turn out to be 'significantly' correlated with each other. One now famous example is between the price level and cumulative rainfall. Because they both rise year after year, it is easy to see why they are correlated, yet it is hard to think of a plausible reason why they should be causally related to each other.

Apart from spurious correlation there are four possible reasons for a non-zero value of r.

(1) X influences Y.
(2) Y influences X.
(3) X and Y jointly influence each other.
(4) Another variable, Z, influences both X and Y.

Correlation alone does not allow us to distinguish between these alternatives. For example, wages (X) and prices (Y) are highly correlated. Some people believe this is due to cost–push inflation, i.e. that wage rises lead to price rises. This is case (1) above. Others believe that wages rise to keep up with the cost of living (i.e. rising prices), which is (2). Perhaps a more convincing explanation is (3), a wage–price spiral where each feeds upon the other. Others would suggest that it is the growth of the money supply, Z, which allows both wages and prices to rise. To distinguish between these alternatives is important for the control of inflation, but correlation alone does not allow that distinction to be made.

Correlation is best used therefore as a suggestive and descriptive piece of analysis, rather than a technique which gives definitive answers. It is often a preparatory piece of analysis, which gives some clues to what the data might yield, to be followed by more sophisticated techniques such as regression.

The coefficient of rank correlation

On occasion it is inappropriate or impossible to calculate the correlation coefficient as described above and an alternative approach is required. Sometimes the original data are unavailable but the ranks are. For example, schools may be ranked in terms of their exam results, but the actual pass rates are not available. Similarly, they may be ranked in terms of spending per pupil, with actual spending levels unavailable. Although the original data are missing, one can still test for an association between spending and exam success by calculating the correlation between the ranks. If extra spending improves exam performance, schools ranked higher on spending should also be ranked higher on exam success, leading to a positive correlation.

Second, even if the raw data are available, they may be highly skewed and hence the correlation coefficient may be influenced heavily by a few outliers. In this case, the hypothesis test for correlation may be misleading as it is based on the assumption of underlying Normal distributions for the data. In this case we could transform the values to ranks, and calculate the correlation of the ranks. In a similar manner to the median, described in Chapter 1, this can effectively deal with heavily skewed distributions.

In these cases, it is **Spearman's coefficient of rank correlation** that is calculated. (The 'standard' correlation coefficient described above is more fully known as **Pearson's product-moment correlation coefficient**, to distinguish it.) The formula to be applied is the same as before, though there are a few tricks to be learned about constructing the ranks and also the hypothesis test is conducted in a different manner.

Table 7.3 Calculation of Spearman's rank correlation coefficient

Country	Birth rate Y	Growth rate X	Rank Y	Rank X	Y^2	X^2	XY
Brazil	30	5.1	7	3	49	9	21
Colombia	29	3.2	9	6	81	36	54
Costa Rica	30	3.0	7	7	49	49	49
India	35	1.4	4	10	16	100	40
Mexico	36	3.8	2.5	5	6.25	25	12.5
Peru	36	1.0	2.5	11	6.25	121	27.5
Philippines	34	2.8	5	8	25	64	40
Senegal	48	−0.3	1	12	1	144	12
South Korea	24	6.9	11	1	121	1	11
Sri Lanka	27	2.5	10	9	100	81	90
Taiwan	21	6.2	12	2	144	4	24
Thailand	30	4.6	7	4	49	16	28
Totals	–	–	78	78	647.5	650	409

Note: The country with the highest growth rate (South Korea) is ranked 1 for variable X; Taiwan, the next fastest growth nation, is ranked 2, etc. For the birth rate, Senegal is ranked 1, having the highest birth rate, 48. Taiwan has the lowest birth rate and so is ranked 12 for variable Y.

Using the ranks is generally less efficient than using the original data, because one is effectively throwing away some of the information (e.g. by *how much* do countries' growth rates differ). However, there is a trade-off: the rank correlation coefficient is more robust, i.e. it is less influenced by outliers or highly skewed distributions. If one suspects this is a risk, it may be better to use the ranks. This is similar to the situation where the median can prove superior to the mean as a measure of central tendency.

We will calculate the rank correlation coefficient for the data on birth and growth rates, to provide a comparison with the ordinary correlation coefficient calculated earlier. It is unlikely that the distributions of birth or of growth rates is particularly skewed (and we have too few observations to reliably tell) so the Pearson measure might generally be preferred, but we calculate the Spearman coefficient for comparison. Table 7.3 presents the data for birth and growth rates in the form of ranks. Calculating the ranks is fairly straightforward, though there are a couple of points to note.

The country with the highest birth rate has the rank of 1, the next highest 2, and so on. Similarly, the country with the highest growth rate ranks 1, etc. One could reverse a ranking, so the lowest birth rate ranks 1, for example; the direction of ranking can be somewhat arbitrary. This would leave the rank correlation coefficient unchanged in value, but the sign would change (e.g. 0.5 would become −0.5). This could be confusing as we would now have a 'negative' correlation rather than a positive one (though the birth rate variable would now have to be redefined). It is better to use the 'natural' order of ranking for each variable.

Where two or more observations are the same, as are the birth rates of Mexico and Peru, then they are given the same rank, which is the average of the relevant ranking values. For example, both countries are given the rank of 2.5, which is the average of 2 and 3. Similarly, Brazil, Costa Rica and Thailand are all given the rank of 7, which is the average of 6, 7 and 8. The next country, Colombia, is then given the rank of 9.

Excel warning

Microsoft Excel has a *rank()* function built in, which takes a variable and calculates a new variable consisting of the ranks, similar to the above table. However, note that it deals with tied values in a different way. In the example above, Brazil, Costa Rica and Thailand would all be given a rank of 6 by *Excel*, not 7. This then gives a different correlation coefficient to that calculated here. *Excel's* method can be shown to be problematic since, if the rankings are reversed (e.g. the highest growth country is numbered 12 rather than 1) *Excel* gives a different numerical result.

We now apply formula (7.1) to the ranked data, giving

$$r_s = \frac{n\Sigma XY - \Sigma X \Sigma Y}{\sqrt{(n\Sigma X^2 - (\Sigma X)^2)(n\Sigma Y^2 - (\Sigma Y)^2)}}$$

$$= \frac{12 \times 409 - 78 \times 78}{\sqrt{(12 \times 650 - 78^2)(12 \times 647.5 - 78^2)}} = -0.691$$

This indicates a negative rank correlation between the two variables, as with the standard correlation coefficient ($r = -0.824$), but with a slightly smaller absolute value.

To test the significance of the result a hypothesis test can be performed on the value of ρ_s, the corresponding population parameter

H_0: $\rho_s = 0$
H_1: $\rho_s \neq 0$

This time the *t* distribution cannot be used (because we are no longer relying on the parent distribution being Normal), but prepared tables of the critical values for ρ_s itself may be consulted; these are given in Table A6 (see page **426**), and an excerpt is given in Table 7.4.

The critical value at the 5% significance level, for $n = 12$, is 0.591. Hence the null hypothesis is rejected if the rank correlation coefficient falls outside the

Table 7.4 Excerpt from Table A6: Critical values of the rank correlation coefficient

n	10%	5%	2%	1%
5	0.900			
6	0.829	0.886	0.943	
⋮	⋮	⋮	⋮	⋮
11	0.523	0.623	0.763	0.794
12	0.497	0.591	0.703	0.780
13	0.475	0.566	0.673	0.746

Note: The critical value is given at the intersection of the shaded row and column.

range [−0.591, 0.591], which it does in this case. Thus the null can be rejected with 95% confidence; the data do support the hypothesis of a relationship between the birth rate and growth. This critical value shown in the table is for a two-tail test. For a one-tail test, the significance level given in the top row of the table should be halved.

Exercise 7.3

(a) Rank the observations for the income ratio across countries (highest = 1) and calculate the coefficient of rank correlation with the birth rate.

(b) Test the hypothesis that $\rho_s = 0$.

(c) Reverse the rankings for both variables and confirm that this does not affect the calculated test statistic.

Worked example 7.1

To illustrate all the calculations and bring them together without distracting explanation, we work through a simple example with the following data on X and Y:

Y	17	18	19	20	27	18
X	3	4	7	6	8	5

An XY graph of the data reveals the following picture, which suggests positive correlation:

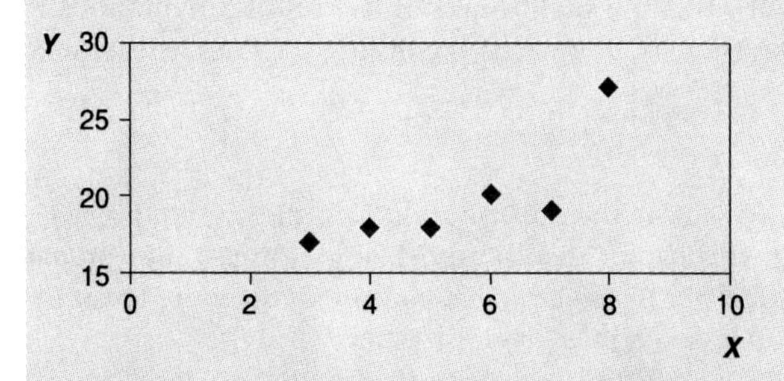

Note that one point appears to be something of an outlier. All the calculations for correlation may be based on the following table:

Obs	Y	X	Y^2	X^2	XY	Rank Y R_Y	Rank X R_X	R_Y^2	R_X^2	$R_X R_Y$
1	17	3	289	9	51	6	6	36	36	36
2	18	4	324	16	72	4.5	5	20.25	25	22.5
3	19	7	361	49	133	3	2	9	4	6
4	20	6	400	36	120	2	3	4	9	6
5	27	8	729	64	216	1	1	1	1	1
6	18	5	324	25	90	4.5	4	20.25	16	18
Totals	119	33	2427	199	682	21	21	90.5	91	89.5

The (Pearson) correlation coefficient r is therefore:

$$r = \frac{n\Sigma XY - \Sigma X \Sigma Y}{\sqrt{(n\Sigma X^2 - (\Sigma X)^2)(n\Sigma Y^2 - (\Sigma Y)^2)}}$$

$$= \frac{6 \times 682 - 33 \times 119}{\sqrt{(6 \times 199 - 33^2)(6 \times 2427 - 119^2)}} = 0.804$$

The hypothesis H_0: $\rho = 0$ versus H_1: $\rho \neq 0$ can be tested using the t test statistic:

$$t = \frac{r\sqrt{n-2}}{\sqrt{1-r^2}} = \frac{0.804 \times \sqrt{6-2}}{\sqrt{1-0.804^2}} = 2.7$$

which is compared to a critical value of 2.776, so the null hypothesis is not rejected, narrowly. This is largely attributable to the small number of observations and anyway it may be unwise to use the t-distribution on such a small sample. The rank correlation coefficient is calculated as

$$r = \frac{n\Sigma XY - \Sigma X \Sigma Y}{\sqrt{(n\Sigma X^2 - (\Sigma X)^2)(n\Sigma Y^2 - (\Sigma Y)^2)}}$$

$$= \frac{6 \times 89.5 - 21 \times 21}{\sqrt{(6 \times 91 - 21^2)(6 \times 90.5 - 21^2)}} = 0.928$$

The critical value at the 5% significance level is 0.886, so the rank correlation coefficient *is* significant, in contrast to the previous result. Not too much should be read into this, however; with few observations the ranking process can easily alter the result substantially.

A simpler formula

When the ranks occur without any ties, equation (7.1) simplifies to the following formula:

$$r_s = 1 - \frac{6 \times \Sigma d^2}{n(n^2 - 1)} \tag{7.4}$$

where d is the difference in the ranks. An example of the use of this formula is given below, using the following data for calculation

Rank Y	Rank X	d	d^2
1	5	−4	16
4	1	3	9
5	2	3	9
6	3	3	9
3	4	−1	1
2	6	−4	16
		Total	60

The differences d and their squared values are shown in the final columns of the table and from these we obtain

$$r_s = 1 - \frac{6 \times 60}{6 \times (6^2 - 1)} = -0.714 \tag{7.5}$$

This is the same answer as would be obtained using the conventional formula (7.1). The verification is left as an exercise. Remember, this formula can only be used if there are no ties in either variable.

Regression analysis

Regression analysis is a more sophisticated way of examining the relationship between two (or more) variables than is correlation. The major differences between correlation and regression are the following:

- Regression can investigate the relationships between two *or more* variables.
- A *direction* of causality is asserted, from the explanatory variable (or variables) to the dependent variable.
- The *influence* of each explanatory variable upon the dependent variable is measured.
- The *significance* of each explanatory variable can be ascertained.

Thus regression permits answers to such questions as:

- Does the growth rate influence a country's birth rate?
- If the growth rate increases, by how much might a country's birth rate be expected to fall?
- Are other variables important in determining the birth rate?

In this example we assert that the direction of causality is from the growth rate (X) to the birth rate (Y) and not vice versa. The growth rate is therefore the **explanatory** variable (also referred to as the **independent** or **exogenous** variable) and the birth rate is the **dependent** variable (also called the **explained** or **endogenous** variable).

Regression analysis describes this causal relationship by fitting a straight line drawn through the data, which best summarises them. It is sometimes called 'the line of best fit' for this reason. This is illustrated in Figure 7.4 for the birth rate and growth rate data. Note that (by convention) the explanatory variable is placed on the horizontal axis, the explained on the vertical. This regression line is downward sloping (its derivation will be explained shortly) for the same

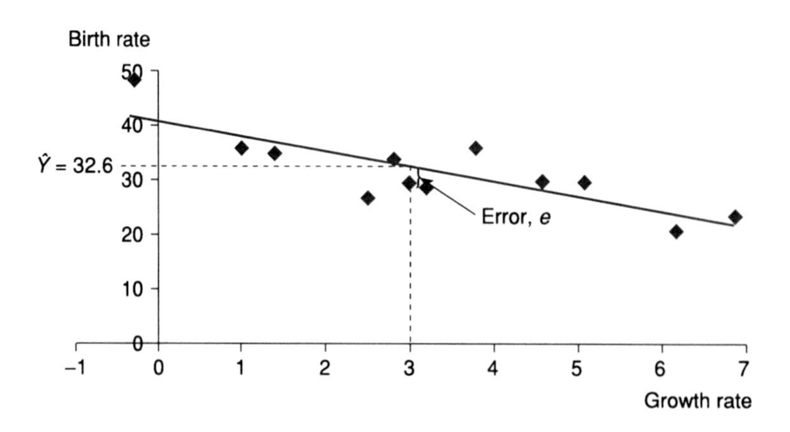

Figure 7.4
The line of best fit

reason that the correlation coefficient is negative, i.e. high values of Y are generally associated with low values of X and vice versa.

Since the regression line summarises knowledge of the relationship between X and Y, it can be used to predict the value of Y given any particular value of X. In Figure 7.4 the value of $X = 3$ (the observation for Costa Rica) is related via the regression line to a value of Y (denoted by $\hat{Y}$) of 32.6. This predicted value is close (but not identical) to the actual birth rate of 30. The difference reflects the absence of perfect correlation between the two variables.

The difference between the actual value, Y, and the predicted value, $\hat{Y}$, is called the **error** or **residual**. It is labelled e in Figure 7.4. (*Note*: The italic e denoting the error term should not be confused with the roman letter e, used as the base for natural logarithms (see Appendix 1C to Chapter 1, page **78**). Why should such errors occur? The relationship is never going to be an exact one for a variety of reasons. There are bound to be other factors besides growth which affect the birth rate (e.g. the education of women) and these effects are all subsumed into the error term. There might additionally be simple measurement error (of Y) and, of course, people do act in a somewhat random fashion rather than follow rigid rules of behaviour.

All of these factors fall into the error term and this means that the observations lie around the regression line rather than on it. If there are many of these factors, none of which is predominant, and they are independent of each other, then these errors may be assumed to be Normally distributed about the regression line.

Why not include these factors explicitly? On the face of it this would seem to be an improvement, making the model more realistic. However, the costs of doing this are that the model becomes more complex, calculation becomes more difficult (not so important now with computers) and it is generally more difficult for the reader (or researcher) to interpret what is going on. If the main interest is the relationship between the birth rate and growth, why complicate the model unduly? There is a virtue in simplicity, as long as the simplified model still gives an undistorted view of the relationship. In Chapter 10 on multiple regression the trade-off between simplicity and realism will be further discussed, particularly with reference to the problems which can arise if relevant explanatory variables are omitted from the analysis.

Calculation of the regression line

The equation of the sample regression line may be written

$$\hat{Y}_i = a + bX_i \tag{7.6}$$

where

$\hat{Y}_i$ is the predicted value of Y for observation (country) i
X_i is the value of the explanatory variable for observation i, and
a, b are fixed coefficients to be estimated; a measures the intercept of the regression line on the Y axis, b measures its slope.

This is illustrated in Figure 7.5.

The first task of regression analysis is to find the values of a and b so that the regression line may be drawn. To do this we proceed as follows. The difference between the actual value, Y_i, and its predicted value, $\hat{Y}_i$, is e_i, the error. Thus

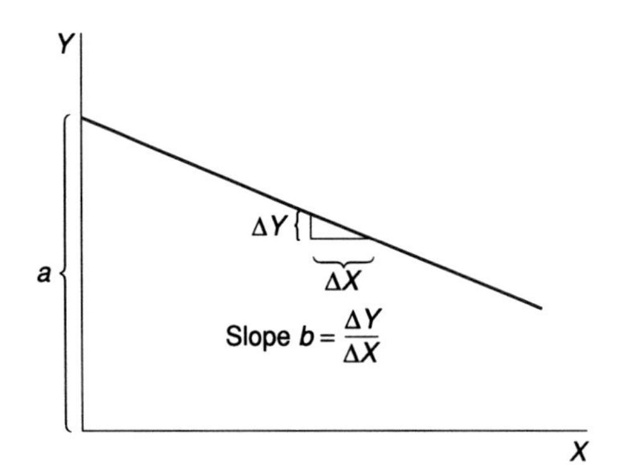

Figure 7.5
Intercept and slope of the regression line

$$Y_i = \hat{Y}_i + e_i \tag{7.7}$$

Substituting equation (7.6) into equation (7.7) the regression equation can be written

$$Y_i = a + bX_i + e_i \tag{7.8}$$

Equation (7.8) shows that observed birth rates are made up of two components:

(1) that part explained by the growth rate, $a + bX_i$, and
(2) an error component, e_i.

In a good model, part (1) should be large relative to part (2) and the regression line is based upon this principle. The line of best fit is therefore found by finding the values of a and b which *minimise the sum of squared errors* ($\sum e_i^2$) from the regression line. For this reason, this method is known as 'the method of least squares' or simply 'ordinary least squares' (OLS). The use of this criterion will be justified later on, but it can be said in passing that the sum of the errors is not minimised because that would not lead to a unique answer for the values a and b. In fact, there is an infinite number of possible regression lines which all yield a sum of errors equal to zero. Minimising the sum of *squared* errors does yield a unique answer.

The task is therefore to

$$\text{minimise } \sum e_i^2 \tag{7.9}$$

by choice of a and b.

Rearranging equation (7.8) the error is given by

$$e_i = Y_i - a - bX_i \tag{7.10}$$

so equation (7.9) becomes

$$\text{minimise } \sum (Y_i - a - bX_i)^2 \tag{7.11}$$

by choice of a and b.

Finding the solution to equation (7.11) requires the use of differential calculus, and is not presented here. The resulting formulae for a and b are

$$b = \frac{n\sum XY - \sum X \sum Y}{n\sum X^2 - (\sum X)^2} \tag{7.12}$$

and

$$a = \bar{Y} - b\bar{X} \quad (7.13)$$

where $\bar{X}$ and $\bar{Y}$ are the mean values of X and Y respectively. The values necessary to evaluate equations (7.12) and (7.13) can be obtained from Table 7.2 which was used to calculate the correlation coefficient. These values are repeated for convenience

$$\Sigma Y = 380 \qquad \Sigma Y^2 = 12\,564$$
$$\Sigma X = 40.2 \qquad \Sigma X^2 = 184.04$$
$$\Sigma XY = 1139.70 \qquad n = 12$$

Using these values we obtain

$$b = \frac{12 \times 1139.70 - 40.2 \times 380}{12 \times 184.04 - 40.2^2} = -2.700$$

and

$$a = \frac{380}{12} - (-2.700) \times \frac{40.2}{12} = 40.711$$

Thus the regression equation can be written, to two decimal places for clarity, as

$$Y_i = 40.71 - 2.70X_i + e_i$$

Interpretation of the slope and intercept

The most important part of the result is the slope coefficient $b = -2.7$ since it measures the effect of X upon Y. This result implies that a unit increase in the growth rate (e.g. from 2% to 3% p.a.) would lower the birth rate by 2.7, for example from 30 births per 1000 population to 27.3. Given that the growth data refer to a 20-year period (1961 to 1981), this increase in the growth rate would have to be sustained over such a time, not an easy task. How big is the effect upon the birth rate? The average birth rate in the sample is 31.67, so a reduction of 2.7 for an average country would be a fall of 8.5% (2.7/31.67 × 100). This is reasonably substantial (although not enough to bring the birth rate down to developed country levels) but would need a considerable, sustained increase in the growth rate to bring it about.

The value of a, the intercept, may be interpreted as the predicted birth rate of a country with zero growth (since $\hat{Y}_i = a$ at $X = 0$). This value of 40.71 is fairly close to that of Senegal, which actually had negative growth over the period and whose birth rate was 48, a little higher than the intercept value. Although a has a sensible interpretation in this case, this is not always so. For example, in a regression of the demand for a good on its price, a would represent demand at zero price, which is unlikely ever to be observed.

Exercise 7.4

(a) Calculate the regression line relating the birth rate to the income ratio.

(b) Interpret the coefficients of this equation.

Measuring the goodness of fit of the regression line

Having calculated the regression line we now ask whether it provides a good fit for the data, i.e. do the observations tend to lie close to, or far away from, the line? If the fit is poor, perhaps the effect of X upon Y is not so strong after all. Note that even if X has *no* effect upon Y we can still calculate a regression line and its slope coefficient b. Although b is likely to be small, it is unlikely to be exactly zero. Measuring the goodness of fit of the data to the line helps us to distinguish between good and bad regressions.

We proceed by comparing the three competing models explaining the birth rate. Which of them fits the data best? Using the income ratio and the GNP variable gives the following regressions (calculations not shown) to compare with our original model:

for the income ratio (IR):	$B = 26.44 + 1.045 \times \text{IR} + e$
for GNP:	$B = 34.72 - 0.003 \times \text{GNP} + e$
for growth:	$B = 40.71 - 2.70 \times \text{GROWTH} + e$

How can we decide which of these three is 'best' on the basis of the regression equations alone? From Figure 7.1 it is evident that some relationships appear stronger than others, yet this is not revealed by examining the regression equation alone. More information is needed. (You cannot choose the best equation simply by looking at the size of the coefficients. Try to think why.)

The goodness of fit is calculated by comparing two lines: the regression line and the 'mean line' (i.e. a horizontal line drawn at the mean value of Y). The regression line *must* fit the data better (if the mean line were the best fit, that is also where the regression line would be) but the question is how much better? This is illustrated in Figure 7.6, which demonstrates the principle behind the calculation of the **coefficient of determination**, denoted by R^2 and usually more simply referred to as 'R squared'.

The figure shows the mean value of Y, the calculated sample regression line and an arbitrarily chosen sample observation (X_i, Y_i). The difference between Y_i and $\bar{Y}$ (length $Y_i - \bar{Y}$) can be divided up into:

(1) That part 'explained' by the regression line, $\hat{Y}_i - \bar{Y}$ (i.e. explained by the value of X_i).
(2) The error term $e_i = Y_i - \hat{Y}_i$.

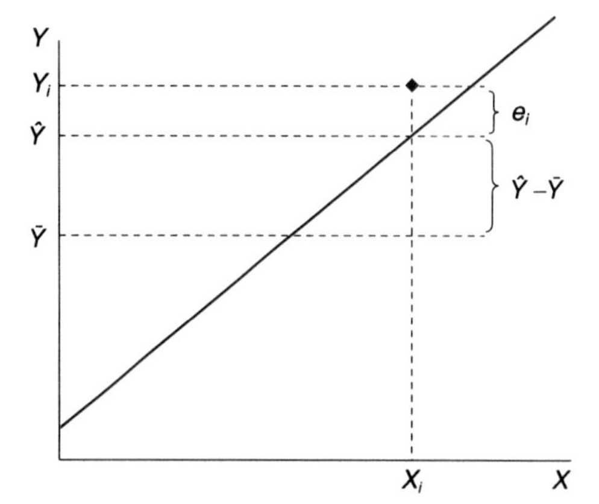

Figure 7.6
The calculation of R^2

In algebraic terms

$$Y_i - \bar{Y} = (Y - \hat{Y}_i) + (\hat{Y}_i - \bar{Y}) \tag{7.14}$$

A good regression model should 'explain' a large part of the differences between the Y_i values and $\bar{Y}$, i.e. the length $\hat{Y}_i - \bar{Y}$ should be large relative to $Y_i - \bar{Y}$. A measure of fit would therefore be $(\hat{Y}_i - \bar{Y})/(Y_i - \bar{Y})$. We need to apply this to all observations, not just a single one. Hence we need to sum this expression over all the sample observations. A problem is that some of the terms would take a negative value and offset the positive terms. To get round this problem we square each of the terms in equation (7.14) to make them all positive, and then sum over the observations. This gives

$\Sigma(Y_i - \bar{Y})^2$, known as the total sum of squares (TSS)
$\Sigma(\hat{Y}_i - \bar{Y})^2$, the regression sum of squares (RSS), and
$\Sigma(Y_i - \hat{Y}_i)^2$, the error sum of squares (ESS)

The measure of goodness of fit, R^2, is then defined as the ratio of the regression sum of squares to the total sum of squares, i.e.

$$R^2 = \frac{\text{RSS}}{\text{TSS}} \tag{7.15}$$

The better the divergences between Y_i and $\bar{Y}$ are explained by the regression line, the better the goodness of fit, and the higher the calculated value of R^2. Further, it is true that

$$\text{TSS} = \text{RSS} + \text{ESS} \tag{7.16}$$

From equations (7.15) and (7.16) we can then see that R^2 must lie between 0 and 1 (note that since each term in equation (7.16) is a sum of squares, none of them can be negative). Thus

$$0 \leqslant R^2 \leqslant 1$$

A value of $R^2 = 1$ indicates that all the sample observations lie exactly on the regression line (equivalent to perfect correlation). If $R^2 = 0$ then the regression line is of no use at all – X does not influence Y (linearly) at all, and to try to predict a value of Y_i one might as well use the mean $\bar{Y}$ rather than the value X_i inserted into the sample regression equation.

To calculate R^2, alternative formulae to those above make the task easier. Instead we use

$$\begin{aligned}
\text{TSS} &= \Sigma(Y_i - \bar{Y})^2 = \Sigma Y_i^2 - n\bar{Y}^2 = 12\,564 - 12 \times 31.67^2 = 530.667 \\
\text{ESS} &= \Sigma(Y_i - \hat{Y})^2 = \Sigma Y_i^2 - a\,\Sigma Y_i - b\,\Sigma X_i Y_i \\
&= 12\,564 - 40.711 \times 380 - (-2.7) \times 1139.70 = 170.754 \\
\text{RSS} &= \text{TSS} - \text{ESS} = 530.667 - 170.754 = 359.913
\end{aligned}$$

This gives the result

$$R^2 = \frac{\text{RSS}}{\text{TSS}} = \frac{359.913}{530.667} = 0.678$$

This is interpreted as follows. Countries' birth rates vary around the overall mean value of 31.67. 67.8% of this variation is explained by variation in countries' growth rates. This is quite a respectable figure to obtain, leaving only 32.8% of the variation in Y left to be explained by other factors (or pure random variation). The regression seems to make a worthwhile contribution to explaining why birth rates differ.

It turns out that in simple regression (i.e. where there is only one explanatory variable), R^2 is simply the square of the correlation coefficient between X and Y. Thus for the income ratio and for GNP we have

for IR: $R^2 = 0.35^2 = 0.13$
for GNP: $R^2 = -0.26^2 = 0.07$

This shows, once again, that these other variables are not terribly useful in explaining why birth rates differ. Each of them only explains a small proportion of the variation in Y.

It should be emphasised at this point that R^2 is not the only criterion (or even an adequate one in all cases) for judging the quality of a regression equation and that other statistical measures, set out below, are also required.

Exercise 7.5

(a) Calculate the R^2 value for the regression of the birth rate on the income ratio, calculated in Exercise 7.4.

(b) Confirm that this result is the same as the square of the correlation coefficient between these two variables, calculated in Exercise 7.1.

Inference in the regression model

So far, regression has been used as a descriptive technique, to measure the relationship between the two variables. We now go on to draw inferences from the analysis about what the *true* regression line might look like. As with correlation, the estimated relationship is in fact a *sample* regression line, based upon data for 12 countries. The estimated coefficients a and b are random variables, since they would differ from sample to sample. What can be inferred about the true (but unknown) regression equation?

The question is best approached by first writing down a true or population regression equation, in a form similar to the sample regression equation

$$Y_i = \alpha + \beta X_i + \varepsilon_i \tag{7.17}$$

As usual, Greek letters denote true, or population, values. Thus α and β are the population *parameters*, of which a and b are (point) estimates, using the method of least squares, and ε is the population error term. If we could observe the individual error terms ε_i then we would be able to get exact values of α and β (even from a sample), rather than just estimates.

Given that a and b are estimates, we can ask about their properties: whether they are unbiased and how precise they are, compared to alternative estimators. Under reasonable assumptions (e.g. see Maddala (2001), Chapter 3) it can be shown that the OLS estimates of the coefficients are unbiased. Thus OLS provides useful point estimates of the parameters (the true values α and β). This is one reason for using the least squares method. It can also be shown that, among the class of linear unbiased estimators, OLS has the minimum variance, i.e. the method provides the most precise estimates. This is another, powerful justification for the use of OLS. So, just as the sample mean provides a more precise estimate of the population mean than does a single observation, the least squares estimates of α and β are the most precise.

Analysis of the errors

To find confidence intervals for α and β we need to know which statistical distribution we should be using, i.e. the distributions of a and b. These can be derived, based on the assumptions that the error term ε in equation (7.17) above is Normally distributed and that the errors are statistically independent of each other. Since we are using cross-section data from countries which are different geographically, politically and socially it seems reasonable to assume the errors are independent.

To check the Normality assumption we can graph the residuals calculated from the sample regression line. If the true errors are Normal it seems likely that these residuals should be approximately Normal also. The residuals are calculated according to equation (7.10) above. For example, to calculate the residual for Brazil we subtract the fitted value from the actual value. The fitted value is calculated by substituting the growth rate into the estimated regression equation, yielding $\hat{Y} = 40.712 - 2.7 \times 5.1 = 26.9$. Subtracting this from the actual value gives $Y_i - \hat{Y} = 30 - 26.9 = 3.1$. Other countries' residuals are calculated in similar manner, yielding the results shown in Table 7.5.

These residuals may then be gathered together in a frequency table (as in Chapter 1) and graphed. This is shown in Figure 7.7.

Although the number of observations is small (and therefore the graph is not a smooth curve) the chart does have the greater weight of frequencies in the centre as one would expect, with less weight as one moves into the tails of the distribution. The assumption that the true error term is Normally distributed does not seem unreasonable.

If the residuals from the sample regression equation appeared distinctly non-Normal (heavily skewed, for example) then one should be wary of constructing confidence intervals using the formulae below. Instead, one might consider transforming the data (see below) before continuing. There are more formal tests for Normality of the residuals but they are beyond the scope of this book. Drawing a graph is an informal alternative, which can be useful, but remember that graphical methods can be misinterpreted.

Table 7.5 Calculation of residuals

	Actual birth rate	Fitted values	Residuals
Brazil	30	26.9	3.1
Colombia	29	32.1	−3.1
Costa Rica	30	32.6	−2.6
⋮	⋮	⋮	⋮
Sri Lanka	27	34.0	−7.0
Taiwan	21	24.0	−3.0
Thailand	30	28.3	1.7

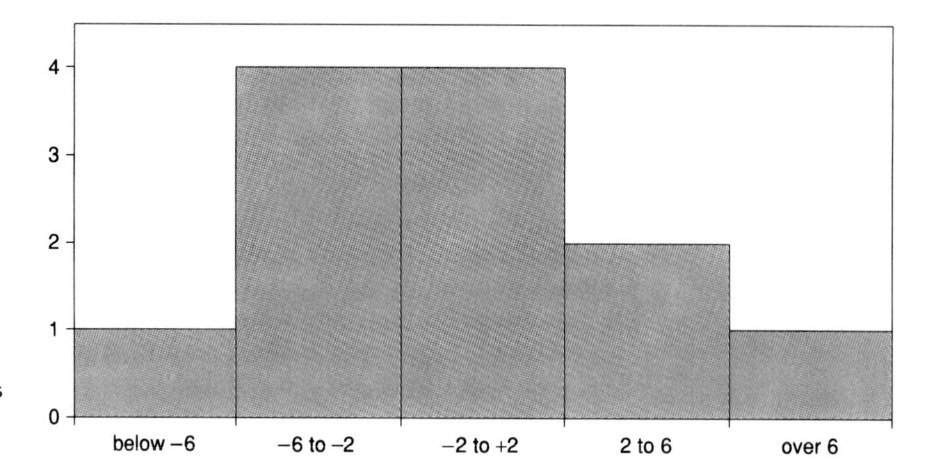

Figure 7.7
Bar chart of residuals from the regression equation

If one were using time-series data one should also check the residuals for **autocorrelation** at this point. This occurs when the error in period t is dependent in some way on the error in the previous period(s) and implies that the method of least squares may not be the best way of estimating the relationship. In this example we have cross-section data, so it is not appropriate to check for autocorrelation, since the ordering of the data does not matter. Chapter 8, on multiple regression, covers this topic.

Confidence interval estimates of α and β

Having checked that the residuals appear reasonably Normal we can proceed with inference. This means finding interval estimates of the parameters α and β and, later on, conducting hypothesis tests. As usual, the 95% confidence interval is obtained by adding and subtracting approximately two standard errors from the point estimate. We therefore need to calculate the standard error of a and of b and we also need to look up tables to find the precise number of standard errors to add and subtract. The principle is just the same as for the confidence interval estimate of the sample mean, covered in Chapter 4.

The estimated sampling variance of b, the slope coefficient, is given by

$$s_b^2 = \frac{s_e^2}{\Sigma(X_i - \bar{X})^2} \tag{7.18}$$

where

$$s_e^2 = \frac{\Sigma e_i^2}{n-2} = \frac{\text{ESS}}{n-2} \tag{7.19}$$

is the **estimated variance of the error term**, ε.

The sampling variance of b measures the uncertainty associated with the estimate. Note that the uncertainty is greater (i) the larger the error variance s_e^2 (i.e. the more scattered the points around the regression line) and (ii) the lower the dispersion of the X observations. When X does not vary much it is then more difficult to measure the effect of changes in X upon Y, and this is reflected in the formula.

First we need to calculate s_e^2. The value of this is

$$s_e^2 = \frac{170.754}{10} = 17.0754 \quad (7.20)$$

and so the estimated variance of b is

$$s_b^2 = \frac{17.0754}{49.37} = 0.346 \quad (7.21)$$

(Use $\Sigma(X_i - \bar{X})^2 = \Sigma X_i^2 - n\bar{X}^2$ in calculating (7.21) – it makes the calculation easier.) The estimated standard error of b is the square root of (7.21),

$$s_b = \sqrt{0.346} = 0.588 \quad (7.22)$$

To construct the confidence interval around the point estimate, $b = -2.7$, the t distribution is used (in regression this applies to all sample sizes, not just small ones). The 95% confidence interval is thus given by

$$[b - t_v s_b, b + t_v s_b] \quad (7.23)$$

where t_v is the (two-tail) critical value of the t distribution at the appropriate significance level (5% in this case), with $v = n - 2$ degrees of freedom. The critical value is 2.228. Thus the confidence interval evaluates to

$$[-2.7 - 2.228 \times 0.588, -2.7 + 2.228 \times 0.588] = [-4.01, -1.39]$$

Thus we can be 95% confident that the true value of β lies within this range. Note that the interval only includes negative values: we can rule out an upwards-sloping regression line.

For the intercept a, the estimate of the variance is given by

$$s_a^2 = s_e^2 \times \left(\frac{1}{n} + \frac{\bar{X}^2}{\Sigma(X_i - \bar{X})^2}\right) = 17.0754 \times \left(\frac{1}{12} + \frac{3.35^2}{49.37}\right) = 5.304 \quad (7.24)$$

and the estimated standard error of a is the square root of this, 2.303. The 95% confidence interval for α, again using the t distribution, is

$$[40.71 - 2.228 \times 2.303, 40.71 + 2.228 \times 2.303] = [35.57, 45.84]$$

The results so far can be summarised as follows

$$Y_i = 40.711 - 2.70X_i + e_i$$
s.e. (2.30) (0.59)
$R^2 = 0.678$ $n = 12$

This conveys, at a glance, all the necessary information to the reader, who can then draw the inferences deemed appropriate. Any desired confidence interval (not just the 95% one) can be quickly calculated with the aid of a set of t tables.

Testing hypotheses about the coefficients

As well as calculating confidence intervals, one can use hypothesis tests as the basis for statistical inference in the regression model. These tests are quickly and easily explained given the information already assembled. Consider the following hypothesis

H_0: $\beta = 0$
H_1: $\beta \neq 0$

This null hypothesis is interesting because it implies no influence of X upon Y at all (i.e. the slope of the true regression line is flat and Y_i can be equally well predicted by $\bar{Y}$). The alternative hypothesis asserts that X does in fact influence Y.

The procedure is in principle the same as in Chapter 5 on hypothesis testing. We measure how many standard deviations separate the observed value of b from the hypothesised value. If this is greater than an appropriate critical value we reject the hypothesis. The test statistic is calculated using the formula

$$t = \frac{b - \beta}{s_b} = \frac{-2.7 - 0}{0.588} = -4.59 \qquad (7.25)$$

Thus the sample coefficient b differs by 4.59 standard errors from its hypothesised value $\beta = 0$. This is compared to the critical value of the t distribution, using $n - 2$ degrees of freedom. Since $t < -t^*_{10}$ (= −2.228), in this case the null hypothesis is rejected with 95% confidence. X does have some influence on Y. Similar tests using the income ratio and GDP to attempt to explain the birth rate show that in neither case is the slope coefficient significantly different from zero, i.e. neither of these variables appears to influence the birth rate.

Rule of thumb for hypothesis tests

A quick and reasonably accurate method for establishing whether a coefficient is significantly different from zero is to see if it is at least twice its standard error. If so, it is significant. This works because the critical value (at 95%) of the t distribution for reasonable sample sizes is about 2.

Sometimes regression results are presented with the t statistic (as calculated above), rather than the standard error, below each coefficient. This implicitly assumes that the hypothesis of interest is that the coefficient is zero. This is not always appropriate: in the consumption function a test for the marginal propensity to consume being equal to 1 might be of greater relevance, for example. In a demand equation, one might want to test for unit elasticity. For this reason, it is better to present the standard errors rather than the t statistics.

Note that the test statistic $t = -4.59$ is exactly the same result as in the case of testing the correlation coefficient. This is no accident, for the two tests are equivalent. A non-zero slope coefficient means there is a relationship between X and Y which also means the correlation coefficient is non-zero. Both null hypotheses are rejected.

Testing the significance of R^2: the F test

Another check of the quality of the regression equation is to test whether the R^2 value, calculated earlier, is significantly greater than zero. This is a test using the F distribution and turns out once again to be equivalent to the two previous tests H_0: $\beta = 0$ and H_0: $\rho = 0$, conducted in previous sections, using the t distribution.

The null hypothesis for the test is H_0: $R^2 = 0$, implying once again that X does not influence Y (hence equivalent to $\beta = 0$). The test statistic is

$$F = \frac{R^2/1}{(1 - R^2)/(n - 2)} \tag{7.26}$$

or equivalently

$$F = \frac{\text{RSS}/1}{\text{ESS}/(n - 2)} \tag{7.27}$$

The F statistic is therefore the ratio of the regression sum of squares to the error sum of squares, each divided by their degrees of freedom (for the RSS there is one degree of freedom because of the one explanatory variable, for the ESS there are $n - 2$ degrees of freedom). A high value of the F statistic rejects H_0 in favour of the alternative hypothesis, H_1: $R^2 > 0$. Evaluating (7.26) gives

$$F = \frac{0.678/1}{(1 - 0.678)/10} = 21.078 \tag{7.28}$$

The critical value of the F distribution at the 5% significance level, with $\nu_1 = 1$ and $\nu_2 = 10$, is $F^*_{1,10} = 4.96$. The test statistic exceeds this, so the regression as a whole is significant. It is better to use the regression model to explain the birth rate than to use the simpler model which assumes all countries have the same birth rate (the sample average).

As stated before, this test is equivalent to those carried out before using the t distribution. The F statistic is, in fact, the square of the t statistic calculated earlier ($-4.59^2 = 21.078$) and reflects the fact that, in general

$$F_{1,n-2} = t^2_{n-2}$$

The Prob-value associated with both statistics is the same (approximately 0.001 in this case) so both tests reject the null at the same level of significance. However, in multiple regression with more than one explanatory variable, the relationship no longer holds and the tests do fulfil different roles, as we shall see in the next chapter.

Exercise 7.6

(a) For the regression of the birth rate on the income ratio, calculate the standard errors of the coefficients and hence construct 95% confidence intervals for both.

(b) Test the hypothesis that the slope coefficient is zero against the alternative that it is not zero.

(c) Test the hypothesis H_0: $R^2 = 0$.

Interpreting computer output

Having shown how to use the appropriate formulae to derive estimates of the parameters, their standard errors and to test hypotheses, we now present all these results as they would be generated by a computer software package, in this case *Excel*. This removes all the effort of calculation and allows us to concentrate on more important issues such as the interpretation of the results. Table 7.6 shows the computer output.

Table 7.6 Regression analysis output using *Excel*

Regression Statistics	
Multiple R	0.824
R square	0.678
Adjusted R square	0.646
Standard error	4.132
Observations	12

ANOVA

	df	SS	MS	F	Significance F
Regression	1	359.913	359.913	21.078	0.001
Residual	10	170.754	17.075		
Total	11	530.667			

	Coefficients	Standard Error	t Stat	P-value	Lower 95%	Upper 95%
Intercept	40.71	2.30	17.68	7.15E-09	35.58	45.84
GR	-2.70	0.59	-4.59	0.001	-4.01	-1.39

The table presents all the results we have already derived, plus a few more.

- The regression coefficients, standard errors and t ratios are given at the bottom of the table, suitably labelled. The column headed '*P value*' (this is how *Excel* refers to the *Prob-value*, discussed in Chapter 5) gives some additional information – it shows the significance level of the t statistic. For example, the slope coefficient is significant at the level of 0.1%,[2] i.e. there is this probability of getting such a sample estimate by chance. This is much less than our usual 5% criterion, so we conclude that the sample evidence did not arise by chance.
- The program helpfully calculates the 95% confidence interval for the coefficients also, which were derived above in equation (7.23).
- Moving up the table, there is a section headed ANOVA (Analysis of Variance). This is similar to the ANOVA covered in Chapter 6. This table provides the sums of squares values (RSS, ESS and TSS, in that order) and their associated degrees of freedom in the '*df*' column. The '*MS*' ('mean square') column calculates the sums of squares each divided by their degrees of freedom, whose ratio gives the F statistic in the next column. This is the value calculated in equation (7.28). The '*Significance F*' value is similar to the *P value* discussed previously: it shows the level at which the F statistic is significant (0.1% in this case) and saves us looking up the F tables.
- At the top of the table is given the R^2 value and the standard error of the error term, s_e, labelled 'Standard Error', which we have already come across. 'Multiple R' is simply the square root of R^2; 'Adjusted R^2' (sometimes called 'R-bar squared' and written $\bar{R}^2$) adjusts the R^2 value for the degrees of freedom. This is an alternative measure of fit, which is not affected by the number of explanatory variables, unlike R^2. See Maddala (2001) Chapter 4 for a more detailed explanation.

[2] This is the area in *both* tails, so it is for a two-tail test.

Prediction

Earlier we showed that the regression line could be used for prediction, using the figures for Costa Rica. The point estimate of Costa Rica's birth rate is calculated simply by putting its growth rate into the regression equation and assuming a zero value for the error, i.e.

$$\hat{Y} = 40.711 - 2.7 \times 3 + 0 = 32.6$$

This is a point estimate, which is unbiased, around which we can build a confidence interval. There are, in fact, two confidence intervals we can construct, the first for the position of the *regression line* at $X = 3$, the second for an *individual observation* (on Y) at $X = 3$. Using the 95% confidence level, the first interval is given by the formula

$$\left[\hat{Y} - t_{n-2} \times s_e\sqrt{\frac{1}{n} + \frac{(X_P - \bar{X})^2}{\Sigma(X - \bar{X})^2}},\ \hat{Y} + t_{n-2} \times s_e\sqrt{\frac{1}{n} + \frac{(X_P - \bar{X})^2}{\Sigma(X - \bar{X})^2}}\right] \quad (7.29)$$

where X_P is the value of X for which the prediction is made. t_{n-2} denotes the critical value of the t distribution at the 5% significance level (for a two-tail test) with $n - 2$ degrees of freedom. This evaluates to

$$\left[32.6 - 2.228 \times 4.132\sqrt{\frac{1}{12} + \frac{(3 - 3.35)^2}{49.37}},\right.$$

$$\left.32.6 + 2.228 \times 4.132\sqrt{\frac{1}{12} + \frac{(3 - 3.35)^2}{49.37}}\right]$$

$$= [29.90, 35.30]$$

This means that we predict with 95% confidence that the *average* birth rate of all countries growing at 3% p.a. is between 29.9 and 35.3.

The second type of interval, for the value of Y itself at $X_P = 3$, is somewhat wider, because there is an additional element of uncertainty: individual countries do not lie on the regression line, but around it. This is referred to as the 95% **prediction interval**. The formula for this interval is

$$\left[\hat{Y} - t_{n-2} \times s_e\sqrt{1 + \frac{1}{n} + \frac{(X_P - \bar{X})^2}{\Sigma(X - \bar{X})^2}},\right.$$

$$\left.\hat{Y} + t_{n-2} \times s_e\sqrt{1 + \frac{1}{n} + \frac{(X_P - \bar{X})^2}{\Sigma(X - \bar{X})^2}}\right] \quad (7.30)$$

Note the extra '1' inside the square root sign. When evaluated, this gives a 95% prediction interval of [23.01, 42.19]. Thus we are 95% confident that an individual country growing at 3% p.a. will have a birth rate within this range.

The two intervals are illustrated in Figure 7.8. The smaller confidence interval is shown in a darker shade, with the wider prediction interval being about twice as big. Note from the formulae that the prediction is more precise (the interval is smaller)

- the closer the sample observations lie to the regression line (smaller s_e);
- the greater the spread of sample X values (larger $\Sigma(X - \bar{X})^2$);

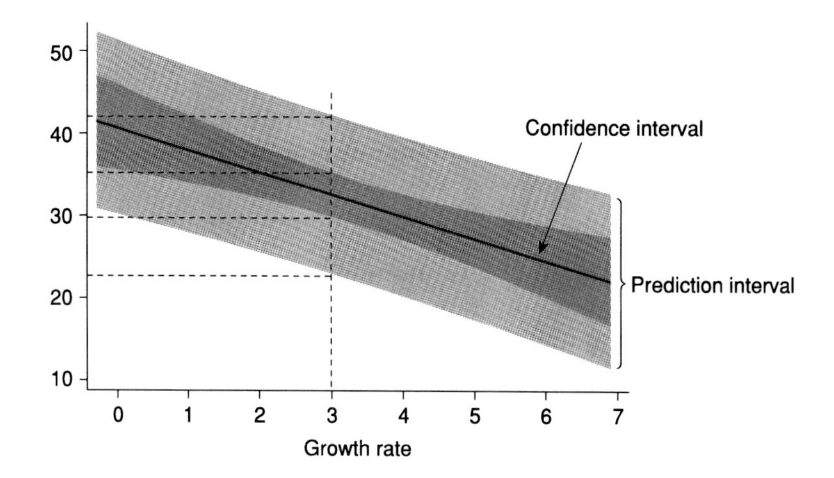

Figure 7.8
Confidence and prediction intervals

- the larger the sample size;
- the closer to the mean of X the prediction is made (smaller $X_p - \bar{X}$).

This last characteristic is evident in the diagram, where the intervals are narrower towards the centre of the diagram.

There is an additional danger of predicting far outside the range of sample X values, if the true regression line is not linear as we have assumed. The linear sample regression line might be close to the true line within the range of sample X values but diverge substantially outside. Figure 7.9 illustrates this point.

In the birth rate sample, we have a fairly wide range of X values; few countries grow more slowly than Senegal or faster than Korea.

Exercise 7.7

Use *Excel*'s regression tool to confirm your answers to Exercises 7.4 to 7.6.

Exercise 7.8

?

(a) Predict (point estimate) the birth rate for a country with an income ratio of 10.

(b) Find the 95% confidence interval prediction for a typical country with IR = 10.

(c) Find the 95% confidence interval prediction for an individual country with IR = 10.

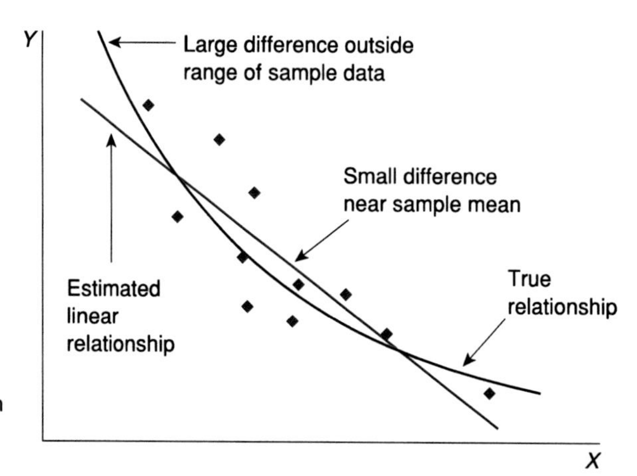

Figure 7.9
The danger of prediction outside the range of sample data

Worked example 7.2

We continue the previous worked example, completing the calculations needed for regression. The previous table contains most of the preliminary calculations. To find the regression line we use

$$b = \frac{n\sum XY - \sum X \sum Y}{n\sum X^2 - (\sum X)^2} = \frac{6 \times 682 - 33 \times 119}{6 \times 199 - 33^2} = 1.57$$

and

$$a = 19.83 - 1.57 \times 5.5 = 11.19$$

Hence we obtain the equation

$$Y_i = 11.19 + 1.57X_i + e_i$$

For inference, we start with the sums of squares:

$$\begin{aligned}
\text{TSS} &= \sum(Y_i - \bar{Y})^2 = \sum Y_i^2 - n\bar{Y}^2 = 2427 - 6 \times 19.83^2 = 66.83 \\
\text{ESS} &= \sum(Y_i - \hat{Y}_i)^2 = \sum Y_i^2 - a\sum Y_i - b\sum X_i Y_i \\
&= 2427 - 11.19 \times 119 - 1.57 \times 682 = 23.62 \\
\text{RSS} &= \text{TSS} - \text{ESS} = 66.83 - 23.62 = 43.21
\end{aligned}$$

We then obtain R^2 = RSS/TSS = 43.21/66.83 = 0.647 or 64.7% of the variation in Y explained by variation in X.

To obtain the standard errors of the coefficients, we first calculate the error variance as $s_e^2 = \text{ESS}/(n-2) = 23.62/4 = 5.905$ and the estimated variance of the slope coefficient is

$$s_b^2 = \frac{s_e^2}{\sum(X - \bar{X})^2} = \frac{5.905}{17.50} = 0.338$$

and the standard error of b is therefore $\sqrt{0.338} = 0.581$.

Similarly for a we obtain

$$s_a^2 = s_e^2 \times \left(\frac{1}{n} + \frac{\bar{X}^2}{\sum(X - \bar{X})^2}\right) = 5.905 \times \left(\frac{1}{6} + \frac{5.5^2}{17.50}\right) = 11.19$$

and the standard error of a is therefore 3.34.

Confidence intervals for a and b can be constructed using the critical value of the t distribution, 2.776 (5%, $\nu = 4$), yielding $1.57 \pm 2.776 \times 0.581$ = [−0.04, 3.16] for b and [1.90, 20.47] for a. Note that zero is inside the confidence interval for b. This is also reflected in the test of H_0: $\beta = 0$ which is

$$t = \frac{1.57 - 0}{0.581} = 2.71$$

which falls short of the two-tailed critical value, 2.776. Hence H_0 cannot be rejected.

The F statistic, to test H_0: $R^2 = 0$ is

$$F = \frac{\text{RSS}/1}{\text{ESS}/(n-2)} = \frac{43.21/1}{23.62/(6-2)} = 7.32$$

which compares to a critical value of $F(1,4)$ of 7.71 so, again, the null cannot be rejected (remember that this and the t test on the slope coefficient are equivalent in simple regression).

We shall predict the value of Y for a value of $X = 10$, yielding $\hat{Y} = 11.19 + 1.57 \times 10 = 26.90$. The 95% confidence interval for this prediction is calculated using equation (7.29), which gives

$$\left[\begin{array}{l} 26.90 - 2.776 \times 2.43\sqrt{\frac{1}{6} + \frac{(10 - 5.5)^2}{17.50}}, \\ \qquad 26.90 + 2.776 \times 2.43\sqrt{\frac{1}{6} + \frac{(10 - 5.5)^2}{17.50}} \end{array}\right] = [19.14, 34.66].$$

The 95% prediction interval for an actual observation at $X = 10$ is given by (7.30), resulting in

$$\left[\begin{array}{l} 26.90 - 2.776 \times 2.43\sqrt{1 + \frac{1}{6} + \frac{(10 - 5.5)^2}{17.50}}, \\ \qquad 26.90 + 2.776 \times 2.43\sqrt{1 + \frac{1}{6} + \frac{(10 - 5.5)^2}{17.50}} \end{array}\right] = [16.62, 37.18].$$

Units of measurement

The measurement and interpretation of the regression coefficients depends upon the units in which the variables are measured. For example, suppose we had measured the birth rate in births per *hundred* (not *thousand*) of population; what would be the implications? Obviously nothing fundamental is changed; we ought to obtain the same qualitative result, with the same interpretation. However, the regression coefficients cannot remain the same: if the slope coefficient remained $b = -2.7$, this would mean that an increase in the growth rate of one percentage point reduces the birth rate by 2.7 births *per hundred*, which is clearly wrong. The right answer should be 0.27 births per hundred (equivalent to 2.7 per thousand) so the coefficient should change to $b = -0.27$. Thus, in general, the sizes of the coefficients depend upon the units in which the variables are measured. This is why one cannot judge the importance of a regression equation from the size of the coefficients alone.

It is easiest to understand this in graphical terms. A graph of the data will look exactly the same, except that the scale on the Y-axis will change; it will be divided by 10. The intercept of the regression line will therefore change to $a = 4.0711$ and the slope to $b = -0.27$. Thus the regression equation becomes

$$Y_i = 4.0711 - 0.27X_i + e_i'$$
$$(e_i' = e_i/10)$$

Since nothing fundamental has altered, any hypothesis test must yield the same test statistic. Thus t and F statistics are unaltered by changes in the units of measurement; nor is R^2 altered. However, standard errors will be divided by 10 (they have to be to preserve the t statistics; see equation (7.25) for example). Table 7.7 sets out the effects of changes in the units of measurement upon the

Table 7.7 The effects of data transformations

Factor (*k*) multiplying . . .		Effect upon			
Y	X	a	s_a	b	s_b
k	1	All multiplied by k			
1	k	Unchanged		Divided by k	
k	k	Multiplied by k		Unchanged	

coefficients and standard errors. In the table it is assumed that the variables have been multiplied by a constant k; in the above case $k = 1/10$ was used.

It is important to be aware of the units in which the variables are measured. If not, it is impossible to know how large is the effect of X upon Y. It may be statistically significant but you have no idea of how important it is. This may occur if, for instance, one of the variables is presented as an index number (see Chapter 10) rather than in the original units.

How to avoid measurement problems: calculating the elasticity

A neat way of avoiding the problems of measurement is to calculate the **elasticity**, i.e. the *proportionate* change in Y divided by the *proportionate* change in X. The proportionate changes are the same whatever units the variables are measured in. The proportionate change in X is given by $\Delta X/X$, where ΔX indicates the *change* in X. Thus if X changes from 100 to 110, the proportionate change is $\Delta X/X = 10/100 = 0.1$ or 10%. The elasticity, η, is therefore given by

$$\eta = \frac{\Delta Y/Y}{\Delta X/X} = \frac{\Delta Y}{\Delta X} \times \frac{X}{Y} \tag{7.31}$$

The second form of the equation is more useful, since $\Delta Y/\Delta X$ is simply the slope coefficient b. We simply need to multiply this by the ratio X/Y, therefore. But what values should be used for X and Y? The convention is to use the means, so we obtain the following formula for the elasticity, from a linear regression equation

$$\eta = b \times \frac{\bar{X}}{\bar{Y}} \tag{7.32}$$

This evaluates to $-2.7 \times 3.35/31.67 = -0.29$. This is interpreted as follows: a 1% increase in the growth rate would lead to a 0.29% decrease in the birth rate. Equivalently, and perhaps a little more usefully, a 10% rise in growth (from say 3% to 3.3% p.a.) would lead to a 2.9% decline in the birth rate (e.g. from 30 to 29.13). This result is the same whatever units the variables X and Y are measured in.

Note that this elasticity is measured at the means; it would have a different value at different points along the regression line. Later on we show an alternative method for estimating the elasticity, in this case the elasticity of demand which is familiar in economics.

Non-linear transformations

So far only *linear* regression has been dealt with, that is fitting a straight line to the data. This can sometimes be restrictive, especially when there is good reason

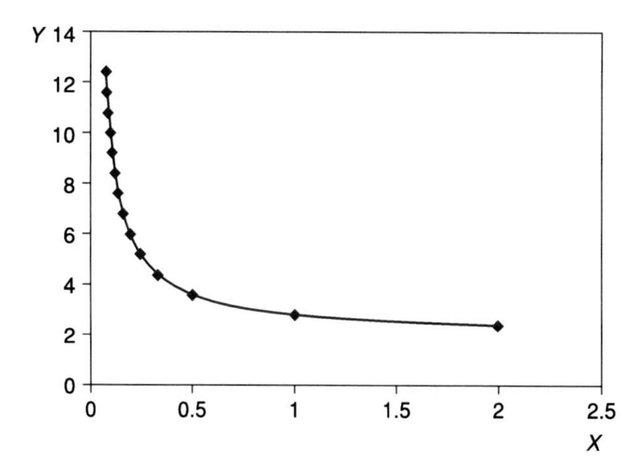

Figure 7.10
Graph of Y against X

to believe that the true relationship is non-linear (e.g. the labour supply curve). Poor results would be obtained by fitting a straight line through the data in Figure 7.10, yet the shape of the relationship seems clear at a glance.

Fortunately this problem can be solved by transforming the data, so that when graphed a linear relationship between the two variables appears. Then a straight line can be fitted to these transformed data. This is equivalent to fitting a curved line to the original data. All that is needed is to find a suitable transformation to 'straighten out' the data. Given the data represented in Figure 7.10, if Y were graphed against $1/X$ the relationship shown in Figure 7.11 would appear.

Thus, if the regression line

$$Y_i = a + b\frac{1}{X_i} + e_i \tag{7.33}$$

were fitted, this would provide a good representation of the data in Figure 7.10. The procedure is straightforward. First, calculate the reciprocal of each of the X values and then use these (together with the original data for Y), using exactly the same methods as before. This transformation appears inappropriate for the birth rate data (see Figure 7.1) but serves as an illustration. The transformed X

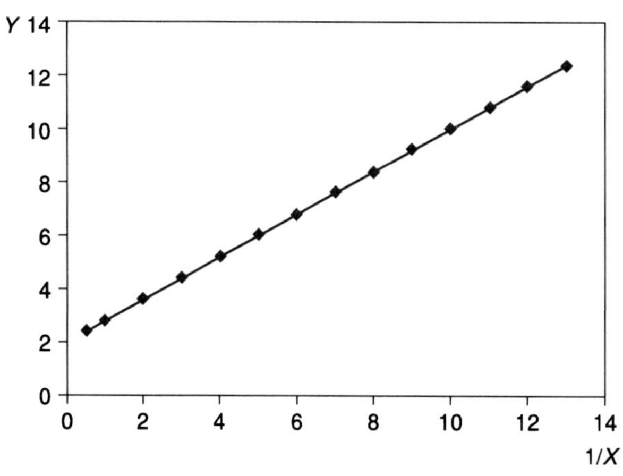

Figure 7.11
Figure 7.10 transformed: Y against $1/X$

values are 0.196 (= 1/5.1) for Brazil, 0.3125 (= 1/3.2) for Colombia, etc. The resulting regression equation is

$$Y_i = 31.92 - 3.96\frac{1}{X_i} + e_i \quad (7.34)$$

s.e. (1.64) (1.56)
$R^2 = 0.39$, $F = 6.44$, $n = 12$

This appears worse than the original specification (the R^2 is low and the slope coefficient is not significantly different from zero) so the transformation does not appear to be a good one. Note also that it is difficult to calculate the effect of X upon Y in this equation. We can see that a unit increase in $1/X$ reduces the birth rate by 3.96, but we do not have an intuitive feel for the inverse of the growth rate. This latest result also implies that a *fall* in the growth rate (hence $1/X$ rises) lowers the birth rate – the converse of our previous result. In the next chapter, we deal with a different example where a non-linear transformation does improve matters.

Table 7.8 presents a number of possible shapes for data, with suggested data transformations which will allow the relationship to be estimated using linear regression. In each case, once the data have been transformed, the methods and formulae used above can be applied.

It is sometimes difficult to know which transformation (if any) to apply. A graph of the data is unlikely to be as tidy as the diagrams in Table 7.8.

Table 7.8 Data transformations

Name	Graph of relationship	Original relationship	Transformed relationship	Regression
Double log	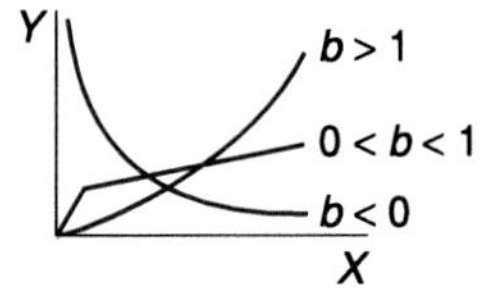 Y, X; $b > 1$; $0 < b < 1$; $b < 0$	$Y = aX^b e$	$\ln Y = \ln a + b \ln X + \ln e$	$\ln Y$ on $\ln X$
Reciprocal	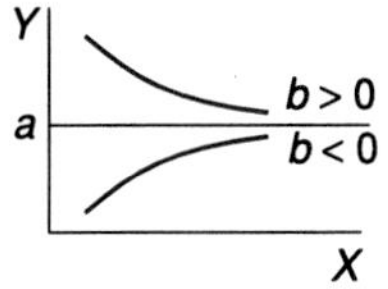 Y, X, a; $b > 0$; $b < 0$	$Y = a + b/X + e$	$Y = a + b\frac{1}{X} + e$	Y on $\frac{1}{X}$
Semi-log	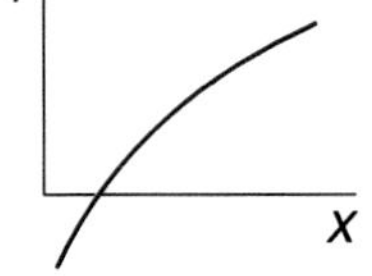 Y, X	$e^Y = aX^b e$	$Y = \ln a + b \ln X + \ln e$	Y on $\ln X$
Exponential	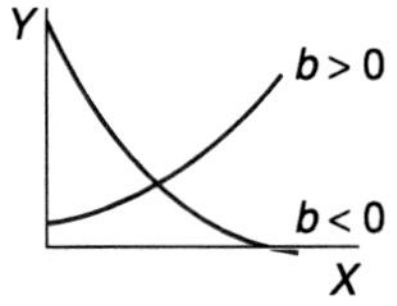 Y, X; $b > 0$; $b < 0$	$Y = e^{a+bX+e}$	$\ln Y = a + bX + e$	$\ln Y$ on X

Economic theory rarely suggests the form which a relationship should follow, and there are no simple statistical tests for choosing alternative formulations. The choice can sometimes be made after visual inspection of the data, or on the basis of convenience. The double log transformation is often used in economics as it has some very convenient properties. Unfortunately it cannot be used with the growth rate data here because Senegal's growth rate was negative. It is impossible to take the logarithm of a negative number. We therefore postpone the use of the log transformation in regression until the next chapter.

Exercise 7.9

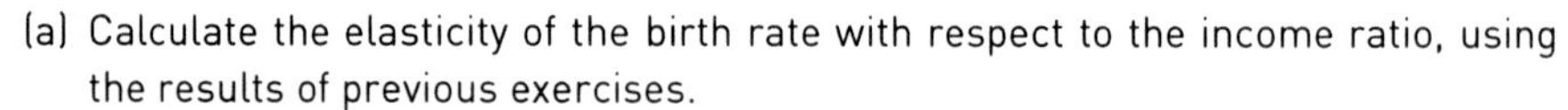

(a) Calculate the elasticity of the birth rate with respect to the income ratio, using the results of previous exercises.

(b) Give a brief interpretation of the meaning of this figure.

Exercise 7.10

Calculate a regression relating the birth rate to the inverse of the income ratio $1/IR$.

Summary

- Correlation refers to the extent of association between two variables. The (sample) correlation coefficient is a measure of this association, extending from $r = -1$ to $r = +1$.
- Positive correlation ($r > 0$) exists when high values of X tend to be associated with high values of Y and low X values with low Y values.
- Negative correlation ($r < 0$) exists when high values of X tend to be associated with low values of Y and vice versa.
- Values of r around 0 indicate an absence of correlation.
- As the sample correlation coefficient is a random variable we can test for its significance, i.e. test whether the true value is zero or not. This test is based upon the t distribution.
- The existence of correlation (even if 'significant') does not necessarily imply causality. There can be other reasons for the observed association.
- Regression analysis extends correlation by asserting a causality from X to Y and then measuring the relationship between the variables via the regression line, the 'line of best fit'.
- The regression line $Y = a + bX$ is defined by the intercept a and slope coefficient b. Their values are found by minimising the sum of squared errors around the regression line.
- The slope coefficient b measures the responsiveness of Y to changes in X.
- A measure of how well the regression line fits the data is given by the coefficient of determination, R^2, varying between 0 (very poor fit) and 1 (perfect fit).
- The coefficients a and b are unbiased point estimates of the true values of the parameters. Confidence interval estimates can be obtained, based on the t distribution. Hypothesis tests on the parameters can also be carried out using the t distribution.

- A test of the hypothesis $R^2 = 0$ (implying the regression is no better at predicting Y than simply using the mean of Y) can be carried out using the F distribution.
- The regression line may be used to predict Y for any value of X by assuming the residual to be zero for that observation.
- The measured response of Y to X (given by b) depends upon the units of measurement of X and Y. A better measure is often the elasticity, which is the proportionate response of Y to a proportionate change in X.
- Data are often transformed prior to regression (e.g. by taking logs) for a variety of reasons (e.g. to fit a curve to the original data).

Key terms and concepts

autocorrelation
correlation coefficient
coefficient of determination (R^2)
coefficient of rank correlation
dependent (endogenous) variable
elasticity
error sum of squares
error term (or residual)
independent (exogenous) variable
intercept
prediction
regression line or equation
regression sum of squares
slope
standard error
t ratio
total sum of squares

References

G. S. Maddala, *Introduction to Econometrics*, 2001, 3rd edn., Wiley.
M. P. Todaro, *Economic Development for a Developing World*, 1992, 3rd edn., Financial Times Prentice Hall.

Problems

Some of the more challenging problems are indicated by highlighting the problem number in colour.

7.1 The other data which Todaro might have used to analyse the birth rate were:

Country	Birth rate	GNP	Growth	Income ratio
Bangladesh	47	140	0.3	2.3
Tanzania	47	280	1.9	3.2
Sierra Leone	46	320	0.4	3.3
Sudan	47	380	−0.3	3.9
Kenya	55	420	2.9	6.8
Indonesia	35	530	4.1	3.4
Panama	30	1910	3.1	8.6
Chile	25	2560	0.7	3.8
Venezuela	35	4220	2.4	5.2
Turkey	33	1540	3.5	4.9
Malaysia	31	1840	4.3	5.0
Nepal	44	150	0.0	4.7
Malawi	56	200	2.7	2.4
Argentina	20	2560	1.9	3.6

For *one* of the three possible explanatory variables (in class, different groups could examine each of the variables):

(a) Draw an *XY* chart of the data above and comment upon the result.

(b) Would you expect a line of best fit to have a positive or negative slope? Roughly, what would you expect the slope to be?

(c) What would you expect the correlation coefficient to be?

(d) Calculate the correlation coefficient, and comment.

(e) Test to see if the correlation coefficient is different from zero. Use the 95% confidence level.

(Analysis of this problem continues in Problem 7.5.)

7.2 The data below show consumption of margarine (in ounces per person per week) and its real price, for the UK.

Year	Consumption	Price	Year	Consumption	Price
1970	2.86	125.6	1980	3.83	104.2
1971	3.15	132.9	1981	4.11	95.5
1972	3.52	126.0	1982	4.33	88.1
1973	3.03	119.6	1983	4.08	88.9
1974	2.60	138.8	1984	4.08	97.3
1975	2.60	141.0	1985	3.76	100.0
1976	3.06	122.3	1986	4.10	86.7
1977	3.48	132.7	1987	3.98	79.8
1978	3.54	126.7	1988	3.78	79.9
1979	3.63	115.7			

(a) Draw an XY plot of the data and comment.

(b) From the chart, would you expect the line of best fit to slope up or down? *In theory*, which way should it slope?

(c) What would you expect the correlation coefficient to be, approximately?

(d) Calculate the correlation coefficient between margarine consumption and its price.

(e) Is the coefficient significantly different from zero? What is the implication of the result?

(The following totals will reduce the burden of calculation: $\sum Y = 67.52$; $\sum X = 2101.70$; $\sum Y^2 = 245.055$; $\sum X^2 = 240\,149.27$; $\sum XY = 7299.638$; Y is consumption, X is price. If you wish, you could calculate a logarithmic correlation. The relevant totals are: $\sum y = 23.88$; $\sum x = 89.09$; $\sum y^2 = 30.45$; $\sum x^2 = 418.40$; $\sum xy = 111.50$, where $y = \ln Y$ and $x = \ln X$.)

(Analysis of this problem continues in Problem 7.6.)

7.3 What would you expect to be the correlation coefficient between the following variables? Should the variables be measured contemporaneously or might there be a lag in the effect of one upon the other?

(a) Nominal consumption and nominal income.

(b) GDP and the imports/GDP ratio.

(c) Investment and the interest rate.

7.4 As Problem 7.3, for:

(a) real consumption and real income;

(b) individuals' alcohol and cigarette consumption;

(c) UK and US interest rates.

7.5 Using the data from Problem 7.1, calculate the rank correlation coefficient between the variables and test its significance. How does it compare with the ordinary correlation coefficient?

7.6 Calculate the rank correlation coefficient between price and quantity for the data in Problem 7.2. How does it compare with the ordinary correlation coefficient?

7.7 (a) For the data in Problem 7.1, find the estimated regression line and calculate the R^2 statistic. Comment upon the result. How does it compare with Todaro's findings?

(b) Calculate the standard error of the estimate and the standard errors of the coefficients. Is the slope coefficient significantly different from zero? Comment upon the result.

(c) Test the overall significance of the regression equation and comment.

(d) Taking your own results and Todaro's, how confident do you feel that you understand the determinants of the birth rate?

(e) What do you think will be the result of estimating your equation using all 26 countries' data? Try it! What do you conclude?

7.8 (a) For the data given in Problem 7.2, estimate the sample regression line and calculate the R^2 statistic. Comment upon the results.

(b) Calculate the standard error of the estimate and the standard errors of the coefficients. Is the slope coefficient significantly different from zero? Is demand inelastic?

(c) Test the overall significance of the regression and comment upon your result.

7.9 From your results for the birth rate model, predict the birth rate for a country with *either* (a) GNP equal to $3000, (b) a growth rate of 3% p.a. *or* (c) an income ratio of 7. How does your prediction compare with one using Todaro's results? Comment.

7.10 Predict margarine consumption given a price of 70. Use the 99% confidence level.

7.11 **(Project)** Update Todaro's study using more recent data.

7.12 Try to build a model of the determinants of infant mortality. You should use cross-section data for 20 countries or more and should include both developing and developed countries in the sample.

Write up your findings in a report which includes the following sections: discussion of the problem; data gathering and transformations; estimation of the model; interpretation of results. Useful data may be found in the Human Development Report (use Google to find it online).

Answers to exercises

Exercise 7.1

(a) The calculation is:

	Birth rate Y	Income ratio X	Y^2	X^2	XY
Brazil	30	9.5	900	90.25	285
Colombia	29	6.8	841	46.24	197.2
Costa Rica	30	4.6	900	21.16	138
India	35	3.1	1225	9.61	108.5
Mexico	36	5	1296	25	180
Peru	36	8.7	1296	75.69	313.2
Philippines	34	3.8	1156	14.44	129.2
Senegal	48	6.4	2304	40.96	307.2
South Korea	24	2.7	576	7.29	64.8
Sri Lanka	27	2.3	729	5.29	62.1
Taiwan	21	3.8	441	14.44	79.8
Thailand	30	3.3	900	10.89	99
Totals	380	60	12 564	361.26	1964

$$r = \frac{12 \times 1964 - 60 \times 380}{\sqrt{(12 \times 361.26 - 60^2)(12 \times 12\,564 - 380^2)}} = 0.355$$

(c) As for (a) except $\Sigma X = 0.6$, $\Sigma Y = 38$, $\Sigma X^2 = 0.036126$, $\Sigma Y^2 = 125.64$, $\Sigma XY = 1.964$. Hence

$$r = \frac{12 \times 1.964 - 0.6 \times 38}{\sqrt{(12 \times 0.036126 - 0.6^2)(12 \times 125.64 - 38^2)}} = 0.355$$

Exercise 7.2

(a) $t = \dfrac{0.355\sqrt{12-2}}{\sqrt{1-(0.355)^2}} = 1.20$

(b) The Prob-value, for a two-tailed test is 0.257 or 25%, so we do not reject the null of no correlation.

Exercise 7.3

(a) The calculation is:

	Birth rate Y	Income ratio X	Rank of Y	Rank of X	Y^2	X^2	XY
Brazil	30	9.5	7	1	49	1	7
Colombia	29	6.8	9	3	81	9	27
Costa Rica	30	4.6	7	6	49	36	42
India	35	3.1	4	10	−16	100	40
Mexico	36	5	2.5	5	−6.25	25	12.5
Peru	36	8.7	2.5	2	6.25	4	5
Philippines	34	3.8	5	7.5	−25	56.25	37.5
Senegal	48	6.4	1	4	−1	16	4
South Korea	24	2.7	11	11	121	121	121
Sri Lanka	27	2.3	10	12	−100	144	120
Taiwan	21	3.8	12	7.5	144	56.25	90
Thailand	30	3.3	7	9	−49	81	63
Totals			78	78	647.5	649.5	569

$$r_s = \frac{12 \times 569 - 78^2}{\sqrt{(12 \times 649.5 - 78^2)(12 \times 647.5 - 78^2)}} = 0.438$$

(b) This is less than the critical value of 0.591 so the null of no rank correlation cannot be rejected.

(c) Reversing the rankings should not alter the result of the calculation.

Exercise 7.4

(a) Using the data and calculations in the answer to Exercise 7.1 we obtain:

$$b = \frac{12 \times 1964 - 60 \times 380}{12 \times 361.26 - 60^2} = 1.045$$

$$a = \frac{380}{12} - (1.045) \times \frac{60}{12} = 26.443$$

(b) A unit increase in the measure of inequality leads to approximately one additional birth per 1000 mothers. The constant has no useful interpretation. The income ratio cannot be zero (in fact, it cannot be less than 0.5).

Exercise 7.5

(a) $TSS = \Sigma(Y_i - \bar{Y})^2 = \Sigma Y_i^2 - n\bar{Y}^2 = 12\,564 - 12 \times 31.67^2 = 530.667$

$ESS = \Sigma(Y_i - \hat{Y}_i)^2 = \Sigma Y_i^2 - a\,\Sigma Y_i - b\,\Sigma X_i Y_i$

$= 12\,564 - 26.443 \times 380 - 1.045 \times 1139.70 = 463.804$

$RSS = TSS - ESS = 530.667 - 463.804 = 66.863$

$R^2 = 0.126.$

(b) This is the square of the correlation coefficient, calculated earlier as 0.355.

Exercise 7.6

(a) $s_e^2 = \dfrac{463.804}{10} = 46.3804$

and so

$$s_b^2 = \frac{46.3804}{61.26} = 0.757$$

and

$s_b = \sqrt{0.757} = 0.870$

For a the estimated variance is

$$s_a^2 = s_e^2 \times \left(\frac{1}{n} + \frac{\bar{X}^2}{\Sigma(X_i - \bar{X})^2}\right) = 46.3804 \times \left(\frac{1}{12} + \frac{5^2}{61.26}\right) = 22.793$$

and hence $s_a = 4.774$. The 95% CIs are therefore $1.045 \pm 2.228 \times 0.87 = [-0.894, 2.983]$ for b and $26.443 \pm 2.228 \times 4.774 = [15.806, 37.081]$.

(b) $t = \dfrac{1.045 - 0}{0.870} = 1.201$

Not significant.

(c) $F = \dfrac{RSS/1}{ESS/(n-2)} = \dfrac{66.863/1}{463.804/(12-2)} = 1.44$

Exercise 7.7

Excel should give the same answers.

Exercise 7.8

(a) $\hat{BR} = 26.44 + 1.045 \times 10 = 36.9.$

(b) $\left[36.9 - 2.228 \times 6.81\sqrt{\frac{1}{12} + \frac{(10-5)^2}{61.26}}, 36.9 + 2.228 \times 6.81\sqrt{\frac{1}{12} + \frac{(10-5)^2}{61.26}}\right]$

$= [26.3, 47.5]$

(c) $\left[36.9 - 2.228 \times 6.81\sqrt{1+\frac{1}{12} + \frac{(10-5)^2}{61.26}}, 36.9 + 2.228 \times 6.81\sqrt{1+\frac{1}{12} + \frac{(10-5)^2}{61.26}}\right]$

$= [18.4, 55.4]$

Exercise 7.9

(*a*) $e = 1.045 \times \frac{5}{31.67} = 0.165$

(*b*) A 10% rise in the inequality measure (e.g. from 4 to 4.4) raises the birth rate by 1.65% (e.g. from 30 to 30.49).

Exercise 7.10

$$BR = 38.82 - 29.61 \times \frac{1}{IR} + e$$

s.e. (19.0)

$R^2 = 0.19$, $F(1,10) = 2.43$.

The regression is rather poor and the *F* statistic is not significant.

2 Probability

Contents

Learning outcomes

By the end of this chapter you should be able to:

- understand the essential concept of the probability of an event occurring;
- appreciate that the probability of a combination of events occurring can be calculated using simple arithmetic rules (the addition and multiplication rules);
- understand that a probability can depend upon the outcome of other events (conditional probability);
- know how to make use of probability theory to help make decisions in situations of uncertainty.

Complete your diagnostic test for Chapter 2 now to create your personal study plan. Exercises with an icon ? are also available for practice in MathXL with additional supporting resources.

Probability theory and statistical inference

In October 1985 Mrs Evelyn Adams of New Jersey, USA, won $3.9 m in the State lottery at odds of 1 in 3 200 000. In February 1986 she won again, although this time only (!) $1.4 m at odds of 1 in 5 200 000. The odds against both these wins were calculated at about 1 in 17 300 bn. Mrs Adams is quoted as saying 'They say good things come in threes, so . . .'.

The above story illustrates the principles of probability at work. The same principles underlie the theory of **statistical inference**, which is the task of drawing conclusions (inferences) about a population from a sample of data drawn from that population. For example, we might have a survey which shows that 30% of a sample of 100 families intend to take a holiday abroad next year. What can we conclude from this about *all* families? The techniques set out in this and subsequent chapters show how to accomplish this.

Why is knowledge of probability necessary for the study of statistical inference? In order to be able to say something about a population on the basis of some sample evidence we must first examine how the sample data are collected. In many cases, the sample is a random one, i.e. the observations making up the sample are chosen at random from the population. If a second sample were selected it would almost certainly be different from the first. Each member of the population has a particular probability of being in the sample (in simple random sampling the probability is the same for all members of the population). To understand sampling procedures, and the implications for statistical inference, we must therefore first examine the theory of probability.

As an illustration of this, suppose we wish to know if a coin is fair, i.e. equally likely to fall heads or tails. The coin is tossed 10 times and 10 heads are recorded. This constitutes a random sample of tosses of the coin. What can we infer about the coin? *If* it is fair, the probability of getting ten heads is 1 in 1024, so a fairly unlikely event seems to have happened. We might reasonably infer therefore that the coin is biased towards heads.

The definition of probability

The first task is to define precisely what is meant by probability. This is not as easy as one might imagine and there are a number of different schools of thought on the subject. Consider the following questions:

- What is the probability of 'heads' occurring on the toss of a coin?
- What is the probability of a driver having an accident in a year of driving?
- What is the probability of a country such as Peru defaulting on its international loan repayments (as Mexico did in the 1980s)?

We shall use these questions as examples when examining the different schools of thought on probability.

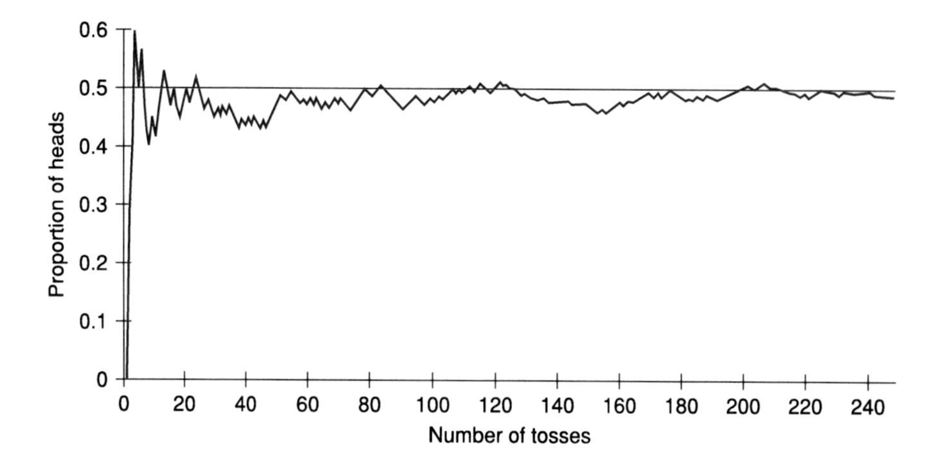

Figure 2.1
The proportion of heads in 250 tosses of a fair coin

The frequentist view

Considering the first question above, the **frequentist view** would be that the probability is equal to the **proportion** of heads obtained from a coin in the long run, i.e. if the coin were tossed many times. The first few results of such an experiment might be

H, T, T, H, H, H, T, H, T, . . .

After a while, the proportion of heads settles down at some particular fraction and subsequent tosses will individually have an insignificant effect upon the value. Figure 2.1 shows the result of tossing a coin 250 times and recording the proportion of heads (actually, this was simulated on a computer: life is too short to do it for real).

This shows the proportion settling down at a value of about 0.50, which indicates an unbiased coin (or rather, an unbiased computer in this case!). This value is the probability, according to the frequentist view. To be more precise, the probability is defined as the proportion of heads obtained as the number of tosses *approaches infinity*. In general we can define $\Pr(H)$, the probability of event H (in this case heads) occurring, as

$$\Pr(H) = \frac{\text{number of occurrences of } H}{\text{number of trials}}, \textit{ as the number of trials approaches infinity.}$$

In this case, each toss of the coin constitutes a trial.

This definition gets round the obvious question of how many trials are needed before the probability emerges, but means that the probability of an event cannot strictly be obtained in finite time.

Although this approach appears attractive in theory, it does have its problems. One could not actually toss the coin an infinite number of times. Or, what if one took a different coin, would the results from the first coin necessarily apply to the second?

Perhaps more seriously, the definition is of less use for the second and third questions posed above. Calculating the probability of an accident is not too problematic: it may be defined as the proportion of all drivers having an accident during the year. However, this may not be relevant for a *particular* driver, since drivers vary so much in their accident records. And how would you answer the third question? There is no long run that we can appeal to. We cannot re-run history over and over again to see in what proportion of cases the country defaults. Yet this is what lenders want to know and credit-rating agencies have to assess. Maybe another approach is needed.

The subjective view

According to the subjective view, probability is a **degree of belief** that someone holds about the likelihood of an event occurring. It is inevitably subjective and therefore some argue that it should be the degree of belief that it is *rational* to hold, but this just shifts the argument to what is meant by 'rational'. Some progress can be made by distinguishing between **prior** and **posterior** beliefs. The former are those held before any evidence is considered; the latter are the modified probabilities in the light of the evidence. For example, one might initially believe a coin to be fair (the prior probability of heads is one-half), but not after seeing only five heads in fifty tosses (the posterior probability would be less than a half).

Although it has its attractions, this approach (which is the basis of **Bayesian** statistics) also has its drawbacks. It is not always clear how one should arrive at the prior beliefs, particularly when one really has no prior information. Also, these methods often require the use of sophisticated mathematics, which may account for the limited use made of them. The development of more powerful computers and user-friendly software may increase the popularity of the Bayesian approach.

There is not universal agreement therefore as to the precise definition of probability. We do not have space here to explore the issue further, so we will ignore the problem! The probability of an event occurring will be defined as a certain value and we will not worry about the precise origin or meaning of that value. This is an **axiomatic** approach: we simply state what the probability is, without justifying it, and then examine the consequences.

Exercise 2.1

(a) Define the probability of an event according to the frequentist view.

(b) Define the probability of an event according to the subjective view.

Exercise 2.2

For the following events, suggest how their probability might be calculated. In each case, consider whether you have used the frequentist or subjective view of probability (or possibly some mixture).

(a) The Republican party winning the next US election.

(b) The number 5 being the first ball drawn in next week's lottery.

(c) A repetition of the 2004 Asian tsunami.

(d) Your train home being late.

Probability theory: the building blocks

We start with a few definitions, to establish a vocabulary that we will subsequently use.

- An **experiment** is an action such as flipping a coin, which has a number of possible **outcomes** or **events**, such as heads or tails.
- A **trial** is a single performance of the experiment, with a single outcome.
- The **sample space** consists of all the possible outcomes of the experiment. The outcomes for a single toss of a coin are {heads, tails}, for example, and these constitute the sample space for a toss of a coin. The outcomes in the sample space are **mutually exclusive**, which means that the occurrence of one rules out all the others. One cannot have both heads and tails in a single toss of a coin. As a further example, if a single card is drawn at random from a pack, then the sample space may be drawn as in Figure 2.2. Each point represents one card in the pack and there are 52 points altogether. (The sample space could be set out in alternative ways. For instance, one could write a list of all the cards: ace of spades, king of spades, . . . , two of clubs. One can choose the representation most suitable for the problem at hand.)
- With each outcome in the sample space we can associate a **probability**, which is the chance of that outcome occurring. The probability of heads is one-half; the probability of drawing the ace of spades from a pack of cards is one in 52, etc.

There are restrictions upon the probabilities we can associate with the outcomes in the sample space. These are needed to ensure that we do not come up with self-contradictory results; for example, it would be odd to arrive at the conclusion that we could expect heads more than half the time *and* tails more than half the time. To ensure our results are always consistent, the following rules apply to probabilities:

- The probability of an event must lie between 0 and 1, i.e.

$$0 \leqslant \Pr(A) \leqslant 1, \text{ for any event } A \tag{2.1}$$

  The explanation is straightforward. If A is certain to occur it occurs in 100% of all trials and so its probability is 1. If A is certain not to occur then its probability is 0, since it never happens however many trials there are. As one cannot be more certain than certain, probabilities of less than 0 or more than 1 can never occur, and equation (2.1) follows.
- The sum of the probabilities associated with all the outcomes in the sample space is 1. Formally

$$\Sigma P_i = 1 \tag{2.2}$$

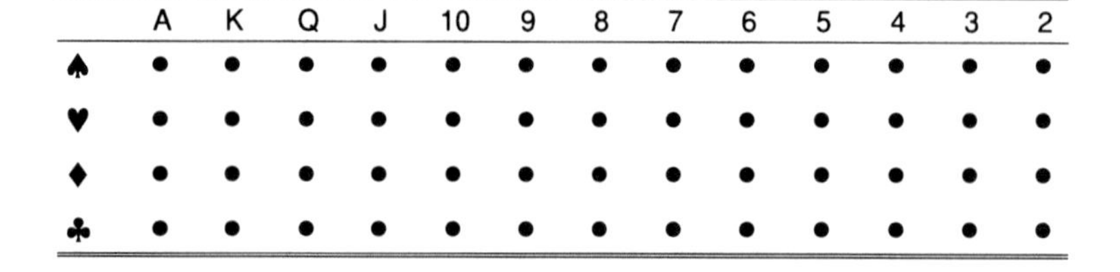

Figure 2.2
The sample space for drawing from a pack of cards

where P_i is the probability of event i occurring. This follows from the fact that one, and only one, of the outcomes *must* occur, since they are mutually exclusive and also **exhaustive**, i.e. they define all the possibilities.

- Following on from equation (2.2) we may define the **complement** of an event as everything in the sample space apart from that event. The complement of heads is tails, for example. If we write the complement of A as not-A then it follows that $\Pr(A) + \Pr(\text{not-}A) = 1$ and hence

$$\Pr(\text{not-}A) = 1 - \Pr(A) \tag{2.3}$$

Compound events

Most practical problems require the calculation of the probability of a set of outcomes rather than just a single one, or the probability of a series of outcomes in separate trials. For example, the probability of drawing a spade at random from a pack of cards encompasses 13 points in the sample space (one for each spade). This probability is 13 out of 52, or one-quarter, which is fairly obvious; but for more complex problems the answer is not immediately evident. We refer to such sets of outcomes as **compound events**. Some examples are getting a five *or* a six on a throw of a die or drawing an ace *and* a queen to complete a 'straight' in a game of poker.

It is sometimes possible to calculate the probability of a compound event by examining the sample space, as in the case of drawing a spade above. However, in many cases this is not so, for the sample space is too complex or even impossible to write down. For example, the sample space for three draws of a card from a pack consists of over 140 000 points! (A typical point might be, for example, the ten of spades, eight of hearts and three of diamonds.) An alternative method is needed. Fortunately there are a few simple rules for manipulating probabilities which help us to calculate the probabilities of compound events.

If the previous examples are examined closely it can be seen that outcomes are being compounded using the words 'or' and 'and': '. . . five *or* six on a single throw . . .'; '. . . an ace *and* a queen . . .'. 'And' and 'or' act as **operators**, and compound events are made up of simple events compounded by these two operators. The following rules for manipulating probabilities show how to use these operators and thus how to calculate the probability of a compound event.

The addition rule

This rule is associated with 'or'. When we want the probability of one outcome *or* another occurring, we add the probabilities of each. More formally, the probability of A or B occurring is given by

$$\Pr(A \text{ or } B) = \Pr(A) + \Pr(B) \tag{2.4}$$

So, for example, the probability of a five or a six on a roll of a die is

$$\Pr(5 \text{ or } 6) = \Pr(5) + \Pr(6) = 1/6 + 1/6 = 1/3 \tag{2.5}$$

This answer can be verified from the sample space, as shown in Figure 2.3. Each dot represents a simple event (one to six). The compound event is made up of two of the six points, shaded in Figure 2.3, so the probability is 2/6 or 1/3.

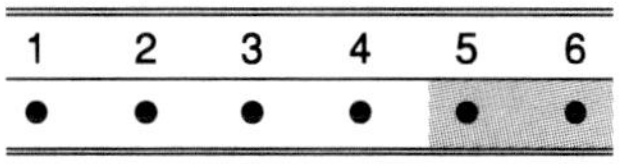

Figure 2.3
The sample space for rolling a die

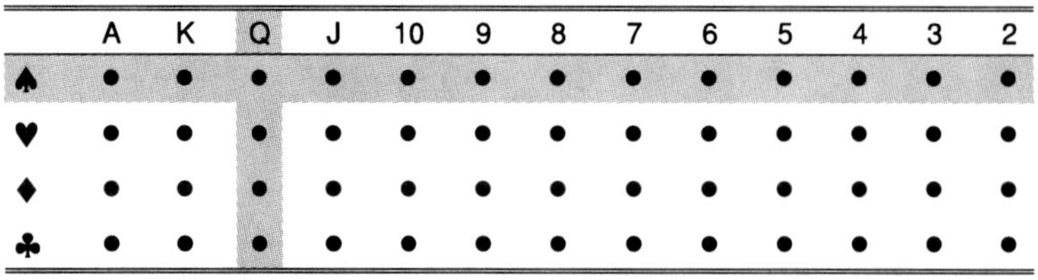

1	2	3	4	5	6
•	•	•	•	•	•

Figure 2.4
The sample space for drawing a queen or a spade

	A	K	Q	J	10	9	8	7	6	5	4	3	2
♠	•	•	•	•	•	•	•	•	•	•	•	•	•
♥	•	•	•	•	•	•	•	•	•	•	•	•	•
♦	•	•	•	•	•	•	•	•	•	•	•	•	•
♣	•	•	•	•	•	•	•	•	•	•	•	•	•

However, equation (2.4) is not a general solution to this type of problem, i.e. it does not *always* work, as can be seen from the following example. What is the probability of a queen or a spade in a single draw from a pack of cards? $\Pr(Q) = 4/52$ (four queens in the pack) and $\Pr(S) = 13/52$ (13 spades), so applying equation (2.4) gives

$$\Pr(Q \text{ or } S) = \Pr(Q) + \Pr(S) = 4/52 + 13/52 = 17/52 \tag{2.6}$$

However, if the sample space is examined, the correct answer is found to be 16/52, as in Figure 2.4. The problem is that one point in the sample space (the one representing the queen of spades) is double-counted, once as a queen and again as a spade. The event 'drawing a queen *and* a spade' is possible, and gets double-counted. Equation (2.4) has to be modified by subtracting the probability of getting a queen *and* a spade, to eliminate this double counting. The correct answer is obtained from

$$\begin{aligned}\Pr(Q \text{ or } S) &= \Pr(Q) + \Pr(S) - \Pr(Q \text{ and } S) \\ &= 4/52 + 13/52 - 1/52 \\ &= 16/52\end{aligned} \tag{2.7}$$

The general rule is therefore

$$\Pr(A \text{ or } B) = \Pr(A) + \Pr(B) - \Pr(A \text{ and } B) \tag{2.8}$$

Rule (2.4) worked for the die example because $\Pr(5 \text{ and } 6) = 0$ since a five and a six cannot simultaneously occur. The double counting did not affect the calculation of the probability.

In general, therefore, one should use equation (2.8), but when two events are mutually exclusive the rule simplifies to equation (2.4).

The multiplication rule

The multiplication rule is associated with use of the word 'and' to combine events. Consider a mother with two children. What is the probability that they are both boys? This is really a compound event: a boy on the first birth *and* a boy on the second. Assume that in a single birth a boy or girl is equally likely, so $\Pr(\text{boy}) = \Pr(\text{girl}) = 0.5$. Denote by $\Pr(B1)$ the probability of a boy on the first birth and by $\Pr(B2)$ the probability of a boy on the second. Thus the question asks for $\Pr(B1 \text{ and } B2)$ and this is given by

$$\Pr(B1 \text{ and } B2) = \Pr(B1) \times \Pr(B2) = 0.5 \times 0.5 = 0.25 \quad (2.9)$$

Intuitively, the multiplication rule can be understood as follows. One-half of mothers have a boy on their first birth and of these, one-half will again have a boy on the second. Therefore a quarter (a half of one-half) of mothers have two boys.

Like the addition rule, the multiplication rule requires slight modification before it can be applied generally and give the right answer in all circumstances. The example assumes first and second births to be **independent events**, i.e. that having a boy on the first birth does not affect the probability of a boy on the second. This assumption is not always valid.

Write $\Pr(B2|B1)$ to indicate the probability of the event $B2$ *given* that the event $B1$ has occurred. (This is known as the **conditional probability**, more precisely the probability of $B2$ conditional upon $B1$.) Let us drop the independence assumption and suppose the following

$$\Pr(B1) = \Pr(G1) = 0.5 \quad (2.10)$$

i.e. boys and girls are equally likely on the first birth, and

$$\Pr(B2|B1) = \Pr(G2|G1) = 0.6 \quad (2.11)$$

i.e. a boy is more likely to be followed by another boy, and a girl by another girl. (It is easy to work out $\Pr(B2|G1)$ and $\Pr(G2|B1)$. What are they?)

Now what is the probability of two boys? Half of all mothers have a boy first, and of these, 60% have another boy. Thus 30% (60% of 50%) of mothers have two boys. This is obtained from the rule

$$\begin{aligned}\Pr(B1 \text{ and } B2) &= \Pr(B1) \times \Pr(B2|B1) \\ &= 0.5 \times 0.6 \\ &= 0.3\end{aligned} \quad (2.12)$$

Thus in general we have

$$\Pr(A \text{ and } B) = \Pr(A) \times \Pr(B|A) \quad (2.13)$$

which simplifies to

$$\Pr(A \text{ and } B) = \Pr(A) \times \Pr(B) \quad (2.14)$$

if A and B are independent.

Independence may therefore be defined as follows: two events, A and B, are independent if the probability of one occurring is not influenced by the fact of the other having occurred. Formally, if A and B are independent then

$$\Pr(B|A) = \Pr(B|\text{not } A) = \Pr(B) \quad (2.15)$$

and

$$\Pr(A|B) = \Pr(A|\text{not } B) = \Pr(A) \quad (2.16)$$

The concept of independence is an important one in statistics, as it usually simplifies problems considerably. If two variables are known to be independent then we can analyse the behaviour of one without worrying about what is happening to the other variable. For example, sales of computers are independent of temperature, so if one is trying to predict sales next month one does not need to worry about the weather. In contrast, ice cream sales do depend on the weather, so predicting sales accurately requires one to forecast the weather first.

Intuition does not always work with probabilities!

Counter-intuitive results frequently arise in probability, which is why it is wise to use the rules to calculate probabilities in tricky situations, rather than rely on intuition. Take the following questions:

- What is the probability of obtaining two heads (HH) in two tosses of a coin?
- What is the probability of obtaining tails followed by heads (TH)?
- If a coin is tossed until either HH or TH occurs, what are the probabilities of each sequence occurring first?

The answers to the first two are easy: $\frac{1}{2} \times \frac{1}{2} = \frac{1}{4}$ in each case. You might therefore conclude that each sequence is equally likely to be the first observed, but you would be wrong!

Unless HH occurs on the first two tosses, then TH *must* occur first. HH is therefore the first sequence *only* if it occurs on the first two tosses, which has a probability of $\frac{1}{4}$. The probability that TH is first is therefore $\frac{3}{4}$. The probabilities are unequal, a strange result. Now try the same thing with HHH and THH and three tosses of a coin.

Combining the addition and multiplication rules

More complex problems can be solved by suitable combinations of the addition and multiplication formulae. For example, what is the probability of a mother having one child of each sex? This could occur in one of two ways: a girl followed by a boy or a boy followed by a girl. It is important to note that these are two different routes to the same outcome. Therefore we have (assuming non-independence according to equation (2.11))

$$\begin{aligned}\Pr(1 \text{ girl}, 1 \text{ boy}) &= \Pr((G1 \text{ and } B2) \text{ or } (B1 \text{ and } G2)) \\ &= \Pr(G1) \times \Pr(B2|G1) + \Pr(B1) \times \Pr(G2|B1) \\ &= (0.5 \times 0.4) + (0.5 \times 0.4) \\ &= 0.4\end{aligned}$$

The answer can be checked if we remember equation (2.2) stating that probabilities must sum to 1. We have calculated the probability of two boys (0.3) and of a child of each sex (0.4). The only other possibility is of two girls. This probability must be 0.3, the same as two boys, since boys and girls are treated symmetrically in this problem (even with the non-independence assumption). The sum of the three possibilities (two boys, one of each or two girls) is therefore $0.3 + 0.4 + 0.3 = 1$, as it should be. This is often a useful check to make, especially if one is unsure that one's calculations are correct.

Note that the problem would have been different if we had asked for the probability of the mother having one girl with a younger brother.

Tree diagrams

The preceding problem can be illustrated using a **tree diagram**, which often helps to clarify a problem. A tree diagram is an alternative way of enumerating

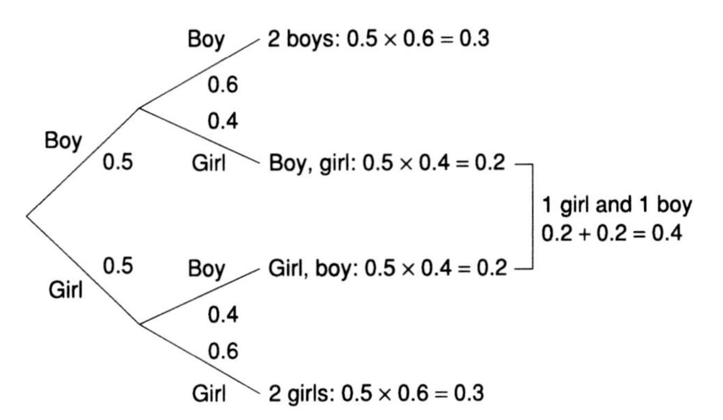

Figure 2.5
Tree diagram for a family with two children

all possible outcomes in the sample space, with the associated probabilities. The diagram for two children is shown in Figure 2.5.

The diagram begins at the left and the first node shows the possible alternatives (boy, girl) at that point and the associated probabilities (0.5, 0.5). The next two nodes show the alternatives and probabilities for the second birth, given the sex of the first child. The final four nodes show the possible results: {boy, boy}; {boy, girl}; {girl, boy}; and {girl, girl}.

To find the probability of two girls, using the tree diagram, follow the lowest path, multiplying the probabilities along it to give $0.5 \times 0.6 = 0.3$. To find the probability of one child of each sex it is necessary to follow all the routes which lead to such an outcome. There are two in this case: leading to boy, girl and to girl, boy. Each of these has a probability of 0.2, obtained by multiplying the probabilities along that branch of the tree. Adding these together (since either one *or* the other leads to the desired outcome) yields the answer, giving $0.2 + 0.2 = 0.4$. This provides a graphical alternative to the formulae used above and may help comprehension.

The tree diagram can obviously be extended to cover third and subsequent children although the number of branches rapidly increases (in geometric progression). The difficulty then becomes not just the calculation of the probability attached to each outcome, but sorting out which branches should be taken into account in the calculation. Suppose we consider a family of five children of whom three are girls. To simplify matters we again assume independence of probabilities. The appropriate tree diagram has $2^5 = 32$ end-points, each with probability 1/32. How many of these relate to families with three girls and two boys, for example? One can draw the diagram and count them, yielding the answer 10, but it takes considerable time and is prone to error. Far better would be to use a formula. To develop this, we use the ideas of **combinations** and **permutations**.

Combinations and permutations

How can we establish the number of ways of having three girls and two boys in a family of five children? One way would be to write down all the possible orderings:

GGGBB GGBGB GGBBG GBGGB GBGBG
GBBGG BGGGB BGGBG BGBGG BBGGG

This shows that there are 10 such orderings, so the probability of three girls and two boys in a family of five children is 10/32. In more complex problems this soon becomes difficult or impossible. The record number of children born to a British mother is 39 (!) of whom 32 were girls. The appropriate tree diagram has over five thousand billion 'routes' through it, and drawing one line (i.e. for one child) per second would imply 17 433 years to complete the task! Rather than do this, we use the **combinatorial** formula to find the answer. Suppose there are n children, r of them girls, then the number of orderings, denoted nCr, is obtained from[1]

$$nCr = \frac{n!}{r!(n-r)!}$$

$$= \frac{n \times (n-1) \times \ldots \times 1}{\{r \times (r-1) \times \ldots \times 1\} \times \{(n-r) \times (n-r-1) \times \ldots \times 1\}} \tag{2.17}$$

In the above example $n = 5$, $r = 3$ so the number of orderings is

$$5C3 = \frac{5 \times 4 \times 3 \times 2 \times 1}{\{3 \times 2 \times 1\} \times \{2 \times 1\}} = 10 \tag{2.18}$$

If there were four girls out of five children then the number of orderings or combinations would be

$$5C4 = \frac{5 \times 4 \times 3 \times 2 \times 1}{\{4 \times 3 \times 2 \times 1\} \times 1} = 5 \tag{2.19}$$

This gives five possible orderings, i.e. the single boy could be the first, second, third, fourth or fifth born.

Why does this formula work? Consider five empty places to fill, corresponding to the five births in chronological order. Take the case of three girls (call them Amanda, Bridget and Caroline for convenience) who have to fill three of the five places. For Amanda there is a choice of five empty places. Having 'chosen' one, there remain four for Bridget, so there are $5 \times 4 = 20$ possibilities (i.e. ways in which these two could choose their places). Three remain for Caroline, so there are 60 ($= 5 \times 4 \times 3$) possible orderings in all (the two boys take the two remaining places). Sixty is the number of **permutations** of three *named* girls in five births. This is written $5P3$ or in general nPr. Hence

$$5P3 = 5 \times 4 \times 3$$

or in general

$$nPr = n \times (n-1) \times \ldots \times (n-r+1) \tag{2.20}$$

A simpler formula is obtained by multiplying and dividing by $(n-r)!$

$$nPr = \frac{n \times (n-r) \times \ldots \times (n-r+1) \times (n-r)!}{(n-r)!} = \frac{n!}{(n-r)!} \tag{2.21}$$

---

[1] $n!$ is read 'n factorial' and is defined as the product of all the integers up to and including n. Thus, for example, $3! = 3 \times 2 \times 1 = 6$.

What is the difference between nPr and nCr? The latter does not distinguish between the girls; the two cases Amanda, Bridget, Caroline, boy, boy and Bridget, Amanda, Caroline, boy, boy are effectively the same (three girls followed by two boys). So nPr is larger by a factor representing the number of ways of ordering the three girls. This factor is given by $r! = 3 \times 2 \times 1 = 6$ (any of the three girls could be first, either of the other two second, and then the final one). Thus to obtain nCr one must divide nPr by $r!$, giving (2.17).

Exercise 2.3

(a) A dart is thrown at a dartboard. What is the sample space for this experiment?

(b) An archer has a 30% chance of hitting the bull's eye on the target. What is the complement to this event and what is its probability?

(c) What is the probability of two mutually exclusive events both occurring?

(d) A spectator reckons there is a 70% probability of an American rider winning the Tour de France and a 40% probability of Frenchman winning. Comment.

Exercise 2.4

(a) For the archer in Exercise 2.3(b) what is the probability that she hits the target with one (and only one) of two arrows?

(b) What is the probability that she hits the target with both arrows?

(c) Explain the importance of the assumption of independence for the answers to both parts (a) and (b) of this exercise.

(d) If the archer becomes more confident after a successful shot (i.e. her probability of a shot on target rises to 50%) and less confident (probability falls to 20%) after a miss, how would this affect the answers to parts (a) and (b)?

Exercise 2.5

(a) Draw the tree diagrams associated with Exercise 2.4. You will need one for the case of independence of events, one for non-independence.

(b) Extend the diagram (assuming independence) to a third arrow. Use this to mark out the paths with two successful shots out of three. Calculate the probability of two hits out of three shots.

(c) Repeat part (b) for the case of non-independence. For this you may assume that a hit raises the problem of success with the next arrow to 50%. A miss lowers it to 20%.

Exercise 2.6

(a) Show how the answer to Exercise 2.5(b) may be arrived at using algebra, including the use of the combinatorial formula.

(b) Repeat part (a) for the non-independence case.

Bayes' theorem

Bayes' theorem is a factual statement about probabilities, which in itself is uncontroversial. However, the use and interpretation of the result is at the heart of the difference between **classical** and **Bayesian** statistics. The theorem itself is easily derived from first principles. Equation (2.22) is similar to equation (2.13) covered earlier when discussing the multiplication rule

$$\Pr(A \text{ and } B) = \Pr(A|B) \times \Pr(B) \tag{2.22}$$

hence

$$\Pr(A|B) = \frac{\Pr(A \text{ and } B)}{\Pr(B)} \tag{2.23}$$

Expanding both top and bottom of the right-hand side

$$\Pr(A|B) = \frac{\Pr(B|A) \times \Pr(A)}{\Pr(B|A) \times \Pr(A) + \Pr(B|\text{not } A) \times \Pr(\text{not } A)} \tag{2.24}$$

Equation (2.24) is known as **Bayes' theorem** and is a statement about the probability of the event A, conditional upon B having occurred. The following example demonstrates its use.

Two bags contain red and yellow balls. Bag A contains six red and four yellow balls, bag B has three red and seven yellow balls. A ball is drawn at random from one bag and turns out to be red. What is the probability that it came from bag A? Since bag A has relatively more red balls to yellow balls than does bag B, it seems bag A ought to be favoured. The probability should be more than 0.5. We can check if this is correct.

Denoting

$\Pr(A) = 0.5$ (the probability of choosing bag A at random) $= \Pr(B)$

$\Pr(R|A) = 0.6$ (the probability of selecting a red ball from bag A), etc.

we have

$$\Pr(A|R) = \frac{\Pr(R|A) \times \Pr(A)}{\Pr(R|A) \times \Pr(A) + \Pr(R|B) \times \Pr(B)} \tag{2.25}$$

using Bayes' theorem. Evaluating this gives

$$\Pr(A|R) = \frac{0.6 \times 0.5}{0.6 \times 0.5 + 0.3 \times 0.5} \tag{2.26}$$

$$= {}^{2}/_{3}$$

(You can check that $\Pr(B|R) = {}^{1}/_{3}$ so that the sum of the probabilities is 1.) As expected, this result is greater than 0.5.

Bayes' theorem can be extended to cover more than two bags: if there are five bags, for example, labelled A to E, then

$$\Pr(A|R) = \frac{\Pr(R|A) \times \Pr(A)}{\Pr(R|A) \times \Pr(A) + \Pr(R|B) \times \Pr(B) + \ldots + \Pr(R|E) \times \Pr(E)} \tag{2.27}$$

In Bayesian language, $\Pr(A)$, $\Pr(B)$, etc., are known as the **prior** (to the drawing of the ball) probabilities, $\Pr(R|A)$, $\Pr(R|B)$, etc., are the **likelihoods** and $\Pr(A|R)$, $\Pr(B|R)$, etc., are the **posterior** probabilities. Bayes' theorem can alternatively be expressed as

$$\text{posterior probability} = \frac{\text{likelihood} \times \text{prior probability}}{\Sigma(\text{likelihoods} \times \text{prior probabilites})} \tag{2.28}$$

This is illustrated below, by reworking the above example.

	Prior probabilities	Likelihoods	Prior × likelihood	Posterior probabilities
A	0.5	0.6	0.30	0.30/0.45 = 2/3
B	0.5	0.3	0.15	0.15/0.45 = 1/3
Total			0.45	

The general version of Bayes' theorem may be stated as follows. If there are n events labelled $E_1, \ldots, E_n$ then the probability of the event E_i occurring, given the sample evidence S, is

$$\Pr(E_i|S) = \frac{\Pr(S|E_i) \times \Pr(E_i)}{\Sigma(\Pr(S|E_i) \times \Pr(E_i))} \tag{2.29}$$

As stated earlier, dispute arises over the interpretation of Bayes' theorem. In the above example there is no difficulty because the probability statements can be interpreted as relative frequencies. If the experiment of selecting a bag at random and choosing a ball from it were repeated many times, then of those occasions when a red ball is selected, in two-thirds of them bag A will have been chosen. However, consider an alternative interpretation of the symbols:

A: a coin is fair;
B: a coin is unfair;
R: the result of a toss is a head.

Then, given a toss (or series of tosses) of a coin, this evidence can be used to calculate the probability of the coin being fair. But this makes no sense according to the frequentist school: either the coin is fair or not; it is not a question of probability. The calculated value must be interpreted as a degree of belief and be given a subjective interpretation.

Exercise 2.7

(a) Repeat the 'balls in the bag' exercise from the text, but with bag A containing five red and three yellow balls, bag B containing one red and two yellow balls. The single ball drawn is red. Before doing the calculation, predict which bag is more likely to be the source of the drawn ball. Explain why.

(b) Bag A now contains 10 red and six yellow balls (i.e. twice as many as before, but in the same proportion). Does this alter the answer you obtained in part (a)?

(c) Set out your answer to part (b) in the form of prior probabilities and likelihoods, in order to obtain the posterior probability.

Decision analysis

The study of probability naturally leads on to the analysis of decision making where risk is involved. This is the realistic situation facing most firms and the use of probability can help to illuminate the problem. To illustrate the topic, we use the example of a firm facing a choice of three different investment projects. The uncertainty that the firm faces concerns the interest rate at which to discount the future flows of income. If the interest/discount rate is high then projects which have income far in the future become less attractive relative to

Table 2.1 Data for decision analysis: present values of three investment projects at different interest rates (£000)

Project	Future interest rate			
	4%	5%	6%	7%
A	1475	1363	1200	1115
B	1500	1380	1148	1048
C	1650	1440	1200	810
Probability	0.1	0.4	0.4	0.1

projects with more immediate returns. A low rate reverses this conclusion. The question is: which project should the firm select? As we shall see, there is no unique, right answer to the question but, using probability theory we can see why the answer might vary.

Table 2.1 provides the data required for the problem. The three projects are imaginatively labelled *A*, *B* and *C*. There are four possible **states of the world**, i.e. future scenarios, each with a different interest rate, as shown across the top of the table. This is the only source of uncertainty, otherwise the states of the world are identical. The figures in the body of the table show the present value of each income stream at the given discount rate.

Present value

The present value of future income is its value today and is obtained using the interest rate. For example, if the interest rate is 10%, the present value (i.e. today) of £110 received in one year's time is £100. In other words, one could invest £100 today at 10% and have £110 in one year's time. £100 today and £110 next year are equivalent.

The present value of £110 received in two years' time is smaller since one has to wait longer to receive it. It is calculated as $£110/1.1^2 = 90.91$. Again, £90.91 invested at 10% per annum will yield £110 in two years' time. After one year it is worth $£90.91 \times 1.1 = 100$ and after a second year that £100 becomes £110. Notice that, if the interest rate rises, the present value falls. For example, if the interest rate is 20%, £110 next year is worth only $£110/1.2 = 91.67$ today.

The present value of £110 in one year's time and another £110 in two years' time is $£110/1.1 + £110/1.1^2 = £190.91$. The present value of more complicated streams of income can be calculated by extension of this principle. In the example used in the text you do not need to worry about how the present value is arrived at. Before reading on you may wish to do Exercise 2.8 to practise calculation of present value.

Thus, for example, if the interest rate turns out to be 4% then project *A* has a present value of £1 475 000 while *B*'s is £1 500 000. If the discount rate turns out to be 5% the *PV* for *A* is £1 363 000 while for *B* it has changed to £1 380 000. Obviously, as the discount rate rises, the present value of the return falls. (Alternatively, we could assume that a higher interest rate increases the cost of borrowing to finance the project, which reduces its profitability.) We assume that each project requires a (certain) initial outlay of £1 100 000 with which the *PV* should be compared.

The final row of the table shows the probabilities which the firm attaches to each interest rate. These are obviously someone's subjective probabilities and are symmetric around a central value of 5.5%.

Exercise 2.8

(a) At an interest or discount rate of 10%, what is the present value of £1200 received in one year's time?

(b) If the interest rate rises to 15%, how is the present value altered? The interest rate has risen by 50% (from 10% to 15%): how has the present value changed?

(c) At an interest rate of 10% what is the present value of £1200 received in (i) two years' time and (ii) five years' time?

(d) An income of £500 is received at the end of years one, two and three (i.e. £1500 in total). What is its present value? Assume r = 10%.

(e) Project *A* provides an income of £300 after one year and another £600 after two years. Project *B* provides £400 and £488 at the same times. At a discount rate of 10% which project has the higher present value? What happens if the discount rate rises to 20%?

Decision criteria: maximising the expected value

We need to decide how a decision is to be made on the basis of these data. The first criterion involves the **expected value** of each project. Because of the uncertainty about the interest rate there is no single present value for each project. We therefore calculate the expected value, using the E operator which was introduced in Chapter 1. In other words, we find the expected present value of each project, by taking a weighted average of the *PV* figures, the weights being the probabilities. The project with the highest expected return is chosen.

The expected values are calculated in Table 2.2. The highest expected present value is £1 302 000, associated with project *C*. On this criterion therefore, *C* is chosen. Is this a wise choice? If the business always uses this rule to evaluate many projects then in the long run it will earn the maximum profits. However, you may notice that if the interest rate turns out to be 7% then *C* would be the *worst* project to choose in this case and the firm would make a substantial loss in such circumstances. Project *C* is the most sensitive to the discount rate (it has the greatest *variance* of *PV* values of the three projects) and therefore the firm faces more risk by opting for *C*. There is a trade-off between risk and return.

Table 2.2 Expected values of the three projects

Project	Expected value
A	1284.2
B	1266.0
C	1302.0

Note: 1284.2 is calculated as 1475 × 0.1 + 1363 × 0.4 + 1200 × 0.4 + 1115 × 0.1. This is the weighted average of the four *PV* values. A similar calculation is performed for the other projects.

Table 2.3 The maximin criterion

Project	Minimum
A	1115
B	1048
C	810
Maximum	1115

Perhaps some alternative criteria should be looked at. These we look at next, in particular the **maximin**, **maximax** and **minimax regret** strategies.

Maximin, maximax and minimax regret

The **maximin** criterion looks at the worst-case scenario for each project and then selects the project which does best in these circumstances. It is inevitably a pessimistic or cautious view therefore. Table 2.3 illustrates the calculation. This time we observe that project *A* is preferred. In the worst case (which occurs when $r = 7\%$ for all projects) then *A* does best, with a *PV* of £1 115 000 and therefore a slight profit. The maximin criterion may be a good one in business where managers tend to over-optimism. Calculating the maximin may be a salutary exercise, even if it is not the ultimate deciding factor.

The opposite criterion is the optimistic one where the **maximax** criterion is used. In this case one looks at the *best* circumstances for each project and chooses the best-performing project. Each project does best when the interest rate is at its lowest level, 3%. Examining the first column of Table 2.1 shows that project *C* (*PV* = 1650) performs best and is therefore chosen. Given the earlier warning about over-optimistic managers, this may not be suitable as the sole criterion for making investment decisions.

A final criterion is that of **minimax regret**. If project *B* were chosen but the interest rate turns out to be 7% then we would regret not having chosen *A*, the best project under these circumstances. Our *regret* would be the extent of the difference between the two, a matter of 1115 – 1048 = 67. Similarly, the regret if we had chosen *C* would be 1115 – 810 = 305. We can calculate these regrets at the other interest rates too, always comparing the *PV* of a project with the best *PV* given that interest rate. This gives us Table 2.4.

The final column of the table shows the maximum regret for each project. The minimax regret criterion is to choose the minimum of these figures. This is

Table 2.4 The costs of taking the wrong decision

Project	4%	5%	6%	7%	Maximum
A	175	77	0	0	175
B	150	60	52	67	150
C	0	0	0	305	305
Minimum					150

given at the bottom of the final column; it is 150 which is associated with project *B*. A justification for using this criterion might be that you do not want to fall too far behind your competitors. If other firms are facing similar investment decisions, then the regret table shows the difference in *PV* (and hence profits) if they choose the best project while you do not. Choosing the minimax regret solution ensures that you will not fall too far behind. During the internet bubble of the 1990s it was important to gain market share and keep up with, or surpass, your competitors. The minimax regret strategy might be a useful tool during such times.

You will probably have noticed that we have managed to find a justification for choosing all three projects! No one project comes out best on all criteria. Nevertheless, the analysis might be of some help: if the investment project is one of many small, independent investments the firm is making, then this would justify use of the expected value criterion. On the other hand, if this is a big, one-off project which could possibly bankrupt the firm if it goes wrong, then the maximin criterion would be appropriate.

The expected value of perfect information

Often a firm can improve its knowledge about future possibilities via research, which costs money. This effectively means buying information about the future state of the world. The question arises: how much should a firm pay for such information? **Perfect information** would reveal the future state of the world with certainty – in this case, the future interest rate. In that case you could be sure of choosing the right project given each state of the world. If interest rates turn out to be 4%, the firm would invest in *C*, if 7% in *A*, and so on.

In such circumstances, the firm would expect to earn

$$(0.1 \times 1650) + (0.4 \times 1440) + (0.4 \times 1200) + (0.1 \times 1115) = 1332.5$$

i.e. the probability of each state of the world is multiplied by the *PV* of the *best* project for that state. This gives a figure which is greater than the expected value calculated earlier, without perfect information, 1302. The **expected value of perfect information** is therefore the difference between these two, 30.5. This sets a *maximum* to the value of information, for it is unlikely in the real world that any information about the future is going to be perfect.

Exercise 2.9

(a) Evaluate the three projects detailed in the table below, using the criteria of expected value, maximin, maximax and minimax regret. The probability of a 4% interest rate is 0.3, of 6% is 0.4 and of 8% is 0.3.

Project	4%	6%	8%
A	100	80	70
B	90	85	75
C	120	60	40

(b) What would be the value of perfect information about the interest rate?

Summary

- The theory of probability forms the basis of statistical inference: the drawing of inferences on the basis of a random sample of data. The reason for this is the probability basis of random sampling.
- A convenient definition of the probability of an event is the number of times the event occurs divided by the number of trials (occasions when the event could occur).
- For more complex events, their probabilities can be calculated by combining probabilities, using the addition and multiplication rules.
- The probability of events A or B occurring is calculated according to the addition rule.
- The probability of A and B occurring is given by the multiplication rule.
- If A and B are not independent, then $\Pr(A \text{ and } B) = \Pr(A) \times \Pr(B|A)$, where $\Pr(B|A)$ is the probability of B occurring given that A has occurred (the conditional probability).
- Tree diagrams are a useful technique for enumerating all the possible paths in series of probability trials, but for large numbers of trials the huge number of possibilities makes the technique impractical.
- For experiments with a large number of trials (e.g. obtaining 20 heads in 50 tosses of a coin) the formulae for combinations and permutations can be used.
- The combinatorial formula nCr gives the number of ways of combining r similar objects among n objects, e.g. the number of orderings of three girls (and hence implicitly two boys also) in five children.
- The permutation formula nPr gives the number of orderings of r distinct objects among n, e.g. three named girls among five children.
- Bayes' theorem provides a formula for calculating a conditional probability, e.g. the probability of someone being a smoker, given they have been diagnosed with cancer. It forms the basis of Bayesian statistics, allowing us to calculate the probability of a hypothesis being true, based on the sample evidence and prior beliefs. Classical statistics disputes this approach.
- Probabilities can also be used as the basis for decision making in conditions of uncertainty, using as decision criteria expected value maximisation, maximin, maximax or minimax regret.

Key terms and concepts

addition rule
Bayes' theorem
combinations
complement
compound event
conditional probability
exhaustive
expected value of perfect information
frequentist approach
independent events
maximin
minimax
minimax regret
multiplication rule
mutually exclusive
outcome or event
permutations
probability experiment
probability of an event
sample space
subjective approach
tree diagram

Problems

Some of the more challenging problems are indicated by highlighting the problem number in colour.

2.1 Given a standard pack of cards, calculate the following probabilities:

(a) drawing an ace;

(b) drawing a court card (i.e. jack, queen or king);

(c) drawing a red card;

(d) drawing three aces without replacement;

(e) drawing three aces with replacement.

2.2 The following data give duration of unemployment by age, in July 1986.

Age	Duration of unemployment (weeks)				Total (000s)	Economically active (000s)
	≤8	8–26	26–52	>52		
	(Percentage figures)					
16–19	27.2	29.8	24.0	19.0	273.4	1270
20–24	24.2	20.7	18.3	36.8	442.5	2000
25–34	14.8	18.8	17.2	49.2	531.4	3600
35–49	12.2	16.6	15.1	56.2	521.2	4900
50–59	8.9	14.4	15.6	61.2	388.1	2560
≥60	18.5	29.7	30.7	21.4	74.8	1110

The 'economically active' column gives the total of employed (not shown) plus unemployed in each age category.

(a) In what sense may these figures be regarded as probabilities? What does the figure 27.2 (top-left cell) mean following this interpretation?

(b) Assuming the validity of the probability interpretation, which of the following statements are true?

(i) The probability of an economically active adult aged 25–34, drawn at random, being unemployed is 531.4/3600.

(ii) If someone who has been unemployed for over one year is drawn at random, the probability that they are aged 16–19 is 19%.

(iii) For those aged 35–49 who became unemployed before July 1985, the probability of their still being unemployed is 56.2%.

(iv) If someone aged 50–59 is drawn at random from the economically active population, the probability of their being unemployed for eight weeks or less is 8.9%.

(v) The probability of someone aged 35–49 drawn at random from the economically active population being unemployed for between 8 and 26 weeks is $0.166 \times 521.2/4900$.

(c) A person is drawn at random from the population and found to have been unemployed for over one year. What is the probability that they are aged between 16 and 19?

2.3 'Odds' in horserace betting are defined as follows: 3/1 (three-to-one against) means a horse is expected to win once for every three times it loses; 3/2 means two wins out of five races; 4/5 (five to four *on*) means five wins for every four defeats, etc.

(a) Translate the above odds into 'probabilities' of victory.

(b) In a three-horse race, the odds quoted are 2/1, 6/4, and 1/1. What makes the odds different from probabilities? Why are they different?

(c) Discuss how much the bookmaker would expect to win in the long run at such odds, assuming each horse is backed equally.

2.4 (a) Translate the following odds to 'probabilities': 13/8, 2/1 *on*, 100/30.

(b) In the 2.45 race at Plumpton on 18/10/94 the odds for the five runners were:

Philips Woody	1/1
Gallant Effort	5/2
Satin Noir	11/2
Victory Anthem	9/1
Common Rambler	16/1

Calculate the 'probabilities' and their sum.

(c) Should the bookmaker base his odds on the true probabilities of each horse winning, or on the amount bet on each horse?

2.5 How might you estimate the probability of Peru defaulting on its debt repayments next year?

2.6 How might you estimate the probability of a corporation reneging on its bond payments?

2.7 Judy is 33, unmarried and assertive. She is a graduate in political science, and involved in union activities and anti-discrimination movements. Which of the following statements do you think is more probable?

(a) Judy is a bank clerk.

(b) Judy is a bank clerk, active in the feminist movement.

2.8 In March 1994 a news item revealed that a London 'gender' clinic (which reportedly enables you to choose the sex of your child) had just set up in business. Of its first six births, two were of the 'wrong' sex. Assess this from a probability point of view.

2.9 A newspaper advertisement reads 'The sex of your child predicted, or your money back!' Discuss this advertisement from the point of view of (a) the advertiser and (b) the client.

2.10 'Roll six sixes to win a Mercedes!' is the announcement at a fair. You have to roll six dice. If you get six sixes you win the car, valued at £20 000. The entry ticket costs £1. What is your expected gain or loss on this game? The organisers of the fair have to take out insurance against the car being won. This costs £250 for the day. Does this seem a fair premium? If not, why not?

2.11 At another stall, you have to toss a coin numerous times. If a head does not appear in 20 tosses you win £1 bn. The entry fee for the game is £100.

(a) What are your expected winnings?

(b) Would you play?

2.12 A four-engine plane can fly as long as at least two of its engines work. A two-engine plane flies as long as at least one engine works. The probability of an individual engine failure is 1 in 1000.

(a) Would you feel safer in a four- or two-engine plane, and why? Calculate the probabilities of an accident for each type.

(b) How much safer is one type than the other?

(c) What crucial assumption are you making in your calculation? Do you think it is valid?

2.13 Which of the following events are independent?

(a) Two flips of a fair coin.

(b) Two flips of a biased coin.

(c) Rainfall on two successive days.

(d) Rainfall on St Swithin's day and rain one month later.

2.14 Which of the following events are independent?

(a) A student getting the first two questions correct in a multiple-choice exam.

(b) A driver having an accident in successive years.

(c) IBM and Dell earning positive profits next year.

(d) Arsenal Football Club winning on successive weekends.

How is the answer to (b) reflected in car insurance premiums?

2.15 Manchester United beat Liverpool 4–2 at soccer, but you do not know the order in which the goals were scored. Draw a tree diagram to display all the possibilities and use it to find (a) the probability that the goals were scored in the order L, MU, MU, MU, L, MU, and (b) the probability that the score was 2–2 at some stage.

2.16 An important numerical calculation on a spacecraft is carried out independently by three computers. If all arrive at the same answer, it is deemed correct. If one disagrees, it is overruled. If there is no agreement then a fourth computer does the calculation and, if its answer agrees with any of the others, it is deemed correct. The probability of an individual computer getting the answer right is 99%. Use a tree diagram to find:

(a) the probability that the first three computers get the right answer;

(b) the probability of getting the right answer;

(c) the probability of getting no answer;

(d) the probability of getting the wrong answer.

2.17 The French national lottery works as follows. Six numbers from the range 0 to 49 are chosen at random. If you have correctly guessed all six you win the first prize. What are your chances of winning if you are only allowed to choose six numbers? A single entry like this costs €1. For €210 you can choose 10 numbers and you win if the six selected numbers are among them. Is this better value than the single entry?

2.18 The UK national lottery works as follows. You choose six (different) numbers in the range 1 to 49. If all six come up in the draw (in any order) you win the first prize, expected to be around £2m (which could be shared if someone else chooses the six winning numbers).

(a) What is your chance of winning with a single ticket?

(b) You win a second prize if you get five out of six right, *and* your final chosen number matches the 'bonus' number in the draw (also in the range 1 to 49). What is the probability of winning a second prize?

(c) Calculate the probabilities of winning a third, fourth or fifth prize, where a third prize is won by matching five out of the six numbers, a fourth prize by matching four out of six and a fifth prize by matching three out of six.

(d) What is the probability of winning a prize?

(e) The prizes are as follows:

Prize	Value	
First	£2 m	(expected, possibly shared)
Second	£100 000	(expected, for each winner)
Third	£1500	(expected, for each winner)
Fourth	£65	(expected, for each winner)
Fifth	£10	(guaranteed, for each winner)

Comment upon the distribution of the fund between first, second, etc., prizes.

(f) Why is the fifth prize guaranteed whereas the others are not?

(g) In the first week of the lottery, 49 million tickets were sold. There were 1 150 000 winners, of which 7 won (a share of) the jackpot, 39 won a second prize, 2139 won a third prize and 76 731 a fourth prize. Are you surprised by these results or are they as you would expect?

2.19 A coin is either fair or has two heads. You initially assign probabilities of 0.5 to each possibility. The coin is then tossed twice, with two heads appearing. Use Bayes' theorem to work out the posterior probabilities of each possible outcome.

2.20 A test for AIDS is 99% successful, i.e. if you are HIV+ it will detect it in 99% of all tests, and if you are not, it will again be right 99% of the time. Assume that about 1% of the population are HIV+. You take part in a random testing procedure, which gives a positive result. What is the probability that you are HIV+? What implications does your result have for AIDS testing?

2.21 (a) Your initial belief is that a defendant in a court case is guilty with probability 0.5. A witness comes forward claiming he saw the defendant commit the crime. You know the witness is not totally reliable and tells the truth with probability p. Use Bayes' theorem to calculate the posterior probability that the defendant is guilty, based on the witness's evidence.

(b) A second witness, equally unreliable, comes forward and claims she saw the defendant commit the crime. Assuming the witnesses are not colluding, what is your posterior probability of guilt?

(c) If $p < 0.5$, compare the answers to (a) and (b). How do you account for this curious result?

2.22 A man is mugged and claims that the mugger had red hair. In police investigations of such cases, the victim was able correctly to identify the assailant's hair colour 80% of the time. Assuming that 10% of the population have red hair, what is the probability that the assailant in this case did, in fact, have red hair? Guess the answer first, then find the right answer using Bayes' theorem. What are the implications of your results for juries' interpretation of evidence in court, particularly in relation to racial minorities?

2.23 A firm has a choice of three projects, with profits as indicated below, dependent upon the state of demand.

Project	**Demand**		
	Low	**Middle**	**High**
A	100	140	180
B	130	145	170
C	110	130	200
Probability	0.25	0.45	0.3

(a) Which project should be chosen on the expected value criterion?

(b) Which project should be chosen on the maximin and maximax criteria?

(c) Which project should be chosen on the minimax regret criterion?

(d) What is the expected value of perfect information to the firm?

2.24 A firm can build a small, medium or large factory, with anticipated profits from each dependent upon the state of demand, as in the table below.

Factory	**Demand**		
	Low	**Middle**	**High**
Small	300	320	330
Medium	270	400	420
Large	50	250	600
Probability	0.3	0.5	0.2

(a) Which project should be chosen on the expected value criterion?

(b) Which project should be chosen on the maximin and maximax criteria?

(c) Which project should be chosen on the minimax regret criterion?

(d) What is the expected value of perfect information to the firm?

2.25 There are 25 people at a party. What is the probability that there are at least two with a birthday in common?

Hint: the *complement* is (much) easier to calculate.

2.26 This problem is tricky, but amusing. Three gunmen, A, B and C, are shooting at each other. The probabilities that each will hit what they aim at are respectively 1, 0.75, 0.5. They take it in turns to shoot (in alphabetical order) and continue until only one is left alive. Calculate the probabilities of each winning the contest. (Assume they draw lots for the right to shoot first.)

Hint 1: Start with one-on-one gunfights, e.g. the probability of A beating B, or of B beating C.

Hint 2: You'll need the formula for the sum of an infinite series, given in Chapter 1.

2.27 The BMAT test (see http://www.ucl.ac.uk/lapt/bmat/) is an on-line test for prospective medical students. It uses 'certainty based marking'. After choosing your answer from the alternatives available, you then have to give your level of confidence that your answer is correct: low, medium or high. If you choose low, you get one mark for the correct answer, zero if it is wrong. For medium confidence you get +2 or −2 marks for correct or incorrect answers. If you choose high, you get +3 or −6.

(a) If you are 60% confident your answer is correct (i.e. you think there is a 60% probability you are right), which certainty level should you choose?

(b) Over what range of probabilities is 'medium' the best choice?

(c) If you were 85% confident, how many marks would you expect to lose by opting for one of the wrong choices?

2.28 A multiple choice test involves 20 questions, with four choices for each answer.

(a) If you guessed the answers to all questions at random, what mark out of 20 would you expect to get?

(b) If you know the correct answer to eight of the questions, what is your expected score out of 20?

(c) The examiner wishes to correct the bias due to students guessing answers. They decide to award a negative mark for incorrect answers (with 1 for a correct answer and 0 for no answer given). What negative mark would ensure that the overall mark out of 20 is a true reflection of the student's ability?

Answers to exercises

Exercise 2.1

Answer in text.

Exercise 2.2

(a) A subjective view would have to be taken, informed by such things as opinion polls.

(b) 1/49, a frequentist view. Some people do add their own subjective evaluations (e.g. that 5 must come up as it has not been drawn for several weeks) but these are often unwarranted according to the frequentist approach.

(c) A mixture of objective and subjective criteria might be used here. Historical data on the occurrence of tsunamis might give a (frequentist) baseline figure, to which might be added subjective considerations such as the amount of recent seismic activity.

(d) A mixture again. Historical data give a benchmark (possibly of little relevance) while immediate factors such as the weather might alter one's subjective judgement. (As I write it is snowing outside, which seems to have a huge impact on British trains!)

Exercise 2.3

(a) 1, 2, 3, . . . , 20, 21 (a triple seven), 22 (double eleven), 24, 25 (outer bull), 26, 27, 28, 30, 32, 33, 34, 36, 38, 39, 40, 42, 45, 48, 50, 51, 54, 57, 60. Or it could miss altogether!

(b) The complement is missing the target, with probability $1 - 0.3 = 70\%$.

(c) Zero, it is impossible.

(d) Impossible, the probabilities sum to more than one.

Exercise 2.4

(a) $0.3 \times 0.7 + 0.7 \times 0.3 = 0.42$. This is a hit followed by a miss or a miss followed by a hit.

(b) $0.3 \times 0.3 = 0.09$.

(c) It is assumed that the probability of the second arrow hitting the target is the same as the first. Altering this assumption would affect both answers.

(d) Part (a) becomes $0.3 \times (1 - 0.5) + 0.7 \times 0.2 = 0.29$. Part (b) becomes $0.3 \times 0.5 = 0.15$.

Exercise 2.5

(a) Independent case:

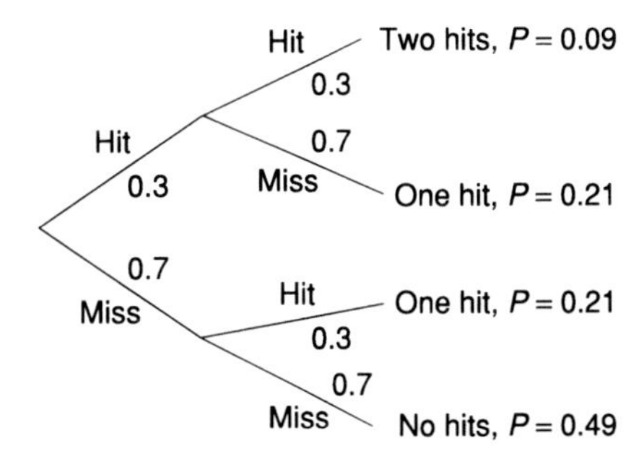

Dependent case:

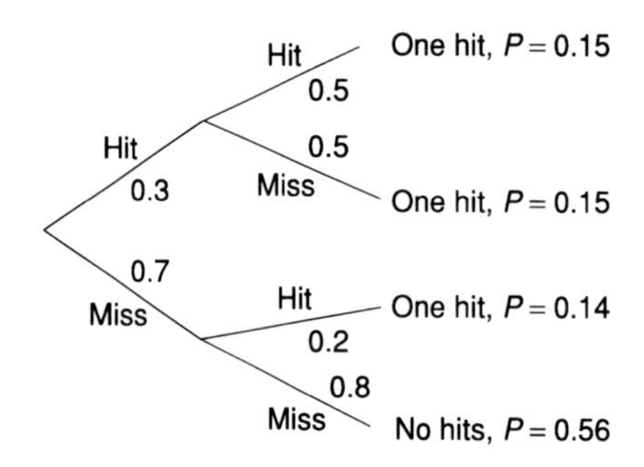

(b)

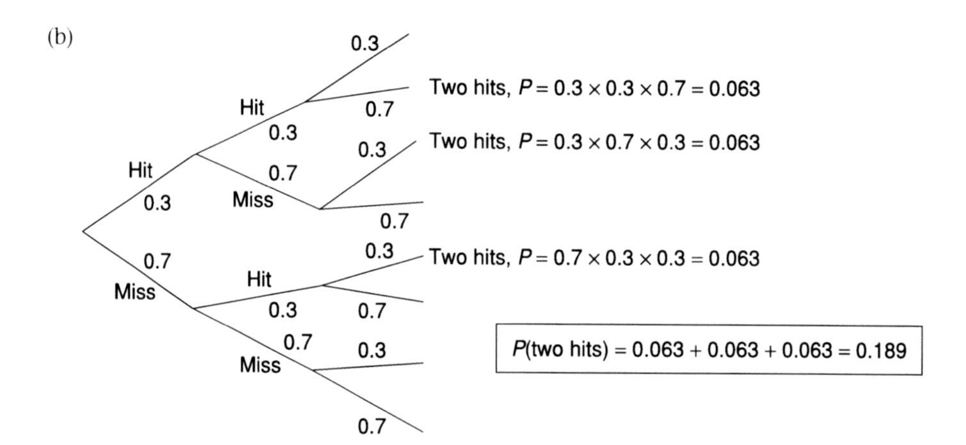

(c)

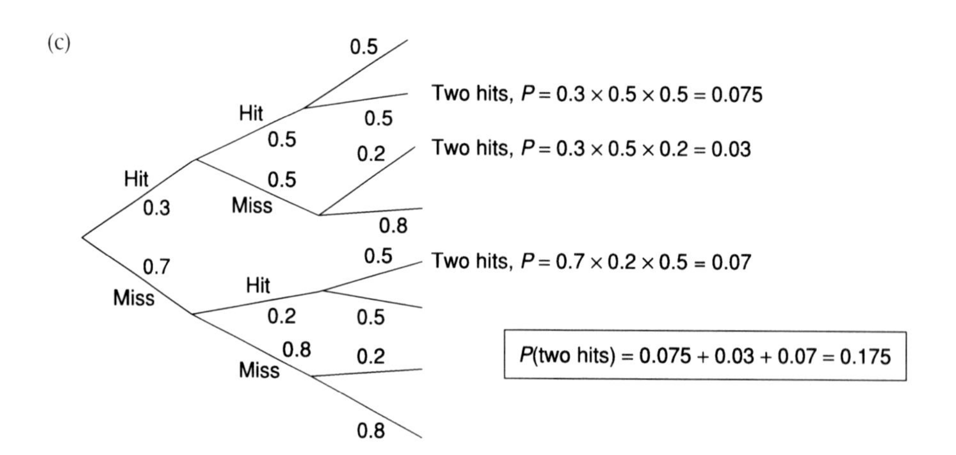

Exercise 2.6

(a) Pr(2 hits) = Pr(H and H and M) × 3C2 = 0.3 × 0.3 × 0.7 × 3 = 0.189.

(b) This cannot be done using the combinatorial formula, because of the non-independence of probabilities. Instead one has to calculate Pr(H and H and M) + Pr(H and M and H) + Pr(M and H and H), yielding the answer 0.175.

Exercise 2.7

(a) Bag A has proportionately more red balls than bag B, hence should be the favoured bag from which the single red ball was drawn. Performing the calculation

$$\Pr(A|R) = \frac{\Pr(R|A) \times \Pr(A)}{\Pr(R|A) \times \Pr(A) + \Pr(R|B) \times \Pr(B)}$$

$$= \frac{0.625 \times 0.5}{0.625 \times 0.5 + 0.5 \times 0.5} = 0.556$$

(b) The result is the same, as $\Pr(R|A) = 0.625$ as before. The number of balls does not enter the calculation.

(c)

	Prior probabilities	Likelihoods	Prior × likelihood	Posterior probabilities
A	0.5	0.625	0.3125	0.3125/0.5625 = 0.556
B	0.5	0.5	0.25	0.25/0.5625 = 0.444
Total			0.5625	

Exercise 2.8

(a) 1200/1.1 = 1090.91.

(b) 1200/1.15 = 1043.48. The *PV* has only changed by 4.3%. This is calculated as 1.1/1.15 – 1 = –0.043.

(c) $1200/1.1^2 = 991.74$; $1200/1.1^5 = 745.11$.

(d) $PV = 500/1.1 + 500/1.1^2 + 500/1.1^3 = 1243.43$.

(e) At 10%: project A yields a *PV* of $300/1.1 + 600/1.1^2 = 768.6$. Project B yields $400/1.1 + 488/1.1^2 = 766.9$. At 20% the *PVs* are 666.7 and 672.2, reversing the rankings. *A*'s large benefits in year 2 are penalised by the higher discount rate.

Exercise 2.9

(a)

Project	Expected value	Minimum	Maximum
A	0.3 × 100 + 0.4 × 80 + 0.3 × 70 = 83	70	100
B	0.3 × 90 + 0.4 × 85 + 0.3 × 75 = 83.5	75	90
C	0.3 × 120 + 0.4 × 60 + 0.3 × 40 = 72	40	120

The maximin is 75, associated with project B and the maximax is 120, associated with project C. The regret values are given by

	4%	6%	8%	Max
A	20	5	5	20
B	30	0	0	30
C	0	25	35	35
			Min	20

The minimax regret is 20, associated with project A.

(b) With perfect information the firm could eam 0.3 × 120 + 0.4 × 85 + 0.3 × 75 = 92.5. The highest expected value is 83.5, so the value of perfect information is 92.5 – 83.5 = 9.

3 Probability distributions

Contents

Learning outcomes

By the end of this chapter you should be able to:

- recognise that the result of most probability experiments (e.g. the score on a die) can be described as a random variable;
- appreciate how the behaviour of a random variable can often be summarised by a probability distribution (a mathematical formula);
- recognise the most common probability distributions and be aware of their uses;
- solve a range of probability problems using the appropriate probability distribution.

Complete your diagnostic test for Chapter 3 now to create your personal study plan. Exercises with an icon ? are also available for practice in MathXL with additional supporting resources.

Introduction

In this chapter the probability concepts introduced in Chapter 2 are generalised by using the idea of a **probability distribution**. A probability distribution lists, in some form, all the possible outcomes of a probability experiment and the probability associated with each one. For example, the simplest experiment is tossing a coin, for which the possible outcomes are heads or tails, each with probability one-half. The probability distribution can be expressed in a variety of ways: in words, or in a graphical or mathematical form. For tossing a coin, the graphical form is shown in Figure 3.1, and the mathematical form is

$$\Pr(H) = \tfrac{1}{2}$$
$$\Pr(T) = \tfrac{1}{2}$$

The different forms of presentation are equivalent, but one might be more suited to a particular purpose.

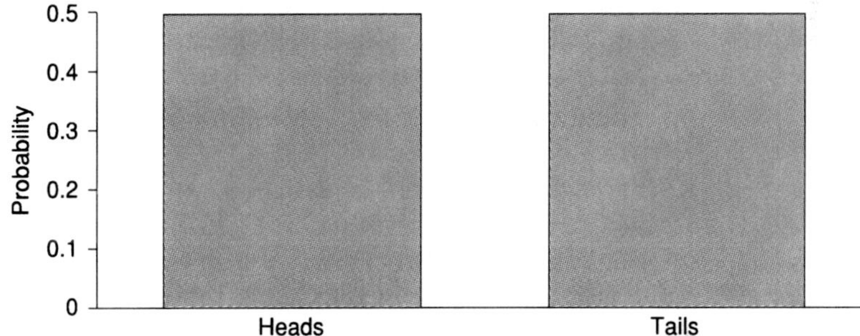

Figure 3.1
The probability distribution for the toss of a coin

Some probability distributions occur often and so are well known. Because of this they have names so we can refer to them easily; for example, the **Binomial distribution** or the **Normal distribution**. In fact, each constitutes a *family* of distributions. A single toss of a coin gives rise to one member of the Binomial distribution family; two tosses would give rise to another member of that family. These two distributions differ in the number of tosses. If a biased coin were tossed, this would lead to yet another Binomial distribution, but it would differ from the previous two because of the different probability of heads.

Members of the Binomial family of distributions are distinguished either by the number of tosses or by the probability of the event occurring. These are the two **parameters** of the distribution and tell us all we need to know about the distribution. Other distributions might have different numbers of parameters, with different meanings. Some distributions, for example, have only one parameter. We will come across examples of different types of distribution throughout the rest of this book.

In order to understand fully the idea of a probability distribution a new concept is first introduced, that of a **random variable**. As will be seen later in the chapter, an important random variable is the sample mean, and to understand how to draw inferences from the sample mean it is important to recognise it as a random variable.

Random variables

Examples of random variables have already been encountered in Chapter 2, for example, the result of the toss of a coin, or the number of boys in a family of five children. A random variable is one whose outcome or value is the result of chance and is therefore unpredictable, although the range of possible outcomes and the probability of each outcome may be known. It is impossible to know in advance the outcome of a toss of a coin for example, but it must be either heads or tails, each with probability one-half. The number of heads in 250 tosses is another random variable, which can take any value between zero and 250, although values near 125 are the most likely. You are very unlikely to get 250 heads from tossing a fair coin!

Intuitively, most people would 'expect' to get 125 heads from 250 tosses of the coin, since heads comes up half the time on average. This suggests we could use the expected value notation introduced in Chapter 1 and write $E(X) = 125$, where X represents the number of heads obtained from 250 tosses. This usage is indeed valid and we will explore this further below. It is a very convenient shorthand notation.

The time of departure of a train is another example of a random variable. It may be timetabled to depart at 11.15, but it probably (almost certainly!) will not leave at exactly that time. If a sample of ten basketball players were taken, and their average height calculated, this would be a random variable. In this latter case, it is the process of taking a sample that introduces the variability which makes the resulting average a random variable. If the experiment were repeated, a different sample and a different value of the random variable would be obtained.

The above examples can be contrasted with some things which are *not* random variables. If one were to take *all* basketball players and calculate their average height, the result would not be a random variable. This time there is no sampling procedure to introduce variability into the result. If the experiment were repeated the same result would be obtained, since the same people would be measured the second time (this assumes that the population does not change, of course). Just because the value of something is unknown does not mean it qualifies as a random variable. This is an important distinction to bear in mind, since it is legitimate to make probability statements about random variables ('the probability that the average height of a sample of basketball players is over 195 cm is 60%') but not about parameters ('the probability that the Pope is over six feet is 60%'). Here again there is a difference of opinion between frequentist and subjective schools of thought. The latter group would argue that it is possible to make probability statements about the Pope's height. It is a way of expressing lack of knowledge about the true value. The frequentists would say the Pope's height is a fact that we do not happen to know; that does not make it a random variable.

The Binomial distribution

One of the simplest distributions which a random variable can have is the Binomial. The Binomial distribution arises whenever the underlying probability experiment has just two possible outcomes, for example heads or tails from the toss of a coin. Even if the coin is tossed many times (so one could end up with one, two, three . . . , etc., heads in total) the *underlying* experiment has only two outcomes, so the Binomial distribution should be used. A counter-example would be the rolling of die, which has six possible outcomes (in this case the Multinomial distribution, not covered in this book, would be used). Note, however, that if we were interested only in rolling a six or not, we *could* use the Binomial by defining the two possible outcomes as 'six' and 'not-six'. It is often the case in statistics that by suitable transformation of the data we can use different distributions to tackle the same problem. We will see more of this later in the chapter.

The Binomial distribution can therefore be applied to the type of problem encountered in the previous chapter, concerning the sex of children. It provides a general formula for calculating the probability of r boys in n births or, in more general terms, the probability of r 'successes' in n trials.[1] We shall use it to calculate the probabilities of 0, 1, . . . , 5 boys in five births.

For the Binomial distribution to apply we first need to assume independence of successive events and we shall assume that, for any birth

$$\Pr(\text{boy}) = P = \tfrac{1}{2}$$

It follows that

$$\Pr(\text{girl}) = 1 - \Pr(\text{boy}) = 1 - P = \tfrac{1}{2}$$

Although we have $P = \frac{1}{2}$ in this example, the Binomial distribution can be applied for any value of P between 0 and 1.

First we consider the case of $r = 5$, $n = 5$, i.e. five boys in five births. This probability is found using the multiplication rule

$$\Pr(r = 5) = P \times P \times P \times P \times P = P^5 = (\tfrac{1}{2})^5 = 1/32$$

The probability of four boys (and then implicitly one girl) is

$$\Pr(r = 4) = P \times P \times P \times P \times (1 - P) = 1/32$$

But this gives only one possible ordering of the four boys and one girl. Our original statement of the problem did not specify a particular ordering of the children. There are five possible orderings (the single girl could be in any of five positions in rank order). Recall that we can use the combinatorial formula nCr to calculate the number of orderings, giving $5C4 = 5$. Hence the probability of four boys and one girl in any order is 5/32. Summarising, the formula for four boys and one girl is

$$\Pr(r = 4) = 5C4 \times P^4 \times (1 - P)$$

[1] The identification of a boy with 'success' is a purely formal one and is not meant to be pejorative!

For three boys (and two girls) we obtain

$$\Pr(r = 3) = 5C3 \times P^3 \times (1 - P)^2 = 10 \times 1/8 \times 1/4 = 10/32$$

In a similar manner

$$\Pr(r = 2) = 5C2 \times P^2 \times (1 - P)^3 = 10/32$$

$$\Pr(r = 1) = 5C1 \times P^1 \times (1 - P)^4 = 5/32$$

$$\Pr(r = 0) = 5C0 \times P^0 \times (1 - P)^5 = 1/32$$

As a check on our calculations we may note that the sum of the probabilities equals 1, as they should do, as we have enumerated all possibilities.

A fairly clear pattern emerges. The probability of r boys in n births is given by

$$\Pr(r) = nCr \times P^r \times (1 - P)^{n-r}$$

and this is known as the Binomial formula or distribution. The Binomial distribution is appropriate for analysing problems with the following characteristics:

- There is a number (n) of trials.
- Each trial has only two possible outcomes, 'success' (with probability P) and 'failure' (probability $1 - P$) and the outcomes are independent between trials.
- The probability P does not change between trials.

The probabilities calculated by the Binomial formula may be illustrated in a diagram, as shown in Figure 3.2. This is very similar to the relative frequency distribution which was introduced in Chapter 1. That distribution was based on empirical data (to do with wealth) while the Binomial probability distribution is a theoretical construction, built up from the basic principles of probability theory.

As stated earlier, the Binomial is, in fact, a family of distributions and each member of this family is distinguished by two **parameters**, n and P. The Binomial is thus a distribution with two parameters, and once their values are known the distribution is completely determined (i.e. $\Pr(r)$ can be calculated for all values of r). To illustrate the difference between members of the family of the Binomial distribution, Figure 3.3 presents three other Binomial distributions, for different values of P and n. It can be seen that for the value of $P = \frac{1}{2}$ the

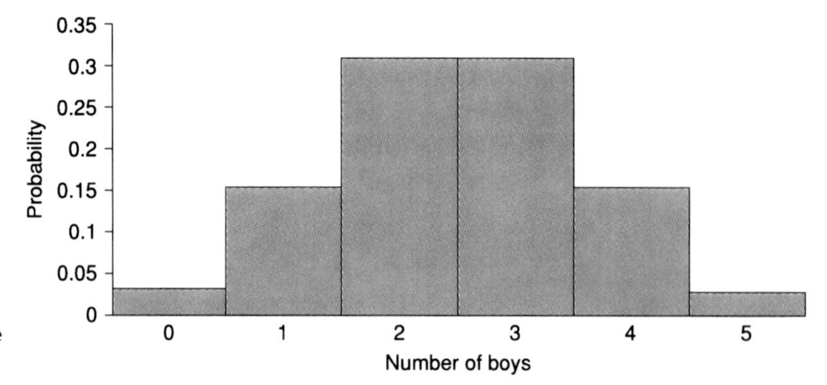

Figure 3.2
Probability distribution of the number of boys in five children

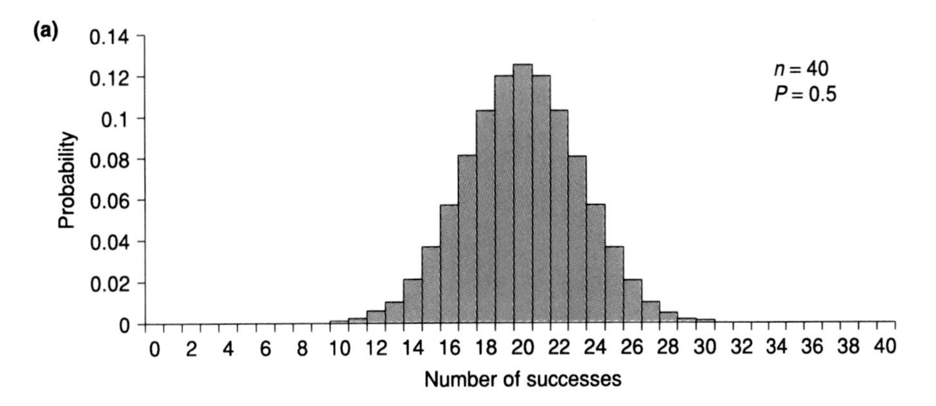

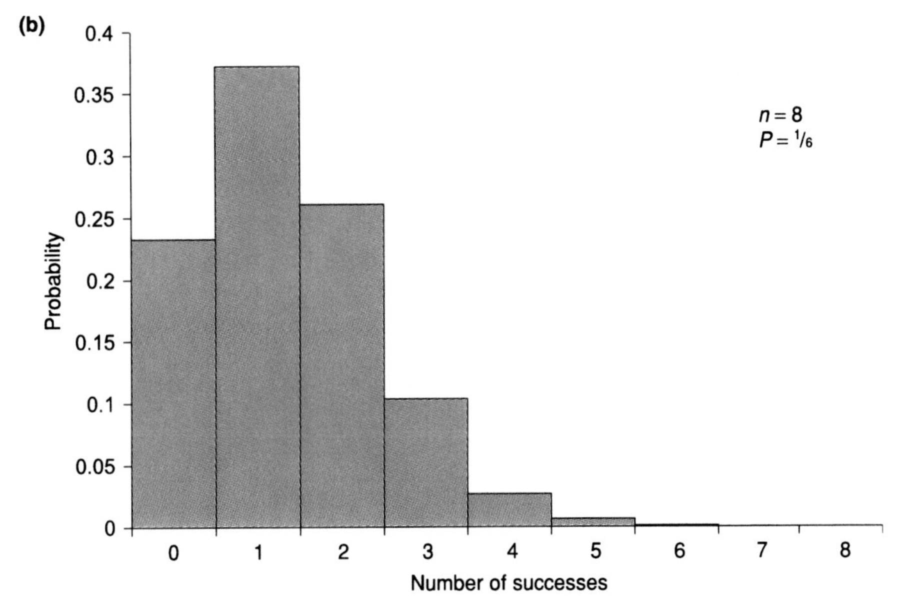

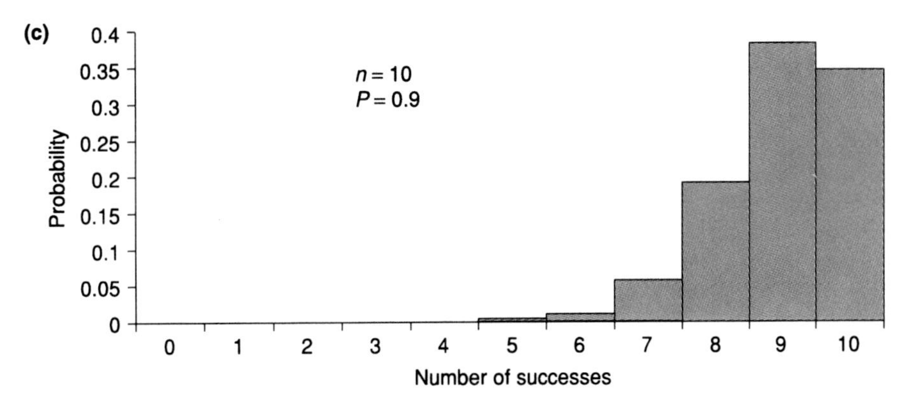

Figure 3.3
Binomial distributions with different parameter values

distribution is symmetric, while for all other values it is skewed to either the left or the right. Part (b) of the figure illustrates the distribution relating to the worked example of rolling a die, described below.

Since the Binomial distribution depends only upon the two values n and P, a shorthand notation can be used, rather than using the formula itself. A random variable r, which has a Binomial distribution with the parameters n and P, can be written in general terms as

$$r \sim B(n, P) \tag{3.1}$$

Thus for the previous example of children, where r represents the number of boys

$$r \sim B(5, \tfrac{1}{2})$$

This is simply a brief and convenient way of writing down the information available; it involves no new problems of a conceptual nature. Writing

$$r \sim B(n, P)$$

is just a shorthand for

$$\Pr(r) = nCr \times P^r \times (1 - P)^{n-r}$$

Teenage weapons

This is a nice example of how knowledge of the Binomial distribution can help our interpretation of events in the news.

> *'One in five teens carry weapon'.* (link on main BBC news web site 23 July 2007)

Following the link to the text of the story, we read:

> *'One in five young teenagers say that their friends are carrying knives and weapons, says a major annual survey of schoolchildren's health and wellbeing'.*

With concerns about knife crime among teenagers, this survey shows that a fifth of youngsters are 'fairly sure' or 'certain' that their male friends are carrying a weapon.'

Notice, incidentally, how the story subtly changes. The headline suggests 20% of teenagers carry a weapon. The text then says this is what *young* teenagers report of their *friends*. It then reveals that some are only 'fairly sure' and that it applies to boys, not girls. By now our suspicions should be aroused. What is the truth?

Note that you are more likely to know someone who carries a weapon than to carry one yourself. Let p be the proportion who truly carry a weapon. Assume also that each person has 10 friends. What is the probability that a person, selected at random, has no friends who carry a weapon? Assuming independence, this is given by $(1 - p)^{10}$. Hence the probability of at least one friend with a weapon is $1 - (1 - p)^{10}$. This is proportion of people who will report having at least one friend with a weapon. How does this vary with p? This is set out in the following table:

p	P(≥ 1 friend with weapon) $1 - (1 - p)^{10}$
0.0%	0%
0.5%	5%
1.0%	10%
1.5%	14%
2.0%	18%
2.5%	22%
3.0%	26%
3.5%	30%
4.0%	34%

Thus a true proportion of just over 2% carrying weapons will generate a report suggesting 20% know someone carrying a weapon! This is much less alarming (and less newsworthy) than in the original story.

You might like to test the assumptions. What happens if there are more than 10 friends assumed? What happens if events are not independent, i.e. having one friend with a weapon increases the probability of another friend with a weapon?

The mean and variance of the Binomial distribution

In Chapter 1 we calculated the mean and variance of a set of data, of the distribution of wealth. The picture of that distribution (Figure 1.9) looks not too dissimilar to one of the Binomial distributions shown in Figure 3.3 above. This suggests that we can calculate the mean and variance of a Binomial distribution, just as we did for the empirical distribution of wealth. Calculating the mean would provide the answer to a question such as 'If we have a family with five children, how many do we expect to be boys?'. Intuitively the answer seems clear, 2.5 (even though such a family could not exist!). The Binomial formula allows us to confirm this intuition.

The mean and variance are most easily calculated by drawing up a relative frequency table based on the Binomial frequencies. This is shown in Table 3.1 for the values $n = 5$ and $P = \frac{1}{2}$. Note that r is equivalent to x in our usual notation and $\Pr(r)$, the relative frequency, is equivalent to $f(x)/\Sigma f(x)$. The mean of this distribution is given by

$$E(r) = \frac{\Sigma r \times \Pr(r)}{\Sigma \Pr(r)} = \frac{80/32}{32/32} = 2.5 \qquad (3.2)$$

Table 3.1 Calculating the mean and variance of the Binomial distribution

r	$\Pr(r)$	$r \times \Pr(r)$	$r^2 \times \Pr(r)$
0	1/32	0	0
1	5/32	5/32	5/32
2	10/32	20/32	40/32
3	10/32	30/32	90/32
4	5/32	20/32	80/32
5	1/32	5/32	25/32
Totals	32/32	80/32	240/32

and the variance is given by

$$V(r) = \frac{\Sigma r^2 \times \Pr(r)}{\Sigma \Pr(r)} - \mu^2 = \frac{240/32}{32/32} - 2.5^2 = 1.25 \quad (3.3)$$

The mean value tells us that in a family of five children we would expect, on average, two and a half boys. Obviously no single family can be like this; it is the average over all such families. The variance is more difficult to interpret intuitively, but it tells us something about how the number of boys in different families will be spread around the average of 2.5.

There is a quicker way to calculate the mean and variance of the Binomial distribution. It can be shown that the mean can be calculated as nP, i.e. the number of trials times the probability of success. For example, in a family with five children and an equal probability that each child is a boy or a girl, then we expect $nP = 5 \times 1/2 = 2.5$ to be boys.

The variance can be calculated as $nP(1 - P)$. This gives $5 \times 1/2 \times 1/2 = 1.25$, as found above by extensive calculation.

Worked example 3.1 Rolling a die

If a die is thrown four times, what is the probability of getting two or more sixes? This is a problem involving repeated experiments (rolling the die) with but two types of outcome for each roll: success (a six) or failure (anything but a six). Note that we combine several possibilities (scores of 1, 2, 3, 4 or 5) together and represent them all as failure. The probability of success (one-sixth) does not vary from one experiment to another, and so use of the Binomial distribution is appropriate. The values of the parameters are $n = 4$ and $P = 1/6$. Denoting by r the random variable 'the number of sixes in four rolls of the die' then

$$r \sim B(4, \tfrac{1}{6})$$

Hence

$$\Pr(r) = nCr \times P^r(1 - P)^{(n-r)}$$

where $P = \frac{1}{6}$ and $n = 4$. The probabilities of two, three and four sixes are then given by

$$\Pr(r = 2) = 4C2(\tfrac{1}{6})^2(\tfrac{5}{6})^2 = 0.116$$

$$\Pr(r = 3) = 4C3(\tfrac{1}{6})^3(\tfrac{5}{6})^1 = 0.015$$

$$\Pr(r = 4) = 4C4(\tfrac{1}{6})^4(\tfrac{5}{6})^0 = 0.00077$$

Since these events are mutually exclusive, the probabilities can simply be added together to achieve the desired result, which is 0.132, or 13.2%. This is the probability of two or more sixes in four rolls of a die.

This result can be illustrated diagrammatically as part of the area under the appropriate Binomial distribution, shown in Figure 3.4.

The shaded areas represent the probabilities of two or more sixes and together their area represents 13.2% of the whole distribution. This illustrates an important principle: that probabilities can be represented by areas under an appropriate probability distribution. We shall see more of this later.

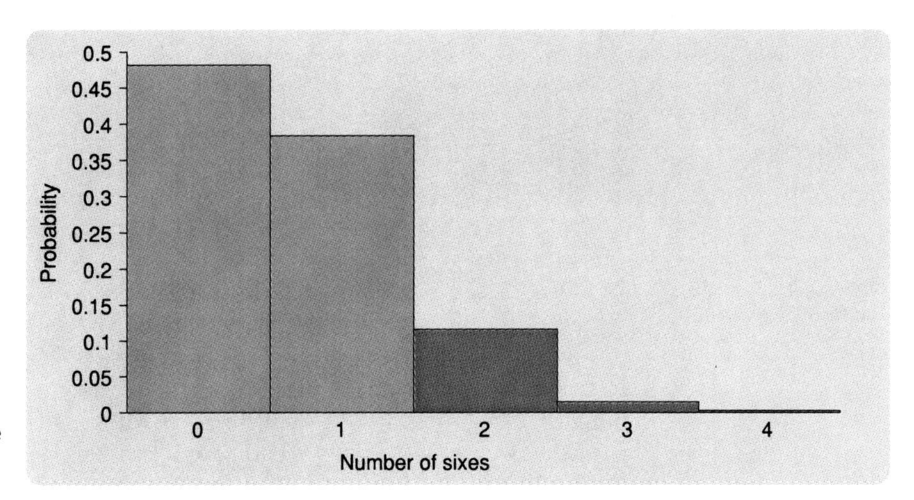

Figure 3.4
Probability of two or more sixes in four rolls of a die

Exercise 3.1

(a) The probability of a randomly drawn individual having blue eyes is 0.6. What is the probability that four people drawn at random all have blue eyes?

(b) What is the probability that two of the sample of four have blue eyes?

(c) For this particular example, write down the Binomial formula for the probability of r blue-eyed individuals, for $r = 0 \ldots 4$. Confirm that the probabilities sum to one.

Exercise 3.2

(a) Calculate the mean and variance of the number of blue-eyed individuals in the previous exercise.

(b) Draw a graph of this Binomial distribution and on it mark the mean value and the mean value +/– one standard deviation.

Having introduced the concept of probability distributions using the Binomial, we now move on to the most important of all probability distributions – the Normal.

The Normal distribution

The Binomial distribution applies when there are two possible outcomes to an experiment, but not all problems fall into this category. For instance, the (random) arrival time of a train is a continuous variable and cannot be analysed using the Binomial. There are many probability distributions in statistics, developed to analyse different types of problem. Several of them are covered in this book and the most important of them is the Normal distribution, which we now turn to. It was discovered by the German mathematician Gauss in the nineteenth century (hence it is also known as the Gaussian distribution), in the course of his work on regression (see Chapter 7).

Many random variables turn out to be Normally distributed. Men's (or women's) heights are Normally distributed. IQ (the measure of intelligence) is also Normally distributed. Another example is of a machine producing (say) bolts with a nominal length of 5 cm which will actually produce bolts of slightly varying length (these differences would probably be extremely small) due to

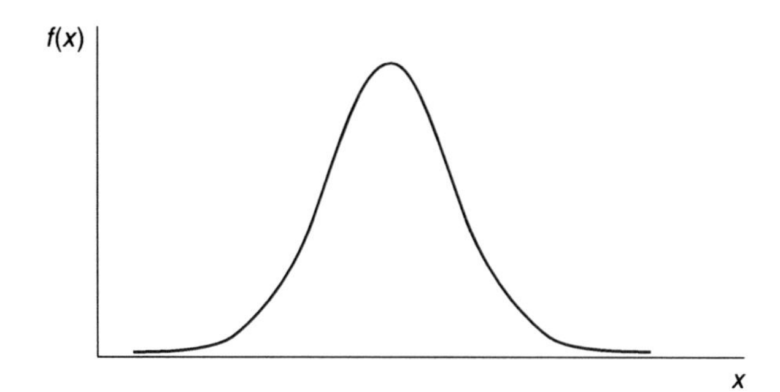

Figure 3.5
The Normal distribution

factors such as wear in the machinery, slight variations in the pressure of the lubricant, etc. These would result in bolts whose length varies, in accordance with the Normal distribution. This sort of process is extremely common, with the result that the Normal distribution often occurs in everyday situations.

The Normal distribution tends to arise when a random variable is the result of many independent, random influences added together, none of which dominates the others. A man's height is the result of many genetic influences, plus environmental factors such as diet, etc. As a result, height is Normally distributed. If one takes the height of men and women together, the result is not a Normal distribution, however. This is because there is one influence which dominates the others: gender. Men are, on average, taller than women. Many variables familiar in economics are not Normal however – incomes, for example (although the logarithm of income is approximately Normal). We shall learn techniques to deal with such circumstances in due course.

Having introduced the idea of the Normal distribution, what does it look like? It is presented below in graphical and then mathematical forms. Unlike the Binomial, the Normal distribution applies to continuous random variables such as height and a typical Normal distribution is illustrated in Figure 3.5. Since the Normal distribution is a continuous one it can be evaluated for all values of x, not just for integers. The figure illustrates the main features of the distribution:

- It is unimodal, having a single, central peak. If this were men's heights it would illustrate the fact that most men are clustered around the average height, with a few very tall and a few very short people.
- It is symmetric, the left and right halves being mirror images of each other.
- It is bell-shaped.
- It extends continuously over all the values of x from minus infinity to plus infinity, although the value of $f(x)$ becomes extremely small as these values are approached (the pages of this book being of only finite width, this last characteristic is not faithfully reproduced!). This also demonstrates that most empirical distributions (such as men's heights) can only be an approximation to the theoretical ideal, although the approximation is close and good enough for practical purposes.

Note that we have labelled the y-axis '$f(x)$' rather than 'Pr(x)' as we did for the Binomial distribution. This is because it is *areas under the curve* that represent probabilities, not the heights. With the Binomial, which is a discrete distribution, one can legitimately represent probabilities by the heights of the bars. For the Normal, although $f(x)$ does not give the probability per se, it does give an indication: you are more likely to encounter values from the middle of the distribution (where $f(x)$ is greater) than from the extremes.

In mathematical terms the formula for the Normal distribution is (x is the random variable)

$$f(x) = \frac{1}{\sigma\sqrt{2\pi}} e^{-\frac{1}{2}\left(\frac{x-\mu}{\sigma}\right)^2} \tag{3.4}$$

The mathematical formulation is not so formidable as it appears. μ and σ are the parameters of the distribution, such as n and P for the Binomial (though they have different meanings); π is 3.1416 and e is 2.7183. If the formula is evaluated using different values of x the values of $f(x)$ obtained will map out a Normal distribution. Fortunately, as we shall see, we do not need to use the mathematical formula in most practical problems.

Like the Binomial, the Normal is a family of distributions differing from one another only in the values of the parameters μ and σ. Several Normal distributions are drawn in Figure 3.6 for different values of the parameters.

Whatever value of μ is chosen turns out to be the centre of the distribution. As the distribution is symmetric, μ is its mean. The effect of varying σ is to narrow (small σ) or widen (large σ) the distribution. σ turns out to be the standard deviation of the distribution. The Normal is another two-parameter family of distributions like the Binomial, and once the mean μ and the standard deviation σ (or equivalently the variance, σ^2) are known the whole of the distribution can be drawn.

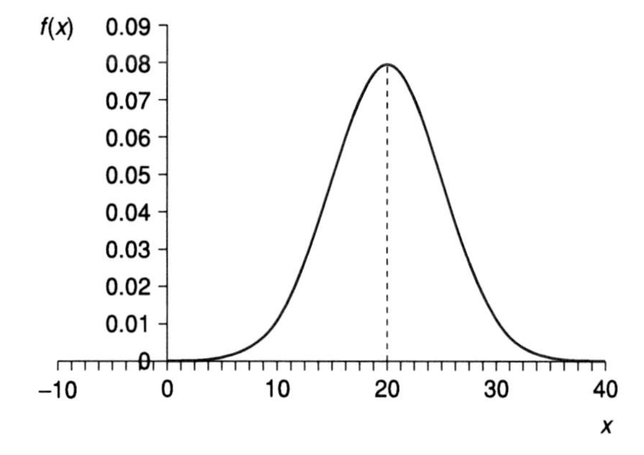

Figure 3.6(a)
The Normal distribution, $\mu = 20$, $\sigma = 5$

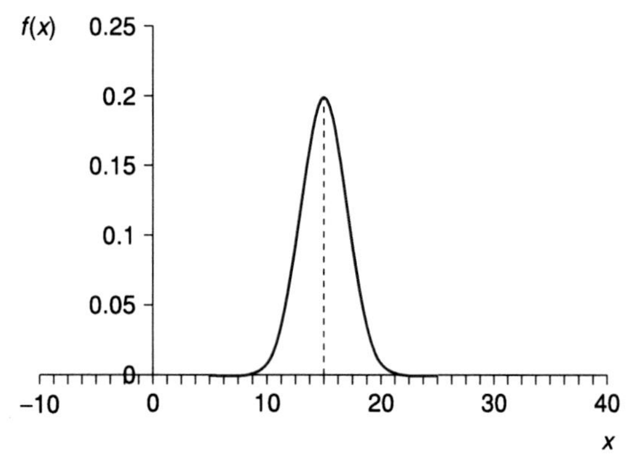

Figure 3.6(b)
The Normal distribution, $\mu = 15$, $\sigma = 2$

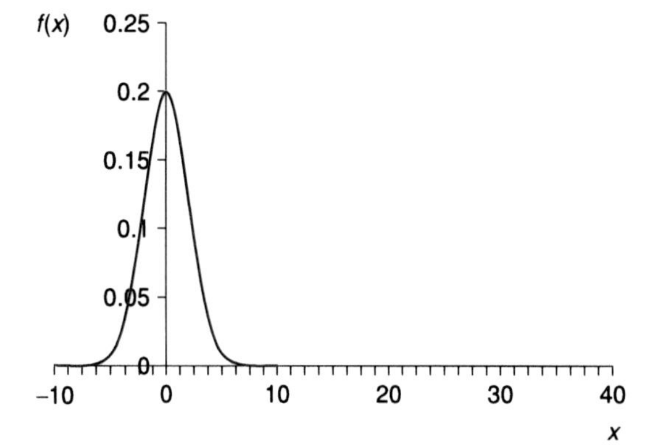

Figure 3.6(c)
The Normal distribution, $\mu = 0$, $\sigma = 4$

The shorthand notation for a Normal distribution is

$$x \sim N(\mu, \sigma^2) \tag{3.5}$$

meaning 'the variable x is Normally distributed with mean μ and variance σ^2'. This is similar in form to the expression for the Binomial distribution, though the meanings of the parameters are different.

Use of the Normal distribution can be illustrated using a simple example. The height of adult males is Normally distributed with mean height $\mu = 174$ cm and standard deviation $\sigma = 9.6$ cm. Let x represent the height of adult males; then

$$x \sim N(174, 92.16) \tag{3.6}$$

and this is illustrated in Figure 3.7. Note that equation (3.6) contains the variance rather than the standard deviation.

What is the probability that a randomly selected man is taller than 180 cm? If all men are equally likely to be selected, this is equivalent to asking what proportion of men are over 180 cm in height. This is given by the area under the Normal distribution, to the right of $x = 180$, i.e. the shaded area in Figure 3.7. The further from the mean of 174, the smaller the area in the tail of the distribution. One way to find this area would be to make use of equation (3.4), but this requires the use of sophisticated mathematics.

Since this is a frequently encountered problem, the answers have been set out in the tables of the **standard Normal distribution**. We can simply look up the solution. However, since there is an infinite number of Normal distributions (one for every combination of μ and σ^2) it would be an impossible task to tabulate

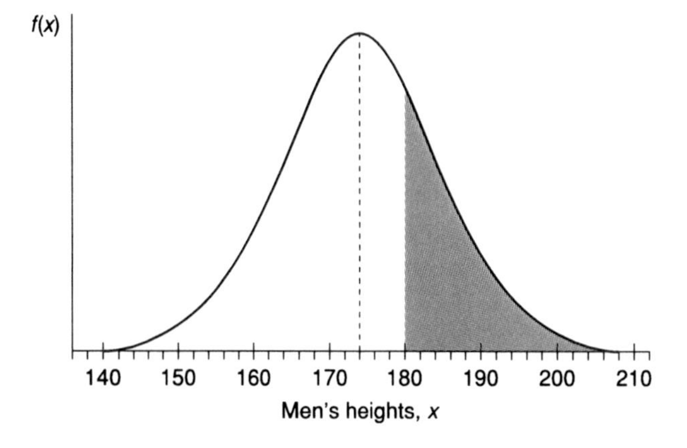

Figure 3.7
Illustration of men's height distribution

them all. The standard Normal distribution, which has a mean of zero and variance of one, is therefore used to represent all Normal distributions. Before the table can be consulted, therefore, the data have to be transformed so that they accord with the standard Normal distribution.

The required transformation is the z score, which was introduced in Chapter 1. This measures the distance between the value of interest (180) and the mean, measured in terms of standard deviations. Therefore we calculate

$$z = \frac{x - \mu}{\sigma} \tag{3.7}$$

and z is a Normally distributed random variable with mean 0 and variance 1, i.e. $z \sim N(0, 1)$. This transformation shifts the original distribution μ units to the left and then adjusts the dispersion by dividing through by σ, resulting in a mean of 0 and variance 1. z is Normally distributed because x is Normally distributed. The transformation in equation (3.7) retains the Normal distribution shape, despite the changes to mean and variance. If x followed some other distribution then z would not be Normal either.

It is easy to verify the mean and variance of z using the rules for E and V operators encountered in Chapter 1

$$\mathrm{E}(z) = \mathrm{E}\left(\frac{x - \mu}{\sigma}\right) = \frac{1}{\sigma}(\mathrm{E}(x) - \mu) = 0 \quad (\text{since } \mathrm{E}(x) = \mu)$$

$$\mathrm{V}(z) = \mathrm{V}\left(\frac{x - \mu}{\sigma}\right) = \frac{1}{\sigma^2}\mathrm{V}(x) = \frac{\sigma^2}{\sigma^2} = 1$$

Evaluating the z score from our data we obtain

$$z = \frac{180 - 174}{9.6} = 0.63 \tag{3.8}$$

This shows that 180 is 0.63 standard deviations above the mean, 174, of the distribution. This is a measure of how far 180 is from 174 and allows us to look up the answer in tables. The task now is to find the area under the standard Normal distribution to the right of 0.63 standard deviations above the mean. This answer can be read off directly from the table of the standard Normal distribution, included as Table A2 in the appendix to this book. An excerpt from Table A2 (see page **414**) is presented in Table 3.2.

The left-hand column gives the z score to one place of decimals. The appropriate row of the table to consult is the one for $z = 0.6$, which is shaded. For the second place of decimals (0.03) we consult the appropriate column, also shaded. At their intersection we find the value 0.2643, which is the desired area and

Table 3.2 Areas of the standard Normal distribution (excerpt from Table A2)

z	0.00	0.01	0.02	0.03	...	0.09
0.0	0.5000	0.4960	0.4920	0.4880	...	0.4641
0.1	0.4602	0.4562	0.4522	0.4483	...	0.4247
⋮	⋮	⋮	⋮	⋮	...	⋮
0.5	0.3085	0.3050	0.3015	0.2981	...	0.2776
0.6	0.2743	0.2709	0.2676	0.2643	...	0.2451
0.7	0.2420	0.2389	0.2358	0.2327	...	0.2148

therefore probability, i.e. 26.43% of the distribution lies to the right of 0.63 standard deviations above the mean. Therefore 26.43% of men are over 180 cm in height.

Use of the standard Normal table is possible because, although there is an infinite number of Normal distributions, they are all fundamentally the same,so that the area to the right of 0.63 standard deviations above the mean is the same for all of them. As long as we measure the distance in terms of standard deviations then we can use the standard Normal table. The process of standardisation turns all Normal distributions into a standard Normal distribution with a mean of zero and a variance of one. This process is illustrated in Figure 3.8.

The area in the right-hand tail is the same for both distributions. It is the standard Normal distribution in Figure 3.8(b), which is tabulated in Table A2. To demonstrate how standardisation turns all Normal distributions into the standard Normal, the earlier problem is repeated but taking all measurements in inches. The answer should obviously be the same. Taking 1 inch = 2.54 cm the figures are

$$x = 70.87 \quad \sigma = 3.78 \quad \mu = 68.50$$

What proportion of men are over 70.87 inches in height? The appropriate Normal distribution is now

$$x \sim N(68.50, 3.78^2) \tag{3.9}$$

The z score is

$$z = \frac{70.87 - 68.50}{3.78} = 0.63 \tag{3.10}$$

which is the same z score as before and therefore gives the same probability.

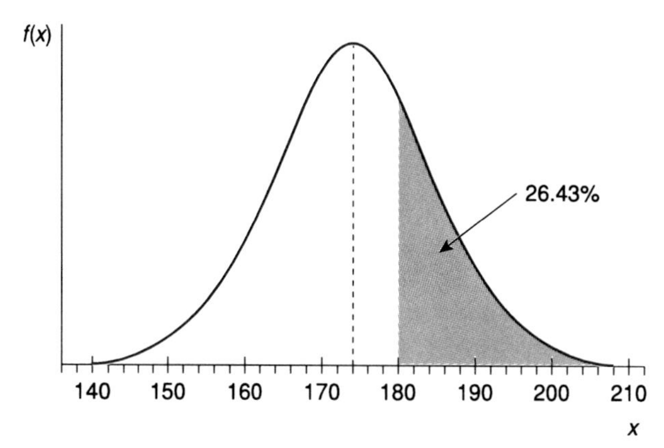

Figure 3.8(a)
The Normal distribution

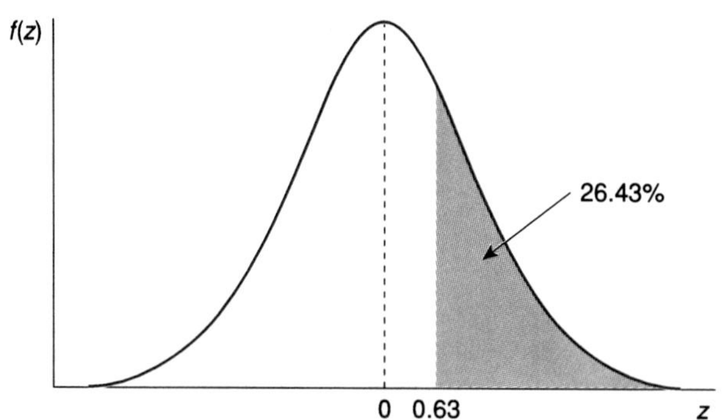

Figure 3.8(b)
The standard Normal distribution corresponding to Figure 3.8(a)

Worked example 3.2

Packets of cereal have a nominal weight of 750 grams, but there is some variation around this as the machines filling the packets are imperfect. Let us assume that the weights follow a Normal distribution. Suppose that the standard deviation around the mean of 750 is 5 grams. What proportion of packets weigh more than 760 grams?

Summarising our information, we have $x \sim N(750, 25)$, where x represents the weight. We wish to find $\Pr(x > 760)$. To be able to look up the answer, we need to measure the distance between 760 and 750 in terms of standard deviations. This is

$$z = \frac{760 - 750}{5} = 2.0$$

Looking up $z = 2.0$ in Table A2 reveals an area of 0.0228 in the tail of the distribution. Thus 2.28% of packets weigh more than 760 grams.

Since a great deal of use is made of the standard Normal tables, it is worth working through a couple more examples to reinforce the method. We have so far calculated that $\Pr(z > 0.63) = 0.2643$. Since the total area under the graph equals one (i.e. the sum of probabilities must be one), the area to the left of $z = 0.63$ must equal 0.7357, i.e. 73.57% of men are under 180 cm. It is fairly easy to manipulate areas under the graph to arrive at any required area. For example, what proportion of men are between 174 and 180 cm in height? It is helpful to refer to Figure 3.9 at this point.

The size of area A is required. Area B has already been calculated as 0.2643. Since the distribution is symmetric the area A + B must equal 0.5, since 174 is at the centre (mean) of the distribution. Area A is therefore 0.5 – 0.2643 = 0.2357. 23.57% is the desired result.

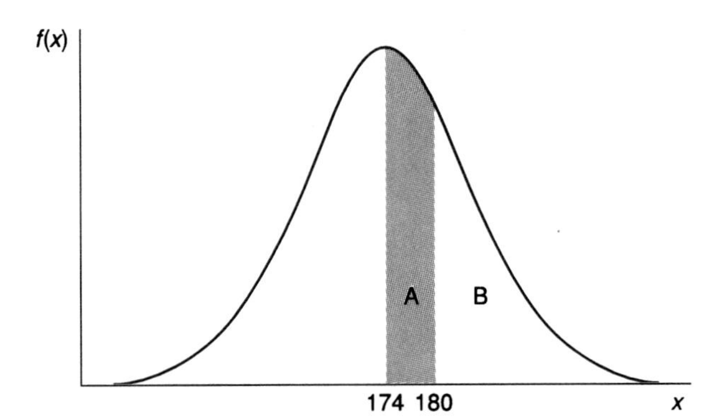

Figure 3.9
The proportion of men between 174 cm and 180 cm in height

Using software to find areas under the standard Normal distribution

If you use a spreadsheet program you can look up the z-distribution directly and hence dispense with tables. In *Excel*, for example, the function '=*NORMSDIST*(0.63)' gives the answer 0.7357, i.e. the area to the *left* of the z score. The area in the right-hand tail is then obtained by subtracting this value from 1, i.e. 1 − 0.7357 = 0.2643. Entering the formula '= 1 − *NORMSDIST*(0.63)' in a cell will give the area in the right-hand tail directly.

As a final exercise consider the question of what proportion of men are between 166 and 178 cm tall. As shown in Figure 3.10 area C + D is wanted. The only way to find this is to calculate the two areas separately and then add them together. For area D the z score associated with 178 is

$$z_D = \frac{178 - 174}{9.6} = 0.42 \tag{3.11}$$

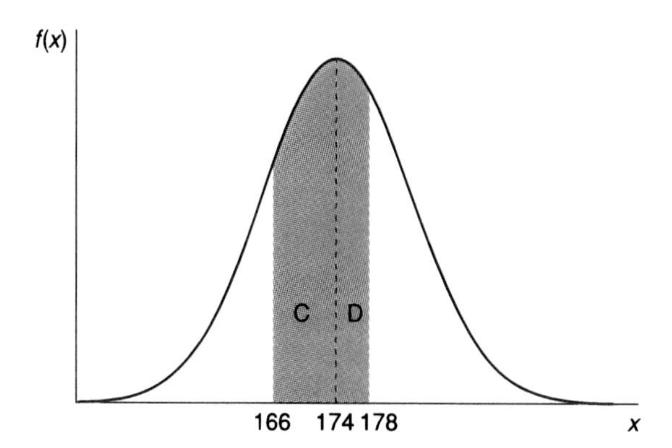

Figure 3.10
The proportion of men between 166 cm and 178 cm in height

Table A2 (see page **414**) indicates that the area in the right-hand tail, beyond $z = 0.42$, is 0.3372, so area D = 0.5 − 0.3372 = 0.1628. For C, the z score is

$$z_C = \frac{166 - 174}{9.6} = -0.83 \tag{3.12}$$

The minus sign indicates that it is the left-hand tail of the distribution, below the mean, which is being considered. Since the distribution is symmetric, it is the same as if it were the right-hand tail, so the minus sign may be ignored when consulting the table. Looking up $z = 0.83$ in Table A2 gives an area of 0.2033 in the tail, so area C is therefore 0.5 − 0.2033 = 0.2967. Adding areas C and D gives 0.1628 + 0.2967 = 0.4595. So nearly half of all men are between 166 and 178 cm in height.

An alternative interpretation of the results obtained above is that if a man is drawn at random from the adult population, the probability that he is over 180 cm tall is 26.43%. This is in line with the frequentist school of thought. Since 26.43% of the population is over 180 cm in height, that is the probability of a man over 180 cm being drawn at random.

Exercise 3.3

(a) The random variable x is distributed Normally, with $x \sim N(40, 36)$. Find the probability that $x > 50$.

(b) Find $\Pr(x < 45)$.

(c) Find $\Pr(36 < x < 44)$.

Exercise 3.4

The mean +/– 0.67 standard deviations cuts off 25% in each tail of the Normal distribution. Hence the middle 50% of the distribution lies within +/– 0.67 standard deviations of the mean. Use this fact to calculate the inter-quartile range for the distribution $x \sim N(200, 256)$.

Exercise 3.5

?

As suggested in the text, the logarithm of income is approximately Normally distributed. Suppose the log (to the base 10) of income has the distribution $x \sim N(4.18, 2.56)$. Calculate the inter-quartile range for x and then take anti-logs to find the inter-quartile range of income.

The sample mean as a Normally distributed variable

One of the most important concepts in statistical inference is the probability distribution of the mean of a random sample, since we often use the sample mean to tell us something about an associated population. Suppose that, from the population of adult males, a random sample of size $n = 36$ is taken, their heights measured and the mean height of the sample calculated. What can we infer from this about the true average height of the population? To do this, we need to know about the statistical properties of the sample mean. The sample mean is a random variable because of the chance element of random sampling (different samples would yield different values of the sample mean). Since the sample mean is a random variable it must have associated with it a probability distribution.

We therefore need to know, first, what is the appropriate distribution and, second, what are its parameters. From the definition of the sample mean we have

$$\bar{x} = \frac{1}{n}(x_1 + x_2 + \ldots + x_n) \quad (3.13)$$

where each observation, x_i, is itself a Normally distributed random variable, with $x_i \sim N(\mu, \sigma^2)$, because each comes from the parent distribution with such characteristics. (We stated earlier that men's heights are Normally distributed.) We now make use of the following theorem to demonstrate that $\bar{x}$ is Normally distributed:

Theorem

Any linear combination of independent, Normally distributed random variables is itself Normally distributed.

A linear combination of two variables x_1 and x_2 is of the form $w_1x_1 + w_2x_2$ where w_1 and w_2 are constants. This can be generalised to any number of x values. It is clear that the sample mean satisfies these conditions and is a linear combination of the individual x values (with the weight on each observation equal to $1/n$). As long as the observations are independently drawn, therefore, the sample mean is Normally distributed.

We now need the parameters (mean and variance) of the distribution. For this we use the E and V operators once again

$$E(\bar{x}) = \frac{1}{n}(E(x_1) + E(x_2) + \ldots + E(x_n)) = \frac{1}{n}(\mu + \mu + \ldots + \mu) = \frac{1}{n}n\mu = \mu \quad (3.14)$$

$$
\begin{aligned}
V(\bar{x}) &= V\left(\frac{1}{n}[x_1 + x_2 + \ldots + x_n]\right) \\
&= \frac{1}{n^2}(V(x_1) + V(x_2) + \ldots + V(x_n)) \\
&= \frac{1}{n^2}(\sigma^2 + \sigma^2 + \ldots + \sigma^2) \\
&= \frac{1}{n^2}n\sigma^2 = \frac{\sigma^2}{n}
\end{aligned}
\tag{3.15}
$$

Putting all this together, we have[2]

$$\bar{x} \sim N\left(\mu, \frac{\sigma^2}{n}\right) \tag{3.16}$$

This we may summarise in the following theorem:

Theorem

The sample mean, $\bar{x}$, drawn from a population which has a Normal distribution with mean μ and variance σ^2, has a sampling distribution which is Normal, with mean μ and variance σ^2/n, where n is the sample size.

The meaning of this theorem is as follows. First of all it is assumed that the population from which the samples are to be drawn is itself Normally distributed (this assumption will be relaxed in a moment), with mean μ and variance σ^2. From this population many samples are drawn, each of sample size n, and the mean of each sample is calculated. The samples are independent, meaning that the observations selected for one sample do not influence the selection of observations in the other samples. This gives many sample means, $\bar{x}_1$, $\bar{x}_2$, etc. If these sample means are treated as a new set of observations, then the probability distribution of these observations can be derived. The theorem states that this distribution is Normal, with the sample means centred around μ, the population mean, and with variance σ^2/n. The argument is set out diagrammatically in Figure 3.11.

Intuitively this theorem can be understood as follows. If the height of adult males is a Normally distributed random variable with mean μ = 174 cm and

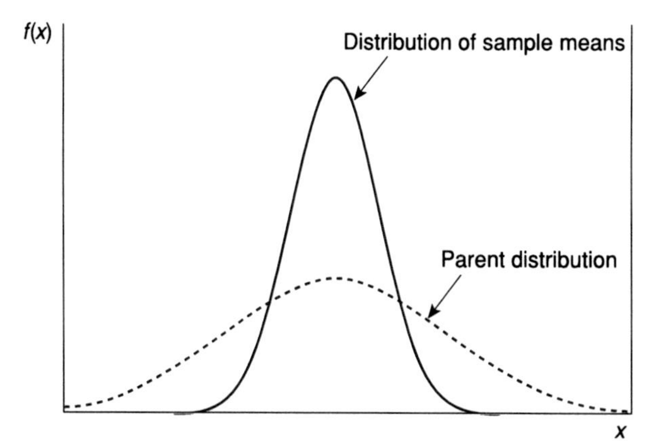

Figure 3.11 The parent distribution and the distribution of sample means

Note: The distribution of $\bar{x}$ is drawn for a sample size of $n = 9$. A larger sample size would narrow the $\bar{x}$ distribution; a smaller sample size would widen it.

[2] Don't worry if you didn't follow the derivation of this formula, just accept that it is correct.

variance $\sigma^2 = 92.16$, then it would be expected that a random sample of (say) nine males would yield a sample mean height of around 174 cm, perhaps a little more, perhaps a little less. In other words, the sample mean is centred around 174 cm, or the mean of the distribution of sample means is 174 cm.

The larger is the size of the individual samples (i.e. the larger n), the closer the sample mean would tend to be to 174 cm. For example, if the sample size is only two, a sample of two very tall people is quite possible, with a high sample mean as a result, well over 174 cm, e.g. 182 cm. But if the sample size were 20, it is very unlikely that 20 very tall males would be selected and the sample mean is likely to be much closer to 174. This is why the sample size n appears in the formula for the variance of the distribution of the sample mean, σ^2/n.

Note that, once again, we have transformed one (or more) random variables, the x_i values, with a particular probability distribution into another random variable, $\bar{x}$, with a (slightly) different distribution. This is common practice in statistics: transforming a variable will often put it into a more useful form, for example one whose probability distribution is well known.

The above theorem can be used to solve a range of statistical problems. For example, what is the probability that a random sample of nine men will have a mean height greater than 180 cm? The height of all men is known to be Normally distributed with mean $\mu = 174$ cm and variance $\sigma^2 = 92.16$. The theorem can be used to derive the probability distribution of the sample mean. For the population we have

$$\bar{x} \sim N(\mu, \sigma^2), \text{ i.e. } \bar{x} \sim N(174, 92.16)$$

Hence for the sample mean

$$\bar{x} \sim N(\mu, \sigma^2/n), \text{ i.e. } \bar{x} \sim N(174, 92.16/9)$$

This is shown diagrammatically in Figure 3.12.

To answer the question posed, the area to the right of 180, shaded in Figure 3.11, has to be found. This should by now be a familiar procedure. First the z score is calculated

$$z = \frac{\bar{x} - \mu}{\sqrt{\sigma^2/n}} = \frac{180 - 174}{\sqrt{92.16/9}} = 1.88 \tag{3.17}$$

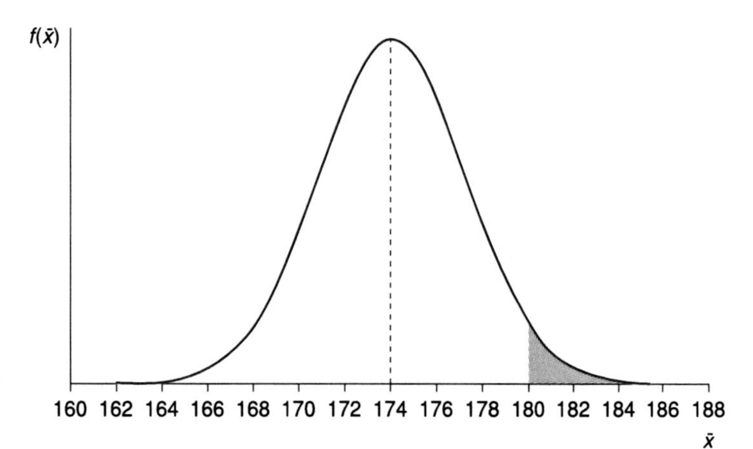

Figure 3.12
The proportion of sample means greater than $\bar{x} = 180$

Note that the z score formula is subtly different because we are dealing with the sample mean $\bar{x}$ rather than x itself. In the numerator we use $\bar{x}$ rather than x and in the denominator we use σ^2/n, not σ^2. This is because $\bar{x}$ has a variance σ^2/n, not σ^2, which is the population variance. $\sqrt{\sigma^2/n}$ is known as the **standard error**, to distinguish it from σ, the standard deviation of the population. The principle behind the z score is the same however: it measures how far is a sample mean of 180 from the population mean of 174, measured in terms of standard deviations.

Looking up the value of $z = 1.88$ in Table A2 gives an area of 0.0311 in the right-hand tail of the Normal distribution. Thus 3.11% of sample means will be greater than or equal to 180 cm when the sample size is nine. The desired probability is therefore 3.11%.

As this probability is quite small, we might consider the reasons for this. There are two possibilities:

(a) through bad luck, the sample collected is not very representative of the population as a whole;
(b) the sample is representative of the population, but the population mean is not 174 cm after all.

Only one of these two possibilities can be correct. How to decide between them will be taken up later on, in Chapter 5 on hypothesis testing.

It is interesting to examine the difference between the answer for a sample size of nine (3.11%) and the one obtained earlier for a single individual (26.43%). The latter may be considered as a sample of size one from the population. The examples illustrate the fact that the larger the sample size, the closer the sample mean is likely to be to the population mean. Thus larger samples tend to give better estimates of the population mean.

Oil reserves

An interesting application of probability distributions is to the estimation of oil reserves. The quantity of oil in an oil field is not known for certain, but is subject to uncertainty. The *proven* oil reserve of a field is the amount recoverable with probability of 90% (known as P90 in the oil industry). One can then add up the proven oil reserves around the world to get a total of proven reserves.

However, using probability theory we can see this might be misleading. Suppose we have 50 fields, where the recoverable quantity of oil is distributed as $x \sim N(100, 81)$ in each. From tables we note that $\bar{x} - 1.28s$ cuts off the bottom 10% of the Normal distribution, 88.48 in this case. This is the proven reserve for a field. Summing across the 50 fields gives 4424 as total reserves. But is there a 90% probability of recovering at least this amount?

Using the first theorem above, the total quantity of oil y is distributed Normally, with mean $E(y) = E(x_1) + \ldots + E(x_{50}) = 5000$ and variance $V(y) = V(x_1) + \ldots + V(x_{50}) = 4050$, assuming independence of the oil fields. Hence we have $y \sim N(5000, 4050)$. Again, the bottom 10% is cut off by $\bar{y} - 1.28s$, which is 4919. This is 11% larger than the 4424 calculated above. Adding up the proven reserves of each field individually underestimates the true total proven reserves. In fact, the probability of total proven reserves being greater than 4424 is almost 100%.

Note that the numbers given here are for illustration purposes and don't reflect the actual state of affairs. The principle of the calculation is correct however.

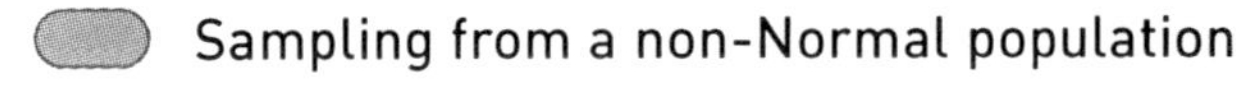

Sampling from a non-Normal population

The previous theorem and examples relied upon the fact that the population followed a Normal distribution. But what happens if it is not Normal? After all, it is not known for certain that the heights of all adult males are exactly Normally distributed, and there are many populations which are not Normal (e.g. wealth, as shown in Chapter 1). What can be done in these circumstances? The answer is to use another theorem about the distribution of sample means (presented without proof). This is known as the **Central Limit Theorem**:

Theorem

The sample mean $\bar{x}$, drawn from a population with mean μ and variance σ^2, has a sampling distribution which approaches a Normal distribution with mean μ and variance σ^2/n, as the sample size approaches infinity.

This is very useful, since it drops the assumption that the population is Normally distributed. Note that the distribution of sample means is only Normal as long as the sample size is infinite; for any finite sample size the distribution is only approximately Normal. However, the approximation is close enough for practical purposes if the sample size is larger than 25 or so observations. If the population distribution is itself nearly Normal then a smaller sample size would suffice. If the population distribution is particularly skewed then more than 25 observations would be desirable. Twenty-five observations constitutes a rule of thumb that is adequate in most circumstances. This is another illustration of statistics as an inexact science. It does not provide absolutely clear-cut answers to questions but, used carefully, helps us to arrive at sensible conclusions.

As an example of the use of the Central Limit Theorem, we return to the wealth data of Chapter 1. Recall that the mean level of wealth was 146.984 (measured in £000) and the variance 56 803. Suppose that a sample of $n = 50$ people were drawn from this population. What is the probability that the sample mean is greater than 160 (i.e. £160 000)?

On this occasion we know that the parent distribution is highly skewed so it is fortunate that we have 50 observations. This should be ample for us to justify applying the Central Limit Theorem. The distribution of $\bar{x}$ is therefore

$$\bar{x} \sim N(\mu, \sigma^2/n) \tag{3.18}$$

and, inserting the parameter values, this gives[3]

$$\bar{x} \sim N(146.984, 56\,803/50) \tag{3.19}$$

To find the area beyond a sample mean of 160, the z score is first calculated

$$z = \frac{160 - 146.984}{\sqrt{56\,803/50}} = 0.39 \tag{3.20}$$

Referring to the standard Normal tables, the area in the tail is then found to be 34.83%. This is the desired probability. So there is a probability of 34.83% of finding a mean of £160 000 or greater with a sample of size 50. This demonstrates

[3] Note that if we used 146 984 for the mean we would have 56 803 000 000 as the variance. Using £000 keeps the numbers more manageable. The z score is the same in both cases.

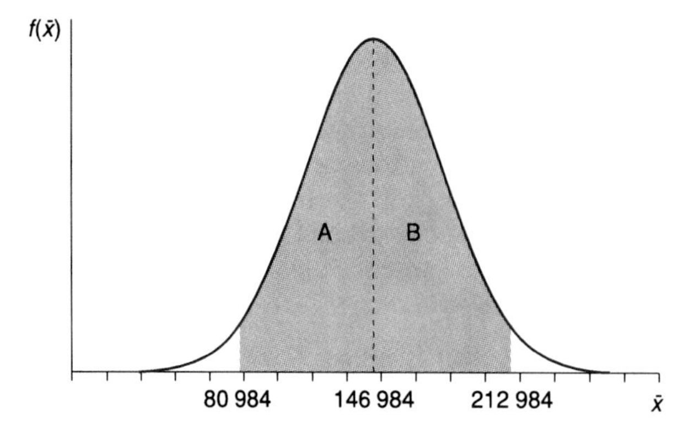

Figure 3.13
The probability of $\bar{x}$ lying within £66 000 either side of £146 984

that there is quite a high probability of getting a sample mean which is a relatively long way from £146 984. This is a consequence of the high degree of dispersion in the distribution of wealth.

Extending this example, we can ask what is the probability of the sample mean lying within, say, £66 000 either side of the true mean of £146 984 (i.e. between £80 984 and £212 984)? Figure 3.13 illustrates the situation, with the desired area shaded. By symmetry, areas A and B must be equal, so we only need find one of them. For B, we calculate the z score

$$z = \frac{212.984 - 146.984}{\sqrt{56\,803/50}} = 1.958 \tag{3.21}$$

From the standard Normal table, this cuts off approximately 2.5% in the upper tail, so area B = 0.475. Areas A and B together make up 95% of the distribution, therefore. There is thus a 95% probability of the sample mean falling within the range [80 984, 212 984] and we call this the 95% **probability interval** for the sample mean. We write this

$$\Pr(80\,984 \leqslant \bar{x} \leqslant 212\,984) = 0.95 \tag{3.22}$$

or, in terms of the formulae we have used[4]

$$\Pr(\mu - 1.96\sqrt{\sigma^2/n} \leqslant \bar{x} \leqslant \mu + 1.96\sqrt{\sigma^2/n}) = 0.95 \tag{3.23}$$

The 95% probability interval and the related concept of the 95% confidence interval (which will be introduced in Chapter 4) play important roles in statistical inference. We deliberately designed the example above to arrive at an answer of 95% for this reason.

Exercise 3.6

(a) If x is distributed as $x \sim N(50, 64)$ and samples of size $n = 25$ are drawn, what is the distribution of the sample mean $\bar{x}$?

(b) If the sample size doubles to 50, how is the standard error of $\bar{x}$ altered?

(c) Using the sample size of 25, (i) what is the probability of $\bar{x} > 51$? (ii) What is $\Pr(\bar{x} < 48)$? (iii) What is $\Pr(49 < \bar{x} < 50.5)$?

[4] 1.96 is the precise value cutting off 2.5% in each tail.

The relationship between the Binomial and Normal distributions

Many statistical distributions are related to one another in some way. This means that many problems can be solved by a variety of different methods (using different distributions), though usually one is more convenient or more accurate than the others. This point may be illustrated by looking at the relationship between the Binomial and Normal distributions.

Recall the experiment of tossing a coin repeatedly and noting the number of heads. We said earlier that this can be analysed via the Binomial distribution. But note that the number of heads, a random variable, is influenced by many independent random events (the individual tosses) added together. Furthermore, each toss counts equally, none dominates. These are just the conditions under which a Normal distribution arises, so it looks like there is a connection between the two distributions.

This idea is correct. Recall that if a random variable r follows a Binomial distribution then

$$r \sim B(n, P)$$

and the mean of the distribution is nP and the variance $nP(1 - P)$. It turns out that as n increases, the Binomial distribution becomes approximately the same as a Normal distribution with mean nP and variance $nP(1 - P)$. This approximation is sufficiently accurate as long as $nP > 5$ and $n(1 - P) > 5$, so the approximation may not be very good (even for large values of n) if P is very close to zero or one. For the coin tossing experiment, where $P = 0.5$, 10 tosses should be sufficient. Note that this approximation is good enough with only 10 observations even though the underlying probability distribution is nothing like a Normal distribution.

To demonstrate, the following problem is solved using both the Binomial and Normal distributions. Forty students take an exam in statistics which is simply graded pass/fail. If the probability, P, of any individual student passing is 60%, what is the probability of at least 30 students passing the exam?

The sample data are

$$P = 0.6$$
$$1 - P = 0.4$$
$$n = 40$$

Binomial distribution method

To solve the problem using the Binomial distribution it is necessary to find the probability of exactly 30 students passing, plus the probability of 31 passing, plus the probability of 32 passing, etc., up to the probability of 40 passing (the fact that the events are mutually exclusive allows this). The probability of 30 passing is

$$\begin{aligned}\Pr(r = 30) &= nCr \times P^r(1 - P)^{n-r} \\ &= 40C^{30} \times 0.6^{30} \times 0.4^{10} \\ &= 0.020\end{aligned}$$

(*Note*: This calculation assumes that the probabilities are independent, i.e. no copying!) This by itself is quite a tedious calculation, but Pr(31), Pr(32), etc., still have to be calculated. Calculating these and summing them gives the result of 3.52% as the probability of at least 30 passing. (It would be a useful exercise for you to do, if only to appreciate how long it takes.)

Normal distribution method

As stated above, the Binomial distribution can be approximated by a Normal distribution with mean nP and variance $nP(1 - P)$. nP in this case is 24 (40×0.6) and $n(1 - P)$ is 16, both greater than 5, so the approximation can be safely used. Thus

$$r \sim N(nP, nP(1 - P))$$

and inserting the parameter values gives

$$r \sim N(24, 9.6)$$

The usual methods are then used to find the appropriate area under the distribution. However, before doing so, there is one adjustment to be made (this only applies when approximating the Binomial distribution by the Normal). The Normal distribution is a continuous one while the Binomial is discrete. Thus 30 in the Binomial distribution is represented by the area under the Normal distribution between 29.5 and 30.5. 31 is represented by 30.5 to 31.5, etc. Thus it is the area under the Normal distribution to the right of 29.5, not 30, which must be calculated. This is known as the **continuity correction**. Calculating the z score gives

$$z = \frac{29.5 - 24}{\sqrt{9.6}} = 1.78 \tag{3.24}$$

This gives an area of 3.75%, not far off the correct answer as calculated by the Binomial distribution. The time saved and ease of calculation would seem to be worth the slight loss in accuracy.

Other examples can be constructed to test this method, using different values of P and n. Small values of n, or values of nP or $n(1 - P)$ less than 5, will give poor results, i.e. the Normal approximation to the Binomial will not be very good.

Exercise 3.7

(a) A coin is tossed 20 times. What is the probability of more than 14 heads? Perform the calculation using both the Binomial and Normal distributions, and compare results.

(b) A biased coin, for which Pr(H) = 0.7 is tossed 6 times. What is the probability of more than 4 heads? Compare Binomial and Normal methods in this case. How accurate is the Normal approximation?

(c) Repeat part (b) but for more than 5 heads.

The Poisson distribution

The section above showed how the Binomial distribution could be approximated by a Normal distribution under certain circumstances. The approximation does not work particularly well for very small values of P, when nP is less than 5. In these circumstances the Binomial may be approximated instead by the Poisson distribution, which is given by the formula

$$\Pr(x) = \frac{\mu^x e^{-\mu}}{x!} \tag{3.25}$$

where μ is the mean of the distribution (similar to μ for the Normal distribution and nP for the Binomial). Like the Binomial, but unlike the Normal, the Poisson is a discrete probability distribution, so that equation (3.25) is only defined for integer values of x. Furthermore, it is applicable to a series of trials which are independent, as in the Binomial case.

The use of the Poisson distribution is appropriate when the probability of 'success' is very small and the number of trials large. Its use is illustrated by the following example. A manufacturer gives a two-year guarantee on the TV screens it makes. From past experience it knows that 0.5% of its screens will be faulty and fail within the guarantee period. What is the probability that of a consignment of 500 screens (a) none will be faulty, (b) more than three are faulty?

The mean of the Poisson distribution in this case is $\mu = 2.5$ (0.5% of 500). Therefore

$$\Pr(x = 0) = \frac{2.5^0 e^{-2.5}}{0!} = 0.082 \tag{3.26}$$

giving a probability of 8.2% of no failures. The answer to this problem via the Binomial method is

$$\Pr(r = 0) = 0.995^{500} = 0.0816$$

Thus the Poisson method gives a reasonably accurate answer. The Poisson approximation to the Binomial is satisfactory if nP is less than about 7.

The probability of more than three screens expiring is calculated as

$$\Pr(x > 3) = 1 - \Pr(x = 0) - \Pr(x = 1) - \Pr(x = 2) - \Pr(x = 3)$$

$$\Pr(x = 1) = \frac{2.5^1 e^{-2.5}}{1!} = 0.205$$

$$\Pr(x = 2) = \frac{2.5^2 e^{-2.5}}{2!} = 0.256$$

$$\Pr(x = 3) = \frac{2.5^3 e^{-2.5}}{3!} = 0.214$$

So

$$\Pr(x > 3) = 1 - 0.082 - 0.205 - 0.256 - 0.214 = 0.242$$

Thus there is a probability of about 24% of more than three failures. The Binomial calculation is much more tedious, but gives an answer of 24.2% also.

The Poisson distribution is also used in problems where events occur over time, such as goals scored in a football match (see Problem 3.25) or queuing-type problems (e.g. arrivals at a bank cash machine). In these problems, there is no natural 'number' of trials but it is clear that, if we take a short interval of time, the probability of an event occurring is small. We can then consider the number of trials to be the number of time intervals. This is illustrated by the following example. A football team scores, on average, two goals every game (you can vary the example by using your own favourite team plus their scoring record!). What is the probability of the team scoring zero or one goal during a game?

The mean of the distribution is 2, so we have, using the Poisson distribution

$$\Pr(x = 0) = \frac{2^0 e^{-2}}{0!} = 0.135$$

$$\Pr(x = 1) = \frac{2^1 e^{-2}}{1!} = 0.271$$

You should continue to calculate the probabilities of 2 or more goals and verify that the probabilities sum to 1.

A queuing-type problem is the following. If a shop receives, on average, 20 customers per hour, what is the probability of no customers within a five-minute period while the owner takes a coffee break?

The average number of customers per five-minute period is $20 \times 5/60 = 1.67$. The probability of a free five-minute spell is therefore

$$\Pr(x = 0) = \frac{1.67^0 e^{-1.67}}{0!} = 0.189$$

a probability of about 19%. Note that this problem cannot be solved by the Binomial method since n and P are not known separately, only their product.

Exercise 3.8

(a) The probability of winning a prize in a lottery is 1 in 50. If you buy 50 tickets, what is the probability that (i) 0 tickets win, (ii) 1 ticket wins, (iii) 2 tickets win. (iv) What is the probability of winning at least one prize?

(b) On average, a person buys a lottery ticket in a supermarket every 5 minutes. What is the probability that 10 minutes will pass with no buyers?

Railway accidents

Andrew Evans of University College, London, used the Poisson distribution to examine the numbers of fatal railway accidents in Britain between 1967 and 1997. Since railway accidents are, fortunately, rare, the probability of an accident in any time period is very small and so use of the Poisson distribution is appropriate. He found that the average number of accidents has been falling over time and by 1997 had reached 1.25 per annum. This figure is therefore used as the mean μ of the Poisson distribution, and we can calculate the probabilities of 0, 1, 2, etc., accidents each year. Using $\mu = 1.25$ and inserting this into equation 3.26 we obtain the following table:

Number of accidents	0	1	2	3	4	5	6
Probability	0.287	0.358	0.224	0.093	0.029	0.007	0.002

and this distribution can be graphed:

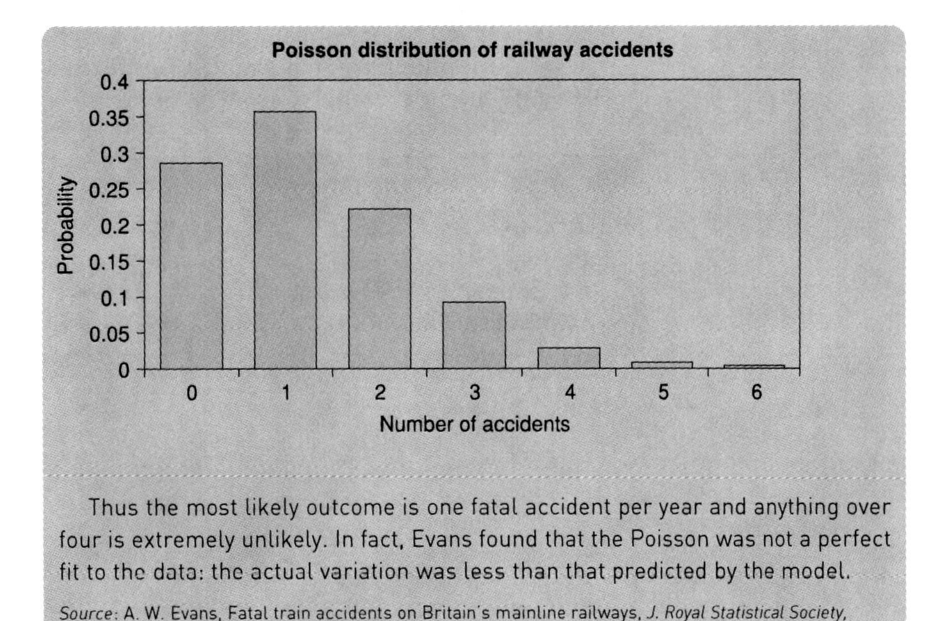

Thus the most likely outcome is one fatal accident per year and anything over four is extremely unlikely. In fact, Evans found that the Poisson was not a perfect fit to the data: the actual variation was less than that predicted by the model.

Source: A. W. Evans, Fatal train accidents on Britain's mainline railways, *J. Royal Statistical Society, Series A*, 2000, **163** (1), 99–119.

Summary

- The behaviour of many random variables (e.g. the result of the toss of a coin) can be described by a probability distribution (in this case, the Binomial distribution).
- The Binomial distribution is appropriate for problems where there are only two possible outcomes of a chance event (e.g. heads/tails, success/failure) and the probability of success is the same each time the experiment is conducted.
- The Normal distribution is appropriate for problems where the random variable has the familiar bell-shaped distribution. This often occurs when the variable is influenced by many, independent factors, none of which dominates the others. An example is men's heights, which are Normally distributed.
- The Poisson distribution is used in circumstances where there is a very low probability of 'success' and a high number of trials.
- Each of these distributions is actually a family of distributions, differing in the parameters of the distribution. Both the Binomial and Normal distributions have two parameters: n and P in the former case, μ and σ^2 in the latter. The Poisson distribution has one parameter, its mean μ.
- The mean of a random sample follows a Normal distribution, because it is influenced by many independent factors (the sample observations), none of which dominates in the calculation of the mean. This statement is always true if the population from which the sample is drawn follows a Normal distribution.

- If the population is not Normally distributed then the Central Limit Theorem states that the sample mean is Normally distributed in large samples. In this case 'large' means a sample of about 25 or more.

Key terms and concepts

Binomial distribution
Central Limit Theorem
Normal distribution
parameters of a distribution
Poisson distribution
probability distribution
random variable
standard error
standard Normal distribution

Problems

Some of the more challenging problems are indicated by highlighting the problem number in colour.

3.1 Two dice are thrown and the sum of the two scores is recorded. Draw a graph of the resulting probability distribution of the sum and calculate its mean and variance. What is the probability that the sum is 9 or greater?

3.2 Two dice are thrown and the absolute difference of the two scores recorded. Graph the resulting probability distribution and calculate its mean and variance. What is the probability that the absolute difference is 4 or more?

3.3 Sketch the probability distribution for the likely time of departure of a train. Locate the timetabled departure time on your chart.

3.4 A train departs every half hour. You arrive at the station at a completely random moment. Sketch the probability distribution of your waiting time. What is your expected waiting time?

3.5 Sketch the probability distribution for the number of accidents on a stretch of road in one day.

3.6 Sketch the probability distribution for the number of accidents on the same stretch of road in one year. How and why does this differ from your previous answer?

3.7 Six dice are rolled and the number of sixes is noted. Calculate the probabilities of 0, 1, . . ., 6 sixes and graph the probability distribution.

3.8 If the probability of a boy in a single birth is $\frac{1}{2}$ and is independent of the sex of previous babies then the number of boys in a family of 10 children follows a Binomial distribution with mean 5 and variance 2.5. In each of the following instances, describe how the distribution of the number of boys differs from the Binomial described above.

(a) The probability of a boy is $\frac{6}{10}$.

(b) The probability of a boy is $\frac{1}{2}$ but births are not independent. The birth of a boy makes it more than an even chance that the next child is a boy.

(c) As (b) above, except that the birth of a boy makes it less than an even chance that the next child will be a boy.

(d) The probability of a boy is $\frac{6}{10}$ on the first birth. The birth of a boy makes it a more than even chance that the next baby will be a boy.

3.9 A firm receives components from a supplier in large batches, for use in its production process. Production is uneconomic if a batch containing 10% or more defective components is used. The firm checks the quality of each incoming batch by taking a sample of 15 and rejecting the whole batch if more than one defective component is found.

(a) If a batch containing 10% defectives is delivered, what is the probability of its being accepted?

(b) How could the firm reduce this probability of erroneously accepting bad batches?

(c) If the supplier produces a batch with 3% defective, what is the probability of the firm sending back the batch?

(d) What role does the assumption of a 'large' batch play in the calculation?

3.10 The UK record for the number of children born to a mother is 39, 32 of them girls. Assuming the probability of a girl in a single birth is 0.5 and that this probability is independent of previous births:

(a) Find the probability of 32 girls in 39 births (you'll need a scientific calculator or a computer to help with this!).

(b) Does this result cast doubt on the assumptions?

3.11 Using equation (3.5) describing the Normal distribution and setting $\mu = 0$ and $\sigma^2 = 1$, graph the distribution for the values $x = -2, -1.5, -1, -0.5, 0, 0.5, 1, 1.5, 2$.

3.12 Repeat the previous Problem for the values $\mu = 2$ and $\sigma^2 = 3$. Use values of x from -2 to $+6$ in increments of 1.

3.13 For the standard Normal variable z, find

(a) $\Pr(z > 1.64)$

(b) $\Pr(z > 0.5)$

(c) $\Pr(z > -1.5)$

(d) $\Pr(-2 < z < 1.5)$

(e) $\Pr(z = -0.75)$.

For (a) and (d), shade in the relevant areas on the graph you drew for Problem 3.11.

3.14 Find the values of z which cut off

(a) the top 10%

(b) the bottom 15%

(c) the middle 50%

of the standard Normal distribution.

3.15 If $x \sim N(10, 9)$ find

(a) $\Pr(x > 12)$

(b) $\Pr(x < 7)$

(c) $\Pr(8 < x < 15)$

(d) $\Pr(x = 10)$.

3.16 IQ (the intelligence quotient) is Normally distributed with mean 100 and standard deviation 16.

(a) What proportion of the population has an IQ above 120?

(b) What proportion of the population has IQ between 90 and 110?

(c) In the past, about 10% of the population went to university. Now the proportion is about 30%. What was the IQ of the 'marginal' student in the past? What is it now?

3.17 Ten adults are selected at random from the population and their IQ measured. (Assume a population mean of 100 and s.d. of 16 as in Problem 3.16.)

(a) What is the probability distribution of the sample average IQ?

(b) What is the probability that the average IQ of the sample is over 110?

(c) If many such samples were taken, in what proportion would you expect the average IQ to be over 110?

(d) What is the probability that the average IQ lies within the range 90 to 110? How does this answer compare to the answer to part (b) of Problem 16? Account for the difference.

(e) What is the probability that a random sample of ten university students has an average IQ greater than 110?

(f) The first adult sampled has an IQ of 150. What do you expect the average IQ of the sample to be?

3.18 The average income of a country is known to be £10 000 with standard deviation £2500. A sample of 40 individuals is taken and their average income calculated.

(a) What is the probability distribution of this sample mean?

(b) What is the probability of the sample mean being over £10 500?

(c) What is the probability of the sample mean being below £8000?

(d) If the sample size were 10, why could you not use the same methods to find the answers to (a)–(c)?

3.19 A coin is tossed 10 times. Write down the distribution of the number of heads:

(a) exactly, using the Binomial distribution;

(b) approximately, using the Normal distribution;

(c) Find the probability of four or more heads, using both methods. How accurate is the Normal method, with and without the continuity correction?

3.20 A machine producing electronic circuits has an average failure rate of 15% (they're difficult to make). The cost of making a batch of 500 circuits is £8400 and the good ones sell for £20 each. What is the probability of the firm making a loss on any one batch?

3.21 An experienced invoice clerk makes an error once in every 100 invoices, on average.

(a) What is the probability of finding a batch of 100 invoices without error?

(b) What is the probability of finding such a batch with more than two errors?

Calculate the answers using both the Binomial and Poisson distributions. If you try to solve the problem using the Normal method, how accurate is your answer?

3.22 A firm employing 100 workers has an average absenteeism rate of 4%. On a given day, what is the probability of (a) no workers, (b) one worker, (c) more than six workers being absent?

3.23 **(Computer project)** This problem demonstrates the Central Limit Theorem at work. In your spreadsheet, use the =*RAND*() function to generate a random sample of 25 observations (I suggest entering this function in cells A4:A28, for example). Copy these cells across 100 columns, to generate 100 samples. In row 29, calculate the mean of each sample. Now examine the distribution of these sample means.

(*Hint*: you will find the *RAND*() function recalculates automatically every time you perform an operation in the spreadsheet. This makes it difficult to complete the analysis. The solution is to copy and then use 'Edit, Paste Special, Values' to create a copy of the values of the sample means. These will remain stable.)

(a) What distribution would you expect them to have?

(b) What is the parent distribution from which the samples are drawn?

(c) What are the parameters of the parent distribution and of the sample means?

(d) Do your results accord with what you would expect?

(e) Draw up a frequency table of the sample means and graph it. Does it look as you expected?

(f) Experiment with different sample sizes and with different parent distributions to see the effect that these have.

3.24 **(Project)** An extremely numerate newsagent (with a spreadsheet program, as you will need) is trying to work out how many copies of a newspaper he should order. The cost to him per copy is 15p, which he then sells at 45p. Sales are distributed Normally with an average daily sale of 250 and variance 625. Unsold copies cannot be returned for credit or refund; he has to throw them away, losing 15p per copy.

(a) What do you think the seller's objective should be?

(b) How many copies should he order?

(c) What happens to the *variance* of profit as he orders more copies?

(d) Calculate the probability of selling *more than X* copies. (Create an extra column in the spreadsheet for this.) What is the value of this probability at the optimum number of copies ordered?

(e) What would the price–cost ratio have to be to justify the seller ordering *X* copies?

(f) The wholesaler offers a sale or return deal, but the cost per copy is 16p. Should the seller take up this new offer?

(g) Are there other considerations which might influence the seller's decision?

Hints:

Set up your spreadsheet as follows:

Col. A: (cells A10:A160) 175, 176, . . . up to 325 in unit increments (to represent sales levels).

Col. B: (cells B10:B160) the probability of sales falling between 175 and 176, between 176 and 177, etc., up to 325 – 326. (*Excel* has the '= NORMDIST()' function to do this – see the help facility.)

Col. C: (cells C10:C160) total cost (= 0.15 × number ordered. Put the latter in cell F3 so you can reference it and change its value).

Col. D: (cells D10:D160) total revenue ('=MIN(sales, number ordered) × 0.45').

Col. E: profit (revenue – cost).

Col. F: profit × probability (i.e. col. E × col. B).

Cell F161: the sum of F10:F160 (this is the expected profit).

Now vary the number ordered (cell F3) to find the maximum value in F161. You can also calculate the variance of profit fairly simply, using an extra column.

3.25 **(Project)** Using a weekend's football results from the Premier (or other) league, see if the number of goals per game can be adequately modelled by a Poisson process. First calculate the average number of goals per game for the whole league, then derive the distribution of goals per game using the Poisson distribution. Do the actual numbers of goals per game follow this distribution? You might want to take several weeks' results to obtain more reliable results.

Answers to exercises

Exercise 3.1

(a) $0.6^4 = 0.1296$ or 12.96%.

(b) $0.6^2 \times 0.4^2 \times 4C2 = 0.3456$.

(c) $\Pr(r) = 0.6^r \times 0.4^{4-r}4Cr$. The probabilities of $r = 0 \ldots 4$ are respectively 0.0256, 0.1536, 0.3456, 0.3456, 0.1296, which sum to one.

Exercise 3.2

(a)

r	P(r)	$r \times$ P(r)	$r^2 \times$ P(r)
0	0.0256	0	0
1	0.1536	0.1536	0.1536
2	0.3456	0.6912	1.3824
3	0.3456	1.0368	3.1104
4	0.1296	0.5184	2.0736
Totals	1	2.4	6.72

The mean = 2.4/1 = 2.4 and the variance = $6.72/1 - 2.4^2 = 0.96$. Note that these are equal to nP and $nP(1 - P)$.

(b)

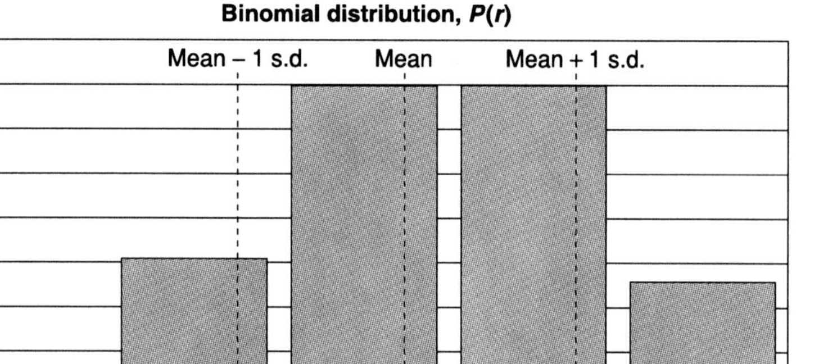

Exercise 3.3

(a) $z = (50 - 40)/\sqrt{36} = 1.67$ and the area beyond $z = 1.67$ is 4.75%.

(b) $z = -0.83$ so area is 20.33%.

(c) This is symmetric around the mean, $z = \pm 0.67$ and the area within these two bounds is 49.72%.

Exercise 3.4

To obtain the IQR we need to go 0.67 s.d.s above and below the mean, giving $200 \pm 0.67 \times 16 = [189.28, 210.72]$.

Exercise 3.5

The IQR (in logs) is within $4.18 \pm 0.67 \times \sqrt{2.56} = [3.11, 5.25]$. Translated out of logs (using 10^x) yields [1288.2, 177 827.9].

Exercise 3.6

(a) $e \sim N(50, 64/25)$.

(b) The s.e. gets smaller. It is $1/\sqrt{2}$ times its previous value.

(c) (i) $z = (51 - 50)/\sqrt{(64/25)} = 0.625$. Hence area in tail = 26.5%. (ii) $z = -1.25$, hence area = 10.56%. (iii) z values are -0.625 and $+0.3125$, giving tail areas of 26.5% and 37.8%, totalling 64.3%. The area between the limits is therefore 35.7%.

Exercise 3.7

(a) Binomial method: $\Pr(r) = 0.5^r \times 0.5^{(20-r)} \times 20Cr$. This gives probabilities of 15, 16, etc., heads of 0.0148, 0.0046, etc., which total 0.0207 or 2.1%. By the Normal approximation, $r \sim N(10, 5)$ and $z = (14.5 - 10)/\sqrt{5} = 2.01$. The area in the tail is then 2.22%, not far off the correct value (a 10% error). Note that $nP = 10 = n(1 - P)$.

(b) Binomial method: Pr(5 or 6 heads) = 0.302 + 0.118 = 0.420 or 42%. By the Normal, $r \sim N(4.2, 1.26)$, $z = 0.267$ and the area is 39.36%, still reasonably close to the correct answer despite the fact that $n(1 - P) = 1.8$.

(c) By similar methods the answers are 11.8% (Binomial) and 12.3% (Normal).

Exercise 3.8

(a) (i) $\mu = 1$ in this case ($1/50 \times 50$) so $\Pr(x = 0) = 1^0 e^{-1}/0! = 0.368$. (ii) $\Pr(x = 1) = 1^1 e^{-1}/1! = 0.368$. (iii) $1^2 e^{-1}/2! = 0.184$. (iv) $1 - 0.368 = 0.632$.

(b) The average number of customer per 10 minutes is 2 (= 10/5). Hence $\Pr(x = 0) = 2^0 e^{-2}/0! = 0.135$.

4 Estimation and confidence intervals

Contents

Learning outcomes

By the end of this chapter you should be able to:

- recognise the importance of probability theory in drawing valid inferences (or deriving estimates) from a sample of data;
- understand the criteria for constructing a good estimate;
- construct estimates of parameters of interest from sample data, in a variety of different circumstances;
- appreciate that there is uncertainty about the accuracy of any such estimate;
- provide measures of the uncertainty associated with an estimate;
- recognise the relationship between the size of a sample and the precision of an estimate derived from it.

Complete your diagnostic test for Chapter 4 now to create your personal study plan. Exercises with an icon ? are also available for practice in MathXL with additional supporting resources.

Introduction

We now come to the heart of the subject of statistical inference. Until now the following type of question has been examined: given the population parameters μ and σ^2, what is the probability of the sample mean $\bar{x}$, from a sample of size n, being greater than some specified value or within some range of values? The parameters μ and σ^2 are assumed to be known and the objective is to try to form some conclusions about possible values of $\bar{x}$. However, in practice it is usually the sample values $\bar{x}$ and s^2 that are known, while the population parameters μ and σ^2 are not. Thus a more interesting question to ask is: given the values of $\bar{x}$ and s^2, what can be said about μ and σ^2? Sometimes the population variance is known, and inferences have to be made about μ alone. For example, if a sample of 50 British families finds an average weekly expenditure on food ($\bar{x}$) of £37.50 with a standard deviation (s) of £6.00, what can be said about the average expenditure (μ) of *all* British families?

Schematically this type of problem is shown as follows:

Sample information		Population parameters
$\bar{x}, s^2$	inferences about $\longrightarrow$	μ, σ^2

This chapter covers the **estimation** of population parameters such as μ and σ^2 while Chapter 5 describes **testing hypotheses** about these parameters. The two procedures are very closely related.

Point and interval estimation

There are basically two ways in which an estimate of a parameter can be presented. The first of these is a **point estimate**, i.e. a single value which is the best estimate of the parameter of interest. The point estimate is the one which is most prevalent in everyday usage; for example, the average Briton surfs the internet for 30 minutes per day. Although this is presented as a fact, it is actually an estimate, obtained from a survey of people's use of personal computers. Since it is obtained from a sample there must be some doubt about its accuracy: the sample will probably not exactly represent the whole population. For this reason **interval estimates** are also used, which give some idea of the likely accuracy of the estimate. If the sample size is small, for example, then it is quite possible that the estimate is not very close to the true value and this would be reflected in a wide interval estimate, for example, that the average Briton spends between 5 and 55 minutes surfing the net per day. A larger sample, or a better method of estimation, would allow a narrower interval to be derived and thus a more precise estimate of the parameter to be obtained, such as an average surfing time of between 20 and 40 minutes. Interval estimates are better for the consumer of the statistics, since they not only show the estimate of the parameter but also give an idea of the confidence which the researcher has in that estimate. The following sections describe how to construct both types of estimate.

Rules and criteria for finding estimates

In order to estimate a parameter such as the population mean, a rule (or set of rules) is required which describes how to derive the estimate of the parameter from the sample data. Such a rule is known as an **estimator**. An example of an estimator for the population mean is 'use the sample mean'. It is important to distinguish between an estimator, a rule and an estimate, which is the value derived as a result of applying the rule to the data.

There are many possible estimators for any parameter, so it is important to be able to distinguish between good and bad estimators. The following examples provide some possible estimators of the population mean:

(1) the sample mean;
(2) the smallest sample observation;
(3) the first sample observation.

A set of criteria is needed for discriminating between good and bad estimators. Which of the above three estimators is 'best'? Two important criteria by which to judge estimators are **bias** and **precision**.

Bias

It is impossible to know if a single estimate of a parameter, derived by applying a particular estimator to the sample data, gives a correct estimate of the parameter or not. The estimate might be too low or too high and, since the parameter is unknown, it is impossible to check this. What *is* possible, however, is to say whether an estimator gives the correct answer *on average*. An estimator which gives the correct answer on average is said to be unbiased. Another way of expressing this is to say that an unbiased estimator does not *systematically* mislead the researcher away from the correct value of the parameter. It is, however, important to remember, that even using an unbiased estimator does not guarantee that a single use of the estimator will yield a correct estimate of the parameter. Bias (or the lack of it) is a theoretical property.

Formally, an estimator is unbiased if its expected value is equal to the parameter being estimated. Consider trying to estimate the population mean using the three estimators suggested above. Taking the sample mean first, we have already learned (see equation (3.15)) that its expected value is μ, i.e.

$$E(\bar{x}) = \mu$$

which immediately shows that the sample mean is an unbiased estimator.

The second estimator (the smallest observation in the sample) can easily be shown to be biased, using the result derived above. Since the smallest sample observation must be less than the sample mean, its expected value must be less than μ. Denote the smallest observation by x_s, then

$$E(x_s) < \mu$$

so this estimator is biased downwards. It underestimates the population mean. The size of the bias is simply the difference between the expected value of the estimator and the value of the parameter, so the bias in this case is

$$\text{Bias} = \mathrm{E}(x_s) - \mu \tag{4.1}$$

For the sample mean $\bar{x}$ the bias is obviously zero.

Turning to the third rule (the first sample observation) this can be shown to be another unbiased estimator. Choosing the first observation from the sample is equivalent to taking a random sample of size one from the population in the first place. Thus the single observation may be considered as the sample mean from a random sample of size one. Since it is a sample mean it is unbiased, as demonstrated earlier.

Precision

Two of the estimators above were found to be unbiased, and, in fact, there are many unbiased estimators (the sample median is another). Some way of choosing between the set of all unbiased estimators is therefore required, which is where the criterion of precision helps. Unlike bias, precision is a relative concept, comparing one estimator to another. Given two estimators A and B, A is more precise than B if the estimates it yields (from all possible samples) are less spread out than those of estimator B. A precise estimator will tend to give similar estimates for all possible samples.

Consider the two unbiased estimators found above: how do they compare on the criteria of precision? It turns out that the sample mean is the more precise of the two, and it is not difficult to understand why. Taking just a single sample observation means that it is quite likely to be unrepresentative of the population as a whole, and thus leads to a poor estimate of the population mean. The sample mean on the other hand is based on all the sample observations and it is unlikely that all of them are unrepresentative of the population. The sample mean is therefore a good estimator of the population mean, being more precise than the single observation estimator.

Just as bias was related to the expected value of the estimator, so precision can be defined in terms of the variance. One estimator is more precise than another if it has a smaller variance. Recall that the probability distribution of the sample mean is

$$\bar{x} \sim N(\mu, \sigma^2/n) \tag{4.2}$$

in large samples, so the variance of the sample mean is

$$V(\bar{x}) = \sigma^2/n$$

As the sample size n becomes larger, the variance of the sample mean becomes smaller, so the estimator becomes more precise. For this reason large samples give better estimates than small samples, and so the sample mean is a better estimator than taking just one observation from the sample. The two estimators can be compared in a diagram (see Figure 4.1) which draws the probability distributions of the two estimators.

It is easily seen that the sample mean yields estimates which are *on average* closer to the population mean.

A related concept is that of **efficiency**. The efficiency of one unbiased estimator, relative to another, is given by the ratio of their sampling variances. Thus the efficiency of the first observation estimator, relative to the sample mean, is given by

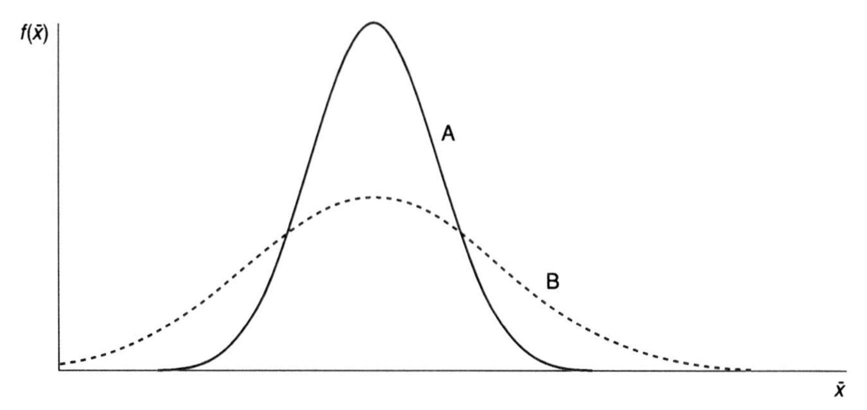

Figure 4.1
The sampling distribution of two estimators

Note: Curve *A* shows the distribution of sample means, which is the more precise estimator. *B* shows the distribution of estimates using a single observation.

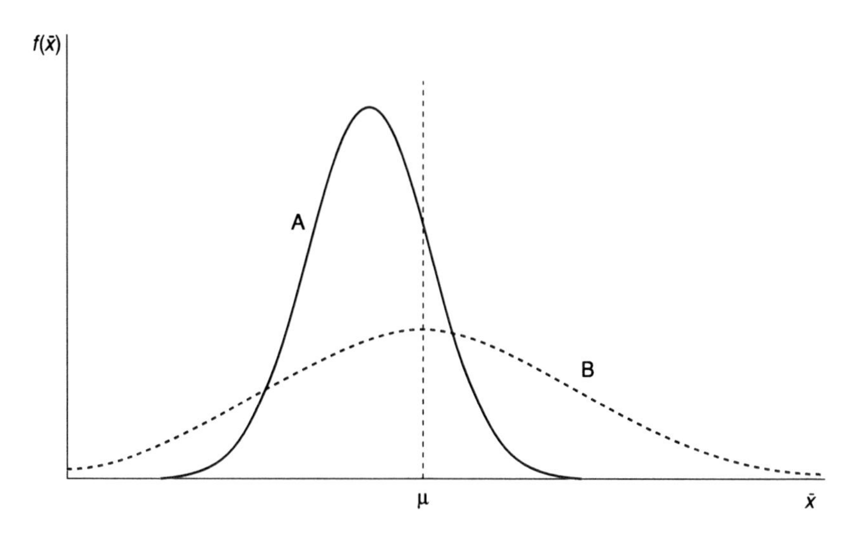

Figure 4.2
The trade-off between bias and precision

$$\text{Efficiency} = \frac{var(\bar{x})}{var(x_1)} = \frac{\sigma^2/n}{\sigma^2} = \frac{1}{n} \tag{4.3}$$

Thus the efficiency is determined by the relative sample sizes in this case. Other things being equal, a more efficient estimator is to be preferred.

Similarly, the variance of the median can be shown to be (for a Normal distribution) $\pi/2 \times \sigma^2/n$. The efficiency of the median is therefore $2/\pi \approx 64\%$.

The trade-off between bias and precision: the Bill Gates effect

It should be noted that just because an estimator is biased does not necessarily mean that it is imprecise. Sometimes there is a trade-off between an unbiased, but imprecise, estimator and a biased, but precise, one. Figure 4.2 illustrates this.

Although estimator *A* is biased it will nearly always yield an estimate which is fairly close to the true value; even though the estimate is expected to be wrong, it is not likely to be far wrong. Estimator *B*, although unbiased, can give estimates which are far away from the true value, so that *A* might be the preferred estimator.

As an example of this, suppose we are trying to estimate the average wealth of the US population. Consider the following two estimators:

(1) use the mean wealth of a random sample of Americans;
(2) use the mean wealth of a random sample of Americans but, if Bill Gates is in the sample, omit him from the calculation.

Bill Gates is the Chairman of Microsoft and one of the world's richest men. Because of this, he is a dollar billionaire (about $50bn according to recent reports – it varies with the stock market). His presence in a sample of, say, 30 observations would swamp the sample and give a highly misleading result. Assuming Bill Gates has $50bn and the others each have $200 000 of wealth, the average wealth would be estimated at about $1.6bn, which is surely wrong.

The first rule could therefore give us a wildly incorrect answer, although the rule is unbiased. The second rule is clearly biased but does rule out the possibility of such an unlucky sample. We can work out the approximate bias. It is the difference between the average wealth of all Americans and the average wealth of all Americans except Bill Gates. If the true average of all 250 million Americans is $200 000, then total wealth is $50 000bn. Subtracting Bill's $50bn leaves $49 950bn shared among the rest, giving $199 800 each, a difference of 0.1%. This is what we would expect the bias to be.

It might seem worthwhile therefore to accept this degree of bias in order to improve the precision of the estimate. Furthermore, if we did use the biased rule, we could always adjust the sample mean upwards by 0.1% to get an approximately unbiased estimate.

Of course, this point applies to any exceptionally rich person, not just Bill Gates. It points to the need to ensure that the rich are not over- (nor under-) represented in the sample. Chapter 9 on sampling methods investigates this point in more detail. In the rest of this book only unbiased estimators are considered, the most important being the sample mean.

Estimation with large samples

For the type of problem encountered in this chapter the method of estimation differs according to the size of the sample. 'Large' samples, by which is meant sample sizes of 25 or more, are dealt with first, using the Normal distribution. Small samples are considered in a later section, where the *t* distribution is used instead of the Normal. The differences are relatively minor in practical terms and there is a close theoretical relationship between the *t* and Normal distributions.

With large samples there are three types of estimation problem we will consider.

(1) The estimation of a mean from a sample of data.
(2) The estimation of a proportion on the basis of sample evidence. This would consider a problem such as estimating the proportion of the population intending to buy an iPhone, based on a sample of individuals. Each person in the sample would simply indicate whether they have bought, or intend to buy, an iPhone. The principles of estimation are the same as in the first case but the formulae used for calculation are slightly different.

(3) The estimation of the difference of two means (or proportions), for example a problem such as estimating the difference between men and women's expenditure on clothes. Once again, the principles are the same, the formulae different.

Estimating a mean

To demonstrate the principles and practice of estimating the population mean, we shall take the example of estimating the average wealth of the UK population, the full data for which were given in Chapter 1. Suppose that we did not have this information but were required to estimate the average wealth from a sample of data. In particular, let us suppose that the sample size is $n = 100$, the sample mean is $\bar{x} = 130$ (in £000) and the sample variance is $s^2 = 50\,000$. Obviously, this sample has got fairly close to the true values (see Chapter 1) but we could not know that from the sample alone. What can we infer about the population mean μ from the sample data alone?

For the point estimate of μ the sample mean is a good candidate since it is unbiased, and it is more precise than other sample statistics such as the median. The point estimate of μ is simply £130 000, therefore.

The point estimate does not give an idea of the uncertainty associated with the estimate. We are not *absolutely* sure that the mean is £130 000 (in fact, it isn't – it is £146 984). The interval estimate gives some idea of the uncertainty. It is centred on the sample mean, but gives a range of values to express the uncertainty.

To obtain the interval estimate we first require the probability distribution of $\bar{x}$, first established in Chapter 3 (equation (3.18))

$$\bar{x} \sim N(\mu, \sigma^2/n) \tag{4.4}$$

From this, it was calculated that there is a 95% probability of the sample mean lying within 1.96 standard errors of μ[1], i.e.

$$\Pr(\mu - 1.96\sqrt{\sigma^2/n} \leq \bar{x} \leq \mu + 1.96\sqrt{\sigma^2/n}) = 0.95$$

We can manipulate each of the inequalities within the brackets to make μ the subject of the expression

$$\mu - 1.96\sqrt{\sigma^2/n} \leq \bar{x} \quad \text{implies} \quad \mu \leq \bar{x} + 1.96\sqrt{\sigma^2/n}$$

Similarly

$$\bar{x} \leq \mu + 1.96\sqrt{\sigma^2/n} \quad \text{implies} \quad \bar{x} - 1.96\sqrt{\sigma^2/n} \leq \mu$$

Combining these two new expressions we obtain

$$[\bar{x} - 1.96\sqrt{\sigma^2/n} \leq \mu \leq \bar{x} + 1.96\sqrt{\sigma^2/n}] \tag{4.5}$$

We have transformed the probability interval. Instead of saying $\bar{x}$ lies within 1.96 standard errors of μ, we now say μ lies within 1.96 standard errors of $\bar{x}$. Figure 4.3 illustrates this manipulation. Figure 4.3(a) shows μ at the centre of a probability interval for $\bar{x}$. Figure 4.3(b) shows a sample mean $\bar{x}$ at the centre of an interval relating to the possible positions of μ.

---

[1] See equation (3.23) in Chapter 3 to remind yourself of this. Remember that ±1.96 is the z score which cuts off 2.5% in each tail of the normal distribution.

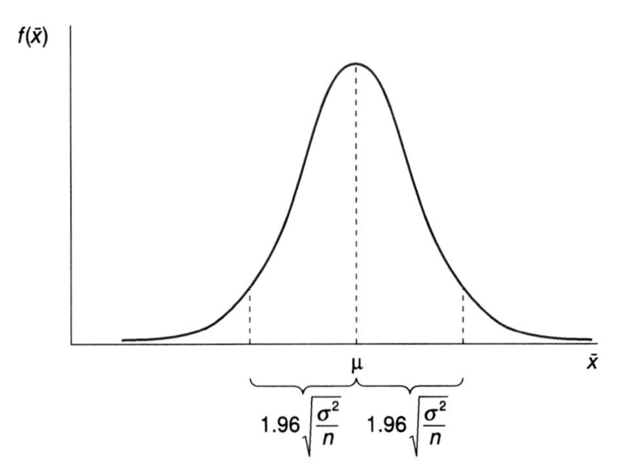

Figure 4.3(a)
The 95% probability interval for $\bar{x}$ around the population mean μ

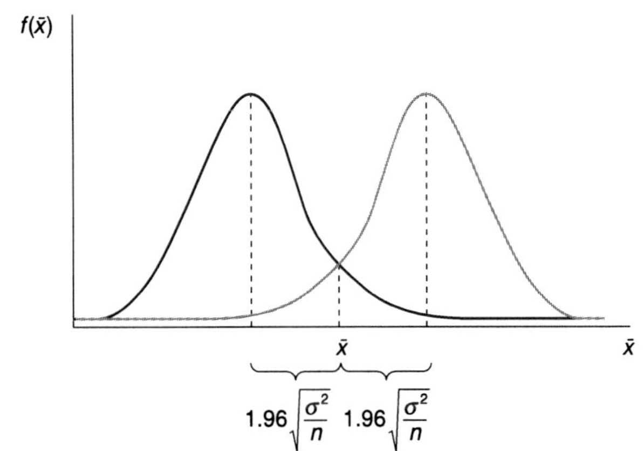

Figure 4.3(b)
The 95% confidence interval for μ around the sample mean $\bar{x}$

The interval shown in equation (4.5) is called the **95% confidence interval** and this is the interval estimate for μ. In this example the value of σ^2 is unknown, but in large ($n \geqslant 25$) samples it can be replaced by s^2 from the sample. s^2 is here used as an estimate of σ^2 which is unbiased and sufficiently precise in large ($n \geqslant 25$ or so) samples. The 95% confidence interval is therefore

$$\begin{aligned}&[\bar{x} - 1.96\sqrt{s^2/n} \leqslant \mu \leqslant \bar{x} + 1.96\sqrt{s^2/n}]\\&= [130 - 1.96\sqrt{50\,000/100},\ 130 + 1.96\sqrt{50\,000/100}]\\&= [86.2,\ 173.8] \qquad (4.6)\end{aligned}$$

Thus we are 95% confident that the true average level of wealth lies between £86 200 and £173 800. It should be noted that £130 000 lies exactly at the centre of the interval[2] (because of the symmetry of the Normal distribution).

By examining equation (4.6) one can see that the confidence interval is wider

- the smaller the sample size;
- the greater the standard deviation of the sample.

---

[2] The two values are the lower and upper limits of the interval, separated by a comma. This is the standard way of writing a confidence interval.

The greater uncertainty which is associated with smaller sample sizes is manifested in a wider confidence interval estimate of the population mean. This occurs because a smaller sample has more chance of being unrepresentative (just because of an unlucky sample).

Greater variation in the sample data also leads to greater uncertainty about the population mean and a wider confidence interval. Greater sample variation suggests greater variation in the population so, again, a given sample could include observations which are a long way off the mean. Note that in this example there is great variation of wealth in the population and hence in the sample also. This means that a sample of 100 is not very informative (the confidence interval is quite wide). We would need a substantially larger sample to obtain a more precise estimate.

Note that the width of the confidence interval does *not* depend upon the population size – a sample of 100 observations reveals as much about a population of 10 000 as it does about a population of 10 000 000. This point will be discussed in more detail in Chapter 9 on sampling methods. This is a result that often surprises people, who generally believe that a larger sample is required if the population is larger.

Worked example 4.1

A sample of 50 school students found that they spent 45 minutes doing homework each evening, with a standard deviation of 15 minutes. Estimate the average time spent on homework by all students.

The sample data are $\bar{x} = 45$, $s = 15$ and $n = 50$. If we can assume the sample is representative we may use $\bar{x}$ as an unbiased estimate of μ, the population mean. The point estimate is therefore 45 minutes.

The 95% confidence interval is given by equation (4.6)

$$[\bar{x} - 1.96\sqrt{s^2/n} \leqslant \mu \leqslant \bar{x} + 1.96\sqrt{s^2/n}]$$
$$= [45 - 1.96\sqrt{15^2/50} \leqslant \mu \leqslant 45 + 1.96\sqrt{15^2/50}]$$
$$= [40.8, 49.2]$$

We are 95% confident the true answer lies between 40.8 and 49.2 minutes.

Exercise 4.1

(a) A sample of 100 is drawn from a population. The sample mean is 25 and the sample standard deviation is 50. Calculate the point and 95% confidence interval estimates for the population mean.

(b) If the sample size were 64, how would this alter the point and interval estimates?

Exercise 4.2

A sample of size 40 is drawn with sample mean 50 and standard deviation 30. Is it likely that the true population mean is 60?

Precisely what is a confidence interval?

There is often confusion over what a confidence interval actually means. This is not really surprising since the obvious interpretation turns out to be wrong. It does *not* mean that there is a 95% chance that the true mean lies within the interval. We cannot make such a probability statement, because of our definition of probability (based on the frequentist view of a probability). That view states that one can make a probability statement about a random variable (such as $\bar{x}$) but not about a parameter (such as μ). μ either lies within the interval or it does not – it cannot lie 95% within it. Unfortunately, we just do not know what the truth is.

It is for this reason that we use the term 'confidence interval' rather than 'probability interval'. Unfortunately, words are not as precise as numbers or algebra, and so most people fail to recognise the distinction. A precise explanation of the 95% confidence interval runs as follows. If we took many samples (all the same size) from a population with mean μ and calculated a confidence interval from each, we would find that μ lies within 95% of the calculated intervals. Of course, in practice we do not take many samples, usually just one. We do not know (and cannot know) if our one sample is one of the 95% or one of the 5% that miss the mean.

Figure 4.4 illustrates the point. It shows 95% confidence intervals calculated from 20 samples drawn from a population with a mean of 5. As expected, we see that 19 of these intervals contain the true mean, while the interval calculated from sample 18 does not contain the true value. This is the expected result, but is not guaranteed. You might obtain all 20 intervals containing the true mean, or fewer than 19. In the long run (with lots of estimates) we would expect 95% of the calculated intervals to contain the true mean.

A second question is, why use a probability (and hence a confidence level) of 95%? In fact, one can choose any confidence level, and thus confidence

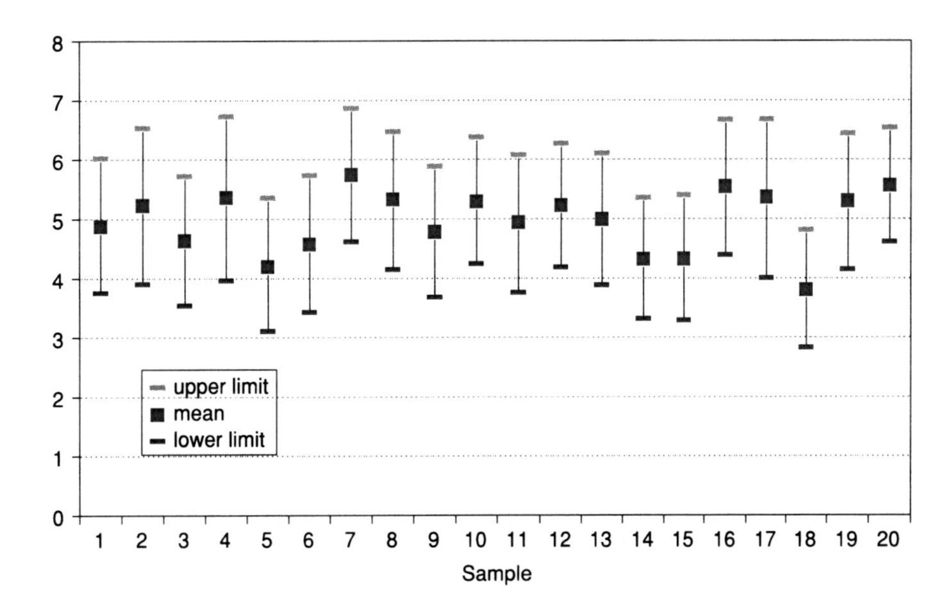

Figure 4.4
Confidence intervals calculated from 20 samples

interval. The 90% confidence interval can be obtained by finding the z score which cuts off 10% of the Normal distribution (5% in each tail). From Table A2 (see page **414**) this is $z = 1.64$, so the 90% confidence interval is

$$[\bar{x} - 1.64\sqrt{s^2/n} \leqslant \mu \leqslant \bar{x} + 1.64\sqrt{s^2/n}] \tag{4.7}$$
$$= [130 - 1.64\sqrt{50\,000/100},\ 130 + 1.64\sqrt{50\,000/100}\,]$$
$$= [93.3,\ 166.7]$$

Notice that this is narrower than the 95% confidence level. The greater the degree of confidence required, the wider the interval has to be. Any confidence level may be chosen, and by careful choice of this level the confidence interval can be made as wide or as narrow as wished. This would seem to undermine the purpose of calculating the confidence interval, which is to obtain some idea of the uncertainty attached to the estimate. This is not the case, however, because the reader of the results can interpret them appropriately, as long as the confidence level is made clear. To simplify matters, the 95% and 99% confidence levels are the most commonly used and serve as conventions. Beware of the researcher who calculates the 76% confidence interval – this may have been chosen in order to obtain the desired answer rather than in the spirit of scientific enquiry! The general formula for the $(100 - \alpha)$% confidence interval is

$$[\bar{x} - z_\alpha\sqrt{s^2/n},\ \bar{x} + z_\alpha\sqrt{s^2/n}\,] \tag{4.8}$$

where z_α is the z score which cuts off the extreme α% of the Normal distribution.

Estimating a proportion

It is often the case that we wish to estimate the **proportion** of the population that has a particular characteristic (e.g. is unemployed), rather than wanting an average. Given what we have already learned this is fairly straightforward and is based on similar principles. Suppose that, following Chapter 1, we wish to estimate the proportion of educated men who are unemployed. We have a random sample of 200 men, of whom 15 are unemployed. What can we infer?

The sample data are

$n = 200$, and
$p = 0.075$ $(= 15/200)$

where p is the (sample) proportion unemployed. We denote the population proportion by the Greek letter π and it is this that we are trying to estimate using data from the sample.

The key to solving this problem is recognising p as a random variable just like the sample mean. This is because its value depends upon the sample drawn and will vary from sample to sample. Once the probability distribution of this random variable is established the problem is quite easy to solve, using the same methods as were used for the mean. The sampling distribution of p is[3]

$$p \sim \mathrm{N}\left(\pi, \frac{\pi(1-\pi)}{n}\right) \tag{4.9}$$

[3] See the Appendix to this chapter (page **170**) for the derivation of this formula.

This tells us that the sample proportion is centred around the true value but will vary around it, varying from sample to sample. This variation is expressed by the variance of p, whose formula is $\pi(1 - \pi)/n$. Having derived the probability distribution of p the same methods of estimation can be used as for the sample mean. Since the expected value of p is π, the sample proportion is an unbiased estimate of the population parameter. The point estimate of π is simply p, therefore. Thus it is estimated that 7.5% of all educated men are unemployed.

Given the sampling distribution for p in equation (4.9) above, the formula for the 95% confidence interval[4] for π can immediately be written down as

$$\left[p - 1.96\sqrt{\frac{\pi(1-\pi)}{n}},\ p + 1.96\sqrt{\frac{\pi(1-\pi)}{n}}\right] \tag{4.10}$$

As the value of π is unknown, the confidence interval cannot yet be calculated, so the sample value of 0.075 has to be used instead of the unknown π. Like the substitution of s^2 for σ^2 in the case of the sample mean above, this is acceptable in large samples. Thus the 95% confidence interval becomes

$$\left[p - 1.96\sqrt{\frac{0.075(1-0.075)}{200}},\ p + 1.96\sqrt{\frac{0.075(1-0.075)}{200}}\right] \tag{4.11}$$

$$= [0.075 - 0.037, 0.075 + 0.037]$$
$$= [0.038, 0.112]$$

We say that we are 95% confident that the true proportion of unemployed, educated men lies between 3.8% and 11.2%.

It can be seen that these two cases apply a common method. The 95% confidence interval is given by the point estimate plus or minus 1.96 standard errors. For a different confidence level, 1.96 would be replaced by the appropriate value from the standard Normal distribution.

With this knowledge two further cases can be swiftly dealt with.

Worked example 4.2 Music down the phone

Do you get angry when you try to phone an organisation and you get an automated reply followed by music while you hang on? Well, you are not alone. Mintel (a consumer survey company) asked 1946 adults what they thought of music played to them while they were trying to get through on the phone. 36% reported feeling angered by the music and more than one in four were annoyed by the automated voice response.

With these data we can calculate a confidence interval for the true proportion of people who dislike the music. First, we assume that the sample is a truly random one. This is probably not strictly true, so our calculated confidence interval will only be an approximate one. With $p = 0.36$ and $n = 1946$ we obtain the following 95% interval

→

[4] As usual, the 95% confidence interval limits are given by the point estimate plus and minus 1.96 standard errors.

$$p \pm 1.96 \times \sqrt{\frac{p(1-p)}{n}} = 0.36 \pm 1.96 \times \sqrt{\frac{0.36(1-0.36)}{1946}}$$
$$= 0.36 \pm 0.021 = [0.339, 0.381]$$

Mintel further estimated that 2800 million calls were made by customers to call centres per year, so we can be (approximately) 95% confident that between 949 million and 1067 million of those calls have an unhappy customer on the line!

Source: *The Times*, 10 July 2000.

Estimating the difference between two means

We now move on to estimating differences. In this case we have two samples and want to know whether there is a difference between their respective populations. One sample might be of men, the other of women, or we could be comparing two different countries, etc. A point estimate of the difference is easy to obtain but once again there is some uncertainty around this figure, because it is based on samples. Hence we measure that uncertainty via a confidence interval. All we require are the appropriate formulae. Consider the following example.

Sixty pupils from school 1 scored an average mark of 62% in an exam, with a standard deviation of 18%; 35 pupils from school 2 scored an average of 70% with standard deviation 12%. Estimate the true difference between the two schools in the average mark obtained.

This is a more complicated problem than those previously treated since it involves two samples rather than one. An estimate has to be found for $\mu_1 - \mu_2$ (the true difference in the mean marks of the schools), in the form of both point and interval estimates. The pupils taking the exams may be thought of as samples of all pupils in the schools who could potentially take the exams.

Notice that this is a problem about sample means, not proportions, even though the question deals in percentages. The point is that each observation in the sample (i.e. each student's mark) can take a value between 0 and 100, and one can calculate the standard deviation of the marks. For this to be a problem of sample proportions the mark for each pupil would each have to be of the pass/fail type, so that one could only calculate the proportion who passed.

It might be thought that the way to approach this problem is to derive one confidence interval for each sample (along the lines set out above), and then to somehow combine them; for example, the degree of overlap of the two confidence intervals could be assessed. This is not the best approach, however. It is sometimes a good strategy, when faced with an unfamiliar problem to solve, to translate it into a more familiar problem and then solve it using known methods. This is the procedure which will be followed here. The essential point is to keep in mind the concept of a random variable and its probability distribution.

Problems involving a single random variable have already been dealt with above. The current problem deals with two samples and therefore there are two random variables to consider, i.e. the two sample means $\bar{x}_1$ and $\bar{x}_2$. Since the aim is to estimate $\mu_1 - \mu_2$, an obvious candidate for an estimator is the difference between the two sample means, $\bar{x}_1 - \bar{x}_2$. We can think of this as a single random variable (even though two means are involved) and use the methods we have already learned. We therefore need to establish the sampling distribution of $\bar{x}_1 - \bar{x}_2$. This is derived in the Appendix to this chapter (see page 170) and results in equation (4.12)

$$\bar{x}_1 - \bar{x}_2 \sim N\left(\mu_1 - \mu_2, \frac{\sigma_1^2}{n_1} - \frac{\sigma_2^2}{n_2}\right) \quad (4.12)$$

This equation states that the difference in sample means will be centred on the difference in the two population means, with some variation around this as measured by the variance. One assumption behind the derivation of equation (4.12) is that the two samples are independently drawn. This is likely in this example; it is difficult to see how the samples from the two schools could be connected. However, one must always bear this possibility in mind when comparing samples. For example, if one were comparing men's and women's heights, it would be dangerous to take samples of men and their wives as they are unlikely to be independent. People tend to marry partners of a similar height to themselves, so this might bias the results.

The distribution of $\bar{x}_1 - \bar{x}_2$ is illustrated in Figure 4.5. Equation (4.12) shows that $\bar{x}_1 - \bar{x}_2$ is an unbiased estimator of $\mu_1 - \mu_2$. The difference between the sample means will therefore be used as the point estimate of $\mu_1 - \mu_2$. Thus the point estimate of the true difference between the schools is

$$\bar{x}_1 - \bar{x}_2 = 62 - 70 = -8\%$$

The 95% confidence interval estimate is derived in the same manner as before, making use of the standard error of the random variable. The formula is[5]

$$\left[(\bar{x}_1 - \bar{x}_2) - 1.96\sqrt{\frac{s_1^2}{n_1} + \frac{s_2^2}{n_2}}, (\bar{x}_1 - \bar{x}_2) + 1.96\sqrt{\frac{s_1^2}{n_1} + \frac{s_2^2}{n_2}}\right] \quad (4.13)$$

As the values of σ^2 are unknown they have been replaced in equation (4.13) by their sample values. As in the single sample case, this is acceptable in large samples. The 95% confidence interval for $\mu_1 - \mu_2$ is therefore

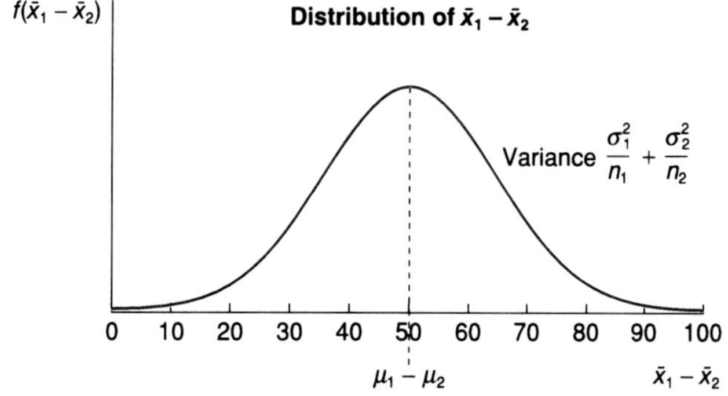

Figure 4.5
The distribution of $\bar{x}_1 - \bar{x}_2$

[5] The term under the square root sign is the standard error for $\bar{x}_1 - \bar{x}_2$

$$\left[(62-70)-1.96\sqrt{\frac{18^2}{60}+\frac{12^2}{35}},\ (62-70)+1.96\sqrt{\frac{18^2}{60}+\frac{12^2}{35}}\right]$$

$$=[-14.05, -1.95]$$

The estimate is that school 2's average mark is between 1.95 and 14.05 percentage points above that of school 1. Notice that the confidence interval does not include the value zero, which would imply equality of the two schools' marks. Equality of the two schools can thus be ruled out with 95% confidence.

Worked example 4.3

A survey of holidaymakers found that on average women spent 3 hours per day sunbathing, men spent 2 hours. The sample sizes were 36 in each case and the standard deviations were 1.1 hours and 1.2 hours respectively. Estimate the true difference between men and women in sunbathing habits. Use the 99% confidence level.

The point estimate is simply one hour, the difference of sample means. For the confidence interval we have

$$\left[(\bar{x}_1-\bar{x}_2)-2.57\sqrt{\frac{s_1^2}{n_1}+\frac{s_2^2}{n_2}}\leqslant\mu\leqslant(\bar{x}_1-\bar{x}_2)+2.57\sqrt{\frac{s_1^2}{n_1}+\frac{s_2^2}{n_2}}\right]$$

$$=\left[(3-2)-2.57\sqrt{\frac{1.1^2}{36}+\frac{1.2^2}{36}}\leqslant\mu\leqslant(3-2)+2.57\sqrt{\frac{1.1^2}{36}+\frac{1.2^2}{36}}\right]$$

$$=[0.30, 1.70]$$

This evidence suggests women do spend more time sunbathing than men (zero is not in the confidence interval). Note that we might worry the samples might not be independent here – it could represent 36 couples. If so, the evidence is likely to underestimate the true difference, if anything, as couples are likely to spend time sunbathing together.

Estimating the difference between two proportions

We move again from means to proportions. We use a simple example to illustrate the analysis of this type of problem. Suppose that a survey of 80 Britons showed that 60 owned personal computers. A similar survey of 50 Swedes showed 30 with computers. Are personal computers more widespread in Britain than Sweden?

Here the aim is to estimate $\pi_1-\pi_2$, the difference between the two population proportions, so the probability distribution of p_1-p_2 is needed, the difference of the sample proportions. The derivation of this follows similar lines to those set out above for the difference of two sample means, so is not repeated. The probability distribution is

$$p_1-p_2\sim N\left(\pi_1-\pi_2,\ \frac{\pi_1(1-\pi_1)}{n_1}+\frac{\pi_2(1-\pi_2)}{n_2}\right) \quad (4.14)$$

Again, the two samples must be independently drawn for this to be correct (it is difficult to see how they could not be in this case).

Since the difference between the sample proportions is an unbiased estimate of the true difference, this will be used for the point estimate. The point estimate is therefore

$$p_1 - p_2 = 60/80 - 30/50$$
$$= 0.15$$

or 15%. The 95% confidence interval is given by

$$\left[p_1 - p_2 - 1.96\sqrt{\frac{\pi_1(1-\pi_1)}{n_1} + \frac{\pi_2(1-\pi_2)}{n_2}},\right.$$
$$\left.p_1 - p_2 + 1.96\sqrt{\frac{\pi_1(1-\pi_1)}{n_1} + \frac{\pi_2(1-\pi_2)}{n_2}}\right] \quad (4.15)$$

π_1 and π_2 are unknown so have to be replaced by p_1 and p_2 for purposes of calculation, so the interval becomes

$$\left[0.75 - 0.60 - 1.96\sqrt{\frac{0.75 \times 0.25}{80} + \frac{0.60 \times 0.40}{50}},\right.$$
$$\left.0.75 - 0.60 + 1.96\sqrt{\frac{0.75 \times 0.25}{80} + \frac{0.60 \times 0.40}{50}}\right]$$
$$= [0.016, 0.316] \quad (4.16)$$

The result is a fairly wide confidence interval due to the relatively small sample sizes. The interval does not include zero, however, so we can be 95% confident there is a difference between the two countries.

Exercise 4.3

(a) Seven people out of a sample of 50 are left-handed. Estimate the true proportion of left-handed people in the population, finding both point and interval estimates.

(b) Repeat part (a) but find the 90% confidence interval. How does the 90% interval compare with the 95% interval?

(c) Calculate the 99% interval and compare to the others.

Exercise 4.4

Given the following data from two samples, calculate the true difference between the means. Use the 95% confidence level.

$\bar{x}_1 = 25 \quad \bar{x}_2 = 30$
$s_1 = 18 \quad s_2 = 25$
$n_1 = 36 \quad n_2 = 49$

Exercise 4.5

A survey of 50 16-year old girls revealed that 40% had a boyfriend. A survey of 100 16-year old boys revealed 20% with a girlfriend. Estimate the true difference in proportions between the sexes.

Estimation with small samples: the *t* distribution

So far only large samples (defined as sample sizes in excess of 25) have been dealt with, which means that (by the Central Limit Theorem) the sampling distribution of $\bar{x}$ follows a Normal distribution, whatever the distribution of the parent population. Remember, from the two theorems of Chapter 3, that:

- if the population follows a Normal distribution, $\bar{x}$ is also Normally distributed; and
- if the population is not Normally distributed, $\bar{x}$ is approximately Normally distributed in large samples ($n \geqslant 25$).

In both cases, confidence intervals can be constructed based on the fact that

$$\frac{\bar{x} - \mu}{\sqrt{\sigma^2/n}} \sim N(0, 1) \qquad (4.17)$$

and so the standard Normal distribution is used to find the values which cut off the extreme 5% of the distribution ($z = \pm 1.96$). In practical examples, we had to replace σ by its estimate, s. Thus the confidence interval was based on the fact that

$$\frac{\bar{x} - \mu}{\sqrt{s^2/n}} \sim N(0, 1) \qquad (4.18)$$

in large samples. For small sample sizes, equation (4.18) is no longer true. Instead, the relevant distribution is the t distribution and we have[6]

$$\frac{\bar{x} - \mu}{\sqrt{s^2/n}} \sim t_{n-1} \qquad (4.19)$$

The random variable defined in equation (4.19) has a t distribution with $n - 1$ degrees of freedom. As the sample size increases, the t distribution approaches the standard Normal, so the latter can be used for large samples.

The t distribution was derived by W.S. Gossett in 1908 while conducting tests on the average strength of Guinness beer (who says statistics has no impact on the real world?). He published his work under the pseudonym 'Student', since the company did not allow its employees to publish under their own names, so the distribution is sometimes also known as the Student distribution.

The t distribution is in many ways similar to the standard Normal, insofar as it is:

- unimodal;
- symmetric;
- centred on zero;
- bell-shaped;
- extends from minus infinity to plus infinity.

---

[6] We also require the assumption that the parent population is Normally distributed for equation (4.19) to be true.

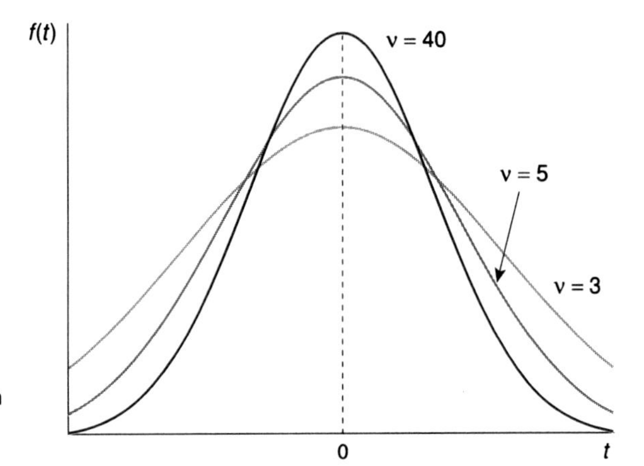

Figure 4.6
The t distribution drawn for different degrees of freedom

The differences are that it is more spread out (has a larger variance) than the standard Normal distribution, and has only one parameter rather than two: the **degrees of freedom**, denoted by the Greek letter ν (pronounced 'nu'[7]). In problems involving the estimation of a sample mean the degrees of freedom are given by the sample size minus one, i.e. $\nu = n - 1$.

The t distribution is drawn in Figure 4.6 for various values of the parameter ν. Note that the fewer the degrees of freedom (smaller sample size) the more dispersed is the distribution.

To summarise the argument so far, when

- the sample size is small, *and*
- the sample variance is used to estimate the population variance,

then the t distribution should be used for constructing confidence intervals, not the standard Normal. This results in a slightly wider interval than would be obtained using the standard Normal distribution, which reflects the slightly greater uncertainty involved when s^2 is used as an estimate of σ^2 if the sample size is small.

Apart from this, the methods are exactly as before and are illustrated by the examples below. We look first at estimating a single mean, then at estimating the difference of two means. The t distribution cannot be used for small sample proportions (explained below) so these cases are not considered.

Estimating a mean

The following would seem to be an appropriate example. A sample of 15 bottles of beer showed an average specific gravity of 1035.6, with standard deviation 2.7. Estimate the true specific gravity of the brew.

The sample information may be summarised as

$$\bar{x} = 1035.6$$
$$s = 2.7$$
$$n = 15$$

[7] Once again, the Greeks pronounce this differently, as 'ni'. They also pronounce π 'pee' rather than 'pie' as in English. This makes statistics lectures in English hard for Greeks to understand!

The sample mean is still an unbiased estimator of μ (this is true regardless of the distribution of the population) and serves as point estimate of μ. The point estimate of μ is therefore 1035.6.

Since σ is unknown, the sample size is small and it can be assumed that the specific gravity of all bottles of beer is Normally distributed (numerous small random factors affect the specific gravity) we should use the t distribution. Thus

$$\frac{\bar{x} - \mu}{\sqrt{s^2/n}} \sim t_{n-1} \tag{4.20}$$

The 95% confidence interval estimate is given by

$$\left[\bar{x} - t_{n-1}\sqrt{s^2/n},\ \bar{x} + t_{n-1}\sqrt{s^2/n}\right] \tag{4.21}$$

where t_{n-1} is the value of the t distribution which cuts off the extreme 5% (2.5% in each tail) of the t distribution with v degrees of freedom. Table A3 (see page **415**) gives percentage points of the t distribution and part of it is reproduced in Table 4.1.

The structure of the t distribution table is different from that of the standard Normal table. The first column of the table gives the degrees of freedom. In this example we want the row corresponding to $v = n - 1 = 14$. The appropriate column of the table is the one headed '0.025' which indicates the area cut off in *each* tail. At the intersection of this row and column we find the appropriate value, $t_{14} = 2.145$. Therefore the confidence interval is given by

$$[1035.6 - 2.145\sqrt{2.7^2/15},\ 1035.6 + 2.145\sqrt{2.7^2/15}]$$

which when evaluated gives

$$[1034.10,\ 1037.10]$$

We can be 95% confident that the true specific gravity lies within this range. If the Normal distribution had (incorrectly) been used for this problem then the t value of 2.145 would have been replaced by a z score of 1.96, giving a confidence interval of

$$[1034.23,\ 1036.97]$$

Table 4.1 Percentage points of the t distribution (excerpt from Table A3)

	Area (α) in each tail						
v	0.4	0.25	0.10	0.05	0.025	0.01	0.005
1	0.325	1.000	3.078	6.314	12.706	31.821	63.656
2	0.289	0.816	1.886	2.920	4.303	6.965	9.925
⋮	⋮	⋮	⋮	⋮	⋮	⋮	⋮
13	0.259	0.694	1.350	1.771	2.160	2.650	3.012
14	0.258	0.692	1.345	1.761	2.145	2.624	2.977
15	0.258	0.691	1.341	1.753	2.131	2.602	2.947

Note: The appropriate t value for constructing the confidence interval is found at the intersection of the shaded row and column.

This underestimates the true confidence interval and gives the impression of a more precise estimate than is actually the case. Use of the Normal distribution leads to a confidence interval which is 8.7% too narrow in this case.

Estimating the difference between two means

As in the case of a single mean the *t*-distribution needs to be used in small samples when the population variances are unknown. Again, both parent populations must be Normally distributed and in addition it must be assumed that the population variances are equal, i.e. $\sigma_1^2 = \sigma_2^2$ (this is required in the mathematical derivation of the *t* distribution). This latter assumption was not required in the large-sample case using the Normal distribution. Consider the following example as an illustration of the method.

A sample of 20 Labour-controlled local authorities shows that they spend an average of £175 per taxpayer on administration with a standard deviation of £25. A similar survey of 15 Conservative-controlled authorities finds an average figure of £158 with standard deviation of £30. Estimate the true difference in expenditure between Labour and Conservative authorities.

The sample information available is

$$\bar{x}_1 = 175 \quad \bar{x}_2 = 158$$
$$s_1 = 25 \quad s_2 = 30$$
$$n_1 = 20 \quad n_2 = 15$$

We wish to estimate $\mu_1 - \mu_2$. The point estimate of this is $\bar{x}_1 - \bar{x}_2$ which is an unbiased estimate. This gives 175 − 158 = 17 as the expected difference between the two sets of authorities.

For the confidence interval, the *t* distribution has to be used since the sample sizes are small and the population variances unknown. It is assumed that the populations are Normally distributed and that the samples have been independently drawn. We also assume that the population variances are equal, which seems justified since s_1 and s_2 do not differ by much (this kind of assumption is tested in Chapter 6). The confidence interval is given by the formula

$$\left[(\bar{x}_1 - \bar{x}_2) - t_v\sqrt{\frac{S^2}{n_1} + \frac{S^2}{n_2}} \leqslant \mu \leqslant (\bar{x}_1 - \bar{x}_2) + t_v\sqrt{\frac{S^2}{n_1} + \frac{S^2}{n_2}}\right] \tag{4.22}$$

where

$$S^2 = \frac{(n_1 - 1)s_1^2 + (n_2 - 1)s_2^2}{n_1 + n_2 - 2} \tag{4.23}$$

is known as the **pooled variance** and

$$v = n_1 + n_2 - 2$$

gives the degrees of freedom associated with the *t* distribution.

S^2 is an estimate of the (common value of) the population variances. It would be inappropriate to have the differing values s_1^2 and s_2^2 in the formula for this *t* distribution, for this would be contrary to the assumption that $\sigma_1^2 = \sigma_2^2$, which is essential for the use of the *t* distribution. The estimate of the common population variance is just the weighted average of the sample variances, using degrees of freedom as weights. Each sample has $n - 1$ degrees of freedom, and the total number of degrees of freedom for the problem is the sum of the degrees of freedom in each sample. The degrees of freedom is thus $20 + 15 - 2 = 33$ and hence the value $t = 2.042$ cuts off the extreme 5% of the distribution. The t table in the Appendix does not give the value for $v = 33$ so we have used $v = 30$ instead, which will give a close approximation.

To evaluate the 95% confidence interval we first calculate S^2

$$S^2 = \frac{(20-1) \times 25^2 + (15-1) \times 30^2}{20 + 15 - 2} = 741.6$$

Inserting this into equation (4.22) gives

$$\left[17 - 2.042\sqrt{\frac{741.6}{20} + \frac{741.6}{15}}, 17 + 2.042\sqrt{\frac{741.6}{20} + \frac{741.6}{15}}\right] = [-1.99, 35.99]$$

Thus the true difference is quite uncertain and the evidence is even consistent with Conservative authorities spending more than Labour authorities. The large degree of uncertainty arises because of the small sample sizes and the quite wide variation within each sample.

One should be careful about the conclusions drawn from this test. The greater expenditure on administration could be either because of inefficiency or because of a higher level of services provided. To find out which is the case would require further investigation. The statistical test carried out here examines the levels of expenditure, but not whether they are productive or not.

Estimating proportions

Estimating proportions when the sample size is small cannot be done with the t distribution. Recall that the distribution of the sample proportion p was derived from the distribution of r (the number of successes in n trials), which followed a Binomial distribution (see the Appendix to this chapter (page **170**)). In large samples the distribution of r is approximately Normal, thus giving a Normally distributed sample proportion. In small samples it is inappropriate to approximate the Binomial distribution with the t distribution, and indeed is unnecessary, since the Binomial itself can be used. Small-sample methods for the sample proportion should be based on the Binomial distribution, therefore, as set out in Chapter 3. These methods are thus not discussed further here.

Exercise 4.6

A sample of size $n = 16$ is drawn from a population which is known to be Normally distributed. The sample mean and variance are calculated as 74 and 121. Find the 99% confidence interval estimate for the true mean.

Exercise 4.7

Samples are drawn from two populations to see if they share a common mean. The sample data are:

$\bar{x}_1 = 45 \quad \bar{x}_2 = 55$
$s_1 = 18 \quad s_2 = 21$
$n_1 = 15 \quad n_2 = 20$

Find the 95% confidence interval estimate of the difference between the two population means.

Summary

- Estimation is the process of using sample information to make good estimates of the value of population parameters, for example using the sample mean to estimate the mean of a population.
- There are several criteria for finding a good estimate. Two important ones are the (lack of) bias and precision of the estimator. Sometimes there is a trade-off between these two criteria – one estimator might have a smaller bias but be less precise than another.
- An estimator is unbiased if it gives a correct estimate of the true value on average. Its expected value is equal to the true value.
- The precision of an estimator can be measured by its sampling variance (e.g. s^2/n for the mean of a sample).
- Estimates can be in the form of a single value (point estimate) or a range of values (confidence interval estimate). A confidence interval estimate gives some idea of how reliable the estimate is likely to be.
- For unbiased estimators, the value of the sample statistic (e.g. $\bar{x}$) is used as the point estimate.
- In large samples the 95% confidence interval is given by the point estimate plus or minus 1.96 standard errors (e.g. $\bar{x} \pm 1.96\sqrt{s^2/n}$ for the mean).
- For small samples the t distribution should be used instead of the Normal (i.e. replace 1.96 by the critical value of the t distribution) to construct confidence intervals of the mean.

Key terms and concepts

bias
confidence level and interval
efficiency
estimator
inference
interval estimate
maximum likelihood
point estimate
precision

Problems

Some of the more challenging problems are indicated by highlighting the problem number in colour.

4.1 (a) Why is an interval estimate better than a point estimate?

(b) What factors determine the width of a confidence interval?

4.2 Is the 95% confidence interval (a) twice as wide, (b) more than twice as wide, (c) less than twice as wide, as the 47.5% interval? Explain your reasoning.

4.3 Explain the difference between an estimate and an estimator. Is it true that a good estimator always leads to a good estimate?

4.4 Explain why an unbiased estimator is not always to be preferred to a biased one.

4.5 A random sample of two observations, x_1 and x_2, is drawn from a population. Prove that $w_1x_1 + w_2x_2$ gives an unbiased estimate of the population mean as long as $w_1 + w_2 = 1$.

Hint: Prove that $E(w_1x_1 + w_2x_2) = \mu$.

4.6 Following the previous question, prove that the most precise unbiased estimate is obtained by setting $w_1 = w_2 = \frac{1}{2}$

(*Hint*: Minimise $V(w_1x_1 + w_2x_2)$ with respect to w_1 after substituting $w_2 = 1 - w_1$. You will need a knowledge of calculus to solve this.)

4.7 Given the sample data

$$\bar{x} = 40 \quad s = 10 \quad n = 36$$

calculate the 99% confidence interval estimate of the true mean. If the sample size were 20, how would the method of calculation and width of the interval be altered?

4.8 A random sample of 100 record shops found that the average weekly sale of a particular CD was 260 copies, with standard deviation of 96. Find the 95% confidence interval to estimate the true average sale for all shops. To compile the CD chart it is necessary to know the correct average weekly sale to within 5% of its true value. How large a sample size is required?

4.9 Given the sample data $p = 0.4$, $n = 50$, calculate the 99% confidence interval estimate of the true proportion.

4.10 A political opinion poll questions 1000 people. Some 464 declare they will vote Conservative. Find the 95% confidence interval estimate for the Conservative share of the vote.

4.11 Given the sample data

$\bar{x}_1 = 25 \quad \bar{x}_2 = 22$
$s_1 = 12 \quad s_2 = 18$
$n_1 = 80 \quad n_2 = 100$

estimate the true difference between the means with 95% confidence.

4.12 (a) A sample of 200 women from the labour force found an average wage of £6000 p.a. with standard deviation £2500. A sample of 100 men found an average wage of £8000 with standard deviation £1500. Estimate the true difference in wages between men and women.

(b) A different survey, of men and women doing similar jobs, obtained the following results:

$\bar{x}_W = £7200 \quad \bar{x}_M = £7600$
$s_W = £1225 \quad s_M = £750$
$n_W = 75 \quad n_M = 50$

Estimate the difference between male and female wages using these new data. What can be concluded from the results of the two surveys?

4.13 67% out of 150 pupils from school A passed an exam; 62% of 120 pupils at school B passed. Estimate the 99% confidence interval for the true difference between the proportions passing the exam.

4.14 (a) A sample of 954 adults in early 1987 found that 23% of them held shares. Given a UK adult population of 41 million and assuming a proper random sample was taken, find the 95% confidence interval estimate for the number of shareholders in the UK.

(b) A 'similar' survey the previous year had found a total of 7 million shareholders. Assuming 'similar' means the same sample size, find the 95% confidence interval estimate of the increase in shareholders between the two years.

4.15 A sample of 16 observations from a Normally distributed population yields a sample mean of 30 with standard deviation 5. Find the 95% confidence interval estimate of the population mean.

4.16 A sample of 12 families in a town reveals an average income of £15 000 with standard deviation £6000. Why might you be hesitant about constructing a 95% confidence interval for the average income in the town?

4.17 Two samples were drawn, each from a Normally distributed population, with the following results

$\bar{x}_1 = 45 \quad s_1 = 8 \quad n_1 = 12$
$\bar{x}_2 = 52 \quad s_2 = 5 \quad n_2 = 18$

Estimate the difference between the population means, using the 95% confidence level.

4.18 The heights of 10 men and 15 women were recorded, with the following results:

	Mean	Variance
Men	173.5	80
Women	162	65

Estimate the true difference between men's and women's heights. Use the 95% confidence level.

4.19 **(Project)** Estimate the average weekly expenditure upon alcohol by students. Ask a (reasonably) random sample of your fellow students for their weekly expenditure on alcohol. From this, calculate the 95% confidence interval estimate of such spending by all students.

Answers to exercises

Exercise 4.1

(a) The point estimate is 25 and the 95% confidence interval is $25 \pm 1.96 \times 50/\sqrt{100} = 25 \pm 9.8 = [15.2, 34.8]$.

(b) The CI becomes larger as the sample size reduces. In this case we would have $25 \pm 1.96 \times 50/\sqrt{64} = 25 \pm 12.25 = [12.75, 37.25]$. Note that the width of the CI is inversely proportional to the square root of the sample size.

Exercise 4.2

The 95% CI is $50 \pm 1.96 \times 30/\sqrt{40} = 50 \pm 9.30 = [40.70, 59.30]$. The value of 60 lies (just) outside this CI so is unlikely to be the true mean.

Exercise 4.3

(a) The point estimate is 14% (7/50). The 95% CI is given by

$$0.14 \pm 1.96 \times \sqrt{\frac{0.14 \times (1 - 0.14)}{50}} = 0.14 \pm 0.096.$$

(b) Use 1.64 instead of 1.96, giving 0.14 ± 0.080.

(c) 0.14 ± 0.126.

Exercise 4.4

$\bar{x}_1 - \bar{x}_2 = 25 - 30 = -5$ is the point estimate. The interval estimate is given by

$$(\bar{x}_1 - \bar{x}_2) \pm 1.96\sqrt{\frac{s_1^2}{n_1} + \frac{s_2^2}{n_2}} = -5 \pm 1.96\sqrt{\frac{18^2}{36} + \frac{25^2}{49}}$$

$$= -5 \pm 9.14 = [-14.14, 4.14]$$

Exercise 4.5

The point estimate is 20%. The interval estimate is

$$0.2 \pm 1.96 \times \sqrt{\frac{0.4 \times 0.6}{50} + \frac{0.2 \times 0.8}{100}} = 0.2 \pm 0.157 = [0.043, 0.357]$$

Exercise 4.6

The 99% CI is given by $74 \pm t^* \times \sqrt{121/16} = 74 \pm 2.947 \times 2.75 = 74 \pm 8.10 = [65.90, 82.10]$.

Exercise 4.7

The pooled variance is given by

$$S^2 = \frac{(15 - 1) \times 18^2 + (20 - 1) \times 21^2}{15 + 20 - 2} = 391.36$$

The 95% CI is therefore

$$(45 - 55) \pm 2.042 \times \sqrt{\frac{391.36}{15} + \frac{391.36}{20}}$$

$$= -10 \pm 13.80 = [-3.8, 23.8]$$

Appendix Derivations of sampling distributions

Derivation of the sampling distribution of p

The sampling distribution of p is fairly straightforward to derive, given what we have already learned. The sampling distribution of p can be easily derived from the distribution of r, the number of successes in n trials of an experiment, since $p = r/n$. The distribution of r for large n is approximately Normal (from Chapter 3)

$$r \sim N(nP, nP(1 - P)) \tag{4.24}$$

Knowing the distribution of r, is it possible to find that of p? Since p is simply r multiplied by a constant, $1/n$, it is also Normally distributed. The mean and variance of the distribution can be derived using the E and V operators. The expected value of p is

$$\mathrm{E}(p) = \mathrm{E}(r/n) = \frac{1}{n}\mathrm{E}(r) = \frac{1}{n}nP = P = \pi \tag{4.25}$$

The expected value of the sample proportion is equal to the population proportion (note that the probability P and the population proportion π are the same thing and may be used interchangeably). The sample proportion therefore gives an unbiased estimate of the population proportion.

For the variance

$$\mathrm{V}(p) = \mathrm{V}\left(\frac{r}{n}\right) = \frac{1}{n^2}\mathrm{V}(r) = \frac{1}{n^2}nP(1 - P) = \frac{\pi(1 - \pi)}{n} \tag{4.26}$$

Hence the distribution of p is given by

$$p \sim \mathrm{N}\left(\pi, \frac{\pi(1 - \pi)}{n}\right) \tag{4.27}$$

Derivation of the sampling distribution of $\bar{x}_1 - \bar{x}_2$

This is the difference between two random variables so is itself a random variable. Since any linear combination of Normally distributed, independent random variables is itself Normally distributed, the difference of sample means follows a Normal distribution. The mean and variance of the distribution can be found using the E and V operators. Letting

$$\mathrm{E}(\bar{x}_1) = \mu_1,\ \mathrm{V}(\bar{x}_1) = \sigma_1^2/n_1 \quad \text{and}$$
$$\mathrm{E}(\bar{x}_2) = \mu_2,\ \mathrm{V}(\bar{x}_2) = \sigma_2^2/n_2$$

then

$$\mathrm{E}(\bar{x}_1 - \bar{x}_2) = \mathrm{E}(\bar{x}_1) - \mathrm{E}(\bar{x}_2) = \mu_1 - \mu_2 \tag{4.28}$$

And

$$\mathrm{V}(\bar{x}_1 - \bar{x}_2) = \mathrm{V}(\bar{x}_1) + \mathrm{V}(\bar{x}_2) = \frac{\sigma_1^2}{n_1} + \frac{\sigma_2^2}{n_2} \tag{4.29}$$

Equation (4.29) assumes $\bar{x}_1$ and $\bar{x}_2$ are independent random variables. The probability distribution of $\bar{x}_1 - \bar{x}_2$ can therefore be summarised as

$$\bar{x}_1 - \bar{x}_2 \sim N\left(\mu_1 - \mu_2, \frac{\sigma_1^2}{n_1} + \frac{\sigma_2^2}{n_2}\right) \tag{4.30}$$

This is equation (4.12) in the text.

5 Hypothesis testing

Contents

Learning outcomes

By the end of this chapter you should be able to:

- understand the philosophy and scientific principles underlying hypothesis testing;
- appreciate that hypothesis testing is about deciding whether a hypothesis is true or false on the basis of a sample of data;
- recognise the type of evidence which leads to a decision that the hypothesis is false;
- carry out hypothesis tests for a variety of statistical problems;
- recognise the relationship between hypothesis testing and a confidence interval;
- recognise the shortcomings of hypothesis testing.

Complete your diagnostic test for Chapter 5 now to create your personal study plan. Exercises with an icon ? are also available for practice in MathXL with additional supporting resources.

Introduction

This chapter deals with issues very similar to those of the previous chapter on estimation, but examines them in a different way. The estimation of population parameters and the testing of hypotheses about those parameters are similar techniques (indeed they are formally equivalent in a number of respects), but there are important differences in the interpretation of the results arising from each method. The process of estimation is appropriate when measurement is involved, such as measuring the true average expenditure on food; hypothesis testing is better when decision making is involved, such as whether to accept that a supplier's products are up to a specified standard. Hypothesis testing is also used to make decisions about the truth or otherwise of different theories, such as whether rising prices are caused by rising wages; and it is here that the issues become contentious. It is sometimes difficult to interpret correctly the results of hypothesis tests in these circumstances. This is discussed further later in this chapter.

The concepts of hypothesis testing

In many ways hypothesis testing is analogous to a criminal trial. In a trial there is a defendant who is *initially presumed innocent*. The *evidence* against the defendant is then presented and, if the jury finds this convincing *beyond all reasonable doubt*, he is found guilty; the presumption of innocence is overturned. Of course, mistakes are sometimes made: an innocent person is convicted or a guilty person set free. Both of these errors involve costs (not only in the monetary sense), either to the defendant or to society in general, and the errors should be avoided if at all possible. The laws under which the trial is held may help avoid such errors. The rule that the jury must be convinced 'beyond all reasonable doubt' helps to avoid convicting the innocent, for instance.

The situation in hypothesis testing is similar. First there is a **maintained** or **null hypothesis** which is initially *presumed* to be true. The empirical evidence, usually data from a random sample, is then gathered and assessed. If the evidence seems inconsistent with the null hypothesis, i.e. it has a low probability of occurring *if* the hypothesis were true, then the null hypothesis is *rejected* in favour of an alternative. Once again there are two types of error one can make, either rejecting the null hypothesis when it is really true, or not rejecting it when in fact it is false. Ideally one would like to avoid both types of error.

An example helps to clarify the issues and the analogy. Suppose that you are thinking of taking over a small business franchise. The current owner claims the weekly turnover of each existing franchise is £5000 and at this level you are willing to take on a franchise. You would be more cautious if the turnover is less than this figure. You examine the books of 26 franchises chosen at random and find that the average turnover was £4900 with standard deviation £280. What do you do?

The null hypothesis in this case is that average weekly turnover is £5000 (or more; that would be even more to your advantage). The **alternative hypothesis** is that turnover is strictly less than £5000 per week. We may write these more succinctly as follows

H_0: $\mu = 5000$
H_1: $\mu < 5000$

H_0 is conventionally used to denote the null hypothesis, H_1 the alternative. Initially, H_0 is presumed to be true and this presumption will be tested using the sample evidence. Note that the sample evidence is *not* used as part of the hypothesis.

You have to decide whether the owner's claim is correct (H_0) or not (H_1). The two types of error you could make are as follows:

- **Type I error** – reject H_0 when it is in fact true. This would mean missing a good business opportunity.
- **Type II error** – not rejecting H_0 when it is in fact false. You would go ahead and buy the business and then find out that it is not as attractive as claimed. You would have overpaid for the business.

The situation is set out in Figure 5.1.

Obviously a good decision rule would give a good chance of making a correct decision and rule out errors as far as possible. Unfortunately it is impossible completely to eliminate the possibility of errors. As the decision rule is changed to reduce the probability of a Type I error, the probability of making a Type II error inevitably increases. The skill comes in balancing these two types of error.

Again a diagram is useful in illustrating this. Assuming that the null hypothesis is true, then the sample observations are drawn from a population with mean 5000 and some variance, which we shall assume is accurately measured by the sample variance. The distribution of $\bar{x}$ is then given by

$$\bar{x} \sim N(\mu, \sigma^2/n) \text{ or} \quad (5.1)$$
$$\bar{x} \sim N(5000, 280^2/26)$$

Under the alternative hypothesis the distribution of $\bar{x}$ would be the same except that it would be centred on a value less than 5000. These two situations are illustrated in Figure 5.2. The distribution of $\bar{x}$ under H_1 is shown by a dashed curve to signify that its exact position is unknown, only that it lies to the left of the distribution under H_0.

A **decision rule** amounts to choosing a point or dividing line on the horizontal axis in Figure 5.2. If the sample mean lies to the left of this point then H_0 is rejected (the sample mean is too far away from H_0 for it to be credible) in favour of H_1 and you do not buy the firm. If $\bar{x}$ lies above this decision point then H_0 is not rejected and you go ahead with the purchase. Such a decision point is

		True situation	
		H_0 true	H_0 false
Decision	Accept H_0	Correct decision	Type II error
	Reject H_0	Type I error	Correct decision

Figure 5.1
The two different types of error

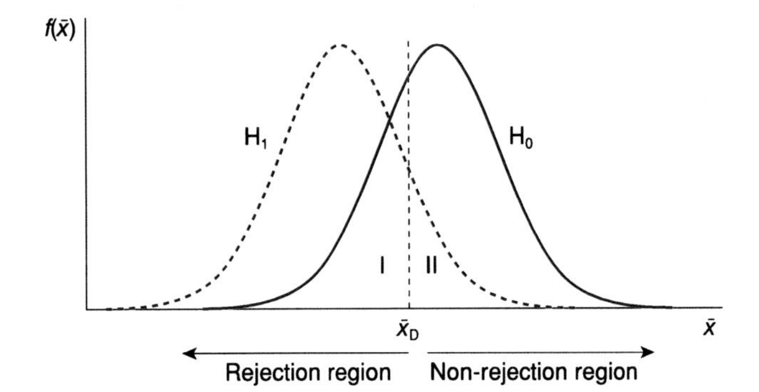

Figure 5.2 The sampling distributions of $\bar{x}$ under H_0 and H_1

shown in Figure 5.2, denoted by $\bar{x}_D$. To the left of $\bar{x}_D$ lies the **rejection** (of H_0) **region**; to the right lies the **non-rejection region**.

Based on this point, we can see the probabilities of Type I and Type II errors. The area under the H_0 distribution to the left of $\bar{x}_D$, labelled I, shows the probability of rejecting H_0 given that it is in fact true: a Type I error. The area under the H_1 distribution to the right of $\bar{x}_D$, labelled II, shows the probability of a Type II error: not rejecting H_0 when it is in fact false (and H_1 is true).

Shifting the decision line to the right or left alters the balance of these probabilities. Moving the line to the right increases the probability of a Type I error but reduces the probability of a Type II error. Moving the line to the left has the opposite effect.

The Type I error probability can be calculated for any value of $\bar{x}_D$. Suppose we set $\bar{x}_D$ to a value of 4950. Using the distribution of $\bar{x}$ given in equation (5.1) above, the area under the distribution to the left of 4950 is obtained using the z score

$$z = \frac{\bar{x}_D - \mu}{\sqrt{s^2/n}} = \frac{4950 - 5000}{\sqrt{280^2/26}} = -0.91 \tag{5.2}$$

From the tables of the standard Normal distribution we find that the probability of a Type I error is 18.1%. Unfortunately, the Type II error probability cannot be established because the exact position of the distribution under H_1 is unknown. Therefore we cannot decide on the appropriate position of $\bar{x}_D$ by some balance of the two error probabilities.

The convention therefore is to set the position of $\bar{x}_D$ by using a Type I error probability of 5%, known as the **significance level**[1] of the test. In other words, we are prepared to accept a 5% probability of rejecting H_0 when it is, in fact, true. This allows us to establish the position of $\bar{x}_D$. From Table A2 (see page **414**) we find that $z = -1.64$ cuts off the bottom 5% of the distribution, so the decision line should be 1.64 standard errors below 5000. The value −1.64 is known as the **critical value** of the test. We therefore obtain

$$\bar{x}_D = 5000 - 1.64\sqrt{280^2/26} = 4910 \tag{5.3}$$

[1] The term **size** of the test is also used, not to be confused with the sample size. We use the term 'significance level' in this text.

Since the sample mean of 4900 lies below 4910 we reject H_0 *at the 5% significance level* or equivalently we reject *with 95% confidence*. The significance level is generally denoted by the symbol α and the complement of this, given by $1 - \alpha$, is known as the confidence level (as used in the confidence interval).

An equivalent procedure would be to calculate the z score associated with the sample mean, known as the **test statistic**, and then compare this to the critical value of the test. This allows the hypothesis testing procedure to be broken down into five neat steps.

(1) Write down the null and alternative hypotheses:

$$H_0: \mu = 5000$$
$$H_1: \mu < 5000$$

(2) Choose the significance level of the test, conventionally $\alpha = 0.05$ or 5%.

(3) Look up the critical value of the test from statistical tables, based on the chosen significance level. $z^* = 1.64$ is the critical value in this case.

(4) Calculate the test statistic

$$z = \frac{\bar{x} - \mu}{\sqrt{s^2/n}} = \frac{-100}{\sqrt{280^2/26}} = -1.82 \qquad (5.4)$$

(5) Decision rule. Compare the test statistic with the critical value: if $z < -z^*$ reject H_0 in favour of H_1. Since $-1.82 < -1.64$ H_0 is rejected with 95% confidence. Note that we use $-z^*$ here (rather than $+z^*$) because we are dealing with the left-hand tail of the distribution.

Worked example 5.1

A sample of 100 workers found the average overtime hours worked in the previous week was 7.8, with standard deviation 4.1 hours. Test the hypothesis that the average for all workers is 5 hours or less.

We can set out the five steps of the answer as follows:

(1) $H_0: \mu = 5$
$H_1: \mu > 5$

(2) Significance level, $\alpha = 5\%$.

(3) Critical value $z^* = 1.64$.

(4) Test statistic

$$z = \frac{\bar{x} - \mu}{\sqrt{s^2/n}} = \frac{7.8 - 5}{\sqrt{4.1^2/100}} = 6.8$$

(5) Decision rule: $6.8 > 1.64$ so we reject H_0 in favour of H_1. Note that in this case we are dealing with the right-hand tail of the distribution (positive values of z and z^*). Only high values of $\bar{x}$ reject H_0.

One-tail and two-tail tests

In the above example the rejection region for the test consisted of one tail of the distribution of $\bar{x}$, since the buyer was only concerned about turnover being less

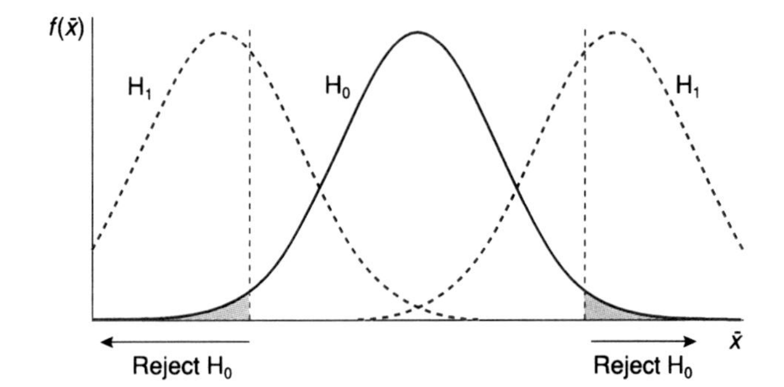

Figure 5.3
A two-tail hypothesis test

than claimed. For this reason it is known as a **one-tail test**. Suppose now that an accountant is engaged to sell the franchise and wants to check the claim about turnover before advertising the business for sale. In this case she would be concerned about turnover being either below *or* above 5000.

This would now become a **two-tail test** with the null and alternative hypotheses being

$$H_0: \mu = 5000$$
$$H_1: \mu \neq 5000$$

Now there are two rejection regions for the test. Either a very low sample mean *or* a very high one will serve to reject the null hypothesis. The situation is presented graphically in Figure 5.3.

The distribution of $\bar{x}$ under H_0 is the same as before, but under the alternative hypothesis the distribution could be shifted either to the left or to the right, as depicted. If the significance level is still chosen to be 5%, then the complete rejection region consist of the *two* extremes of the distribution under H_0, containing 2.5% in each tail (hence 5% in total). This gives a Type I error probability of 5% as before.

The critical value of the test therefore becomes $z^* = 1.96$, the value which cuts off 2.5% in each tail of the standard Normal distribution. Only if the test statistic falls into one of the rejection regions beyond 1.96 standard errors from the mean is H_0 rejected.

Using data from the previous example, the test statistic remains $z = -1.82$ so that the null hypothesis cannot be rejected in this case, as -1.82 does not fall beyond -1.96. To recap, the five steps of the test are:

(1) $H_0: \mu = 5000$
$H_1: \mu \neq 5000$
(2) Choose the significance level: $\alpha = 0.05$.
(3) Look up the critical value: $z^* = 1.96$.
(4) Evaluate the test statistic

$$z = \frac{-100}{\sqrt{280^2/26}} = -1.82$$

(5) Compare test statistic and critical values: if $z < -z^*$ or $z > z^*$ reject H_0 in favour of H_1. In this case $-1.82 > -1.96$ so H_0 cannot be rejected with 95% confidence.

One- and two-tail tests therefore differ only at steps 1 and 3. Note that we have come to different conclusions according to whether a one- or two-tail test was used, with the same sample evidence. There is nothing wrong with this, however, for there are different interpretations of the two results. If the investor always uses his rule, he will miss out on 5% of good investment opportunities, when sales are (by chance) low. He will never miss out on a good opportunity because the investment appears too good (i.e. sales by chance are very high). For the accountant, 5% of the firms with sales averaging £5000 will not be advertised as such, *either* because sales appear too low *or* because they appear too high.

It is tempting on occasion to use a one-tail test because of the sample evidence. For example, the accountant might look at the sample evidence above and decide that the franchise operation can only have true sales less than or equal to 5000. Therefore a one-tail test is used. This is a dangerous practice, since the sample evidence is being used to help formulate the hypothesis, which is then tested on that same evidence. This is going round in circles; the hypothesis should be chosen *independently* of the evidence, which is then used to test it. Presumably the accountant would also use a one-tail test (with H_1: $\mu > 5000$ as the alternative hypothesis) if it was noticed that the sample mean were *above* the hypothesised value. In effect therefore the 10% significance level would be used, not the 5% level, since there would be 5% in each tail of the distribution. A Type I error would be made on 10% of all occasions rather than 5%.

It is acceptable to use a one-tail test when you have *independent* information about what the alternative hypothesis should be, or you are not concerned about one side of the distribution (such as the investor) and can effectively add that into the null hypothesis. Otherwise, it is safer to use a two-tail test.

Exercise 5.1

(a) Two political parties are debating crime figures. One party says that crime has increased compared to the previous year. The other party says it has not. Write down the null and alternative hypotheses.

(b) Explain the two types of error that could be made in this example and the possible costs of each type of error.

Exercise 5.2

(a) We test the hypothesis H_0: $\mu = 100$ against H_1: $\mu > 100$ by rejecting H_0 if our sample mean is greater than 108. If in fact $\bar{x} \sim N(100, 900/25)$, what is the probability of making a Type I error?

(b) If we wanted a 5% Type I error probability, what decision rule should we adopt?

(c) If we knew that μ could only take on the values 100 (under H_0) or 112 (under H_1) what would be the Type II error probability using the decision rule in part (a)?

Exercise 5.3

Test the hypothesis H_0: $\mu = 500$ versus H_1: $\mu \neq 500$ using the evidence $\bar{x} = 530$, $s = 90$ from a sample of size $n = 30$.

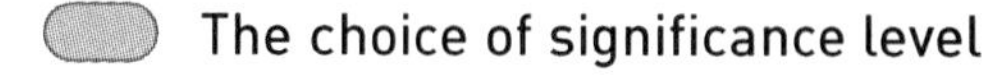

The choice of significance level

We justified the choice of the 5% significance level by reference to convention. This is usually a poor argument for anything, but it does have some justification. In an ideal world we would have precisely specified null *and* alternative hypotheses (e.g. we would test H_0: $\mu = 5000$ against H_1: $\mu = 4500$, these being the only possibilities). Then we could calculate the probabilities of both Type I *and* Type II errors, for any given decision rule. We could then choose the optimal decision rule, which gives the best compromise between the two types of error. This is reflected in a court of law. In criminal cases, the jury must be convinced of the prosecution's case beyond reasonable doubt, because of the cost of committing a Type I error. In a civil case (libel, for example) the jury need only be convinced *on the balance of probabilities*. In a civil case, the costs of Type I and Type II error are more evenly balanced and so the burden of proof is lessened.

However, in practice we usually do not have the luxury of two well-specified hypotheses. As in the example, the null hypothesis is precisely specified (it has to be or the test could not be carried out) but the alternative hypothesis is imprecise (sometimes called a **composite hypothesis** because it encompasses a range of values). Statistical inference is often used not so much as an aid to decision making but to provide evidence for or against a particular theory, to alter one's degree of belief in the truth of the theory. For example, an economic theory might assert that rising prices are caused by rising wages (the cost–push theory of inflation). The null and alternative hypotheses would be:

H_0: there is no connection between rising wages and rising prices;
H_1: there is some connection between rising wages and rising prices.

(Note that the null has 'no connection', since this is a precise statement. 'Some connection' is too vague to be the null hypothesis.) Data could be gathered to test this hypothesis (the appropriate methods will be discussed in the chapters on correlation and regression). But what decision rests upon the result of this test? It could be thought that government might make a decision to impose a prices and incomes policy, but if every academic study of inflation led to the imposition or abandonment of a prices and incomes policy there would have been an awful lot of policies! (In fact, there *were* a lot of such policies, but not as many as the number of studies of inflation.) No single study is decisive ('more research is needed' is a very common phrase) but each does influence the climate of opinion which may eventually lead to a policy decision. But if a hypothesis test is designed to influence opinion, how is the significance level to be chosen?

It is difficult to trade off the costs of Type I and Type II errors and the probability of making those errors. A Type I error in this case means concluding that rising wages do cause rising prices when, in fact, they do not. So what would be the cost of this error, i.e. imposing a prices and incomes policy when, in fact, it is not needed? It is extremely difficult, if not impossible, to put a figure on it. It would depend on what type of prices and incomes policy were imposed – would wages be frozen or allowed to rise with productivity, how fast would prices be allowed to rise, would company dividends be frozen? The costs of the Type II error would also be problematic (not imposing a needed prices and incomes policy), for they would depend, among other things, on what alternative policies might be adopted.

The 5% significance level really does depend upon convention therefore, it cannot be justified by reference to the relative costs of Type I and Type II errors (it is too much to believe that everyone does consider these costs and independently arrives at the conclusion that 5% is the appropriate significance level!). However, the 5% convention does impose some sort of discipline upon research;

it sets some kind of standard which all theories (hypotheses) should be measured against. Beware the researcher who reports that a particular hypothesis is rejected at the 8% significance level; it is likely that the significance level was chosen so that the hypothesis could be rejected, which is what the researcher was hoping for in the first place!

The Prob-value approach

Suppose a result is significant at the 4.95% level (i.e. it just meets the 5% convention and the null hypothesis is rejected). A *very* slight change in the sample data could have meant the result being significant at only the 5.05% level, and the null hypothesis not being rejected. Would we really be happy to alter our belief completely on such fragile results? Most researchers (but not all!) would be cautious if their results were only just significant (or fell just short of significance).

This suggests an alternative approach: the significance level of the test statistic could be reported and the reader could make his own judgements about it. This is known as the **Prob-value** approach, the Prob-value being the significance level of the calculated test statistic. For example, the calculated test statistic for the investor problem was $z = -1.82$ and the associated Prob-value is obtained from Table A2 (see page **414**) as 3.44%, i.e. −1.82 cuts off 3.44% in one tail of the standard Normal distribution. This means that the null hypothesis can be rejected at the 3.44% significance level or, alternatively expressed, with 96.56% confidence.

Notice that Table A2 gives the Prob-value for a one-tail test; for a two-tail test the Prob-value should be doubled. Thus for the accountant, using the two-tail test, the significance level is 6.88% and this is the level at which the null hypothesis can be rejected. Alternatively we could say we reject the null with 93.12% confidence. This does not meet the standard 5% criterion (for the significance level) which is most often used, so would result in non-rejection of the null.

An advantage of using the Prob-value approach is that many statistical software programs routinely provide the Prob-value of a calculated test statistic. If one understands the use of Prob-values then one does not have to look up tables (this applies to any distribution, not just the Normal), which can save a lot of time.

To summarise, one rejects the null hypothesis if either:

- (Method 1) – the test statistic **is greater than** the critical value, i.e. $z > z^*$, or
- (Method 2) – the Prob-value associated with the test statistic **is less than** the significance level, i.e. $P < 0.05$ (if the 5% significance level is used).

I have found that many students initially find this confusing, because of the opposing inequality in the two versions (greater than and less than). For example, a program might calculate a hypothesis test and report the result as '$z = 1.4$ (P value $= 0.162$)'. The first point to note is that most software programs report the Prob-value for a two-tail test by default. Hence, assuming a 5% significance level, in this case we cannot reject H_0 because $z = 1.4 < 1.96$ or equivalently because $0.162 > 0.05$, against a two-tailed alternative (i.e. H_1 contains $\neq$).

If you wish to conduct a one-tailed test you have to halve the reported Prob-value, becoming 0.081 in this example. This is again greater than 5%, so the hypothesis is still accepted, even against a one-sided alternative (H_1 contains > or <). Equivalently, one could compare 1.4 with the one-tail critical value, 1.64, showing non-rejection of the null, but one has to look up the standard Normal table with this method. Computers cannot guess whether a one- or two-sided test is wanted, so take the conservative option and report the two-sided value. The correction for a one-sided test has to be done manually.

Significance, effect size and power

Researchers usually look for 'significant' results. Academic papers report that 'the results are significant' or that 'the coefficient is significantly different from zero at the 5% significance level'. It is vital to realise that the word 'significant' is used here in the *statistical* sense and not in its everyday sense of being *important*. Something can be statistically significant yet still unimportant.

Suppose that we have some more data about the business examined earlier. Data for 100 franchises have been uncovered, revealing an average weekly turnover of £4975 with standard deviation £143. Can we reject the hypothesis that the average weekly turnover is £5000? The test statistic is

$$z = \frac{4975 - 5000}{\sqrt{143^2/100}} = -1.75$$

Since this is less than $-z^* = -1.64$ the null is rejected with 95% confidence. True average weekly turnover is less than £5000. However, the difference is only £25 per week, which is 0.5% of £5000. Common sense would suggest that the difference may be unimportant, even if it is significant in the statistical sense. One should not interpret statistical results in terms of significance alone, therefore; one should also look at the size of the difference (sometimes known as the **effect size**) and ask whether it is important or not. This is a mistake made by even experienced researchers; a review of articles in the prestigious *American Economic Review* reported that 82% of them confused statistical significance for economic significance in some way (McCloskey and Ziliak, 2004).

This problem with hypothesis testing paradoxically grows worse as the sample size increases. For example, if 250 observations reveal average sales of 4985 with standard deviation 143, the null would (just) be rejected at 5% significance. In fact, given a large enough sample size we can virtually guarantee to reject the null hypothesis even before we have gathered the data. This can be seen from equation (5.4) for the z score test statistic: as n grows larger, the test statistic also inevitably increases.

A good way to remember this point is to appreciate that it is the *evidence* which is significant, not the size of the effect. Strictly, it is better to say '. . . there is significant evidence of difference between . . .' than '. . . there is a significant difference between . . .'.

A related way of considering the effect of increasing sample size is via the concept of the **power** of a test. This is defined as

$$\textbf{Power} \text{ of a test} = 1 - \Pr(\text{Type II error}) = 1 - \beta \qquad (5.5)$$

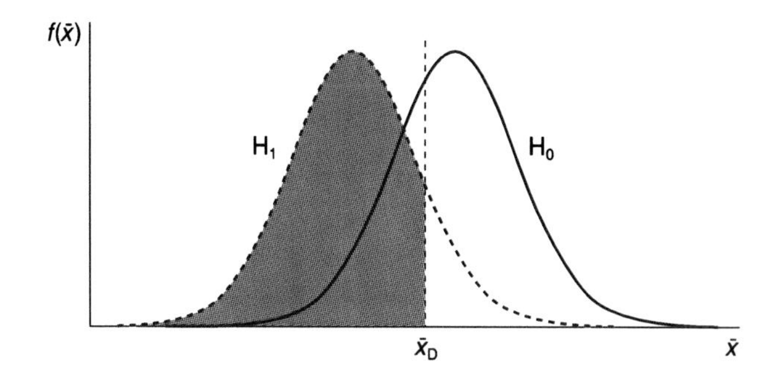

Figure 5.4
The power of a test

where β is the symbol conventionally used to indicate the probability of a Type II error. As a Type II error is defined as not rejecting H_0 when false (equivalent to rejecting H_1 when true), power is the probability of rejecting H_0 when false (if H_0 is false, it must be *either* accepted *or* rejected; hence these probabilities sum to one). This is one of the correct decisions identified earlier, associated with the lower right-hand box in Figure 5.1, that of correctly rejecting a false null hypothesis. The power of a test is therefore given by the area under the H_1 distribution, to the left of the decision line, as illustrated (shaded) in Figure 5.4 (for a one-tail test).

It is generally desirable to maximise the power of a test, as long as the probability of a Type I error is not raised in the process. There are essentially three ways of doing this.

- Avoid situations where the null and alternative hypotheses are very similar, i.e. the hypothesised means are not far apart (a small effect size).
- Use a large sample size. This reduces the sampling variance of $\bar{x}$ (under both H_0 and H_1) so the two distributions become more distinct.
- Use good sampling methods which have small sampling variances. This has a similar effect to increasing the sample size.

Unfortunately, in economics and business the data are very often given in advance and there is little or no control possible over the sampling procedures. This leads to a neglect of consideration of power, unlike in psychology, for example, where the experiment can often be designed by the researcher. The gathering of sample data will be covered in detail in Chapter 9.

Exercise 5.4

If a researcher believes the cost of making a Type I error is much greater than the cost of a Type II error, should they choose a 5% or 1% significance level? Explain why.

Exercise 5.5

(a) A researcher uses *Excel* to analyse data and test a hypothesis. The program reports a test statistic of $z = 1.77$ (P value $= 0.077$). Would you reject the null hypothesis if carrying out (i) a one-tailed test (ii) a two-tailed test? Use the 5% significance level.

(b) Repeat part (a) using a 1% significance level.

Further hypothesis tests

We now proceed to consider a number of different types of hypothesis test, all involving the same principles but differing in details of their implementation. This is similar to the exposition in the last chapter covering, in turn, tests of a proportion, tests of the difference of two means and proportions, and finally problems involving small sample sizes.

Testing a proportion

A car manufacturer claims that no more than 10% of its cars should need repairs in the first three years of their life, the warranty period. A random sample of 50 three-year-old cars found that 8 had required attention. Does this contradict the maker's claim?

This problem can be handled in a very similar way to the methods used for a mean. The key, once again, is to recognise the sample proportion as a random variable with an associated probability distribution. From Chapter 4 (equation (4.9)), the sampling distribution of the sample proportion in large samples is given by

$$p \sim N\left(\pi, \frac{\pi(1-\pi)}{n}\right) \tag{5.6}$$

In this case $\pi = 0.10$ (under the null hypothesis, the maker's claim). The sample data are

$$p = 8/50 = 0.16$$
$$n = 50$$

Thus 16% of the sample required attention within the warranty period. This is substantially higher than the claimed 10%, but is this just because of a bad sample or does it reflect the reality that the cars are badly built? The hypothesis test is set out along the same lines as for a sample mean.

(1) H_0: $\pi = 0.10$
    H_1: $\pi > 0.10$

    (The only concern is the manufacturer not matching its claim.)
(2) Significance level: $\alpha = 0.05$.
(3) The critical value of the one-tail test at the 5% significance level is $z^* = 1.64$, obtained from the standard Normal table.
(4) The test statistic is

$$z = \frac{p - \pi}{\sqrt{\frac{\pi(1-\pi)}{n}}} = \frac{0.16 - 0.10}{\sqrt{\frac{0.1 \times 0.9}{50}}} = 1.41$$

(5) Since the test statistic is less than the critical value, it falls into the non-rejection region. The null hypothesis is not rejected by the data. The manufacturer's claim is not unreasonable.

Note that for this problem, the rejection region lies in the *upper* tail of the distribution because of the 'greater than' inequality in the alternative hypothesis. The null hypothesis is therefore rejected in this case if $z > z^*$.

Do children prefer branded goods only because of the name?

Researchers at Johns Hopkins Bloomberg School of Public Health in Maryland found young children were influenced by the packaging of foods. 63 children were offered two identical meals, save that one was still in its original packaging (from MacDonalds). 76% of the children preferred the branded French fries.

Is this evidence significant? The null hypothesis is H_0: $\pi = 0.5$ versus H_1: $\pi > 0.5$. The test statistic for this hypothesis test is

$$z = \frac{p - \pi}{\sqrt{\frac{\pi(1-\pi)}{n}}} = \frac{0.76 - 0.50}{\sqrt{\frac{0.5 \times 0.5}{63}}} = 4.12$$

which is greater than the critical value of $z^* = 1.64$. Hence we conclude that children are influenced by the packaging or brand name.

(Source: *New Scientist*, 11 August 2007.)

Testing the difference of two means

Suppose a car company wishes to compare the performance of its two factories producing an identical model of car. The factories are equipped with the same machinery but their outputs might differ due to managerial ability, labour relations, etc. Senior management wishes to know if there is any difference between the two factories. Output is monitored for 30 days, chosen at random, with the following results:

	Factory 1	Factory 2
Average daily output	420	408
Standard deviation of daily output	25	20

Does this produce sufficient evidence of a real difference between the factories, or does the difference between the samples simply reflect random differences such as minor breakdowns of machinery? The information at our disposal may be summarised as

$$\bar{x}_1 = 420 \qquad \bar{x}_2 = 408$$
$$s_1 = 25 \qquad s_2 = 20$$
$$n_1 = 30 \qquad n_2 = 30$$

The hypothesis test to be conducted concerns the difference between the factories' outputs, so the appropriate random variable to examine is $\bar{x}_1 - \bar{x}_2$. From Chapter 4 (equation (4.12)), this has the following distribution, in large samples

$$\bar{x}_1 - \bar{x}_2 \sim N\left(\mu_1 - \mu_2, \frac{\sigma_1^2}{n_1} + \frac{\sigma_2^2}{n_2}\right) \tag{5.7}$$

The population variances, σ_1^2 and σ_2^2, may be replaced by their sample estimates, s_1^2 and s_1^2, if the former are unknown, as here. The hypothesis test is therefore as follows.

(1) H_0: $\mu_1 - \mu_2 = 0$
H_1: $\mu_1 - \mu_2 \neq 0$

The null hypothesis posits no real difference between the factories. This is a two-tail test since there is no a priori reason to believe one factory is better than the other, apart from the sample evidence.

(2) Significance level: $\alpha = 1\%$. This is chosen since the management does not want to interfere unless it is really confident of some difference between the factories. In order to favour the null hypothesis, a lower significance level than the conventional 5% is set.

(3) The critical value of the test is $z^* = 2.57$. This cuts off 0.5% in each tail of the standard Normal distribution.

(4) The test statistic is

$$z = \frac{(\bar{x}_1 - \bar{x}_2) - (\mu_1 - \mu_2)}{\sqrt{\frac{s_1^2}{n_1} + \frac{s_2^2}{n_2}}} = \frac{(420 - 408) - 0}{\sqrt{\frac{25^2}{30} + \frac{20^2}{30}}} = 2.05$$

Note that this is of the same form as in the single-sample cases. The hypothesised value of the difference (zero in this case) is subtracted from the sample difference and this is divided by the standard error of the random variable.

(5) Decision rule: $z < z^*$ so the test statistic falls into the non-rejection region. There does not appear to be a significant difference between the two factories.

A number of remarks about this example should be made. First, it should be noted that it is not necessary for the two sample sizes to be equal (although they are in the example). For example, 45 days' output from factory 1 and 35 days' from factory 2 could have been sampled. Second, the values of s_1^2 and s_2^2 do not have to be equal. They are respectively estimates of σ_1^2 and σ_2^2 and, although the null hypothesis asserts that $\mu_1 = \mu_2$, it does not assert that the variances are equal. Management wants to know if the *average* levels of output are the same; it is not concerned about daily fluctuations in output. A test of the hypothesis of equal variances is set out in Chapter 6.

The final point to consider is whether all the necessary conditions for the correct application of this test have been met. The example noted that the 30 days were chosen at random. If the 30 days sampled were consecutive we might doubt whether the observations were truly independent. Low output on one day (e.g. due to a mechanical breakdown) might influence the following day's output (e.g. if a special effort were made to catch up on lost production).

Testing the difference of two proportions

The general method should by now be familiar, so we will proceed by example for this case. Suppose that, in a comparison of two holiday companies' customers, of the 75 who went with Happy Days Tours, 45 said they were satisfied, while 48 of the 90 who went with Fly by Night Holidays were satisfied. Is there a significant difference between the companies?

This problem can be handled by a hypothesis test on the difference of two sample proportions. The procedure is as follows. The sample evidence is

$$p_1 = 45/75 = 0.6 \qquad n_1 = 75$$
$$p_2 = 48/90 = 0.533 \qquad n_2 = 90$$

The hypothesis test is carried out as follows

(1) H_0: $\pi_1 - \pi_2 = 0$
H_1: $\pi_1 - \pi_2 \neq 0$

(2) Significance level: $\alpha = 5\%$.
(3) Critical value: $z^* = 1.96$.
(4) Test statistic: The distribution of $p_1 - p_2$ is

$$p_1 - p_2 \sim N\left(\pi_1 - \pi_2, \frac{\pi_1(1-\pi_1)}{n_1} + \frac{\pi_2(1-\pi_2)}{n_2}\right)$$

so the test statistic is

$$z = \frac{(p_1 - p_2) - (\pi_1 - \pi_2)}{\sqrt{\frac{\pi_1(1-\pi_1)}{n_1} + \frac{\pi_2(1-\pi_2)}{n_2}}} \tag{5.8}$$

However, π_1 and π_2 in the denominator of equation (5.8) have to be replaced by estimates from the samples. They cannot simply be replaced by p_1 and p_2 because these are unequal; to do so would contradict the null hypothesis that they *are* equal. Since the null hypothesis is assumed to be true (for the moment), it doesn't make sense to use a test statistic which explicitly supposes the null hypothesis to be false. Therefore π_1 and π_2 are replaced by an estimate of their common value which is denoted $\hat{\pi}$ and whose formula is

$$\hat{\pi} = \frac{n_1 p_1 + n_2 p_2}{n_1 + n_2} \tag{5.9}$$

i.e. a weighted average of the two sample proportions. This yields

$$\hat{\pi} = \frac{75 \times 0.6 + 90 \times 0.533}{75 + 90} = 0.564$$

This, in fact, is just the proportion of all customers who were satisfied, 93 out of 165. The test statistic therefore becomes

$$z = \frac{0.6 - 0.533 - 0}{\sqrt{\frac{0.564 \times (1 - 0.564)}{75} + \frac{0.564 \times (1 - 0.564)}{90}}} = 0.86$$

(5) The test statistic is less than the critical value so the null hypothesis cannot be rejected with 95% confidence. There is not sufficient evidence to demonstrate a difference between the two companies' performance.

Are women better at multi-tasking?

The conventional wisdom is 'yes'. However, the concept of multi-tasking originated in computing and, in that domain it appears men are more likely to multi-task. Oxford Internet Surveys (http://www.oii.ox.ac.uk/microsites/oxis/) asked a sample of 1578 people if they multi-tasked while on-line (e.g. listening to music, using the phone). 69% of men said they did compared to 57% of women. Is this difference statistically significant?

The published survey does not give precise numbers of men and women respondents for this question, so we will assume equal numbers (the answer is not very sensitive to this assumption). We therefore have the test statistic

$$z = \frac{0.69 - 0.57 - 0}{\sqrt{\frac{0.63 \times (1 - 0.63)}{789} + \frac{0.63 \times (1 - 0.63)}{789}}} = 4.94$$

(0.63 is the overall proportion of multi-taskers.) The evidence is significant and clearly suggests this is a genuine difference: men are the multi-taskers!

Exercise 5.6

A survey of 80 voters finds that 65% are in favour of a particular policy. Test the hypothesis that the true proportion is 50%, against the alternative that a majority is in favour.

Exercise 5.7

A survey of 50 teenage girls found that on average they spent 3.6 hours per week chatting with friends over the internet. The standard deviation was 1.2 hours. A similar survey of 90 teenage boys found an average of 3.9 hours, with standard deviation 2.1 hours. Test if there is any difference between boys' and girls' behaviour.

Exercise 5.8

One gambler on horse racing won on 23 of his 75 bets. Another won on 34 out of 95. Is the second person a better judge of horses, or just luckier?

Hypothesis tests with small samples

As with estimation, slightly different methods have to be employed when the sample size is small ($n < 25$) and the population variance is unknown. When both of these conditions are satisfied the t distribution must be used rather than the Normal, so a t test is conducted rather than a z test. This means consulting tables of the t distribution to obtain the critical value of a test, but otherwise the methods are similar. These methods will be applied to hypotheses about sample means only, since they are inappropriate for tests of a sample proportion, as was the case in estimation.

Testing the sample mean

A large chain of supermarkets sells 5000 packets of cereal in each of its stores each month. It decides to test-market a different brand of cereal in 15 of its stores. After a month the 15 stores have sold an average of 5200 packets each, with a standard deviation of 500 packets. Should all supermarkets switch to selling the new brand?

The sample information is

$$\bar{x} = 5200,\ s = 500,\ n = 15$$

From Chapter 4 the distribution of the sample mean from a small sample when the population variance is unknown is based upon

$$\frac{\bar{x} - \mu}{\sqrt{s^2/n}} \sim t_v \qquad (5.10)$$

with $v = n - 1$ degrees of freedom. The hypothesis test is based on this formula and is conducted as follows

(1) H_0: $\mu = 5000$
H_1: $\mu > 5000$
(Only an improvement in sales is relevant.)
(2) Significance level: $\alpha = 1\%$ (chosen because the cost of changing brands is high).
(3) The critical value of the t distribution for a one-tail test at the 1% significance level with $v = n - 1 = 14$ degrees of freedom is $t^* = 2.62$.
(4) The test statistic is

$$t = \frac{\bar{x} - \mu}{\sqrt{s^2/n}} = \frac{5200 - 5000}{\sqrt{500^2/15}} = 1.55$$

(5) The null hypothesis is not rejected since the test statistic, 1.55, is less than the critical value, 2.62. It would probably be unwise to switch over to the new brand of cereals.

Testing the difference of two means

A survey of 20 British companies found an average annual expenditure on research and development of £3.7m with a standard deviation of £0.6m. A survey of 15 similar German companies found an average expenditure on research and development of £4.2m with standard deviation £0.9m. Does this evidence lend support to the view often expressed that Britain does not invest enough in research and development?

This is a hypothesis about the difference of two means, based on small sample sizes. The test statistic is again based on the t distribution, i.e.

$$\frac{(\bar{x}_1 - \bar{x}_2) - (\mu_1 - \mu_2)}{\sqrt{\dfrac{S^2}{n_1} + \dfrac{S^2}{n_2}}} \sim t_v \qquad (5.11)$$

where S^2 is the pooled variance (as given in equation (4.23)) and the degrees of freedom are given by $v = n_1 + n_2 - 2$.

The hypothesis test procedure is as follows:

(1) H_0: $\mu_1 - \mu_2 = 0$
H_1: $\mu_1 - \mu_2 < 0$
(2) Significance level: $\alpha = 5\%$.

(3) The critical value of the t distribution at the 5% significance level for a one-tail test with $v = n_1 + n_2 - 2 = 33$ degrees of freedom is approximately $t^* = 1.70$.

(4) The test statistic is based on equation (5.11)

$$t = \frac{(\bar{x}_1 - \bar{x}_2) - (\mu_1 - \mu_2)}{\sqrt{\frac{S^2}{n_1} + \frac{S^2}{n_2}}} = \frac{3.7 - 4.2 - 0}{\sqrt{\frac{0.55}{20} + \frac{0.55}{15}}} = -1.97$$

where S^2 is the pooled variance, calculated by

$$S^2 = \frac{(n_1 - 1)s_1^2 + (n_2 - 1)s_2^2}{n_1 + n_2 - 2} = \frac{19 \times 0.6^2 + 14 \times 0.9^2}{33} = 0.55$$

(5) The test statistic falls in the rejection region, $t < -t^*$, so the null hypothesis is rejected. The data do support the view that Britain spends less on R&D than Germany.

Exercise 5.9

?

It is asserted that parents spend, on average, £540 per annum on toys for each child. A survey of 24 parents finds expenditure of £490, with standard deviation £150. Does this evidence contradict the assertion?

Exercise 5.10

A sample of 15 final-year students were found to spend on average 15 hours per week in the university library, with standard deviation 3 hours. A sample of 20 freshers found they spend on average 9 hours per week in the library, standard deviation 5 hours. Is this sufficient evidence to conclude that finalists spend more time in the library?

Are the test procedures valid?

A variety of assumptions underlie each of the tests which we have applied above and it is worth considering in a little more detail whether these assumptions are justified. This will demonstrate that one should not rely upon the statistical tests alone; it is important to retain one's sense of judgement.

The first test concerned the weekly turnover of a series of franchise operations. To justify the use of the Normal distribution underlying the test, the sample observations must be independently drawn. The random errors around the true mean turnover figure should be independent of each other. This might not be the case if, for example, similar events could affect the turnover figures of all franchises.

If one were using time-series data, as in the car factory comparison, similar issues arise. Do the 30 days represent independent observations or might there be an autocorrelation problem (e.g. if the sample days were close together in time)? Suppose that factory 2 suffered a breakdown of some kind which took three days to fix. Output would be reduced on three successive days and factory 2 would almost inevitably appear less efficient than factory 1. A look at the individual sample observations might be worthwhile, therefore, to see if there are unusual patterns. It would have been altogether better if the samples had been collected on randomly chosen days over a longer time period to reduce the danger of this type of problem.

If the two factories both obtain their supplies from a common, but limited, source then the output of one factory might not be independent of the output of the other. A high output of one factory would tend to be associated with a low output from the other, which has little to do with their relative efficiencies. This might leave the average difference in output unchanged but might increase the variance substantially (either a very high positive value of $\bar{x}_1 - \bar{x}_2$ or a very high negative value is obtained). This would lead to a low value of the test statistic and the conclusion of no difference in output. Any real difference in efficiency is masked by the common supplier problem. If the two samples are not independent then the distribution of $\bar{x}_1 - \bar{x}_2$ may not be Normal.

Hypothesis tests and confidence intervals

Formally, two-tail hypothesis tests and confidence intervals are equivalent. Any value that lies within the 95% confidence interval around the sample mean cannot be rejected as the 'true' value using the 5% significance level in a hypothesis test using the same sample data. For example, our by now familiar accountant could construct a confidence interval for the firm's sales. This yields the 95% confidence interval

$$[4792, 5008] \tag{5.12}$$

Notice that the hypothesised value of 5000 is within this interval and that it was not rejected by the hypothesis test carried out earlier. As long as the same confidence level is used for both procedures, they are equivalent.

Having said this, their interpretation is different. The hypothesis test forces us into the reject/do not reject dichotomy, which is rather a stark choice. We have seen how it becomes more likely that the null hypothesis is rejected as the sample size increases. This problem does not occur with estimation. As the sample size increases the confidence interval becomes narrower (around the unbiased point estimate) which is entirely beneficial. The estimation approach also tends to emphasise importance over significance in most people's minds. With a hypothesis test one might know that turnover is significantly different from 5000 without knowing how far from 5000 it actually is.

On some occasions a confidence interval is inferior to a hypothesis test, however. Consider the following case. In the UK only 17 out of 465 judges are women (3.7%).[2] The Equal Opportunities Commission commented that since the appointment system is so secretive it is impossible to tell if there is discrimination or not. What can the statistician say about this? No discrimination (in its broadest sense) would mean half of all judges would be women. Thus the hypotheses are

---

[2] This figure is somewhat out of date now, but it is still a useful example.

H_0: $\pi = 0.5$ (no discrimination)
H_1: $\pi < 0.5$ (discrimination against women)

The sample data are $p = 0.037$, $n = 465$. The z score is

$$z = \frac{p - \pi}{\sqrt{\frac{\pi(1-\pi)}{n}}} = \frac{0.037 - 0.5}{\sqrt{\frac{0.5 \times 0.5}{465}}} = -19.97$$

This is clearly significant (*and* 3.7% is a long way from 50%!) so the null hypothesis is rejected. There is some form of discrimination somewhere against women (unless women choose not to be judges). But a confidence interval estimate of the 'true' proportion of female judges would be meaningless. To what population is this 'true' proportion related?

The lesson from all this is that there exist differences between confidence intervals and hypothesis tests, despite their formal similarity. Which technique is more appropriate is a matter of judgement for the researcher. With hypothesis testing, the rejection of the null hypothesis at some significance level might actually mean a small (and unimportant) deviation from the hypothesised value. It should be remembered that the rejection of the null hypothesis based on a large sample of data is also consistent with the true value and hypothesised value possibly being quite close together.

Independent and dependent samples

The following example illustrates the differences between **independent samples** (as encountered so far) and **dependent samples** where slightly different methods of analysis are required. The example also illustrates how a particular problem can often be analysed by a variety of statistical methods.

A company introduces a training programme to raise the productivity of its clerical workers, which is measured by the number of invoices processed per day. The company wants to know if the training programme is effective. How should it evaluate the programme? There is a variety of ways of going about the task, as follows:

- Take two (random) samples of workers, one trained and one not trained, and compare their productivity.
- Take a sample of workers and compare their productivity before and after training.
- Take two samples of workers, one to be trained and the other not. Compare the improvement of the trained workers with any change in the other group's performance over the same time period.

We shall go through each method in turn, pointing out any possible difficulties.

Two independent samples

Suppose a group of 10 workers is trained and compared to a group of 10 non-trained workers, with the following data being relevant

$\bar{x}_T = 25.5 \qquad \bar{x}_N = 21.0$
$s_T = 2.55 \qquad s_N = 2.91$
$n_T = 10 \qquad n_N = 10$

Thus, trained workers process 25.5 invoices per day compared to only 21 by non-trained workers. The question is whether this is significant, given that the sample sizes are quite small.

The appropriate test here is a t test of the difference of two sample means, as follows:

$H_0: \mu_T - \mu_N = 0$
$H_1: \mu_T - \mu_N > 0$

$$t = \frac{25.5 - 21.0}{\sqrt{\frac{7.49}{10} + \frac{7.49}{10}}} = 3.68$$

(7.49 is S^2, the pooled variance). The t statistic leads to rejection of the null hypothesis; the training programme does seem to be effective.

One problem with this test is that the two samples might not be truly random and thus not properly reflect the effect of the training programme. Poor workers might have been reluctant (and thus refused) to take part in training, departmental managers might have selected better workers for training as some kind of reward, or simply better workers may have volunteered. In a well-designed experiment this should not be allowed to happen, of course, but we do not rule out the possibility. There is also the 5% (significance level) chance of unrepresentative samples being selected and a Type I error occurring.

Paired samples

This is the situation where a sample of workers is tested before and after training. The sample data are as follows:

Worker	1	2	3	4	5	6	7	8	9	10
Before	21	24	23	25	28	17	24	22	24	27
After	23	27	24	28	29	21	24	25	26	28

In this case, the observations in the two samples are paired and this has implications for the method of analysis. One *could* proceed by assuming these are two independent samples and conduct a t test. The summary data and results are

$\bar{x}_B = 23.50 \qquad \bar{x}_A = 25.5$
$s_B = 3.10 \qquad s_A = 2.55$
$n_B = 10 \qquad n_A = 10$

The resulting test statistic is $t_{18} = 1.58$ which is not significant at the 5% level.

There are two problems with this test and its result. First, the two samples are not truly independent, since the before and after measurements refer to the same group of workers. Second, nine out of 10 workers in the sample have shown an improvement, which is odd in view of the result found above, of no significant improvement. If the training programme really has no effect, then the probability of a single worker showing an improvement is $\frac{1}{2}$. The probability of nine or more workers showing an improvement is, by the Binomial method, $(\frac{1}{2})^{10} \times 10C9 + (\frac{1}{2})^{10}$, which is about one in a hundred. A very unlikely event seems to have occurred.

The t test used above is inappropriate because it does not make full use of the information in the sample. It does not reflect the fact, for example, that the before and after scores, 21 and 23, relate to the same worker. The Binomial calculation above does reflect this fact. A re-ordering of the data would not affect the t test result, but would affect the Binomial, since a different number of workers would now show an improvement. Of course, the Binomial does not use all the sample information either – it dispenses with the actual productivity data for each worker and replaces it with 'improvement' or 'no improvement'. It disregards the amount of improvement for each worker.

The best use of the sample data comes by measuring the improvement for each worker, as follows (if a worker had deteriorated, this would be reflected by a negative number):

Worker	1	2	3	4	5	6	7	8	9	10
Improvement	2	3	1	3	1	4	0	3	2	1

These new data can be treated by single sample methods, and account is taken both of the actual data values and of the fact that the original samples were dependent (re-ordering of the data would produce different improvement figures). The summary statistics of the new data are as follows

$\bar{x} = 2.00,\ s = 1.247,\ n = 10$

The null hypothesis of no improvement can now be tested as follows

$H_0\colon \mu = 0$
$H_1\colon \mu > 0$

$$t = \frac{2.0 - 0}{\sqrt{\frac{1.247^2}{10}}} = 5.07$$

This is significant at the 5% level so the null hypothesis of no improvement is rejected. The correct analysis of the sample data has thus reversed the previous conclusion. It is perhaps surprising that treating the same data in different ways leads to such a difference in the results. It does illustrate the importance of using the appropriate method.

Matters do not end here, however. Although we have discovered an improvement, this might be due to other factors apart from the training programme. For example, if the before and after measurements were taken on different days of the week (that Monday morning feeling . . .), or if one of the days were sunnier, making people feel happier and therefore more productive, this would bias the results. These may seem trivial examples but these effects do exist, for example the 'Friday afternoon car', which has more faults than the average.

The way to solve this problem is to use a control group, so called because extraneous factors are controlled for, in order to isolate the effects of the factor under investigation. In this case, the productivity of the control group would be measured (twice) at the same times as that of the training group, though no training would be given to them. Ideally, the control group would be matched on other factors (e.g. age) to the treatment group to avoid other factors influencing the results. Suppose that the average improvement of the control group were 0.5 invoices per day with standard deviation 1.0 (again for a group of 10). This can be compared with the improvement of the training group via the two-sample t test, giving

$$t = \frac{2.0 - 0.5}{\sqrt{\frac{1.13^2}{10} + \frac{1.13^2}{10}}} = 2.97$$

(1.13^2 is the pooled variance). This confirms the finding that the training programme is of value.

Exercise 5.11 A group of students' marks on two tests, before and after instruction, were as follows:

Student	1	2	3	4	5	6	7	8	9	10	11	12
Before	14	16	11	8	20	19	6	11	13	16	9	13
After	15	18	15	11	19	18	9	12	16	16	12	13

Test the hypothesis that the instruction had no effect, using both the independent sample and paired sample methods. Compare the two results.

Discussion of hypothesis testing

The above exposition has served to illustrate how to carry out a hypothesis test and the rationale behind it. However, the methodology has been subject to criticism and it is important to understand this since it gives a greater insight into the meaning of the results of a hypothesis test.

In the previous examples the problem has often been posed as a decision-making one, yet we noted that in many instances no decision is actually taken and therefore it is difficult to justify a particular significance level. Bayesian statisticians would argue that their methods do not suffer from this problem, since the result of their analysis (termed a posterior probability) gives the degree of belief which the researcher has in the truth of the null hypothesis. However, this posterior probability does in part depend upon the prior probability (i.e. before the statistical analysis) that the researcher attaches to the null hypothesis. As noted in Chapter 2, the derivation of the prior probabilities can be difficult.

In practice, most people do not regard the results of a hypothesis test as all-or-nothing proof, but interpret the result on the basis of the quality of the data, the care the researcher has taken in analysing the data, personal experience and a multitude of other factors. Both schools of thought, classical and Bayesian, introduce subjectivity into the analysis and interpretation of data: classical statisticians in the choice of the significance level (and choice of one- or two-tail test), Bayesians in their choice of prior probabilities. It is not clear which method is superior, but classical methods have the advantage of being simpler.

Another criticism of hypothesis testing is that it is based on weak methodological foundations. The philosopher Karl Popper argued that theories should be rigorously tested against the evidence, and that strenuous efforts should be made to try to falsify the theory or hypothesis. This methodology is not strictly followed in hypothesis testing, where the researcher's favoured hypothesis is usually the alternative. A conclusion in favour of the alternative hypothesis is arrived at by default, because of the failure of the null hypothesis to survive the evidence.

Consider the researcher who believes that health standards have changed in the last decade. This may be tested by gathering data on health and testing the null hypothesis of no change in health standards against the alternative hypothesis of some change. The researcher's theory thus becomes the alternative hypothesis and is never actually tested against the data. No attempt is made to falsify the (alternative) hypothesis; it is accepted by default if the null hypothesis falls. *Only* the null hypothesis is ever tested.

A further problem is the asymmetry between the null and alternative hypotheses. The null hypothesis is that there is *exactly* no change in health standards whereas the alternative hypothesis contains all other possibilities, from a large deterioration to a large improvement. The dice seem loaded against the null hypothesis. Indeed, as noted earlier, if a large enough sample is taken the null hypothesis is almost certain to be rejected, because there is bound to have been *some* change, however, small. The large sample size leads to a small standard error (σ^2/n) and thus a large z score. This suggests that the significance level of a test should decrease as the sample size increases.

These particular problems are avoided by the technique of estimation, which measures the size of the change and focuses attention upon that, rather than upon some accept/reject decision. As the sample size increases, the confidence interval narrows and an improved measure of the true change in health standards is obtained. Zero (i.e. no change in health standards) might be in the confidence interval or it might not; it is not the central issue. We might say that an estimate tells us what the value of a population parameter *is*, while a hypothesis test tells us what it is *not*. Thus the techniques of estimation and hypothesis testing put different emphasis upon interpretation of the results, even though they are formally identical.

Summary

- Hypothesis testing is the set of procedures for deciding whether a hypothesis is true or false. When conducting the test we presume the hypothesis, termed the null hypothesis, is true until it is proved false on the basis of some sample evidence.
- If the null is proved false, it is rejected in favour of the alternative hypothesis. The procedure is conceptually similar to a court case, where the defendant is presumed innocent until the evidence proves otherwise.
- Not all decisions turn out to be correct and there are two types of error that can be made. A Type I error is to reject the null hypothesis when it is in fact true. A Type II error is not to reject the null when it is false.

- Choosing the appropriate decision rule (for rejecting the null hypothesis) is a question of trading off Type I and Type II errors. Because the alternative hypothesis is imprecisely specified, the probability of a Type II error usually cannot be specified.
- The rejection region for a test is therefore chosen to give a 5% probability of making a Type I error (sometimes a 1% probability is chosen). The critical value of the test statistic (sometimes referred to as the critical value of the test) is the value which separates the acceptance and rejection regions.
- The decision is based upon the value of a test statistic, which is calculated from the sample evidence and from information in the null hypothesis

$$\left(\text{e.g. } z = \frac{\bar{x} - \mu}{s/\sqrt{n}}\right)$$

- The null hypothesis is rejected if the test statistic falls into the rejection region for the test (i.e. it exceeds the critical value).
- For a two-tail test there are two rejection regions, corresponding to very high and very low values of the test statistic.
- Instead of comparing the test statistic to the critical value, an equivalent procedure is to compare the Prob-value of the test statistic with the significance level. The null is rejected if the Prob-value is less than the significance level.
- The power of a test is the probability of a test correctly rejecting the null hypothesis. Some tests have low power (e.g. when the sample size is small) and therefore are not very useful.

Key terms and concepts

alternative hypothesis
critical value
effect size
independent samples
null or maintained hypothesis
one- and two-tail tests
paired samples
power
Prob-value
rejection region
significance level
Type I and Type II errors

Reference

D. McCloskey, and S. Ziliak, Size matters: the standard error of regressions in the *American Economic Review, Journal of Socio-Economics*, 2004, **33**, 527–546.

Problems

Some of the more challenging problems are indicated by highlighting the problem number in colour.

5.1 Answer true or false, with reasons if necessary.

(a) There is no way of reducing the probability of a Type I error without simultaneously increasing the probability of a Type II error.

(b) The probability of a Type I error is associated with an area under the distribution of $\bar{x}$ assuming the null hypothesis to be true.

(c) It is always desirable to minimise the probability of a Type I error.

(d) A larger sample, *ceteris paribus*, will increase the power of a test.

(e) The significance level is the probability of a Type II error.

(f) The confidence level is the probability of a Type II error.

5.2 Consider the investor in the text, seeking out companies with weekly turnover of at least £5000. He applies a one-tail hypothesis test to each firm, using the 5% significance level. State whether each of the following statements is true or false (or not known) and explain why.

(a) 5% of his investments are in companies with less than £5000 turnover.

(b) 5% of the companies he *fails* to invest in have turnover greater than £5000 per week.

(c) He invests in 95% of all companies with turnover of £5000 or over.

5.3 A coin which is either fair or has two heads is to be tossed twice. You decide on the following decision rule: if two heads occur you will conclude it is a two-headed coin, otherwise you will presume it is fair. Write down the null and alternative hypotheses and calculate the probabilities of Type I and Type II errors.

5.4 In comparing two medical treatments for a disease, the null hypothesis is that the two treatments are equally effective. Why does making a Type I error not matter? What significance level for the test should be set as a result?

5.5 A firm receives components from a supplier, which it uses in its own production. The components are delivered in batches of 2000. The supplier claims that there are only 1% defective components on average from its production. However, production occasionally becomes out of control and a batch is produced with 10% defective components. The firm wishes to intercept these low-quality batches, so a sample of size 50 is taken from each batch and tested. If two or more defectives are found in the sample then the batch is rejected.

(a) Describe the two types of error the firm might make in assessing batches of components.

(b) Calculate the probability of each type of error given the data above.

(c) If instead, samples of size 30 were taken and the batch rejected if one or more rejects were found, how would the error probabilities be altered?

(d) The firm can alter the two error probabilities by choice of sample size and rejection criteria. How should it set the relative sizes of the error probabilities

(i) if the product might affect consumer safety?
(ii) if there are many competitive suppliers of components?
(iii) if the costs of replacement under guarantee are high?

5.6 Computer diskettes, which do not meet the quality required for high-density (1.44 Mb) diskettes, are sold as double-density diskettes (720 kb) for 80p each. High-density diskettes are sold for £1.20 each. A firm samples 30 diskettes from each batch of 1000 and if any fail the quality test the whole batch is sold as double-density diskettes. What are the types of error possible and what is the cost to the firm of a Type I error?

5.7 Testing the null hypothesis that $\mu = 10$ against $\mu > 10$, a researcher obtains a sample mean of 12 with standard deviation 6 from a sample of 30 observations. Calculate the z score and the associated Prob-value for this test.

5.8 Given the sample data $\bar{x} = 45$, $s = 16$, $n = 50$, at what level of confidence can you reject H_0: $\mu = 40$ against a two-sided alternative?

5.9 What is the power of the test carried out in Problem 5.3?

5.10 Given the two hypotheses

H_0: $\mu = 400$
H_1: $\mu = 415$

and $\sigma^2 = 1000$ (for both hypotheses):

(a) Draw the distribution of $\bar{x}$ under both hypotheses.

(b) If the decision rule is chosen to be: reject H_0 if $\bar{x} \geqslant 410$ from a sample of size 40, find the probability of a Type II error and the power of the test.

(c) What happens to these answers as the sample size is increased? Draw a diagram to illustrate.

5.11 Given the following sample data

$\bar{x} = 15 \quad s^2 = 270 \quad n = 30$

test the null hypothesis that the true mean is equal to 12, against a two-sided alternative hypothesis. Draw the distribution of $\bar{x}$ under the null hypothesis and indicate the rejection regions for this test.

5.12 From experience it is known that a certain brand of tyre lasts, on average, 15 000 miles with standard deviation 1250. A new compound is tried and a sample of 120 tyres yields an average life of 15 150 miles. Are the new tyres an improvement? Use the 5% significance level.

5.13 Test H_0: $\pi = 0.5$ against H_0: $\pi \neq 0.5$ using $p = 0.45$ from a sample of size $n = 35$.

5.14 Test the hypothesis that 10% of your class or lecture group are left-handed.

5.15 Given the following data from two independent samples

$\bar{x}_1 = 115 \quad \bar{x}_2 = 105$
$s_1 = 21 \quad s_2 = 23$
$n_1 = 49 \quad n_2 = 63$

test the hypothesis of no difference between the population means against the alternative that the mean of population 1 is greater than the mean of population 2.

5.16 A transport company wants to compare the fuel efficiencies of the two types of lorry it operates. It obtains data from samples of the two types of lorry, with the following results:

Type	Average mpg	Std devn	Sample size
A	31.0	7.6	33
B	32.2	5.8	40

Test the hypothesis that there is no difference in fuel efficiency, using the 99% confidence level.

5.17 A random sample of 180 men who took the driving test found that 103 passed. A similar sample of 225 women found that 105 passed. Test whether pass rates are the same for men and women.

5.18 (a) A pharmaceutical company testing a new type of pain reliever administered the drug to 30 volunteers experiencing pain. Sixteen of them said that it eased their pain. Does this evidence support the claim that the drug is effective in combating pain?

(b) A second group of 40 volunteers were given a placebo instead of the drug. Thirteen of them reported a reduction in pain. Does this new evidence cast doubt upon your previous conclusion?

5.19 (a) A random sample of 20 observations yielded a mean of 40 and standard deviation 10. Test the hypothesis that $\mu = 45$ against the alternative that it is not. Use the 5% significance level.

(b) What assumption are you implicitly making in carrying out this test?

5.20 A photo processing company sets a quality standard of no more than 10 complaints per week on average. A random sample of 8 weeks showed an average of 13.6 complaints, with standard deviation 5.3. Is the firm achieving its quality objective?

5.21 Two samples are drawn. The first has a mean of 150, variance 50 and sample size 12. The second has mean 130, variance 30 and sample size 15. Test the hypothesis that they are drawn from populations with the same mean.

5.22 (a) A consumer organisation is testing two different brands of battery. A sample of 15 of brand *A* shows an average useful life of 410 hours with a standard deviation of 20 hours. For brand *B*, a sample of 20 gave an average useful life of 391 hours with standard deviation 26 hours. Test whether there is any significant difference in battery life.

(b) What assumptions are being made about the populations in carrying out this test?

5.23 The output of a group of 11 workers before and after an improvement in the lighting in their factory is as follows:

Before	52	60	58	58	53	51	52	59	60	53	55
After	56	62	63	50	55	56	55	59	61	58	56

Test whether there is a significant improvement in performance

(a) assuming these are independent samples,

(b) assuming they are dependent.

5.24 Another group of workers were tested at the same times as those in Problem 5.23, although their department *also* introduced rest breaks into the working day.

Before	51	59	51	53	58	58	52	55	61	54	55
After	54	63	55	57	63	63	58	60	66	57	59

Does the introduction of rest days alone appear to improve performance?

5.25 Discuss in general terms how you might 'test' the following:

(a) astrology;

(b) extra-sensory perception;

(c) the proposition that company takeovers increase profits.

5.26 **(Project)** Can your class tell the difference between tap water and bottled water? Set up an experiment as follows: fill r glasses with tap water and $n - r$ glasses with bottled water. The subject has to guess which is which. If they get more than p correct, you conclude they can tell the difference. Write up a report of the experiment including:

(a) a description of the experimental procedure;

(b) your choice of n, r and p, with reasons;

(c) the power of your test;

(d) your conclusions.

5.27 **(Computer project)** Use the = *RAND( )* function in your spreadsheet to create 100 samples of size 25 (which are effectively all from the same population). Compute the mean and standard deviation of each sample. Calculate the z score for each sample, using a hypothesised mean of 0.5 (since the = *RAND( )* function chooses a random number in the range 0 to 1).

(a) How many of the z scores would you expect to exceed 1.96 in absolute value? Explain why.

(b) How many do exceed this? Is this in line with your prediction?

(c) Graph the sample means and comment upon the shape of the distribution. Shade in the area of the graph beyond $z = \pm 1.96$.

Answers to exercises

Exercise 5.1

(a) H_0: crime is the same as last year, H_1: crime has increased.

(b) Type I error – concluding crime has risen, when in fact it has not. Type II – concluding it has not risen, when, in fact, it has. The cost of the former might be employing more police officers which are not in fact warranted; of the latter, not employing more police to counter the rising crime level. (The *Economist* magazine (19 July 2003) reported that 33% of respondents to a survey in the UK felt that crime had risen in the previous two years, only 4% thought that it had fallen. In fact, crime had fallen slightly, by about 2%. A lot of people were making a Type I error, therefore.)

Exercise 5.2

(a) $z = (108 - 100)/\sqrt{36} = 1.33$. The area in the tail beyond 1.33 is 9.18%, which is the probability of a Type I error.

(b) $z = 1.64$ cuts off 5% in the upper tail of the distribution, hence we need the decision rule to be at $\bar{x} + 1.64 \times s/\sqrt{n} = 100 + 1.64 \times \sqrt{36} = 109.84$.

(c) Under H_1: $\mu = 112$, we can write $\bar{x} \sim N(112, 900/25)$. (We assume the same variance under both H_0 and H_1 in this case.) Hence $z = (108 - 112)/\sqrt{36} = -0.67$. This gives an area in the tail of 25.14%, which is the Type II error probability. Usually, however, we do not have a precise statement of the value of μ under H_1 so cannot do this kind of calculation.

Exercise 5.3

$\alpha = 0.05$ (significance level chosen), hence the critical value is $z^* = 1.96$. The test statistic is $z = (530 - 500)/(90/\sqrt{30}) = 1.83 < 1.96$ so H_0 is not rejected at the 5% significance level.

Exercise 5.4

One wants to avoid making a Type I error if possible, i.e. rejecting H_0 when true. Hence set a low significance level (1%) so that H_0 is only rejected by very strong evidence.

Exercise 5.5

(a) (i) Reject. The Prob-value should be halved, to 0.0385, which is less than 5%. Alternatively, $1.77 > 1.64$. (ii) Do not reject, the Prob-value is greater than 5%; equivalently $1.77 < 1.96$.

(b) In this case, the null is not rejected in both cases. In the one-tailed case, $0.0385 > 1\%$, so the null is not rejected.

Exercise 5.6

$$z = \frac{0.65 - 0.5}{\sqrt{\frac{0.5 \times 0.5}{80}}} = 2.68$$ hence the null is decisively rejected.

Exercise 5.7

We have the data: $\bar{x}_1 = 3.6$, $s_1 = 1.2$, $n_1 = 50$; $\bar{x}_2 = 3.9$, $s_2 = 2.1$, $n_2 = 90$. The null hypothesis is H_0: $\mu_1 = \mu_2$ versus H_1: $\mu_1 \neq \mu_2$. The test statistic is

$$z = \frac{(\bar{x}_1 - \bar{x}_2) - (\mu_1 - \mu_2)}{\sqrt{\frac{s_1^2}{n_1} + \frac{s_2^2}{n_2}}} = \frac{(3.6 - 3.9) - 0}{\sqrt{\frac{1.2^2}{50} + \frac{2.1^2}{90}}} = -1.08 < 1.96$$

(absolute value) so the null is not rejected at the 5% significance level.

Exercise 5.8

The evidence is $p_1 = 23/75$, $n_1 = 75$, $p_2 = 34/95$, $n_2 = 95$. The hypothesis to be tested is H_0: $\pi_1 - \pi_2 = 0$ versus H_1: $\pi_1 - \pi_2 < 0$. Before calculating the test statistic we must calculate the pooled variance as

$$\hat{\pi} = \frac{n_1 p_1 + n_2 p_2}{n_1 + n_2} = \frac{75 \times 0.3067 + 95 \times 0.3579}{75 + 95} = 0.3353$$

The test statistic is then

$$z = \frac{0.3067 - 0.3579 - 0}{\sqrt{\frac{0.3353 \times (1 - 0.3353)}{75} + \frac{0.3353 \times (1 - 0.3353)}{95}}} = -0.70$$

This is less in absolute magnitude than 1.64, the critical value of a one tailed test, so the null is not rejected. The second gambler is just luckier than the first, we conclude. We have to be careful about our interpretation, however: one of the gamblers might prefer longer-odds bets, so wins less often but gets more money each time. Hence this may not be a fair comparison.

Exercise 5.9

We shall treat this as a two-tailed test, although a one-tailed test might be justified if there were other evidence that spending had fallen. The hypothesis is H_0: $\mu = 540$ versus H_1: $\mu \neq 540$. Given the sample evidence, the test statistic is

$$t = \frac{\bar{x} - \mu}{\sqrt{s^2/n}} = \frac{490 - 540}{\sqrt{150^2/24}} = -1.63$$

The critical value of the t distribution for 23 degrees of freedom is 2.069, so the null is not rejected.

Exercise 5.10

The hypothesis to test is H_0: $\mu_F - \mu_N = 0$ versus H_1: $\mu_F - \mu_N > 0$ (F indexes finalists, N the new students). The pooled variance is calculated as

$$S^2 = \frac{(n_1 - 1)s_1^2 + (n_2 - 1)s_2^2}{n_1 + n_2 - 2} = \frac{15 \times 3^2 + 20 \times 5^2}{35} = 18.14$$

The test statistic is

$$t = \frac{(\bar{x}_1 - \bar{x}_2) - (\mu_1 - \mu_2)}{\sqrt{\frac{S^2}{n_1} + \frac{S^2}{n_2}}} = \frac{15 - 9 - 0}{\sqrt{\frac{18.14}{15} + \frac{18.14}{20}}} = 4.12$$

The critical value of the t distribution with 15 + 20 – 2 = 33 degrees of freedom is approximately 1.69 (5% significance level, for a one-tailed test). Thus the null is decisively rejected and we conclude finalists do spend more time in the library.

Exercise 5.11

By the method of independent samples we obtain $\bar{x}_1 = 13$, $\bar{x}_2 = 14.5$, $s_1 = 4.29$, $s_2 = 3.12$, with $n = 12$ in both cases. The test statistic is therefore

$$t = \frac{(\bar{x}_1 - \bar{x}_2) - (\mu_1 - \mu_2)}{\sqrt{\frac{S^2}{n_1} + \frac{S^2}{n_2}}} = \frac{13 - 14.5 - 0}{\sqrt{\frac{14.05}{12} + \frac{14.05}{12}}} = -0.98$$

with pooled variance

$$S^2 = \frac{(n_1 - 1)s_1^2 + (n_2 - 1)s_2^2}{n_1 + n_2 - 2} = \frac{11 \times 4.29^2 + 11 \times 3.12^2}{22} = 14.05$$

The null of no effect is therefore accepted. By the method of paired samples, we have a set of improvements as follows:

Student	1	2	3	4	5	6	7	8	9	10	11	12
Improvement	1	2	4	3	−1	−1	3	1	3	0	3	0

The mean of these is 1.5 and the variance is 3. The t statistic is therefore

$$t = \frac{1.5 - 0}{\sqrt{3/12}} = 3$$

This now conclusively rejects the null hypothesis (critical value 1.8), in stark contrast to the former method. The difference arises because 10 out of 12 students have improved or done as well as before, only two have fallen back (slightly). The gain in marks is modest but applies consistently to nearly all candidates.

10 Index numbers

Contents

Learning outcomes

By the end of this chapter you should be able to:

- represent a set of data in index number form;
- understand the role of index numbers in summarising or presenting data;
- recognise the relationship between price, quantity and expenditure index numbers;
- turn a series measured at current prices into one at constant prices (or in volume terms);
- splice separate index number series together;
- measure inequality using index numbers.

Complete your diagnostic test for Chapter 10 now to create your personal study plan. Exercises with an icon ? are also available for practice in MathXL with additional supporting resources.

Introduction

'Consumer price index up 3.8%. Retail price index up 4.6%.' (UK, June 2008)

'Vietnam reports an inflation rate of 27.04%' (July 2008)

'Zimbabwe inflation at 2,200,000%' (July 2008)

The above headlines reveal startling differences between the inflation rates of three different countries. This chapter is concerned with how such measures are constructed and then interpreted. Index numbers are not restricted to measuring inflation, though that is one of the most common uses. There are also indexes of national output, of political support, of corruption in different countries of the world, and even of happiness (Danes are the happiest, it seems).

An **index number** is a descriptive statistic, in the same sense as the mean or standard deviation, which summarises a mass of information into some readily understood statistic. As such, it shares the advantages and disadvantages of other summary statistics: it provides a useful overview of the data but misses out the finer detail. The retail price index (RPI) referred to above is one example, which summarises information about the prices of different goods and services, aggregating them into a single number. We have used index numbers earlier in the book (e.g. in the chapters on regression), without fully explaining their derivation or use. This will now be remedied.

Index numbers are most commonly used for following trends in data over time, such as the RPI measuring the price level or the index of industrial production (IIP) measuring the output of industry. The RPI also allows calculation of the rate of inflation, which is simply the rate of change of the price index; and from the IIP it is easy to measure the rate of growth of output. Index numbers are also used with cross-section data, for example, an index of regional house prices would summarise information about the different levels of house prices in different regions of the country at a particular point in time. There are many other examples of index numbers in use, common ones being the

Financial Times All Share index, the trade weighted exchange rate index, and the index of the value of retail sales.

This chapter will explain how index numbers are constructed from original data and the problems which arise in doing this. There is also a brief discussion of the RPI to illustrate some of these problems and to show how they are resolved in practice. Finally, a different set of index numbers is examined, which are used to measure inequality, such as inequality in the distribution of income, or in the market shares held by different firms competing in a market.

A simple index number

We begin with the simplest case, where we wish to construct an index number series for a single commodity. In this case, we shall construct an index number series representing the price of coal. This is a series of numbers showing, in each year, the price of coal and how it changes over time. More precisely, we measure the cost of coal to industrial users, for the years 2002–2006. Later in the chapter we will expand the analysis to include other fuels and thereby construct an index of the price of energy as a whole. The raw data for coal are given in Table 10.1 (adapted from the Digest of UK Energy Statistics, available on the internet). We assume that the product itself has not changed from year to year, so that the index provides a fair representation of costs. This means, for example, that the quality of coal has not changed during the period.

To construct a price index from these data we choose one year as the **reference year** (2002 in this case) and set the price index in that year equal to 100. The prices in the other years are then measured *relative* to the reference year figure of 100. The index, and its construction, are presented in Table 10.2.

All we have done so far is to change the form in which the information is presented. We have perhaps gained some degree of clarity (for example, it is easy to see that the price in 2006 is 18% higher than in 2002), but we have lost the original information about the actual level of prices. Since it is usually *relative* prices that are of interest, this loss of information about the actual price level is not too serious, and information about relative prices is retained by the price

Table 10.1 The price of coal, 2002–2006

	2002	2003	2004	2005	2006
Price (£/tonne)	36.97	34.03	37.88	44.57	43.63

Table 10.2 The price index for coal, 2002 = 100

Year	Price	Index	
2002	36.97	100.0	(= 36.97/36.97 × 100)
2003	34.03	92.0	(= 34.03/36.97 × 100)
2004	37.88	102.5	(= 37.88/36.97 × 100)
2005	44.57	120.6	Etc.
2006	43.63	118.0	

Table 10.3 The price index for coal, 2004 = 100

Year	Price	Index	
2002	36.97	97.6	(= 36.97/37.88 × 100)
2003	34.03	89.8	(= 34.03/37.88 × 100)
2004	37.88	100.0	(= 37.88/37.88 × 100)
2005	44.57	117.7	Etc.
2006	43.63	115.2	

index. For example, using either the index or actual prices, we can see that the price of coal was 8% lower in 2003 than in 2002.

In terms of a formula we have calculated

$$P^t = \frac{\text{price of coal in year } t}{\text{price of coal in 2002}} \times 100$$

where P^t represents the value of the index in year t.

The choice of reference year is arbitrary and we can easily change it for a different year. If we choose 2004 to be the reference year, then we set the price in that year equal to 100 and again measure all other prices relative to it. This is shown in Table 10.3, which can be derived from Table 10.2 or directly from the original data on prices. You should choose whichever reference year is most convenient for your purposes. Whichever year is chosen, the informational content is the same.

Exercise 10.1

(a) Average house prices in the UK for 2000–2004 were:

Year	2000	2001	2002	2003	2004
Price (£)	86 095	96 337	121 137	140 687	161 940

(a) Turn this into an index with a reference year of 2000.

(b) Recalculate the index with reference year 2003.

(c) Check that the ratio of house prices in 2004 relative to 2000 is the same for both indexes.

A price index with more than one commodity

Constructing an index for a single commodity is a simple process but only of limited use, mainly in terms of presentation. Once there is more than a single commodity, index numbers become more useful but are more difficult to calculate. Industry uses other sources of energy as well as coal, such as gas, petroleum and electricity and managers might wish to know the overall price of energy, which affects their costs. This is a more common requirement in reality, rather than the simple index number series calculated above. If the price of each fuel were rising at the same rate, say at 5% per year, then it is straightforward to say

Table 10.4 Fuel prices to industry, 2002–2006

Year	Coal (£/tonne)	Petroleum (£/tonne)	Electricity (£/MWh)	Gas (£/therm)
2002	36.97	132.24	29.83	0.780
2003	34.03	152.53	28.68	0.809
2004	37.88	153.71	31.26	0.961
2005	44.57	204.28	42.37	1.387
2006	43.63	260.47	55.07	1.804

that the price of energy is also rising at 5% per year. But supposing, as is likely, that the prices are all rising at different rates, as shown in Table 10.4. Is it now possible to say how fast the price of energy is increasing? Several different prices now have to be combined in order to construct an index number, a more complex process than the simple index number calculated above.

From the data presented in Table 10.4 we can calculate that the price of coal has risen by 18% over the five-year period, petrol has risen by 97%, electricity by 85% and gas by 131%. It is fairly clear prices are rising rapidly, but how do we measure this precisely?

Using base-year weights: the Laspeyres index

We tackle the problem by taking a **weighted average** of the price changes of the individual fuels, the weights being derived from the quantities of each fuel used by the industry. Thus, if industry uses relatively more coal than petrol, more weight is given to the rise in the price of coal in the calculation.

We put this principle into effect by constructing a hypothetical 'shopping basket' of the fuels used by industry, and measure how the cost of this basket has risen (or fallen) over time. Table 10.5 gives the quantities of each fuel consumed by industry in 2002 (again from the Digest of UK Energy Statistics) and it is this which forms the shopping basket. 2002 is referred to as the **base year** since it is the quantities consumed in this year which are used to make up the shopping basket.

The cost of the basket in 2002 prices therefore works out as shown in Table 10.6 (using information from Tables 10.4 and 10.5).

The final column of the table shows the expenditure on each of the four energy inputs and the total cost of the basket is 8581.01 (this is in £m, so altogether about £8.58bn was spent on energy by industry). This sum may be written as

$$\sum_i p_{0i} q_{0i} = 8581.01$$

where the summation is calculated over all the four fuels. Here, p refers to prices, q to quantities. The first subscript (0) refers to the year, the second (i) to each

Table 10.5 Quantities of fuel used by industry, 2002

Coal (m. tonnes)	1.81
Petroleum (m. tonnes)	5.70
Electricity (m. MWh)	112.65
Gas (m. therms)	5641

Table 10.6 Cost of the energy basket, 2002

	Price	Quantity	Price × quantity
Coal (£/tonne)	36.97	1.81	66.916
Petroleum (£/tonne)	132.24	5.70	753.768
Electricity (£/MWh)	29.83	112.65	3360.350
Gas (£/million therms)	0.780	5641	4399.980
Total			8581.013

Table 10.7 The cost of the 2002 energy basket at 2003 prices

	2003 Price	2002 Quantity	Price × quantity
Coal (£/tonne)	34.03	1.81	61.594
Petroleum (£/tonne)	152.53	5.70	869.421
Electricity (£/MWh)	28.68	112.65	3230.802
Gas (£/million therms)	0.809	5641	4563.569
Total			8725.386

energy source in turn. We refer to 2002 as year 0, 2003 as year 1, etc., for brevity of notation. Thus, for example, p_{01} means the price of coal in 2002, q_{12} the consumption of petroleum by industry in 2003.

We now need to find what the 2002 basket of energy would cost in each of the subsequent years, using the prices pertaining to those years. For example, for 2003 we value the 2002 basket using the 2003 prices. This is shown in Table 10.7 and yields a cost of £87.25 bn.

Firms would therefore have to spend an extra £144m (= 8725 − 8581) in 2003 to buy the same quantities of energy as in 2002. This amounts to an additional 1.7% over the expenditure in 2002. The sum of £8725m may be expressed as $\Sigma p_{1i}q_{0i}$, since it is obtained by multiplying the prices in year 1 (2003) by quantities in year 0 (2002).

Similar calculations for subsequent years produce the costs of the 2002 basket as shown in Table 10.8.

It can be seen that *if* firms had purchased the same quantities of each energy source in the following years, they would have had to pay more in each subsequent year up to 2006.

To obtain the energy price index from these numbers we measure the cost of the basket in each year relative to its 2002 cost, i.e. we divide the cost of the basket in each successive year by $\Sigma p_{0i}q_{0i}$ and multiply by 100.

Table 10.8 The cost of the energy basket, 2002–2006

	Formula	Cost
2002	$\Sigma p_0 q_0$	8581.01
2003	$\Sigma p_1 q_0$	8725.39
2004	$\Sigma p_2 q_0$	9887.15
2005	$\Sigma p_3 q_0$	13 842.12
2006	$\Sigma p_4 q_0$	17 943.65

Note: For brevity, we have dropped the *i* subscript in the formula.

Table 10.9 The Laspeyres price index

Year	Formula	Index	
2002	$\frac{\sum p_0 q_0}{\sum p_0 q_0} \times 100$	100	(= 8725.39/8581.01 × 100)
2003	$\frac{\sum p_1 q_0}{\sum p_0 q_0} \times 100$	101.68	(= 9887.15/8581.01 × 100)
2004	$\frac{\sum p_2 q_0}{\sum p_0 q_0} \times 100$	115.22	etc.
2005	$\frac{\sum p_3 q_0}{\sum p_0 q_0} \times 100$	161.31	
2006	$\frac{\sum p_4 q_0}{\sum p_0 q_0} \times 100$	209.11	

This index is given in Table 10.9 and is called the **Laspeyres price index** after its inventor. We say that it uses **base-year weights** (i.e. quantities in the base year 2002 form the weights in the basket).

We have set the value of the index to 100 in 2002, i.e. the reference year and the base year coincide, though this is not essential.

The Laspeyres index for year n with the base year as year 0 is given by the following formula

$$P_{\mathrm{L}}^{n} = \frac{\sum p_{ni} q_{0i}}{\sum p_{0i} q_{0i}} \times 100 \tag{10.1}$$

(Henceforth we shall omit the i subscript on prices and quantities in the formulae for index numbers, for brevity.) The index shows that energy prices increased by 109.11% over the period – a rapid rate of increase. The rise amounts to an average increase of 20.25% p.a. in the cost of energy. During the same period, prices in general rose by 12.5% (or 3.0% p.a.) so in relative terms energy became markedly more expensive.

The choice of 2002 as the base year for the index was an arbitrary one; any year will do. If we choose 2003 as the base year then the cost of the 2003 basket is evaluated in each year (including 2002), and this will result in a slightly different Laspeyres index. The calculations are in Table 10.10. The final two columns of the table compare the Laspeyres index constructed using the 2003 and 2002 baskets respectively (the former adjusted to 2002 = 100). A very small

Table 10.10 The Laspeyres price index using the 2003 basket

	Cost of 2003 basket	Laspeyres index 2003 = 100	Laspeyres index 2002 = 100 (2003 basket)	Laspeyres index using 2002 basket
2002	8707.50	98.24	100	100
2003	8863.52	100	101.79	101.68
2004	10 033.45	113.20	115.23	115.22
2005	14 040.80	158.41	161.25	161.31
2006	18 198.34	205.32	209.00	209.11

difference can be seen, which is due to the fact that consumption patterns were very similar in 2002 and 2003. It would not be uncommon to get a larger difference between the series than in this instance.

The Laspeyres price index shows the increase in the price of energy for the 'average' firm, i.e. one which consumes energy in the same proportions as the 2002 basket overall. There are probably very few such firms: most would use perhaps only one or two energy sources. Individual firms may therefore experience price rises quite different from those shown here. For example, a firm depending upon electricity alone would face an 85% price increase over the four years, significantly different from the figure of 109% suggested by the Laspeyres index.

Exercise 10.2

(a) The prices of fuels used by industry 1999–2003 were:

Year	Coal (£/tonne)	Petroleum (£/tonne)	Electricity (£/MWh)	Gas (£/therm)
1999	34.77	104.93	36.23	0.546
2000	35.12	137.90	34.69	0.606
2001	38.07	148.10	31.35	0.816
2002	34.56	150.16	29.83	0.780
2003	34.50	140.00	28.44	0.807

and quantities consumed by industry were:

	Coal (m tonnes)	Petroleum (m tonnes)	Electricity (m MWh)	Gas (m therms)
1999	2.04	5.33	110.98	6039

Calculate the Laspeyres price index of energy based on these data. Use 1999 as the reference year.

(b) Recalculate the index making 2001 the reference year.

(c) The quantities consumed in 2000 were:

	Coal (m tonnes)	Petroleum (m tonnes)	Electricity (m MWh)	Gas (m therms)
2000	0.72	5.52	114.11	6265

Calculate the Laspeyres index using this basket and compare to the answer to part (a).

Using current-year weights: the Paasche index

Firms do not of course consume the same basket of energy every year. One would expect them to respond to changes in the relative prices of fuels and to other factors. Technological progress means that the efficiency with which the fuels can be used changes, causing fluctuations in demand. Table 10.11 shows the quantities consumed in the years after 2002 and indicates that firms did indeed alter their pattern of consumption.

Each of these annual patterns of consumption could be used as the 'shopping basket' for the purpose of constructing the Laspeyres index and each would give a slightly different price index, as we saw with the usage of the 2002 and 2003 baskets. One cannot say that one of these is more correct than the others.

Table 10.11 Quantities of energy used, 2000–2006

	Coal (m tonnes)	Petroleum (m tonnes)	Electricity (m MWh)	Gas (m therms)
2003	1.86	6.27	113.36	5677
2004	1.85	6.45	115.84	5258
2005	1.79	6.57	118.52	5226
2006	1.71	6.55	116.31	4910

One further problem is that whichever basket is chosen remains the same over time and eventually becomes unrepresentative of the current pattern of consumption.

The **Paasche index** (denoted P_{P}^{n} to distinguish it from the Laspeyres index) overcomes these problems by using **current-year weights** to construct the index, in other words the basket is continually changing. Suppose 2002 is to be the reference year, so $P_{\mathrm{P}}^{0} = 100$. To construct the Paasche index for 2003 we use the 2003 weights (or basket), for the 2004 value of the index we use the 2004 weights, and so on. An example will clarify matters.

The Paasche index for 2003 will be the cost of the 2003 basket at 2003 prices relative to its cost at 2002 prices, i.e.

$$P_{\mathrm{P}}^{1} = \frac{\Sigma p_1 q_1}{\Sigma p_0 q_1} \times 100$$

$$P_{\mathrm{P}}^{1} = \frac{8863.52}{8707.50} \times 100 = 101.79$$

The general formula for the Paasche index in year n is given in equation (10.2).

$$P_{\mathrm{P}}^{n} = \frac{\Sigma p_n q_n}{\Sigma p_0 q_n} \times 100 \qquad (10.2)$$

Table 10.12 shows the calculation of this index for the later years.

The Paasche formula gives a slightly different result than does the Laspeyres, as is usually the case. The Paasche should generally give a slower rate of increase than does the Laspeyres index. This is because one would expect profit-maximising firms to respond to changing relative prices by switching their consumption in the direction of the inputs which are becoming relatively cheaper. The Paasche index, by using the current weights, captures this change, but the Laspeyres, assuming fixed weights, does not. This may happen slowly, as it takes time for firms to switch to different fuels, even if technically possible. This is why the Paasche can increase faster than the Laspeyres in some years (e.g. 2003), although in the long run it should increase more slowly.

Table 10.12 The Paasche price index

	Cost of basket at current prices	Cost at 2002 prices	Index
2002	8581.01	8581.01	100
2003	8863.52	8707.50	101.79
2004	9735.60	8478.09	114.83
2005	13 692.05	8546.72	160.20
2006	17 043.52	8228.72	207.12

Is one of the indices more 'correct' than the other? The answer is that neither is definitively correct. It can be shown that the 'true' value lies somewhere between the two, but it is difficult to say exactly where. If all the items which make up the index increase in price at the same rate then the Laspeyres and Paasche indices would give the same answer, so it is the change in *relative* prices and the resultant change in consumption patterns which causes problems.

Units of measurement

It is important that the units of measurement in the price and quantity tables be consistent. Note that in the example, the price of coal was measured in £/tonne and the consumption was measured in millions of tonnes. The other fuels were similarly treated (in the case of electricity, one MWh equals one million watt-hours). But suppose we had measured electricity consumption in kWh instead of MWh (1 MWh = 1000 kWh), but still measured its price in £ per MWh? We would then have 2002 data of 29.83 for price as before, but 112 650 for quantity. It is as if electricity consumption has been boosted 1000-fold, and this would seriously distort the results. The (Laspeyres) energy price index would be (by a similar calculation to the one above):

2002	2003	2004	2005	2006
100	96.2	104.8	142.09	184.68

This is incorrect, and shows a much lower value than the correct Laspeyres index (because electricity is now given too much weight in the calculation, and electricity prices were rising less rapidly than others).

The Human Development Index

One of the more interesting indices to appear in recent years is the Human Development Index (HDI), produced by the United Nations Development Programme (UNDP). The HDI aims to provide a more comprehensive socioeconomic measure of a country's progress than GDP (national output). Output is a measure of how well-off we are in material terms, but makes no allowance for the quality of life and other factors.

The HDI combines a measure of well-being (GDP per capita) with longevity (life expectancy) and knowledge (based on literacy and years of schooling). As a result, each country obtains a score, from 0 (poor) to 1 (good). Some selected values are given in the following table.

Country	HDI 1970	HDI 1980	HDI 2003	Rank (HDI 92)	Rank (GDP)
Canada	0.887	0.911	0.932	1	11
UK	0.873	0.892	0.919	10	19
Hong Kong	0.737	0.830	0.875	24	22
Gabon	0.378	0.468	0.525	114	42
Senegal	0.176	0.233	0.322	143	114

One can see that there is an association between the HDI and GDP, but not a perfect one. Canada has the world's 11th highest GDP per capita but comes top of 

the HDI rankings. In contrast, Gabon, some way up the GDP rankings, is much lower when the HDI is calculated.

So how is the HDI calculated from the initial data? How can we combine life expectancy (which can stretch from 0 to 80 years or more) with literacy (the proportion of the population who can read and write)? The answer is to score all of the variables on a scale from 0 to 100.

The HDI sets a range for (national average) life expectancy between 25 and 85 years. A country with a life expectancy of 52.9 (the case of Gabon) therefore scores 0.465, i.e. 52.9 is 46.5% of the way between 25 and 85.

Adult literacy can vary between 0% and 100% of the population, so needs no adjustment. Gabon's figure is 0.625. The scale used for years of schooling is 0 to 15, so Gabon's very low average of 2.6 yields a score of 0.173. Literacy and schooling are then combined in a weighted average (with a ⅔ weight on literacy) to give a score for knowledge of $\frac{2}{3} \times 0.625 + \frac{1}{3} \times 0.173 = 0.473$.

For income, Gabon's average of $3498 is compared to the global average of $5185 to give a score of 0.636. (Incomes above $5185 are manipulated to avoid scores above 1.)

A simple average of 0.465, 0.473 and 0.636 then gives Gabon's final figure of 0.525. One can see that its average income is brought down by the poorer scores in the two other categories, resulting in a poorer HDI ranking.

The construction of this index number shows how disparate information can be brought together into a single index number for comparative purposes. Further work by UNDP adjusts the HDI on the basis of gender and reveals the stark result that no country treats its women as well as it does its men.

Adapted from: Human Development Report, 1994 and other years. More on the HDI can be found at http://www.undp.org/

It is possible to make some manipulations of the units of measurement (usually to make calculation easier) as long as all items are treated alike. If, for example, all prices were measured in pence rather than pounds (so all prices in Table 10.4 were multiplied by 100) then this would have no effect on the resultant index, as you would expect. Similarly, if all quantity figures were measured in thousands of tonnes, thousands of therms and thousands of MWh there would be no effects on the index, even if prices remained in £/tonne, etc. But if electricity were measured in pence per MWh, while all other fuels were in £/tonne, a wrong answer would again be obtained. Quantities consumed should also be measured over the same time period, for example millions of therms *per annum*. It does not matter what the time period is (days, weeks, months or years) as long as all the items are treated similarly.

Exercise 10.3

The quantities of energy used in subsequent years were:

	Coal (m tonnes)	Petroleum (m tonnes)	Electricity (m MWh)	Gas (m therms)
2001	1.69	6.60	111.34	6142
2002	1.10	5.81	112.37	5650
2003	0.69	6.69	113.93	5880

Calculate the Paasche index for 1999–2003 with 1999 as reference year. Compare this to the Laspeyres index result.

Using expenditures as weights

On occasion the quantities of each commodity consumed are not available, but expenditures are, and a price index can still be constructed using slightly modified formulae. It is often easier to find the expenditure on a good than to know the actual quantity consumed (think of housing as an example). We shall illustrate the method with a simplified example, using the data on energy prices and consumption for the years 2002 and 2003 only. The data are repeated in Table 10.13.

The data for consumption are assumed to be no longer available, but only the expenditure on each energy source as a percentage of total expenditure. Expenditure is derived as the product of price and quantity consumed.

The formula for the Laspeyres index can be easily manipulated to accord with the data as presented in Table 10.13.

The Laspeyres index formula based on expenditure shares is given in equation (10.3)[1]

$$P_{\mathrm{L}}^{n} = \sum \frac{p_n}{p_0} \times s_0 \times 100 \qquad (10.3)$$

Equation (10.3) is made up of two component parts. The first, p_n/p_0, is simply the price in year n relative to the base-year price for each energy source. The second component, $s_0 = p_0q_0/\Sigma p_0q_0$, is the share or proportion of total expenditure spent on each energy source in the base year, the data for which are in Table 10.13. It should be easy to see that the sum of the s_0 values is 1, so that equation (10.3) calculates a weighted average of the individual price increases, the weights being the expenditure shares.

The calculation of the Laspeyres index for 2003 using 2002 as the base year is therefore

$$P_{\mathrm{L}}^{n} = \frac{34.03}{36.97} \times 0.008 + \frac{152.53}{132.24} \times 0.088 + \frac{28.68}{29.83} \times 0.392 + \frac{0.809}{0.780} \times 0.513$$
$$= 1.0168$$

giving the value of the index as 101.68, the same value as derived earlier using the more usual methods. Values of the index for subsequent years are calculated

Table 10.13 Expenditure shares, 2002

	Prices	Quantities	Expenditure	Share
Coal (£/tonne)	36.97	1.81	66.92	0.8%
Petroleum (£/tonne)	132.24	5.7	753.77	8.8%
Electricity (£/MWh)	29.83	112.65	3360.35	39.2%
Gas (£/therm)	0.78	5641	4399.98	51.3%
Total			8581.01	100.0%

Note: The 0.8% share of coal is calculated as (66.92/8581.01) × 100; others are calculated similarly.

[1] See the Appendix to this chapter (page **385**) for the derivation of this formula.

by appropriate application of equation (10.3) above. This is left as an exercise for the reader, who may use Table 10.9 to verify the answers.

The Paasche index may similarly be calculated from data on prices and expenditure shares, as long as these are available for each year for which the index is required. The formula for the Paasche index is

$$P_{\mathrm{P}}^{n} = \frac{1}{\sum \frac{p_0}{p_n} s_n} \times 100 \tag{10.4}$$

The calculation of the Paasche index is also left as an exercise.

Comparison of the Laspeyres and Paasche indices

The advantages of the Laspeyres index are that it is easy to calculate and that it has a fairly clear intuitive meaning, i.e. the cost each year of a particular basket of goods. The Paasche index involves more computation, and it is less easy to envisage what it refers to. As an example of this point, consider the following simple case. The Laspeyres index values for 2004 and 2005 are 115.22 and 161.31. The ratio of these two numbers, 1.40, would suggest that prices rose by 40% between these years. What does this figure actually represent? The 2005 Laspeyres index has been divided by the same index for 2004, i.e.

$$\frac{P_{\mathrm{L}}^{3}}{P_{\mathrm{L}}^{2}} = \frac{\sum p_3 q_0}{\sum p_0 q_0} \bigg/ \frac{\sum p_2 q_0}{\sum p_0 q_0} = \frac{\sum p_3 q_0}{\sum p_2 q_0}$$

which is the ratio of the cost of the 2002 basket at 2005 prices to its cost at 2004 prices. This makes some intuitive sense. Note that it is not the same as the Laspeyres index for 2005 with 2004 as base year, which would require using q_2 in the calculation.

If the same is done with the Paasche index numbers a rise of 39.5% is obtained between 2004 and 2005, virtually the same result. But the meaning of this is not so clear, since the relevant formula is

$$\frac{P_{\mathrm{P}}^{3}}{P_{\mathrm{P}}^{2}} = \frac{\sum p_3 q_3}{\sum p_0 q_3} \bigg/ \frac{\sum p_2 q_2}{\sum p_0 q_2}$$

which does not simplify further. This is a curious mixture of 2004 and 2005 quantities, and 2002, 2004 and 2005 prices!

The major advantage of the Paasche index, however, is that the weights are continuously updated, so that the basket of goods never becomes out of date. In the case of the Laspeyres index the basket remains unchanged over a period, becoming less and less representative of what is being bought by consumers. When revision is finally made there may therefore be a large change in the weighting scheme. The extra complexity of calculation involved in the Paasche index is less important now that computers do most of the work.

Exercise 10.4

(a) Calculate the share of expenditure going to each of the four fuel types in the previous exercises and use this result to recalculate the Laspeyres and Paasche indexes using equations (10.3) and (10.4).

(b) Check that the results are the same as calculated in previous exercises.

The story so far – a brief summary

We have encountered quite a few different concepts and calculations thus far and it might be worthwhile to briefly summarise what we have covered before moving on. In order, we have examined:

- a simple index for a single commodity;
- a Laspeyres price index, which uses base year weights;
- a Paasche price index, which uses current year weights and is an alternative to the Laspeyres formulation;
- the same Laspeyres and Paasche indices, but calculated using the data in a slightly different form, using expenditure shares rather than quantities.

We now move on to examine quantity and expenditure indices, then look at the relationship between them all.

Quantity and expenditure indices

Just as one can calculate price indices, it is also possible to calculate **quantity** and **value** (or **expenditure**) **indices**. We first concentrate on quantity indices, which provide a measure of the total quantity of energy consumed by industry each year. The problem again is that we cannot easily aggregate the different sources of energy. It makes no sense to add together tonnes of coal and petroleum, therms of gas and megawatts of electricity. Some means has to be found to put these different fuels on a comparable basis. To do this, we now reverse the roles of prices and quantities: the quantities of the different fuels are weighted by their different prices (prices represent the value to the firm, at the margin, of each different fuel). As with price indices, one can construct both Laspeyres and Paasche quantity indices.

The Laspeyres quantity index

The Laspeyres quantity index for year n is given by

$$Q_L^n = \frac{\Sigma q_n p_0}{\Sigma q_0 p_0} \times 100 \tag{10.5}$$

i.e. it is the ratio of the cost of the year n basket to the cost of the year 0 basket, both valued at year 0 prices. Note that it is the same as equation (10.1) but with prices and quantities reversed.

Using 2002 as the base year, the cost of the 2003 basket at 2002 prices is

$$\Sigma q_1 p_0 = 1.86 \times 36.97 + 6.27 \times 132.24 + 113.36 \times 29.83 + 5677 \times 0.78$$
$$= 8707.50$$

and the cost of the 2002 basket at 2002 prices is 8581.01 (calculated earlier). The value of the quantity index for 2003 is therefore

$$Q_L^1 = \frac{8707.50}{8581.01} \times 100 = 101.47$$

Table 10.14 Calculation of the Laspeyres quantity index

	$\Sigma p_0 q_n$	Index	
2002	8581.01	100	
2003	8707.50	101.47	(= 8707.5/8581.01 × 100)
2004	8478.09	98.80	(= 8478.09/8581.01 × 100)
2005	8546.72	99.60	
2006	8228.72	95.89	

In other words, if prices had remained constant between 2002 and 2003, industry would have consumed 1.47% more energy (and spent 1.47% more also).

The value of the index for subsequent years is shown in Table 10.14, using the formula given in equation (10.5).

The Paasche quantity index

Just as there are Laspeyres and Paasche versions of the price index, the same is true for the quantity index. The Paasche quantity index is given by

$$Q_P^n = \frac{\Sigma q_n p_n}{\Sigma q_0 p_n} \times 100 \qquad (10.6)$$

and is the analogue of equation (10.2) with prices and quantities reversed. The calculation of this index is shown in Table 10.15, which shows a similar trend to the Laspeyres index in Table 10.14. Normally one would expect the Paasche to show a slower increase than the Laspeyres quantity index: firms should switch to inputs whose relative prices fall; the Paasche gives lesser weight (current prices) to these quantities than does the Laspeyres (base-year prices) and thus shows a slower rate of increase.

Expenditure indices

The **expenditure** or **value index** is simply an index of the cost of the year n basket at year n prices and so it measures how expenditure changes over time. The formula for the index in year n is

$$E^n = \frac{\Sigma p_n q_n}{\Sigma p_0 q_0} \times 100 \qquad (10.7)$$

There is obviously only one value index and one does not distinguish between Laspeyres and Paasche formulations. The index can be easily derived, as shown

Table 10.15 Calculation of the Paasche quantity index

	$\Sigma p_n q_n$	$\Sigma p_n q_0$	Index
2002	8581.01	8581.01	100
2003	8863.52	8725.39	101.58
2004	9735.60	9887.15	98.47
2005	13 692.05	13 842.12	98.92
2006	17 043.52	17 943.65	94.98

Note: The final column is calculated as the ratio of the previous two columns.

Table 10.16 The expenditure index

	$\Sigma p_n q_n$	Index
2002	8581.01	100
2003	8863.52	103.29
2004	9735.60	113.46
2005	13 692.05	159.56
2006	17 043.52	198.62

Note: The expenditure index is a simple index of the expenditures in the previous column.

in Table 10.16. The expenditure index shows how industry's expenditure on energy is changing over time. Thus expenditure in 2006 was 99% higher than in 2002, for example.

The increase in expenditure over time is a consequence of two effects: (i) changes in the prices of energy and (ii) changes in quantities purchased. It should therefore be possible to decompose the expenditure index into price and quantity effects. You many not be surprised to learn that these effects can be measured by the price and quantity indices we have already covered. We look at this decomposition in more detail in the next section.

Relationships between price, quantity and expenditure indices

Just as multiplying a price by a quantity gives total value, or expenditure, the same is true of index numbers. The value index can be decomposed as the product of a price index and a quantity index. In particular, it is the product of a Paasche quantity index and a Laspeyres price index, *or* the product of a Paasche price index and a Laspeyres quantity index. This can be very simply demonstrated using Σ notation

$$E^n = \frac{\Sigma p_n q_n}{\Sigma p_0 q_0} = \frac{\Sigma p_n q_n}{\Sigma p_n q_0} \times \frac{\Sigma p_n q_0}{\Sigma p_0 q_0} = Q_P^n \times P_L^n$$

(Paasche quantity times Laspeyres price index) **(10.8)**

or

$$E^n = \frac{\Sigma p_n q_n}{\Sigma p_0 q_0} = \frac{\Sigma p_n q_n}{\Sigma p_0 q_n} \times \frac{\Sigma p_0 q_n}{\Sigma p_0 q_0} = P_P^n \times Q_L^n$$

(Paasche price times Laspeyres quantity index) **(10.9)**

Thus increases in value or expenditure can be decomposed into price and quantity effects. Two decompositions are possible and give slightly different answers.

It is also evident that a quantity index can be constructed by dividing a value index by a price index, since by simple manipulation of equations (10.8) and (10.9) we obtain

$$Q_P^n = E^n / P_L^n \quad \textbf{(10.10)}$$

and

$$Q_L^n = E^n / P_P^n \quad \textbf{(10.11)}$$

Note that dividing the expenditure index by a Laspeyres price index gives a Paasche quantity index, and dividing by a Paasche price index gives a Laspeyres

Table 10.17 Deflating the expenditure series

	Expenditure at current prices	Laspeyres price index	Expenditure in volume terms	Index
2002	8581.01	100	8581.01	100
2003	8863.52	101.68	8716.86	101.58
2004	9735.60	115.22	8449.49	98.47
2005	13 692.05	161.31	8487.99	98.92
2006	17 043.52	209.11	8150.55	94.98

quantity index. In either case we go from a series of expenditures to one representing quantities, having taken out the effect of price changes. This is known as **deflating** a series and is a widely used and very useful technique. We shall reconsider our earlier data in the light of this. Table 10.17 provides the detail. Column 2 of the table shows the expenditure on fuel at **current prices** or in **cash terms**. Column 3 contains the Laspeyres price index repeated from Table 10.9 above. Deflating (dividing) column 2 by column 3 and multiplying by 100 yields column 4 which shows expenditure on fuel in **quantity** or **volume terms**. The final column turns the volume series in column 4 into an index with 2002 = 100.

This final index is equivalent to a Paasche quantity index, as illustrated by equation (10.7) and can be seen by comparison to Table 10.15 above.

Trap!

A common mistake is to believe that once a series has been turned into an index, it is inevitably in real (or volume) terms. This is *not* the case. One can have an index of a cash (or nominal) series (e.g. in Table 10.16 above) *or* of a real series (the final column of Table 10.17). An index number is really just a change of the units of measurement to something more useful for presentation purposes; it is not the same as deflating the series.

In the example above we used the energy price index to deflate the expenditure series. However, it is also possible to use a *general* price index (such as the retail price index or the GDP deflator) to deflate. This gives a slightly different result, both in numerical terms and in its interpretation. Deflating by a general price index yields a series of expenditures in **constant prices** or in **real terms**. Deflating by a specific price index (e.g. of energy) results in a **quantity** or **volume** series.

An example should clarify this (see Problem 10.11 for data). The government spends billions of pounds each year on the health service. If this cash expenditure series is deflated by a general price index (e.g. the GDP deflator) then we obtain expenditure on health services at constant prices, or real expenditure on the health service. If the NHS pay and prices index is used as a deflator, then the result is an index of the quantity or volume of health services provided. Since the NHS index tends to rise more rapidly than the GDP deflator, the volume series rises more slowly than the series of expenditure at constant prices. This can lead to a vigorous, if pointless, political debate. The government claims it is spending more on the health service, in real terms, while the opposition claims that the health service is getting fewer resources. As we have seen, both can be right.

Exercise 10.5

(a) Use the data from earlier exercises to calculate the Laspeyres quantity index.

(b) Calculate the Paasche quantity index.

(c) Calculate the expenditure index.

(d) Check that dividing the expenditure index by the price index gives the quantity index (remember that there are two ways of doing this).

The real rate of interest

Another example of 'deflating' is calculating the 'real' rate of interest. This adjusts the actual (sometimes called 'nominal') rate of interest for changes in the value of money, i.e. inflation. If you earn a 7% rate of interest on your money over a year, but the price level rises by 5% at the same time, you are clearly not 7% better off. The real rate of interest in this case would be given by

$$\text{real interest rate } \frac{1+0.07}{1+0.05} - 1 = 0.019 = 1.9\% \qquad \textbf{(10.12)}$$

In general, if r is the interest rate and i is the inflation rate, the real rate of interest is given by

$$\text{real interest rate } \frac{1+r}{1+i} - 1 \qquad \textbf{(10.13)}$$

A simpler method is often used in practice, which gives virtually identical results for small values of r and i. This is to subtract the inflation rate from the interest rate, giving 7% − 5% = 2% in this case.

Chain indices

Whenever an index number series over a long period of time is wanted, it is usually necessary to link together a number of separate, shorter indices, resulting in a **chain index**. Without access to the original raw data it is impossible to construct a proper Laspeyres or Paasche index, so the result will be a mixture of different types of index number but it is the best that can be done in the circumstances.

Suppose that the following two index number series are available. Access to the original data is assumed to be impossible.

Laspeyres price index for energy, 2002–2006 (from Table 10.9)

2002	2003	2004	2005	2006
100	101.68	115.22	161.31	209.11

Laspeyres price index for energy, 1999–2003

1998	1999	2000	2001	2002
104.54	100	104.63	116.68	111.87

The two series have different reference years and use different shopping baskets of consumption. The first index measures the cost of the 2002 basket in each of the subsequent years. The second measures the price of the 1999 basket

Table 10.18 A chain index of energy prices, 1998–2006

	'Old' index	'New' index	Chain index
1998	104.54	–	104.54
1999	100	–	100
2000	104.63	–	104.63
2001	116.68	–	116.68
2002	111.87	100	111.87
2003	–	101.68	113.75
2004	–	115.22	128.90
2005	–	161.31	180.46
2006	–	209.11	233.93

Note: After 2001, the chain index values are calculated by multiplying the 'new' index by 1.1187; e.g. 113.75 = 101.68 × 1.1187 for 2003.

in surrounding years. There is an 'overlap' year which is 2002. How do we combine these into one continuous index covering the whole period?

The obvious method is to use the ratio of the costs of the two baskets in 2002, 111.87/100 = 1.1187, to alter one of the series. To base the continuous series on 1999 = 100 requires multiplying each of the post-2002 figures by 1.1187, as is demonstrated in Table 10.18. Alternatively, the continuous series could just as easily be based on 2002 = 100 by dividing the pre-2002 numbers by 1.1187.

The continuous series is not a proper Laspeyres index number as can be seen if we examine the formulae used. We shall examine the 2006 figure, 233.93, by way of example. This figure is calculated as 233.93 = 209.11 × 111.87/100 which in terms of our formulae is

$$\frac{\sum p_{06}q_{02}}{\sum p_{02}q_{02}} \times \frac{\sum p_{02}q_{99}}{\sum p_{02}q_{99}} \; 100 \qquad (10.14)$$

The proper Laspeyres index for 2006 using 1999 weights is

$$\frac{\sum p_{06}q_{99}}{\sum p_{99}q_{99}} \times 100 \qquad (10.15)$$

There is no way that this latter equation can be derived from equation (10.14), proving that the former is not a properly constructed Laspeyres index number. Although it is not a proper index number series it does have the advantage of the weights being revised and therefore more up-to-date.

Similar problems arise when deriving a chain index from two Paasche index number series. Investigation of this is left to the reader; the method follows that outlined above for the Laspeyres case.

The Retail Price Index

As an example, consider the UK **Retail Price Index**, which is one of the more sophisticated of index numbers, involving the recording of the prices of around 550 items each month, and weighting them on the basis of households' expenditure patterns as revealed by the Expenditure and Food Survey (the EFS was explained in more detail in Chapter 9 on sampling methods). The principles involved in the calculation are similar to those set out above, with slight differences due to a variety of reasons.

The RPI is something of a compromise between a Laspeyres and a Paasche index. It is calculated monthly, and within each calendar year the weights used remain constant, so that it takes the form of a Laspeyres index. Each January, however, the weights are updated on the basis of evidence from the EFS, so that the index is in fact a set of chain-linked Laspeyres indices, the chaining taking place in January each year. Despite the formal appearance as a Laspeyres index, the RPI measured over a period of years has the characteristics of a Paasche index, due to the annual change in the weights.

Another departure from principle is the fact that about 14% of households are left out when expenditure weights are calculated. These consist of most pensioner households (10%) and the very rich (4%), because they tend to have significantly different spending patterns from the rest of the population and their inclusion would make the index too unrepresentative. A separate RPI is calculated for pensioners, while the very rich have to do without one.

A change in the quality of goods purchased can also be problematic, as alluded to earlier. If a manufacturer improves the quality of a product and charges more, is it fair to say that the price has gone up? Sometimes it is possible to measure improvement (if the power of a vacuum cleaner is increased, for example), but other cases are more difficult, such as if the punctuality of a train service is improved. By how much has quality improved? In many circumstances the statistician has to make a judgement about the best procedure to adopt. The ONS does make explicit allowance for the increase for quality of personal computers, for example, taking account of such as factors as increased memory and processing speed.

Prices in the long run

Table 10.19 shows how prices have changed over the longer term. The 'inflation-adjusted' column shows what the item would have cost if it had risen in line with the overall retail price index. It is clear that some relative prices have changed substantially and you can try to work out the reasons.

Table 10.19 80 years of prices: 1914–1994

Item	1914 price	Inflation-adjusted price	1994 price
Car	£730	£36 971	£6995
London–Manchester 1st class rail fare	£2.45	£124.08	£130
Pint of beer	1p	53p	£1.38
Milk (quart)	1.5p	74p	70p
Bread	2.5p	£1.21	51p
Butter	6p	£3.06	68p
Double room at Savoy Hotel, London	£1.25	£63.31	£195

→

The Office for National Statistics has gone back even further and shown that, since 1750, prices have increased about 140 times. Most of this occurred after 1938: up till then prices had only risen by about three times (over two centuries, about half a per cent per year on average), since then prices have risen 40-fold, or about 6% per annum.

Exercise 10.6

The index of energy prices for the years 1995–1999 was:

1995	1996	1997	1998	1999
100	86.3	85.5	88.1	88.1

Use these data to calculate a chain index from 1995 to 2006, setting 1995 = 100.

Discounting and present values

Deflating makes expenditures in different years comparable by correcting for the effect of inflation. The future sum is deflated (reduced) because of the increase in the general price level. **Discounting** is a similar procedure for comparing amounts across different years, correcting for **time preference**. For example, suppose that by investing £1000 today a firm can receive £1100 in a year's time. To decide if the investment is worthwhile, the two amounts need to be compared.

If the prevailing interest rate is 12%, then the firm could simply place its £1000 in the bank and earn £120 interest, giving it £1120 at the end of the year. Hence the firm should not invest in this particular project; it does better keeping money in the bank. The investment is not undertaken because

$$£1000 \times (1 + r) > £1100$$

where r is the interest rate, 12% or 0.12. Alternatively, this inequality may be expressed as

$$£1000 > \frac{£1100}{(1 + r)}$$

The expression on the right-hand side of the inequality sign is the **present value** (PV) of £1100 received in one year's time. Here, r is the rate of discount and is equal to the rate of interest in this example because this is the rate at which the firm can transform present into future income, and vice versa. In what follows, we use the terms interest rate and discount rate interchangeably. The term $1/(1 + r)$ is known as the **discount factor**. Multiplying an amount by the discount factor results in the present value of the sum.

We can also express the inequality as follows (by subtracting £1000 from each side):

$$0 > -£1000 + \frac{£1100}{(1 + r)}$$

The right-hand side of this expression is known as the **net present value** (NPV) of the project. It represents the difference between the initial outlay and the present value of the return generated by the investment. Since this is negative the investment is not worthwhile (the money would be better placed on deposit in a bank). The general rule is to invest if the *NPV* is positive.

Similarly, the present value of £1100 to be received in two years' time is

$$PV = \frac{£1100}{(1+r)^2} = \frac{£1100}{(1+0.12)^2} = £876.91$$

when $r = 12\%$. In general, the *PV* of a sum *S* to be received in *t* years is

$$PV = \frac{S}{(1+r)^t}$$

The *PV* may be interpreted as the amount a firm would be prepared to pay today to receive an amount *S* in *t* years' time. Thus a firm would not be prepared to make an outlay of more than £876.91 in order to receive £1100 in two years' time. It would gain more by putting the money on deposit and earning 12% interest per annum.

Most investment projects involve an initial outlay followed by a *series* of receipts over the following years, as illustrated by the figures in Table 10.20. In order to decide if the investment is worthwhile, the present value of the income stream needs to be compared to the initial outlay. The *PV* of the income stream is obtained by adding together the present value of each year's income. Thus we calculate[2]

$$PV = \frac{S_1}{(1+r)} + \frac{S_2}{(1+r)^2} + \frac{S_3}{(1+r)^3} + \frac{S_4}{(1+r)^4} \tag{10.16}$$

or more concisely, using Σ notation

$$PV = \sum \frac{S_t}{(1+r)^t} \tag{10.17}$$

Columns 3 and 4 of the table show the calculation of the present value. The discount factors, $1/(1+r)^t$, are given in column 3. Multiplying column 2 by column 3 gives the individual elements of the *PV* calculation (as in equation (10.16) above) and their sum is 1034.14, which is the present value of the returns. Since the *PV* is greater than the initial outlay of 1000 the investment generates a return of at least 12% and so is worthwhile.

Table 10.20 The cash flows from an investment project

Year	Outlay or income		Discount factor	Discounted income
2001	Outlay	−1000		
2002	Income	300	0.893	267.86
2003		400	0.797	318.88
2004		450	0.712	320.30
2005		200	0.636	127.10
Total				1034.14

Note: The discount factors are calculated as $0.893 = 1/(1.12)$, $0.797 = 1/(1.12)^2$, etc.

[2] This present value example has only four terms but in principle there can be any number of terms stretching into the future.

An alternative investment criterion: the internal rate of return

The investment rule can be expressed in a different manner, using the **internal rate of return** (*IRR*). This is the rate of discount which makes the *NPV* equal to zero, i.e. the present value of the income stream is equal to the initial outlay. An *IRR* of 10% equates £1100 received next year to an outlay of £1000 today. Since the *IRR* is less than the market interest rate (12%) this indicates that the investment is not worthwhile: it only yields a rate of return of 10%. The rule 'invest if the *IRR* is greater than the market rate of interest' is equivalent to the rule 'invest if the net present value is positive, using the interest rate to discount future revenues'.

In general it is mathematically difficult to find the *IRR* of a project with a stream of future income, except by trial and error methods. The *IRR* is the value of r which sets the *NPV* equal to zero, i.e. it is the solution to

$$NPV = -S_0 + \sum \frac{S_t}{(1+r)^t} = 0 \qquad (10.18)$$

where S_0 is the initial outlay. Fortunately, most spreadsheet programs have an internal routine for its calculation. This is illustrated in Figure 10.1 which shows the calculation of the *IRR* for the data in Table 10.20 above.

Cell C13 contains the formula '= *IRR*(C6:C10, 0.1)' – this can be seen just above the column headings – which is the function used in *Excel* to calculate the internal rate of return. The financial flows of the project are in cells C6:C10; the value 0.1 (10%) is an initial guess at the answer – *Excel* starts from this value and then tries to improve upon it. The *IRR* for this project is found to be 13.7% which is indeed above the market interest rate of 12%. The final two columns show that the *PV* of the income stream, when discounted using the internal rate of return, is equal to the initial outlay (as it should be). The discount factors in the penultimate column are calculated using $r = 13.7\%$.

Microsoft Excel - IRR calculation.xls

File Edit View Insert Format Tools Data Window Help

C13 =IRR(C6:C10,0.1)

	A	B	C	D	E	F	G
4							
5	Year	Outlay or income		Discount factor	Discounted income	Discount factor using IRR	Discounted income
6	2001	Outlay	-1000				
7	2002	Income	300	0.893	267.86	0.880	263.96
8	2003		400	0.797	318.88	0.774	309.66
9	2004		450	0.712	320.3	0.681	306.52
10	2005		200	0.636	127.1	0.599	119.86
11	Total				1034.14		1000.00
12							
13	Internal rate of return		13.7%				
14							
15							
16							
17							

Figure 10.1
Calculation of *IRR*

Note: Note that the first term in the series is the initial outlay (cell C4) and that it is entered as a *negative* number. If a positive value is entered, the *IRR* function will not work.

The *IRR* is particularly easy to calculate if the income stream is a constant monetary sum. If the initial outlay is S_0 and a sum S is received each year in perpetuity (like a bond), then the *IRR* is simply

$$IRR = \frac{S}{S_0}$$

For example, if an outlay of £1000 yields a permanent income stream of £120 p.a. then the *IRR* is 12%. This should be intuitively obvious, since investing £1000 at an interest rate of 12% would give you an annual income of £120.

Although the *NPV* and *IRR* methods are identical in the above example, this is not always the case in more complex examples. When comparing two investment projects of different sizes, it is possible for the two methods to come up with different rankings. Delving into this issue is beyond the scope of this book but, in general, the *NPV* method is the more reliable of the two.

Nominal and real interest rates

The above example took no account of possible inflation. If there were a high rate of inflation, part of the future returns to the project would be purely inflationary gains and would not reflect real resources. Is it possible our calculation is misleading under such circumstances?

There are two ways of dealing with this problem:

(1) use the actual cash flows and the nominal (market) interest rate to discount, or

(2) use real (inflation-adjusted) flows and the real interest rate.

These two methods should give the same answer.

If an income stream has already been deflated to real terms then the present value should be obtained by discounting by the **real interest rate**, not the nominal (market) rate. Table 10.21 illustrates the principle. Column (1) repeats the income flows in cash terms from Table 10.20. Assuming an inflation rate of $i = 7\%$ per annum gives the price index shown in column (2), based on 2001 = 100. This is used to deflate the cash series to real terms, shown in column (3). This is in constant (2001) prices. If we were presented only with the real income series and could not obtain the original cash flows we would have to discount the real series by the real interest rate r_r, defined by

Table 10.21 Discounting a real income stream

Year		Cash flows (1)	Price index (2)	Real income (3)	Real discount factor (4)	Discounted sums (5)
2001	Outlay	−1000	100			
2002	Income	300	107.0	280.37	0.955	267.86
2003		400	114.5	349.38	0.913	318.88
2004		450	122.5	367.33	0.872	320.30
2005		200	131.1	152.58	0.833	127.10
Total						1034.14

$$1 + r_r = \frac{1 + r}{1 + i} \tag{10.19}$$

With a (nominal) interest rate of 12% and an inflation rate of 7% this gives

$$1 + r_r = \frac{1 + 0.12}{1 + 0.07} = 1.0467 \tag{10.20}$$

so that the real interest rate is 4.67% and in this example is the same every year. The discount factors used to discount the real income flows are shown in column (4) of the table, based on the real interest rate, the discounted sums are in column (5) and the present value of the real income series is £1034.14. This is the same as was found earlier, by discounting the cash figures by the nominal interest rate. Thus one can discount *either* the nominal (cash) values using the nominal discount rate, *or* the real flows by the real interest rate. Make sure you do not confuse the nominal and real interest rates.

The real interest rate can be approximated by subtracting the inflation rate from the nominal interest rate, i.e. 12% − 7% = 5%. This gives a reasonably accurate approximation for low values of the interest and inflation rates (below about 10% p.a.). Because of the simplicity of the calculation, this method is often preferred.

Exercise 10.7

(a) An investment of £100 000 yields returns of £25 000, £35 000, £30 000 and £15 000 in each of the subsequent four years. Calculate the present value of the income stream and compare to the initial outlay, using an interest rate of 10% per annum.

(b) Calculate the internal rate of return on this investment.

Exercise 10.8

(a) An investment of £50 000 yields cash returns of £20 000, £25 000, £30 000 and £10 000 in each subsequent year. The rate of inflation is a constant 5% and the rate of interest is constant at 9%. Use the rate of inflation to construct a price index and discount the cash flows to real terms.

(b) Calculate the real discount rate.

(c) Use the real discount rate to calculate the present value of the real income flows.

(d) Compare the answer to part (c) to the result where the nominal cash flows and nominal interest rate are used.

Inequality indices

A separate set of index numbers is used specifically in the measurement of inequality, such as inequality in the distribution of income. We have already seen how we can measure the dispersion of a distribution via the variance and standard deviation. This is based upon the deviations of the observations about the mean. An alternative idea is to measure the difference between *every pair* of observations, and this forms the basis of a statistic known as the **Gini coefficient**. This would probably have remained an obscure measure, due to the complexity of calculation, were it not for Konrad Lorenz, who showed that there is an attractive visual interpretation of it, now known as the **Lorenz curve**, and a relatively simple calculation of the Gini coefficient, based on this curve.

We start off by constructing the Lorenz curve, based on data for the UK income distribution in 2006, and proceed then to calculate the Gini coefficient. We then use these measures to look at inequality both over time (in the UK) and across different countries.

We then examine another manifestation of inequality, in terms of market shares of firms. For this analysis we look at the calculation of **concentration ratios** and at their interpretation.

The Lorenz curve

Table 10.22 shows the data for the distribution of income in the UK based on data from the *Family Resources Survey 2006–07*, published by the ONS. The data report the total weekly income of each household, which means that income is recorded after any cash benefits from the state (e.g. a pension) have been received but before any taxes have been paid.

The table indicates a substantial degree of inequality. For example, the poorest 14% of households earn £200 per week or less, while the richest 17% earn more than £1000, five times as much. Although these figures give some idea of the extent of inequality, they relate only to relatively few households at the extremes of the distribution. A **Lorenz curve** is a way of graphically presenting the whole distribution. A typical Lorenz curve is shown in Figure 10.2.

Households are ranked along the horizontal axis, from poorest to richest, so that the median household, for example, is halfway along the axis. On the vertical axis is measured the cumulative share of income, which goes from 0% to 100%. A point such as A on the diagram indicates that the poorest 30% of households earn 5% of total income. Point B shows that the poorest half of

Table 10.22 The distribution of gross income in the UK, 2006–2007

Range of weekly household income	Mid-point of interval	Number of households
0–	50	516
100–	150	3095
200–	250	3869
300–	350	3095
400–	450	2579
500–	550	2063
600–	650	2063
700–	750	1548
800–	850	1290
900–	950	1032
1000–	1250	4385
Total		25 534

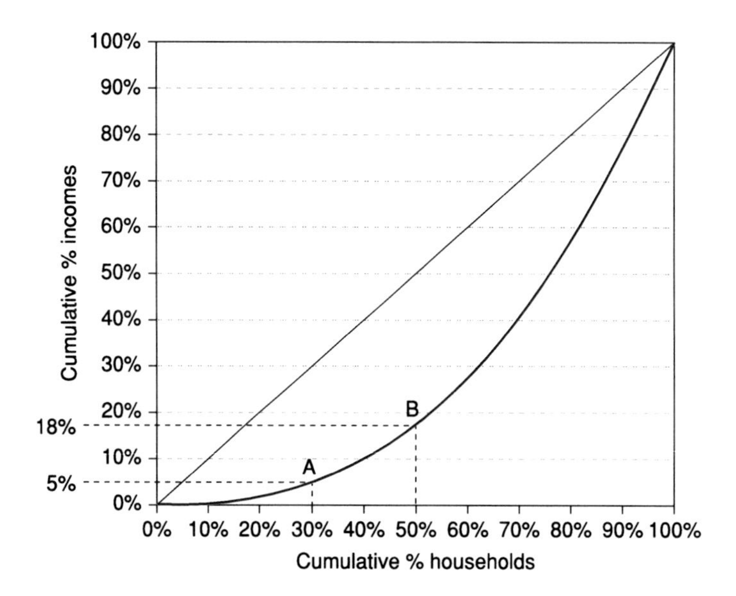

Figure 10.2
Typical Lorenz curve

the population earn only 18% of income (and hence the other half earn 82%). Joining up all such points maps out the Lorenz curve.

A few things are immediately obvious about the Lorenz curve:

- Since 0% of households earn 0% of income, and 100% of households earn 100% of income, the curve must run from the origin up to the opposite corner.
- Since households are ranked from poorest to richest, the Lorenz curve must lie below the 45° line, which is the line representing complete equality. The further away from the 45° line is the Lorenz curve, the greater is the degree of inequality.
- The Lorenz curve must be concave from above: as we move to the right we encounter successively richer individuals, so the cumulative income grows faster.

Table 10.23 shows how to generate a Lorenz curve for the data given in Table 10.22. The task is to calculate the $\{x, y\}$ coordinates for the Lorenz curve. These are given in columns 6 and 8 respectively of the table. Column 5 of the table calculates the proportion of households in each income category (i.e. the relative frequencies, as in Chapter 1), and these are then cumulated in column 6. These are the figures which are used along the horizontal axis. Column 4 calculates the total income going to each income class (by multiplying the class frequency by the mid-point). The proportion of total income going to each class is then calculated in column 7 (class income divided by total income). Column 8 cumulates the values in column 7.

Using columns 6 and 8 of the table we can see, for instance, that the poorest 2% of the population have about 0.2% of total income (one-tenth of their 'fair share'); the poorer half have about 25% of income; and the top 20% have about 40% of total income. Figure 10.3 shows the Lorenz curve plotted, using the data in columns 6 and 8 of the table above.

Table 10.23 Calculation of the Lorenz curve coordinates

Range of income	Mid-point	Number of households	Total income	% Households	% Cumulative households (x)	% Income	% Cumulative income (y)
(1)	(2)	(3)	(4)	(5)	(6)	(7)	(8)
0–	50	516	25 792	2.0%	2.0%	0.2%	0.2%
100–	150	3095	464 256	12.1%	14.1%	3.1%	3.3%
200–	250	3869	967 200	15.2%	29.3%	6.5%	9.8%
300–	350	3095	1 083 264	12.1%	41.4%	7.3%	17.1%
400–	450	2579	1 160 640	10.1%	51.5%	7.8%	24.8%
500–	550	2063	1 134 848	8.1%	59.6%	7.6%	32.5%
600–	650	2063	1 341 184	8.1%	67.7%	9.0%	41.5%
700–	750	1548	1 160 640	6.1%	73.7%	7.8%	49.3%
800–	850	1290	1 096 160	5.1%	78.8%	7.4%	56.6%
900–	950	1032	980 096	4.0%	82.8%	6.6%	63.2%
1000–	1250	4385	5 480 800	17.2%	100.0%	36.8%	100.0%
		25 534	14 894 880	100.0%		100.0%	

Notes:
Column 4 = column 2 × column 3
Column 5 = column 3 ÷ 25 534
Column 6 = column 5 cumulated
Column 7 = column 4 ÷ 14 894 880
Column 8 = column 7 cumulated

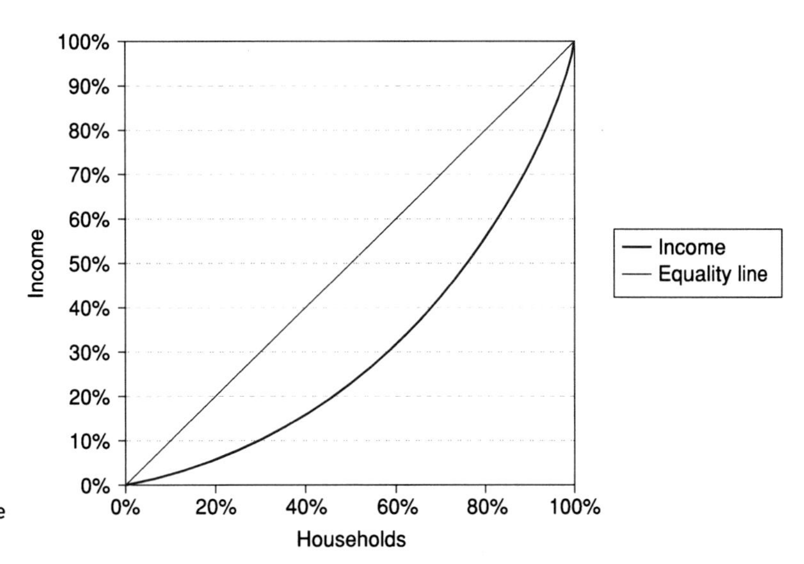

Figure 10.3
Lorenz curve for income data

The Gini coefficient

The Gini coefficient is a numerical representation of the degree of inequality in a distribution and can be derived directly from the Lorenz curve. The Lorenz curve is illustrated once again in Figure 10.4 and the Gini coefficient is simply the ratio of area A to the sum of areas A and B.

Denoting the Gini coefficient by G, we have

$$G = \frac{\text{A}}{\text{A} + \text{B}} \tag{10.21}$$

and it should be obvious that G must lie between 0 and 1. When there is total equality the Lorenz curve coincides with the 45 line, area A then disappears, and $G = 0$. With total inequality (one household having all the income), area B disappears, and $G = 1$. Neither of these extremes is likely to occur in real life; instead one will get intermediate values, but the lower the value of G, the less inequality there is (though see the *caveats* listed below). One could compare two countries, for example, simply by examining the values of their Gini coefficients.

The Gini coefficient may be calculated from the following formulae for areas A and B, using the x and y co-ordinates from Table 10.23

$$\begin{aligned} \text{B} = \tfrac{1}{2}\{&(x_1 - x_0) \times (y_1 + y_0) \\ &+ (x_2 - x_1) \times (y_2 + y_1) \\ &\quad\vdots \\ &+ (x_k - x_{k-1}) \times (y_k + y_{k-1})\} \end{aligned} \tag{10.22}$$

$x_0 = y_0 = 0$ and $x_k = y_k = 100$ represent the two end-points of the Lorenz curve and the other x and y values are the coordinates of the intermediate points. k is the number of classes for income in the frequency table. Area A is then given by[3]

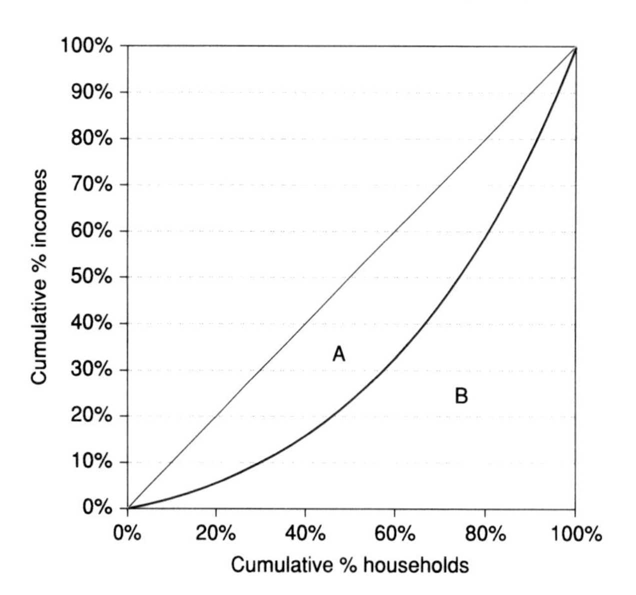

Figure 10.4 Calculation of the Gini coefficient from the Lorenz curve

[3] The value 5000 is correct if one uses percentages, as here (it is $100 \times 100 \times \frac{1}{2}$, the area of the triangle). If one uses percentages expressed as decimals, then $\text{A} = 0.5 - \text{B}$.

$$A = 5000 - B \tag{10.23}$$

and the Gini coefficient is then calculated as

$$G = \frac{A}{A + B} \text{ or } \frac{A}{5000} \tag{10.24}$$

Thus for the data in Table 10.23 we have

$$\begin{aligned} B = \tfrac{1}{2} \times \{&(2.0 - 0) \times (0.2 + 0) \\ &+ (14.1 - 2.0) \times (3.3 + 0.2) \\ &+ (29.3 - 14.1) \times (9.8 + 3.3) \\ &+ (41.4 - 29.3) \times (17.1 + 9.8) \\ &+ (51.5 - 41.4) \times (24.8 + 17.1) \\ &+ (59.6 - 51.5) \times (32.5 + 24.8) \\ &+ (67.7 - 59.6) \times (41.5 + 32.5) \\ &+ (73.7 - 67.7) \times (49.3 + 41.5) \\ &+ (78.8 - 73.7) \times (56.6 + 49.3) \\ &+ (82.8 - 78.8) \times (63.2 + 56.6) \\ &+ (100 - 82.8) \times (100 + 63.2)\} \\ = 3210.5& \end{aligned} \tag{10.25}$$

Therefore area A = 5000 − 3210.5 = 1789.5 and we obtain

$$G = \frac{1789.5}{5000} = 0.3579 \tag{10.26}$$

or approximately 36%.

This method implicitly assumes that the Lorenz curve is made up of straight line segments connecting the observed points, which is in fact not true – it should be a smooth curve. Since the straight lines will lie inside the true Lorenz curve, area B is *over*-estimated and so the calculated Gini coefficient is biased downwards. The true value of the Gini coefficient is slightly greater than 36% therefore. The bias will be greater (a) the fewer the number of observations and (b) the more concave is the Lorenz curve (i.e. the greater is inequality). The bias is unlikely to be substantial, however, so is best left untreated.

An alternative method of calculating G is simply to draw the Lorenz curve on gridded paper and count squares. This has the advantage that you can draw a smooth line joining the observations and avoid the bias problem mentioned above. This alternative method can prove reasonably quick and accurate, but has the disadvantage that you cannot use a computer to do it!

Is inequality increasing?

The Gini coefficient is only useful as a comparative measure, for looking at trends in inequality over time, or for comparing different countries or regions. Table 10.24, taken from the *Statbase* website, shows the value of the Gini coefficient for the UK over the past 10 years and shows how it was affected by the tax system. The results are based on *equivalised* income, i.e. after making a correction for differences in family size.[4] For this reason there is a slight difference from the Gini coefficient calculated above, which uses unadjusted data.

[4] This is because a larger family needs more income to have the same living standard as a smaller one.

Table 10.24 Gini coefficients for the UK, 1995/96–2005/06

	Original income	Gross income	Disposable income	Post-tax income
1995/96	51.9	35.7	32.5	36.5
2000/01	51.3	37.5	34.6	38.9
2005/06	51.9	37.3	33.6	37.3

Note: Gross income is original income plus certain state benefits, such as pensions. Taking off direct taxes gives disposable income and subtracting other taxes gives post-tax income.

Table 10.25 Gini coefficients in past times

Year	Gini
1688	0.55
1801–03	0.56
1867	0.52
1913	0.43–0.63

Using equivalised income appears to make little difference in this case (compare the 'gross income' column with the earlier calculation).

The table shows essentially two things:

(1) The Gini coefficient changes little over time, suggesting that the income distribution is fairly stable.
(2) The biggest reduction in inequality comes through cash benefits paid out by the state, rather than through taxes. In fact, the tax system appears to *increase* inequality rather than to reduce it, primarily because of the effects of indirect taxes.

Recent increases in inequality are a reversal of the historical trend. The figures presented in Table 10.25, from L. Soltow,[5] provide estimates of the Gini coefficient in earlier times. These figures suggest that a substantial decline in the Gini coefficient has occurred in the last century or so, perhaps related to the process of economic development. It is difficult to compare Soltow's figures directly with the modern ones because of such factors as the quality of data and different definitions of income.

A simpler formula for the Gini coefficient

Kravis, Heston and Summers[6] provide estimates of 'world' GDP by decile and these figures, presented in Table 10.26, will be used to illustrate another method of calculating the Gini coefficient.

These figures show that the poorer half of the world population earns only about 10% of world income and that a third of world income goes to the richest 10% of the population. This suggests a higher degree of inequality than for a single country such as the UK, as one might expect.

[5] Long run changes in British income inequality, *Economic History Review*, 1968, **21**, 17–29.
[6] Real GDP per capita for more than 100 countries, *Economic Journal*, 1978, **88**, 215–242.

Table 10.26 The world distribution of income by decile

Decile	1	2	3	4	5	6	7	8	9	10
% GDP	1.5	2.1	2.4	2.4	3.3	5.2	8.4	17.1	24.1	33.5
Cumulative %	1.5	3.6	6.0	8.4	11.7	16.9	25.3	42.4	66.5	100.0

When the class intervals contain equal numbers of households (e.g. when the data are given for deciles of the income distribution, as here) formula (10.22) for area B simplifies to

$$B = \frac{100}{2k}(y_0 + 2y_1 + 2y_2 + \ldots + 2y_{k-1} + y_k) = \frac{100}{k}\left(\sum_{i=0}^{i=k} y_i - 50\right) \qquad (10.27)$$

where k is the number of intervals (e.g. 10 in the case of deciles, 5 for quintiles). Thus you simply sum the y values, subtract 50,[7] and divide by the number of classes k. The y values for the Kravis *et al.* data appear in the final row of Table 10.26, and their sum is 282.3. We therefore obtain

$$B = \frac{100}{10}(282.3 - 50) = 2323$$

Hence

$$A = 5000 - 2323 = 2677$$

and

$$G = \frac{2677}{5000} = 0.5354$$

or about 53%. This is surprisingly similar to the figure for original income in the UK, but, of course, differences in definition, measurement, etc., may make direct comparison invalid. While the Gini coefficient may provide some guidance when comparing inequality over time or across countries, one needs to take care in its interpretation.

Exercise 10.9

(a) The same data as used in the text are presented below, but with fewer class intervals:

Range of income	Mid-point of interval	Number of households
0–	100	3611
200–	300	6964
400–	500	4643
600–	700	3611
800–	900	2321
1000–	1250	4385
Total		25 534

Draw the Lorenz curve for these data.

(b) Calculate the Gini coefficient for these data and compare to that calculated earlier.

[7] If using decimal percentages, subtract 0.5.

Exercise 10.10

Given shares of total income of 8%, 15%, 22%, 25% and 30% by each quintile of a country's population, calculate the Gini coefficient.

Inequality and development

Table 10.27 presents figures for the income distribution in selected countries around the world. They are in approximately ascending order of national income.

Table 10.27 Income distribution figures in selected countries

	Year	Quintiles					Top 10%	Gini
		1	2	3	4	5		
Bangladesh	1981–82	6.6	10.7	15.3	22.1	45.3	29.5	0.36
Kenya	1976	2.6	6.3	11.5	19.2	60.4	45.8	0.51
Côte d'Ivoire	1985–86	2.4	6.2	10.9	19.1	61.4	43.7	0.52
El Salvador	1976–77	5.5	10.0	14.8	22.4	47.3	29.5	0.38
Brazil	1972	2.0	5.0	9.4	17.0	66.6	50.6	0.56
Hungary	1982	6.9	13.6	19.2	24.5	35.8	20.5	0.27
Korea, Rep.	1976	5.7	11.2	15.4	22.4	45.3	27.5	0.36
Hong Kong	1980	5.4	10.8	15.2	21.6	47.0	31.3	0.38
New Zealand	1981–82	5.1	10.8	16.2	23.2	44.7	28.7	0.37
UK	1979	7.0	11.5	17.0	24.8	39.7	23.4	0.31
Netherlands	1981	8.3	14.1	18.2	23.2	36.2	21.5	0.26
Japan	1979	8.7	13.2	17.5	23.1	37.5	22.4	0.27

The table shows that countries have very different experiences of inequality, even for similar levels of income (e.g. compare Bangladesh and Kenya). Hungary, the only (former) communist country, shows the greatest equality, although whether income accurately measures people's access to resources in such a regime is perhaps debatable. Note that countries with fast growth (such as Korea and Hong Kong) do not have to have a high degree of inequality. Developed countries seem to have uniformly low Gini coefficients.

Source: World Development Report 2002.

Concentration ratios

Another type of inequality is the distribution of market shares of the firms in an industry. We all know that Microsoft currently dominates the software market with a large market share. In contrast, an industry such as bakery has many different suppliers and there is little tendency to dominance. The **concentration ratio** is a commonly used measure to examine the distribution of market shares among firms competing in a market. Of course, it would be possible to measure this using the Lorenz curve and Gini coefficient, but the concentration ratio has the advantage that it can be calculated on the basis of less information and also tends to focus attention on the largest firms in the industry. The concentration

Table 10.28 Sales figures for an industry (millions of units)

Firm	A	B	C	D	E	F	G	H	I	J
Sales	180	115	90	62	35	25	19	18	15	10

ratio is often used as a measure of the competitiveness of a particular market but, as with all statistics, it requires careful interpretation.

A market is said to be concentrated if most of the demand is met by a small number of suppliers. The limiting case is monopoly where the whole of the market is supplied by a single firm. We shall measure the degree of concentration by the **five-firm concentration ratio**, which is the proportion of the market held by the largest five firms, and it is denoted C_5. The larger is this proportion, the greater the degree of concentration and potentially the less competitive is that market. Table 10.28 gives the (imaginary) sales figures of the 10 firms in a particular industry.

For convenience the firms have already been ranked by size from A (the largest) to J (smallest). The output of the five largest firms is 482, out of a total of 569, so the five-firm concentration ratio is $C_5 = 84.7\%$, i.e. 84.7% of the market is supplied by the five largest firms.

Without supporting evidence it is hard to interpret this figure. Does it mean that the market is not competitive and the consumer being exploited? Some industries, such as the computer industry, have a very high concentration ratio yet it is hard to deny that they are fiercely competitive. On the other hand, some industries with no large firms have restrictive practices, entry barriers, etc., which mean that they are not very competitive (lawyers might be one example). A further point is that there may be a *threat* of competition from outside the industry which keeps the few firms acting competitively.

Concentration ratios can be calculated for different numbers of largest firms, for example the three-firm or four-firm concentration ratio. Straightforward calculation reveals them to be 67.7% and 78.6% respectively for the data given in Table 10.28. There is little reason in general to prefer one measure to the others, and they may give different pictures of the degree of concentration in an industry.

The concentration ratio calculated above relates to the quantity of output produced by each firm, but it is possible to do the same with sales revenue, employment, investment or any other variable for which data are available. The interpretation of the results will be different in each case. For example, the largest firms in an industry, while producing the majority of output, might not provide the greater part of employment if they use more capital-intensive methods of production. Concentration ratios obviously have to be treated with caution, therefore, and are probably best combined with case studies of the particular industry before conclusions are reached about the degree of competition.

Exercise 10.11

Total sales in an industry are \$400m. The largest five firms have sales of \$180m, \$70m, \$40m, \$25m and \$15m. Calculate the three- and five-firm concentration ratios.

Summary

- An index number summarises the variation of a variable over time or across space in a convenient way.
- Several variables can be combined into one index, providing an average measure of their individual movements. The retail price index is an example.
- The Laspeyres price index combines the prices of many individual goods using base-year quantities as weights. The Paasche index is similar but uses current-year weights to construct the index.
- Laspeyres and Paasche quantity indices can also be constructed, combining a number of individual quantity series using prices as weights. Base-year prices are used in the Laspeyres index, current-year prices in the Paasche.
- A price index series multiplied by a quantity index series results in an index of expenditures. Rearranging this demonstrates that deflating (dividing) an expenditure series by a price series results in a volume (quantity) index. This is the basis of deflating a series in cash (or nominal) terms to one measured in real terms (i.e. adjusted for price changes).
- Two series covering different time periods can be spliced together (as long as there is an overlapping year) to give one continuous chain index.
- Discounting the future is similar to deflating but corrects for the rate of time preference rather than inflation. A stream of future income can thus be discounted and summarised in terms of its present value.
- An investment can be evaluated by comparing the discounted present value of the future income stream to the initial outlay. The internal rate of return of an investment is a similar but alternative way of evaluating an investment project.
- The Gini coefficient is a form of index number that is used to measure inequality (e.g. of incomes). It can be given a visual representation using a Lorenz curve diagram.
- For measuring the inequality of market shares in an industry, the concentration ratio is commonly used.

Key terms and concepts

base year
chain index
concentration ratio
deflating a data series
discounting
Gini coefficient
internal rate of return
Laspeyres index
Lorenz curve
Paasche index
present value
reference year
retail price index
weighted average

References

Kravis, I. B., Heston, A. and Summers, R., Real GDP per capita for more than one hundred countries, *Economic Journal*, 1978, (349) **88**, 215–242.

L. Soltow, Long run changes in British income inequality, *Economic History Review*, 1968, **21**(1), 17–29.

Problems

Some of the more challenging problems are indicated by highlighting the problem number in colour.

10.1 The data below show exports and imports for the UK, 1987–1992, in £bn at current prices.

	1987	1988	1989	1990	1991	1992
Exports	120.6	121.2	126.8	133.3	132.1	135.5
Imports	122.1	137.4	147.6	148.3	140.2	148.3

(a) Construct index number series for exports and imports, setting the index equal to 100 in 1987 in each case.

(b) Is it possible, using only the two indices, to construct an index number series for the balance of trade? If so, do so; if not, why not?

10.2 The following data show the gross trading profits of companies, 1987–1992, in the UK, in £m.

1987	1988	1989	1990	1991	1992
61 750	69 180	73 892	74 405	78 063	77 959

(a) Turn the data into an index number series with 1987 as the reference year.

(b) Transform the series so that 1990 is the reference year.

(c) What increase has there been in profits between 1987 and 1992? Between 1990 and 1992?

10.3 The following data show energy prices and consumption in 1995–1999 (analogous to the data in the chapter for the years 2002–2006).

Prices	Coal (£/tonne)	Petroleum (£/tonne)	Electricity (£/MWh)	Gas (£/therm)
1995	37.27	92.93	40.07	0.677
1996	35.41	98.33	39.16	0.464
1997	34.42	90.86	36.87	0.509
1998	35.16	87.23	36.67	0.560
1999	34.77	104.93	36.23	0.546

Quantities	Coal (m tonnes)	Petroleum (m tonnes)	Electricity (m MWh)	Gas (m therms)
1995	2.91	6.37	102.88	4938
1996	2.22	6.21	105.45	5406
1997	2.14	5.64	107.31	5565
1998	1.81	5.37	107.97	5639
1999	2.04	5.33	110.98	6039

(a) Construct a Laspeyres price index using 1995 as the base year.

(b) Construct a Paasche price index. Compare this result with the Laspeyres index. Do they differ significantly?

(c) Construct Laspeyres and Paasche quantity indices. Check that they satisfy the conditions that $E^n = P_L \times Q_P$, etc.

10.4 The prices of different house types in south-east England are given in the table below:

Year	Terraced houses	Semi-detached	Detached	Bungalows	Flats
1991	59 844	77 791	142 630	89 100	47 676
1992	55 769	73 839	137 053	82 109	43 695
1993	55 571	71 208	129 414	82 734	42 746
1994	57 296	71 850	130 159	83 471	44 092

(a) If the numbers of each type of house in 1991 were 1898, 1600, 1601, 499 and 1702 respectively, calculate the Laspeyres price index for 1991–1994, based on 1991 = 100.

(b) Calculate the Paasche price index, based on the following numbers of dwellings:

Year	Terraced houses	Semi-detached	Detached	Bungalows	Flats
1992	1903	1615	1615	505	1710
1993	1906	1638	1633	511	1714
1994	1911	1655	1640	525	1717

(c) Compare Paasche and Laspeyres price series.

10.5 (a) Using the data in Problem 10.3, calculate the expenditure shares on each fuel in 1995 and the individual price index number series for each fuel, with 1995 = 100.

(b) Use these data to construct the Laspeyres price index using the expenditures shares approach. Check that it gives the same answer as in Problem 10.3(a).

10.6 The following table shows the weights in the retail price index and the values of the index itself, for 1990 and 1994.

	Food	Alcohol and tobacco	Housing	Fuel and light	Household items	Clothing	Personal goods	Travel	Leisure
Weights									
1990	205	111	185	50	111	69	39	152	78
1994	187	111	158	45	123	58	37	162	119
Prices									
1990	121.0	120.7	163.7	115.9	116.9	115.0	122.7	121.2	117.1
1994	139.5	162.1	156.8	133.9	132.4	116.0	152.4	150.7	145.7

(a) Calculate the Laspeyres price index for 1994, based on 1990 = 100.

(b) Draw a bar chart of the expenditure weights in 1990 and 1994 to show how spending patterns have changed. What major changes have occurred? Do individuals seem to be responding to changes in relative prices?

(c) The pensioner price index is similar to the general index calculated above, except that it excludes housing. What effect does this have on the index? What do you think is the justification for this omission?

(d) If consumers spent, on average, £188 per week in 1990 and £240 per week in 1994, calculate the real change in expenditure on food.

(e) Do consumers appear rational, i.e. do they respond as one would expect to relative price changes? If not, why not?

10.7 Construct a chain index from the following data series:

	1998	2002	2000	2001	2002	2006	2004
Series 1	100	110	115	122	125		
Series 2			100	107	111	119	121

What problems arise in devising such an index and how do you deal with them?

10.8 Construct a chain index for 1995–2004 using the following data, setting 1998 = 100.

1995	1996	1997	1998	2002	2000	2001	2002	2006	2004
87	95	100	105						
			98	93	100	104	110		
							100	106	112

10.9 Industry is complaining about the rising price of energy. It demands to be compensated for any rise over 5% in energy prices between 2003 and 2004. How much would this compensation cost? Which price index should be used to calculate the compensation and what difference would it make? (Use the energy price data in the chapter.)

10.10 Using the data in Problem 10.6 above, calculate how much the average consumer would need to be compensated for the rise in prices between 1990 and 1994.

10.11 The following data show expenditure on the National Health Service (in cash terms), the GDP deflator, the NHS pay and prices index, population and population of working age:

Year	NHS expenditure (£m) (1)	GDP deflator 1973 = 100 (2)	NHS pay and price index 1973 = 100 (3)	Population (000) (4)	Population of working age (000) (5)
1987	21 495	442	573	56 930	34 987
1988	23 601	473	633	57 065	35 116
1989	25 906	504	678	57 236	35 222
1990	28 534	546	728	57 411	35 300
1991	32 321	585	792	57 801	35 467

(In all the following answers, set your index to 1987 = 100.)

(a) Turn the expenditure cash figures into an index number series.

(b) Calculate an index of 'real' NHS expenditure using the GDP deflator. How does this alter the expenditure series?

(c) Calculate an index of the volume of NHS expenditure using the NHS pay and prices index. How and why does this differ from the answer arrived at in (b)?

(d) Calculate indices of real and volume expenditure *per capita*. What difference does this make?

(e) Suppose that those not of working age cost twice as much to treat, on average, as those of working age. Construct an index of the need for health care and examine how health care expenditures have changed relative to need.

(f) How do you think the needs index calculated in (e) could be improved?

10.12 (a) If w represents the wage rate and p the price level, what is w/p?

(b) If Δw represents the annual growth in wages and i is the inflation rate, what is $\Delta w - i$?

(c) What does $\ln(w) - \ln(p)$ represent? (ln = natural logarithm)

10.13 A firm is investing in a project and wishes to receive a rate of return of at least 15% on it. The stream of net income is:

Year	1	2	3	4
Income	600	650	700	400

(a) What is the present value of this income stream?

(b) If the investment costs £1600 should the firm invest? What is the net present value of the project?

10.14 A firm uses a discount rate of 12% for all its investment projects. Faced with the following choice of projects, which yields the higher *NPV*?

Project	Outlay	Income stream					
		1	2	3	4	5	6
A	5600	1000	1400	1500	2100	1450	700
B	6000	800	1400	1750	2500	1925	1200

10.15 Calculate the internal rate of return for the project in Problem 10.13. Use either trial and error methods or a computer to solve.

10.16 Calculate the internal rates of return for the projects in Problem 10.14.

10.17 (a) Draw a Lorenz curve and calculate the Gini coefficient for the wealth data in Chapter 1 (Table 1.3).

(b) Why is the Gini coefficient typically larger for wealth distributions than for income distributions?

10.18 (a) Draw a Lorenz curve and calculate the Gini coefficient for the 1979 wealth data contained in Problem 1.5 (Chapter 1). Draw the Lorenz curve on the same diagram as you used in Problem 10.17.

(b) How does the answer compare to 2003?

10.19 The following table shows the income distribution by quintile for the UK in 2006–2007, for various definitions of income:

Quintile	Income measure			
	Original	Gross	Disposable	Post-tax
1 (bottom)	3%	7%	7%	6%
2	7%	10%	12%	11%
3	15%	16%	16%	16%
4	24%	23%	22%	2%
5 (top)	51%	44%	42%	44%

(a) Use equation (10.27) to calculate the Gini coefficient for each of the four categories of income.

(b) For the 'original income' category, draw a smooth Lorenz curve on a piece of gridded paper and calculate the Gini coefficient using the method of counting squares. How does your answer compare to that for part (a)?

10.20 For the Kravis, Heston and Summers data (Table 10.26), combine the deciles into quintiles and calculate the Gini coefficient from the quintile data. How does your answer compare with the answer given in the text, based on deciles? What do you conclude about the degree of bias?

10.21 Calculate the three-firm concentration ratio for employment in the following industry:

Firm	A	B	C	D	E	F	G	H
Employees	3350	290	440	1345	821	112	244	352

10.22 Compare the degrees of concentration in the following two industries. Can you say which is likely to be more competitive?

Firm	A	B	C	D	E	F	G	H	I	J
Sales	337	384	696	321	769	265	358	521	880	334
Sales	556	899	104	565	782	463	477	846	911	227

10.23 **(Project)** The World Development Report contains data on the income distributions of many countries around the world (by quintile). Use these data to compare income distributions across countries, focusing particularly on the differences between poor countries, middle-income and rich countries. Can you see any pattern emerging? Are there countries which do not fit into this pattern? Write a brief report summarising your findings.

Answers to exercises

Exercise 10.1

(a) 100, 111.9, 140.7, 163.4, 188.1.

(b) 61.2, 68.5, 86.1, 100, 115.1.

(c) 115.1/61.2 = 1.881.

Exercise 10.2

	1999	2000	2001	2002	2003
(a) 1999 = 100	100	104.63	116.68	111.87	111.30
(b) 2001 = 100	85.70	89.67	100	95.87	95.39
(c) Using 2000 basket	100	104.69	116.86	112.08	111.52

Exercise 10.3

The Paasche index is:

1999	2000	2001	2002	2003
100	104.69	117.27	111.09	110.93

Exercise 10.4

(a) Expenditure shares in 1999 are:

	Expenditure	Share
Coal	70.93	0.9%
Petroleum	559.28	7.0%
Electricity	4020.81	50.6%
Gas	3297.29	41.5%

giving the Laspeyres index for 2000 as

$$P_1^n = \frac{35.12}{34.77} \times 0.009 + \frac{137.90}{104.93} \times 0.070 + \frac{34.69}{36.23} \times 0.506 + \frac{0.606}{0.546} \times 0.415$$

$$= 1.0463 \text{ or } 104.63.$$

The expenditure shares in 2000 are 0.3%, 8.9%, 46.3%, 44.4% which allows the 2000 Paasche index to be calculated as

$$P_1^n = \frac{1}{\frac{34.77}{35.12} \times 0.003 + \frac{104.93}{137.9} \times 0.089 + \frac{36.23}{34.69} \times 0.463 + \frac{0.546}{0.606} \times 0.444} \times 100$$

$$= 1.0469 \text{ or } 104.69.$$

Later years can be calculated in similar fashion.

Exercise 10.5

(a/b) The Laspeyres and Paasche quantity indexes are:

	Laspeyres index	Paasche index
1999	100	100
2000	102.65	102.71
2001	102.40	102.91
2002	98.18	97.50
2003	101.46	101.12

(c) The expenditure index is 100, 107.46, 120.08, 109.07, 112.55.

(d) The Paasche quantity index times Laspeyres price index (or vice versa) gives the expenditure index.

Exercise 10.6

The full index is (using Laspeyres indexes):

		Chain index		
1995	100			100
1996	86.3			86.3
1997	85.5			85.5
1998	88.1			88.1
1999	88.1	100		88.1
2000		104.63		92.2
2001		116.68		102.8
2002		111.87	100	98.6
2003		111.30	101.68	100.2
2004			115.22	113.6
2005			161.31	159.0
2006			209.11	206.1

Exercise 10.7

(a) The discounted figures are:

Year	Investment/yield	Discount factor	Discounted yield
0	−100 000		
1	25 000	0.9091	22 727.3
2	35 000	0.8264	28 925.6
3	30 000	0.7513	22 539.4
4	15 000	0.6830	10 245.2
Total			84 437.5

The present value is less than the initial outlay.

(b) The internal rate of return is 2.12%.

Exercise 10.8

(a) Deflating to real income gives:

Year	Investment/yield	Price index	Real income
0	–50 000	100	–50 000.0
1	20 000	105	19 047.6
2	25 000	110.250	22 675.7
3	30 000	115.763	25 915.1
4	10 000	121.551	8 227.0

(b) The real discount rate is 1.09/1.05 = 1.038 or 3.8% p.a.

(c/d)

Nominal values	Discount factor	Discounted value	Real values	Discount factor	Discounted value
–50 000			–50 000.0		
20 000	0.917	18 348.6	19 047.6	0.963	18 348.6
25 000	0.842	21 042.0	22 675.7	0.928	21 042.0
30 000	0.772	23 165.5	25 915.1	0.894	23 165.5
10 000	0.708	7 084.3	8 227.0	0.861	7 084.3
Totals		69 640.4			69 640.38

The present value is the same in both cases and exceeds the initial outlay.

Exercise 10.9

(b)

Range of income	Mid-point	Number of households	Total income	% Households	% Cumulative households x	% Income	% Cumulative income y
(1)	(2)	(3)	(4)	(5)	(6)	(7)	(8)
0–	100	3611	361 100	14.1%	14.1%	2.4%	2.4%
200–	300	6964	2 089 200	27.3%	41.4%	14.1%	16.5%
400–	500	4643	2 321 500	18.2%	59.6%	15.6%	32.1%
600–	700	3611	2 527 700	14.1%	73.7%	17.0%	49.1%
800–	900	2321	2 088 900	9.1%	82.8%	14.0%	63.1%
1000–	1250	4385	5 481 250	17.2%	100.0%	36.9%	100.0%
Totals		25 535	14 869 650	100.0%		100.0%	

The Gini coefficient is then calculated as follows: $B = 0.5 \times \{14.1 \times (2.4 + 0) + 27.3 \times (16.5 + 2.4) + 18.2 \times (32.1 + 16.5) + 14.1 \times (49.1 + 32.1) + 9.1 \times (63.1 + 49.1) + 17.2 \times (100 + 63.1)\} = 3201$. Area $A = 5000 - 3301 = 1799$. Hence Gini $= 1799/5000 = 0.360$, very similar to the value in the text using more categories of income.

Exercise 10.10

$B = 100/5 \times (246 - 50) = 3920$. Hence $A = 1080$ and Gini $= 0.216$. 246 is the sum of the cumulative y values.

Exercise 10.11

$C_3 = 290/400 = 72.5\%$ and $C_5 = 82.5\%$.

Appendix Deriving the expenditure share form of the Laspeyres price index

We can obtain the expenditure share version of the formula from the standard formula given in equation (10.1)

$$P_L^n = \frac{\Sigma p_n q_0}{\Sigma p_0 q_0} = \frac{\Sigma \frac{p_n}{p_0} p_0 q_0}{\Sigma \frac{p_0}{p_0} p_0 q_0}$$

$$= \frac{\Sigma \frac{p_n}{p_0} \frac{p_0 q_0}{\Sigma p_0 q_0}}{\Sigma \frac{p_0}{p_0} \frac{p_0 q_0}{\Sigma p_0 q_0}} = \Sigma \frac{p_n}{p_0} \frac{p_0 q_0}{\Sigma p_0 q_0}$$

$$= \Sigma \frac{p_n}{p_0} \times s_0$$

which is equation (10.3) in the text (the × 100 is omitted from this derivation for simplicaty.

9 Data collection and sampling methods

Contents

Learning outcomes

By the end of this chapter you should be able to:

- recognise the distinction between primary and secondary data sources;
- avoid a variety of common pitfalls when using secondary data;
- make use of electronic sources to gather data;
- recognise the main types of random sample and understand their relative merits;
- appreciate how such data are collected;
- conduct a small sample survey yourself.

Complete your diagnostic test for Chapter 9 now to create your personal study plan. Exercises with an icon ? are also available for practice in MathXL with additional supporting resources.

Introduction

It may seem a little odd to look at data collection now, after several chapters covering the analysis of data. Collection of data logically comes first, but the fact is that most people's experience is as a user of data, which determines their priorities. Also, it is difficult to have the motivation for learning about data collection when one does not know what it is subsequently used for. Having spent considerable time learning how to analyse data, it is now time to look at their collection and preparation.

There are two reasons why you might find this chapter useful. First, it will help if you have to carry out some kind of survey yourself. Second, it will help you in your data analysis, even if you are using someone else's data. Knowing the issues involved in data collection can help your judgement of the quality of the data you are using.

When conducting statistical research, there are two ways of obtaining data:

(1) use **secondary data sources**, such as the UN Yearbook, or
(2) collect sample data personally, a **primary data source**.

The first category should nowadays be divided into two subsections: printed and electronic sources. The latter is obviously becoming more important as time progresses, but printed documentation still has its uses. Using secondary data sources sounds simple, but it is easy to waste valuable time by making elementary errors. The first part of this chapter provides some simple advice to help you avoid such mistakes.

Much of this text has been concerned with the analysis of sample evidence and the inferences that can be drawn from it. It has been stressed that this evidence must come from randomly drawn samples and, although the notion of randomness was discussed in Chapter 2, the precise details of random sampling have not been set out.

The second part of this chapter is therefore concerned with the problems of collecting sample survey data prior to their analysis. The decision to collect the data personally depends upon the type of problem faced, the current availability of data relating to the problem and the time and cost needed to conduct a survey. It should not be forgotten that the first question that needs answering is whether the answer obtained is worth the cost of finding it. It is probably not worthwhile for the government to spend £50 000 to find out how many biscuits people eat, on average (although it may be worth biscuit manufacturers doing this). The sampling procedure is always subject to some limit on cost, therefore, and the researcher is trying to obtain the best value for money.

Using secondary data sources

Much of the research in economics and finance is based on secondary data sources, i.e. data which the researcher did not collect herself. The data may be in the form of official statistics such as those published in *Economic Trends* or they may come from unofficial surveys. In either case one has to use the data as presented; there is no control over sampling procedures.

It may seem easy enough to look up some figures in a publication, but there are a number of pitfalls for the unwary. The following advice comes from experience, some of it painful, and it may help you to avoid wasting time and effort. I have also learned much from the experiences of my students, whom I have also watched suffer.

A lot of data are now available online, so the advice given here covers both printed and electronic sources, with a separate section for the latter.

Make sure you collect the right data

This may seem obvious, but most variables can be measured in a variety of different ways. Suppose you want to measure the cost of labour (over time) to firms. Should you use the wage rate or earnings? The latter includes payment for extra hours such as overtime payments and reflects general changes in the length of the working week. Is the wage measured per hour or per week? Does it include part-time workers? If so, a trend in the proportion of part-timers will bias the wage series. Does the series cover all workers, men only, or women only? Again, changes in the composition will influence the wage series. What about tax and social security costs? Are they included? There are many questions one could ask.

One needs to have a clear idea therefore of the precise variable one needs to collect. This will presumably depend upon the issue in question. Economic theory might provide some guidance: for instance, theory suggests that firms care about *real* wage rates (i.e. after taking account of inflation, so related to the price of the goods the firm sells) so this is what one should measure. Check the definition of any series you collect (this is often at the back of the printed publication, or in a separate supplement giving explanatory notes and definitions). Make sure that the definition has not changed over the time period you require: the definition of unemployment used in the UK changed about 20 times in the 1980s, generally with the effect of reducing *measured* unemployment, even if actual unemployment was unaffected. In the UK the geographical coverage of data may vary: one series may relate to the UK, another to Great Britain and yet another to England and Wales. Care should obviously be taken if one is trying to compare such series.

Try to get the most up-to-date figures

Many macroeconomic series are revised as more information becomes available. The balance of payments serves as a good example. The first edition of this book showed the balance of payments (current balance, in £m for the UK) for 1970, as published in successive years, as follows:

1971	1972	1973	1974	1975	1976	1977	1978	...	1986
579	681	692	707	735	733	695	731	...	795

The difference between the largest and smallest figures is of the order of 37%, a wide range. In the third edition of this book the figure was (from the 1999 edition of *Economic Trends Annual Supplement*) £911m which is 57% higher than the initial estimate. The latest figure at the time of writing is £819m. Most series are better than this. The balance of payments is hard to measure because it is the small difference between two large numbers, exports and imports. A 5% increase in measured exports and a 5% decrease in measured imports could thus change the measured balance by 100% or more.

One should always try to get the most up-to-date figures, therefore, which often means working *backwards* through printed data publications, i.e. use the current issue first and get data back as far as is available, then find the previous issue to go back a little further, etc. This can be tedious but it will also give some idea of the reliability of the data from the size of data revisions.

Keep a record of your data sources

You should always keep *precise* details of where you obtained each item of data. If you need to go back to the original publication (e.g. to check on the definition of a series) you will then be able to find it easily. It is easy to spend hours (if not *days*) trying to find the source of some interesting numbers that you wish to update. 'Precise details' means the name of the publication, issue number or date, and table or page number. It also helps to keep the library reference number of the publication if it is obscure. It is best to take a photocopy of the data (but check copyright restrictions) rather than just copy it down, if possible.

Keeping data in *Excel* or another spreadsheet

Spreadsheets are ideal for keeping your data. It is often a good idea to keep the data all together in one worksheet and extract portions of them as necessary and analyse them in another worksheet. Alternatively, it is usually quite easy to transfer data from the spreadsheet to another program (e.g. *SPSS* or *Stata*) for more sophisticated analysis. In most spreadsheets you can attach a comment to any cell, so you can use this to keep a record of the source of each observation, changes of definition, etc. Thus you can retain all the information about your data together in one place.

Check your data

Once you have collected your data you must check them. Once you have done this, you must check them again. Better, persuade someone else to help with the second check. Note that if your data are wrong then all your subsequent calculations could be incorrect and you will have wasted much time. I have known many students who have spent months or even years on a dissertation or thesis who have then found an error in the data they collected earlier.

A useful way to check the data is first to graph them (e.g. a time-series plot). Obvious outliers will show up and you can investigate them for possible errors. Do not just rely on the graphs, however, look through your data and check them against the original source. Do not forget that the original source could be wrong too, so be wary of 'unusual' observations.

Using electronic sources of data

A vast amount of data are now available electronically, usually online, and this is becoming increasingly the norm. Sometimes the data are available free but sometimes they have to be paid for, especially if they have a commercial value.

My experience suggests that many students nowadays *only* consider online resources, which I feel is a mistake. Not everything is online and sometimes, even if it is, it is extremely hard to find. It can sometimes take less time to go to the library, find the appropriate journal and type in the numbers. As an estimate, 100 observations should take no longer than about 10 minutes to type into a computer, which is probably quicker than finding them electronically, converting to the right format, etc. Hence the advantage of online data lies principally with large datasets.

Obtaining data electronically should avoid input errors and provide consistent, up-to-date figures. However, this is not always guaranteed. For example, the UK Office for National Statistics (ONS) online databank provides plenty of information, but some of the series clearly have breaks in them and there is little warning of this in the on-screen documentation. The series for revenue per admission to cinemas (roughly the price of admission) goes:

1963	1964	1965	1966	1967
37.00	40.30	45.30	20.60	21.80

which strongly suggests an artificial break in the series in 1966 (especially as admissions *fell* by 12% between 1965 and 1966!). Later in the series, the observations appear to be divided by 100. The lesson is that even with electronic data you should check the numbers to ensure they are correct.[1]

You need to follow the same advice with electronic sources as with printed ones: make sure you collect the right variables and keep a note of your source. Online sources do not seem to be as good as many printed sources when it comes to providing definitions of the variables. It is often unclear if the data are in real terms, seasonally adjusted, etc. Sometimes you may need to go to the printed document to find the definitions, even if the data themselves come from the internet. Keeping a note of your source means taking down the URL of the site you visit. Remember that some sites generate the page 'on demand' so the web address is not a permanent one and typing it in later on will not take you back to the same source. In these circumstances is may be better to note the 'root' part of the address (e.g. www.imf.org/data/) rather than the complete detail. You should also take a note of the date you accessed the site, this may be needed if you put the source into a bibliography.

Tips on downloading data

- If you are downloading a spreadsheet, save it to your hard disk then include the URL of the source within the spreadsheet itself. You will always know where it came from. You can do the same with Word documents.
- You cannot do this with PDF files, which are read-only. You could save the file to your disk, including the URL within the file name (but avoid putting extra full stops in the file name, that confuses the operating system: replace them with hyphens.).

[1] I wrote this for the previous edition of this book. I can no longer find the same data on Statbase, it seems to have disappeared into the ether!

- You can use the 'Text select tool' within Acrobat to copy items of data from a PDF file and then paste them into a spreadsheet.
- Often, when pasting several columns of such data into *Excel*, all the numbers go into a single column. You can fix this using the Data, Text to Columns menu. Experimentation is required, but it works well.

Since there are now so many online sources (and they are constantly changing) a list of useful data sites rapidly becomes out of date. The following sites seem to have withstood the test of time so far and have a good chance of surviving throughout the life of this edition.

- The UK Office for National Statistics is at http://www.statistics.gov.uk/ and their Statbase service supplies over 1000 datasets online for free. This is tied to information on 13 'themes', such as education, agriculture, etc.
- The Data and Story Library at http://lib.stat.cmu.edu/DASL/ is just that: datasets with accompanying statistical analyses which are useful for learning.
- The IMF's World Economic Database is at http://www.imf.org/ (follow the links to publications, World Economic Outlook, then the database). It has macroeconomic series for most countries for several years. It is easy to download in csv (text) format, for use in spreadsheets.
- The Biz/Ed site at http://www.bized.co.uk/ contains useful material on business (including financial case studies of companies) as well as economic data. There is a link from here to the Penn World Tables, which contain national accounts data for many countries (on a useful, comparable basis) from 1960 onwards. Alternatively, visit the Penn home page at http://pwt.econ.upenn.edu/.
- The World Bank provides a lot of information, particularly relating to developing countries, at http://www.worldbank.org/data/. Much of the data appears to be in .pdf format so, although it is easy to view on-screen, it cannot be easily transferred into a spreadsheet or similar software.
- Bill Goffe's Resources for Economists site (http://rfe.org) contains a data section which is a good starting point for data sources.
- Google. Possibly the most useful website of all. Intelligent use of this search tool is often the best way to find what you want.
- http://davidmlane.com/hyperstat/ has an online textbook and glossary. This is useful if you have a computer handy but not a textbook.
- Financial and business databases are often commercial enterprises and hence are not freely available. Two useful free (or partially free) sites, however, are *The Financial Times* (http://www.ft.com/home/uk) and *Yahoo Finance* (http://finance.yahoo.com/).

Collecting primary data

Primary data are data that you have collected yourself from original sources, often by means of a sample survey. This has the advantage that you can design the questionnaire to include the questions of interest to you and you have total control over all aspects of data collection. You can also choose the size of the sample (as long as you have sufficient funds available) so as to achieve the desired width of any confidence intervals.

Almost all surveys rely upon some method of sampling, whether random or not. The probability distributions which have been used in previous chapters as the basis of the techniques of estimation and hypothesis testing rely upon the samples having been drawn at random from the population. If this is not the case, then the formulae for confidence intervals, hypothesis tests, etc., are incorrect and not strictly applicable (they may be reasonable approximations but it is difficult to know how reasonable). In addition, the results about the bias and precision of estimators will be incorrect. For example, suppose an estimate of the average expenditure on repairs and maintenance by car owners is obtained from a sample survey. A poor estimate would arise if only Rolls-Royce owners were sampled, since they are not representative of the population as a whole. The precision of the estimator (the sample mean, $\bar{x}$) is likely to be poor because the mean of the sample could either be very low (Rolls-Royce cars are very reliable so rarely need repairs) or very high (if they do break down the high quality of the car necessitates a costly repair). This means the confidence interval estimate will be very wide and thus imprecise. It is not immediately obvious if the estimator would be biased upwards or downwards.

Thus some form of random sampling method is needed to be able to use the theory of the probability distributions of random variables. Nor should it be believed that the theory of random sampling can be ignored if a very large sample is taken, as the following cautionary tale shows. In 1936 the *Literary Digest* tried to predict the result of the forthcoming US election by sending out 10 million mail questionnaires. Two million were returned, but even with this enormous sample size Roosevelt's vote was incorrectly estimated by a margin of 19 percentage points. The problem is that those who respond to questionnaires are not a random sample of those who receive them.

The meaning of random sampling

The definition of random sampling is that every element of the population should have a known, non-zero probability of being included in the sample. The problem with the sample of cars used above was that Ford cars (for example) had a zero probability of being included. Many sampling procedures give an equal probability of being selected to each member of the population but this is not an essential requirement. It is possible to adjust the sample data to take account of unequal probabilities of selection. If, for example, Rolls-Royce had a much greater chance of being included than Ford, then the estimate of the population mean would be calculated as a weighted average of the sample observations, with greater weight being given to the few 'Ford' observations than to relatively abundant 'Rolls-Royce' observations. A very simple illustration of this is given below. Suppose that for the population we have the following data:

	Rolls-Royce	Ford
Number in population	20 000	2 000 000
Annual repair bill	£1000	£200

Then the true average repair bill is

$$\mu = \frac{20\ 000 \times 1000 + 2\ 000\ 000 \times 200}{2\ 020\ 000} = 207.92$$

Suppose the sample data are as follows:

	Rolls-Royce	Ford
Number in sample	20	40
Probability of selection	1/1000	1/50 000
Repair bill	£990	£205

To calculate the average repair bill from the sample data we use a weighted average, using the relative population sizes as weights, not the sample sizes

$$\bar{x} = \frac{20\ 000 \times 990 + 2\ 000\ 000 \times 205}{2\ 020\ 000} = 212.77$$

If the sample sizes were used as weights the average would come out at £466.67, which is substantially incorrect.

As long as the probability of being in the sample is known (and hence the relative population sizes must be known), the weight can be derived; but if the probability is zero this procedure breaks down.

Other theoretical assumptions necessary for deriving the probability distribution of the sample mean or proportion are that the population is of infinite size and that each observation is independently drawn. In practice the former condition is never satisfied since no population is of infinite size, but most populations are large enough that it does not matter. For each observation to be independently drawn (i.e. the fact of one observation being drawn does not alter the probability of others in the sample being drawn) strictly requires that sampling be done with replacement, i.e. each observation drawn is returned to the population before the next observation is drawn. Again in practice this is often not the case, sampling being done without replacement, but again this is of negligible practical importance where the population is large relative to the sample.

On occasion the population is quite small and the sample constitutes a substantial fraction of it. In these circumstances the **finite population correction (fpc)** should be applied to the formula for the variance of $\bar{x}$, the fpc being given by

$$\text{fpc} = (1 - n/N) \tag{9.1}$$

where N is the population size and n the sample size. The table below illustrates its usage:

Variance of $\bar{x}$ from infinite population	Variance of $\bar{x}$ from finite population	Example values of fpc			
		$n = 20$	25	50	100
		$N = 50$	100	1000	10 000
σ^2/n	$\sigma^2/n \times (1 - n/N)$	0.60	0.75	0.95	0.99

The finite population correction serves to narrow the confidence interval because a sample size of (say) 25 reveals more about a population of 100 than about a population of 100 000, so there is less uncertainty about population parameters. When the sample size constitutes only a small fraction of the population (e.g. 5% or less) the finite population correction can be ignored in practice. If the whole population is sampled ($n = N$) then the variance becomes zero and there is no uncertainty about the population mean.

A further important aspect of random sampling occurs when there are two samples to be analysed, when it is important that the two samples are independently drawn. This means that the drawing of the first sample does not influence the drawing of the second sample. This is a necessary condition for the derivation of the probability distribution of the difference between the sample means (or proportions).

Types of random sample

The meaning and importance of randomness in the context of sampling has been explained. However, there are various different types of sampling, all of them random, but which have different statistical properties. Some methods lead to greater precision of the estimates, while others can lead to considerable cost savings in the collection of the sample data, but at the cost of lower precision. The aim of sampling is usually to obtain the most precise estimates of the parameter in question, but the best method of sampling will depend on the circumstances of each case. If it is costly to sample individuals, a sampling method which lowers cost may allow a much larger sample size to be drawn and thus good (precise) estimates to be obtained, even if the method is inherently not very precise. These issues are investigated in more detail below, as a number of different sampling methods are examined.

Simple random sampling

This type of sampling has the property that every possible sample that could be obtained from the population has an equal chance of being selected. This implies that each element of the population has an equal probability of being included in the sample, but this is not the defining characteristic of simple random sampling. As will be shown below, there are sampling methods where every member of the population has an equal chance of being selected, but some samples (i.e. certain combinations of population members) can never be selected.

The statistical methods in this book are based upon the assumption of simple random sampling from the population. It leads to the most straightforward formulae for estimation of the population parameters. Although many statistical surveys are not based upon simple random sampling, the use of statistical tests based on simple random sampling is justified since the sampling process is often hypothetical. For example, if one were to compare annual growth rates of two countries over a 30-year period, a z test on the difference of two sample means (i.e. the average annual growth rate in each country) would be conducted. In a sense the data are not a sample since they are the only possible data for those two countries over that time period. Why not therefore just regard the data as constituting the whole population? Then it would just be a case of finding which country had the higher growth rate; there would be no uncertainty about it.

The alternative way of looking at the data would be to suppose that there exists some hypothetical population of annual growth rates and that the data for the two countries were drawn by (simple) random sampling from this population. Is this story consistent with the data available? In other words, could the data we have simply arise by chance? If the answer to this is no (i.e. the z score exceeds the critical value) then there is something causing a difference between the two countries (it may not be clear what that something is). In this case it is reasonable to assume that all possible samples have an equal chance of selection, i.e. that simple random sampling takes place. Since the population is hypothetical one might as well suppose it to have an infinite number of members, again required by sampling theory.

Stratified sampling

Returning to the practical business of sampling, one problem with simple random sampling is that it is possible to collect 'bad' samples, i.e. those which are unrepresentative of the population. An example of this is what we may refer to as the 'basketball player' problem, i.e. in trying to estimate the average height of the population, the sample (by sheer bad luck) contains a lot of basketball players. One way round this problem is to ensure that the proportion of basketball players in the sample accurately reflects the proportion of basketball players in the population (i.e. very small!). The way to do this is to divide up the population into 'strata' (e.g. basketball players and non-players) and then to ensure that each stratum is properly represented in the sample. This is best illustrated by means of an example.

A survey of newspaper readership, which is thought to be associated with age, is to be carried out. Older people are thought to be more likely to read newspapers, as younger people are more likely to use other sources (principally the internet) to obtain the news. Suppose the population is made up of three age strata: old, middle-aged and young, as follows:

Percentage of population in age group		
Old	*Middle aged*	*Young*
20%	50%	30%

Suppose a sample of size 100 is taken. With luck it would contain 20 old people, 50 who are middle-aged and 30 young people, and thus would be representative of the population as a whole. But if, by bad luck (or bad sample design), all 100 people in the sample were middle-aged, poor results might be obtained since newspaper readership differs between age groups.

To avoid this type of problem a stratified sample is taken, which ensures that all age groups are represented in the sample. This means that the survey would have to ask people about their age as well as their reading habits. The simplest form of stratified sampling is equiproportionate sampling, whereby a stratum which constitutes (say) 20% of the population also makes up 20% of the sample. For the example above the sample would be made up as follows:

Class	Old	Middle-aged	Young	Total
Number in sample	20	50	30	100

It should be clear why stratified sampling constitutes an improvement over simple random sampling, since it rules out 'bad' samples, i.e. those not representative of the population. It is simply impossible to get a sample consisting completely of middle-aged people. In fact, it is impossible to get a sample in anything but the proportions 20:50:30, as in the population; this is ensured by the method of collecting the sample.

It is easy to see when stratification leads to large improvements over simple random sampling. If there were no difference between strata (age groups) in reading habits then there would be no gain from stratification. If reading habits were the same regardless of age group there would be no point in dividing up the population according to that factor. On the other hand, if there were large differences between strata, but within strata reading habits were similar, then the gains from stratification would be large. (The fact that reading habits are similar within strata means that even a small sample from a stratum should give an accurate picture of that stratum.)

Stratification is beneficial therefore when

- the between-strata differences are large and
- the within-strata differences are small.

These benefits take the form of greater precision of the estimates, i.e. narrower confidence intervals.[2] The greater precision arises because stratified sampling makes use of supplementary information – i.e. the proportion of the population in each age group. Simple random sampling does not make use of this. Obviously, therefore, if those proportions of the population are unknown, stratified sampling cannot be carried out. However, even if the proportions are only known approximately there could be a gain in precision.

In this example age is a **stratification factor**, i.e. a variable which is used to divide the population into strata. Other factors could, of course, be used, such as income or even height. A good stratification factor is one which is related to the subject of investigation. Income would therefore probably be a good stratification factor, because it is related to reading habits, but height is not since there is probably little difference between tall and short people regarding the newspaper they read. What is a good stratification factor obviously depends upon the subject of study. A bed manufacturer might well find height to be a good stratification factor if conducting an enquiry into preferences about the size of beds. Although good stratification factors improve the precision of estimates, bad factors do not make them worse; there will simply be no gain over simple random sampling. It would be as if there were no differences between the age groups in reading habits, so that ensuring the right proportions in the sample is irrelevant, but it has no detrimental effects.

[2] The formulae for calculating confidence intervals with stratified sampling are not given here, since they merit a whole book to themselves. The interested reader should consult, for example, C. A. Moser and G. Kalton, *Survey Methods in Social Investigation*, 1971, Heinemann.

Proportional allocation of sample observations to the different strata (as done above) is the simplest method but is not necessarily the best. For the optimal allocation there should generally be a divergence from proportional allocation, and the sample should have more observations in a particular stratum (relative to proportional allocation):

- the more diverse the stratum, and
- the cheaper it is to sample the stratum.

Starting from the 20:50:30 proportional allocation derived earlier, suppose that older people all read the same newspaper, but youngsters read a variety of titles. Then the representation of youngsters in the sample should be increased, and that of older people reduced. If it really were true that everyone old person read the same paper then one observation from that class would be sufficient to yield all there is to know about it. Furthermore, if it is cheaper to sample younger readers, perhaps because they are easier to contact than older people, then again the representation of youngsters in the sample should be increased. This is because, for a given budget, it will allow a larger total sample size.

Surveying concert-goers

A colleague and I carried out a survey of people attending a concert in Brighton (by Jamiroquai – hope they're still popular by the time you read this) to find out who they were, how much they spent in the town and how they travelled to the concert. The spreadsheet below gives some of the results.

Microsoft Excel - Book1

	A	B	C	D	E	F
3						
4		Number	%	Cumulative %		
5	Work in Brighton	11	15.1	15.1		
6	Student in Brighton	19	26.0	41.1		
7	Work outside Brighton	42	57.5	98.6		
8	Not working	1	1.4	100.0		
9	Total	73	100.0			
10						
11	**Method of travelling to concert:**					
12		Number	%	Cumulative %		
13	Car	23	31.5	31.5		
14	Car share	24	32.9	64.4		
15	Bus/coach	4	5.5	69.9		
16	On foot	9	12.3	82.2		
17	Train	10	13.7	95.9		
18	Taxi	3	4.1	100.0		
19	Total	73	100.0			
20						

→

The data were collected by face-to-face interviews before the concert. We did not have a sampling frame, so the (student) interviewers simply had to choose the sample themselves on the night. The one important instruction about sampling we gave them was that they should not interview more than one person in any group. People in the same group are likely to be influenced by each other (e.g. travel together) so we would not get independent observations, reducing the effective sample size.

From the results you can see that 41.1% either worked or studied in Brighton and that only one person in the sample was neither working nor studying. The second half of the table shows that 64.4% travelled to the show in a car (obviously adding to congestion in the town), about half of whom shared a car ride. Perhaps surprisingly, Brighton residents were just as likely to use their car to travel as were those from out of town.

The average level of spending was £24.20, predominantly on food (£7.38), drink (£5.97) and shopping (£5.37). The last category had a high variance associated with it – many people spent nothing, one person spent £200 in the local shops.

Cluster sampling

A third form of sampling is cluster sampling which, although intrinsically inefficient, can be much cheaper than other forms of sampling, allowing a larger sample size to be collected. Drawing a simple or a stratified random sample of size 100 from the whole of Britain would be very expensive to collect since the sample observations would be geographically very spread out. Interviewers would have to make many long and expensive journeys simply to collect one or two observations. To avoid this, the population can be divided into 'clusters' (e.g. regions or local authorities) and one or more of these clusters are then randomly chosen. Sampling takes place only within the selected clusters, and is therefore geographically concentrated, and the cost of sampling falls, allowing a larger sample to be collected for the same expenditure of money.

Within each cluster one can have either a 100% sample or a lower sampling fraction, which is called multistage sampling (this is explained further below). Cluster sampling gives unbiased estimates of population parameters but, for a given sample size, these are less precise than the results from simple or stratified sampling. This arises in particular when the clusters are very different from each other, but fairly homogeneous within themselves. In this case, once a cluster is chosen, if it is unrepresentative of the population, a poor (inaccurate) estimate of the population parameter is inevitable. The ideal circumstances for cluster sampling are when all clusters are very similar, since in that case examining one cluster is almost as good as examining the whole population.

Dividing up the population into clusters and dividing it into strata are similar procedures, but the important difference is that sampling is from one or at most a few clusters, but from all strata. This is reflected in the characteristics which make for good sampling. In the case of stratified sampling, it is beneficial if the between-strata differences are large and the within-strata differences small. For cluster sampling this is reversed: it is desirable to have small between-cluster differences but heterogeneity within clusters. Cluster sampling is less efficient (precise) for a given sample size, but is cheaper and so can offset this disadvantage with a larger sample size. In general, cluster sampling needs a much larger sample to be effective, so is only worthwhile where there are significant gains in cost.

Multistage sampling

Multistage sampling was briefly referred to in the previous section and is commonly found in practice. It may consist of a mixture of simple, stratified and cluster sampling at the various stages of sampling. Consider the problem of selecting a random sample of 1000 people from a population of 25 million to find out about voting intentions. A simple random sample would be extremely expensive to collect, for the reasons given above, so an alternative method must be found. Suppose further that it is suspected that voting intentions differ according to whether one lives in the north or south of the country and whether one is a home owner or renter. How is the sample to be selected? The following would be one appropriate method.

First the country is divided up into clusters of counties or regions, and a random sample of these taken, say one in five. This would be the first way of reducing the cost of selection, since only one-fifth of all counties now need to be visited. This one-in-five sample would be stratified to ensure that north and south were both appropriately represented. To ensure that each voter has an equal chance of being in the sample, the probability of a county being drawn should be proportional to its adult population. Thus a county with twice the population of another should have twice the probability of being in the sample.

Having selected the counties, the second stage would be to select a random sample of local authorities within each selected county. This might be a one-in-ten sample from each county and would be a simple random sample within each cluster. Finally a selection of voters from within each local authority would be taken, stratified according to tenure. This might be a one in 500 sample. The sampling fractions would therefore be

$$\frac{1}{5} \times \frac{1}{10} \times \frac{1}{500} = \frac{1}{25\,000}$$

So from the population of 25 million voters a sample of 1000 would be collected. For different population sizes the sampling fractions could be adjusted so as to achieve the goal of a sample size of 1000.

The sampling procedure is a mixture of simple, stratified and cluster sampling. The two stages of cluster sampling allow the selection of 50 local authorities for study and so costs are reduced. The north and south of the country are both adequately represented and housing tenures are also correctly represented in the sample by the stratification at the final stage. The resulting confidence intervals will be complicated to calculate but should give an improvement over the method of simple random sampling.

The UK Time Use Survey

The UK Time Use Survey provides a useful example of the effects of multistage sampling. It uses a mixture of cluster and stratified sampling and the results are weighted to compensate for unequal probabilities of selection into the sample and for the effects of non-response. Together, these act to increase the size of standard errors, relative to those obtained from a simple random sample of the same size. This increase can be measured by the **design factor**, defined as the ratio of the true standard error to the one arising from a simple random sample of the same size. For the time use survey, the design factor is typically 1.5 or more. Thus the standard errors are increased by 50% or more, but a simple random sample of the same size would be much more expensive to collect (e.g. the clustering means that only a minority of geographical areas are sampled).

The following table shows the average amount of time spent sleeping by 16–24 year olds (in minutes per day):

	Mean	True s.e.	95% CI	Design factor	n	Effective sample size
Male	544.6	6.5	[531.9, 557.3]	1.63	1090	412
Female	545.7	4.2	[537.3, 554.0]	1.14	1371	1058

The true standard error, taking account of the sample design, is 6.5 minutes for men. The design factor is 1.63, meaning this standard error is 63% larger than for a similar sized ($n = 1090$) simple random sample. Equivalently, a simple random sample of size $n = 412$ ($= 1090/1.63^2$) would achieve the same precision (but at greater cost).

How the design factor is made up is shown in the following table:

Design factor (deft)	Deft due to stratification	Deft due to clustering	Deft due to weighting
1.63	1.00	1.17	1.26

It can be seen that stratification has no effect on the standard error, but both clustering and the post-sample weighting serve to increase the standard errors.

Source: The UK 2000 Time Use Survey, Technical Report, 2003, Her Majesty's Stationery Office.

Quota sampling

Quota sampling is a non-random method of sampling and therefore it is impossible to use sampling theory to calculate confidence intervals from the sample data, or to find whether or not the sample will give biased results. Quota sampling simply means obtaining the sample information as best one can, for example, by asking people in the street. However, it is by far the cheapest method of sampling and so allows much larger sample sizes. As shown above, large sample sizes can still give biased results if sampling is non-random; but in some cases the budget is too small to afford even the smallest properly conducted random sample, so a quota sample is the only alternative.

Even with quota sampling, where the interviewer is simply told to go out and obtain (say) 1000 observations, it is worth making some crude attempt at stratification. The problem with human interviewers is that they are notoriously non-random, so that when they are instructed to interview every tenth person they see (a reasonably random method), if that person turns out to be a shabbily dressed tramp slightly the worse for drink, they are quite likely to select the eleventh person instead. Shabbily dressed tramps, slightly the worse for drink, are therefore under-represented in the sample. To combat this sort of problem the interviewers are given quotas to fulfil, for example, 20 men and 20 women, 10 old-age pensioners, one shabbily dressed tramp, etc., so that the sample will at least broadly reflect the population under study and give reasonable results.

It is difficult to know how accurate quota samples are, since it is rare for their results to be checked against proper random samples or against the population itself. Probably the most common quota samples relate to voting intentions and so can be checked against actual election results. The 1992 UK general election provides an interesting illustration. The opinion polls predicted a fairly substantial Labour victory but the outcome was a narrow Conservative majority. An enquiry concluded that the erroneous forecast occurred because a substantial number of voters changed their minds at the last moment and that there was 'differential turn-out', i.e. Conservative supporters were more likely to vote than Labour ones. Since then, pollsters have tried to take this factor into account when trying to predict election outcomes.

Can you always believe surveys?

Many surveys are more interested in publicising something than in finding out the facts. One has to be wary of surveys finding that people enjoy high-rise living . . . when the survey is sponsored by an elevator company. In July 2007 a survey of 1000 adults found that 'the average person attends 3.4 weddings each year'. This sounds suspiciously high to me. I've never attended three or more weddings in a year, nor have friends I have asked. Let's do some calculations. There were 283 730 weddings in the UK in 2005. There are about 45m adults, so if they each attend 3.4 weddings, that makes $45 \times 3.4 = 153$ million attendees. This means 540 per wedding. That seems excessively high (remember this excludes children) and probably means the sample design was poor, obtaining an unrepresentative result.

A good way to make a preliminary judgement on the likely accuracy of a survey is to ask 'who paid for this?'

Calculating the required sample size

Before collecting sample data it is obviously necessary to know how large the sample size has to be. The required sample size will depend upon two factors:

- the desired level of precision of the estimate, and
- the funds available to carry out the survey.

The greater the precision required the larger the sample size needs to be, other things being equal. But a larger sample will obviously cost more to collect and

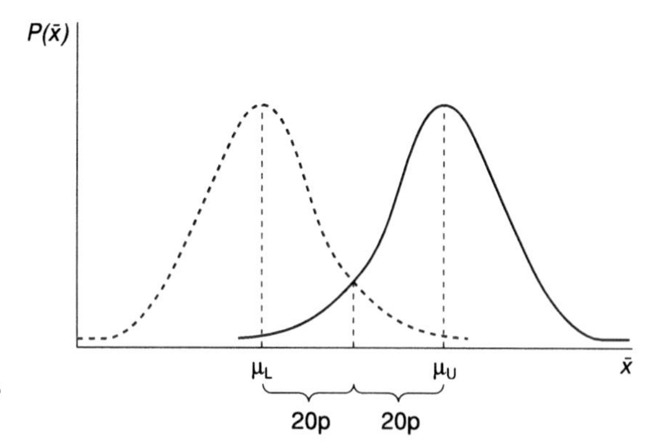

Figure 9.1
The desired width of the confidence interval

this might conflict with a limited amount of funds being available. There is a trade-off therefore between the two desirable objectives of high precision and low cost. The following example shows how these two objectives conflict.

A firm producing sweets wishes to find out the average amount of pocket money children receive per week. It wants to be 99% confident that the estimate is within 20 pence of the correct value. How large a sample is needed?

The problem is one of estimating a confidence interval, turned on its head. Instead of having the sample information $\bar{x}$, s and n, and calculating the confidence interval for μ, the desired width of the confidence interval is given and it is necessary to find the sample size n which will ensure this. The formula for the 99% confidence interval, assuming a Normal rather than t distribution (i.e. it is assumed that the required sample size will be large), is

$$[\bar{x} - 2.58 \times \sqrt{s^2/n},\ \bar{x} + 2.58 \times \sqrt{s^2/n}] \tag{9.2}$$

Diagrammatically this can be represented as in Figure 9.1.

The firm wants the distance between $\bar{x}$ and μ to be no more than 20 pence in either direction, which means that the confidence interval must be 40 pence wide. The value of n which makes the confidence interval 40 pence wide has to be found. This can be done by solving the equation

$$20 = 2.58 \times \sqrt{s^2/n}$$

and hence by rearranging:

$$n = \frac{2.58^2 \times s^2}{20^2} \tag{9.3}$$

All that is now required to solve the problem is the value of s^2, the sample variance; but since the sample has not yet been taken this is not available. There are a number of ways of trying to get round this problem:

- using the results of existing surveys if available;
- conducting a small, preliminary, survey;
- guessing.

These may not seem very satisfactory (particularly the last), but something has to be done and some intelligent guesswork should give a reasonable estimate of s^2. Suppose, for example, that a survey of children's *spending* taken five years previously showed a standard deviation of 30 pence. It might be reasonable to expect that the standard deviation of spending would be similar to the standard deviation of income, so 30 pence (updated for inflation) can be used as an estimate of the standard deviation. Suppose that five years' inflation turns the 30 pence into 50 pence. Using $s = 50$ we obtain

$$n = \frac{2.58^2 \times 50^2}{20^2} = 41.6$$

giving a required sample size of 42 (the sample size has to be an integer). This is a large ($n \geqslant 25$) sample size so the use of the Normal distribution was justified.

Is the firm willing to pay for such a large sample? Suppose it was willing to pay out £1000 in total for the survey, which costs £600 to set up and then £6 per person sampled. The total cost would be £600 + 42 × 6 = £852 which is within the firm's budget. If the firm wished to spend less than this, it would have to accept a smaller sample size and thus a lower precision or a lower level of confidence. For example, if only a 95% confidence level were required, the appropriate z score would be 1.96, yielding

$$n = \frac{1.96^2 \times 50^2}{20^2} = 24.01$$

A sample size of 24 would only cost £600 + 6 × 24 = £804. (At this sample size the assumption that $\bar{x}$ follows a Normal distribution becomes less tenable, so the results should be treated with caution. Use of the t distribution is tricky, because the appropriate t value depends upon the number of degrees of freedom which in turn depends on sample size, which is what is being looked for!)

The general formula for finding the required sample size is

$$n = \frac{z_\alpha^2 \times s^2}{p^2} \tag{9.4}$$

where z_α is the z score appropriate for the $(100 - \alpha)\%$ confidence level and p is the desired accuracy (20 pence in this case).

Collecting the sample

The sampling frame

We now move on to the fine detail of how to select the individual observations which make up the sample. In order to do this it is necessary to have some sort of sampling frame, i.e. a list of all the members of the population from which the sample is to be drawn. This can be a problem if the population is extremely large, for example the population of a country, since it is difficult to manipulate so much information (cutting up 50 million pieces of paper to put into a hat for a random draw is a tedious business). Alternatively the list might not even exist or, if it does, not be in one place convenient for consultation and use. In this case there is often an advantage to multistage sampling, for the selection of regions or even local authorities is fairly straightforward and not too time-consuming. Once at this lower level the sampling frame is more manageable – each local authority has an electoral register, for example – and individual observations can be relatively easily chosen. Thus it is not always necessary to have a complete sampling frame for the entire population in one place.

Choosing from the sampling frame

There is a variety of methods available for selecting a sample of (say) 1000 observations from a sampling frame of 25 000 names, varying from the manual to the electronic. The oldest method is to cut up 25 000 pieces of paper, put them in a (large) hat, shake it (to randomise) and pick out 1000. This is fairly time-consuming, however, and has some pitfalls – if the pieces are not all cut to the same size is the probability of selection the same? It is much better if the population in the sampling frame is numbered in some way, for then one only has to select random numbers. This can be done by using a table of random numbers (see Table A1 on page **412**, for example), or a computer. The use of random number tables for such purposes is an important feature of statistics and in 1955 the Rand Corporation produced a book entitled *A Million Random Digits with 100 000 Normal Deviates*. This book, as the title suggests, contained nothing but pages of random numbers, which allowed researchers to collect random samples. Interestingly, the authors did not bother fully to proofread the text, since a few (random) errors here and there wouldn't matter! These numbers were calculated electronically and nowadays every computer has a facility for rapidly choosing a set of random numbers. (It is an interesting question how a computer, which follows rigid rules of behaviour, can select random numbers which, by definition, are unpredictable by any rule.)

A further alternative, if a 1 in 25 sample is required, is to select a random starting point between 1 and 25 and then select every subsequent 25th observation (e.g. the 3rd, 28th, 53rd, etc.). This is a satisfactory procedure if the sampling frame is randomly sorted to start with, but otherwise there can be problems. For example, if the list is sorted by income (poorest first), a low starting value will almost certainly give an underestimate of the population mean. If all the numbers were randomly selected, this 'error' in the starting value would not be important.

Interviewing techniques

Good training of interviewers is vitally important to the results of a survey. It is very easy to lead an interviewee into a particular answer to a question. Consider the following two sets of questions:

A

(1) Do you know how many people were killed by the atomic bomb at Hiroshima?
(2) Do you think nuclear weapons should be banned?

B

(1) Do you believe in nuclear deterrence?
(2) Do you think nuclear weapons should be banned?

A2 is almost certain to get a higher 'yes' response than *B2*. Even a different ordering of the questions can have an effect upon the answers (consider asking *A2* before *A1*). The construction of the questionnaire has to be done with care, therefore. The manner in which the questions are asked is also important, since it can often suggest the answer. Good interviewers are trained to avoid these problems by sticking precisely to the wording of the question and not to suggest an expected answer.

Telephone surveys

An article by M. Collins in the *Journal of the Royal Statistical Society* reveals some of the difficulties in conducting surveys by telephone. First, the sampling frame is incomplete since, although most people have a telephone, some are not listed in the directory. In the late 1980s this was believed to be around 12% of all numbers, but it has been growing since, to around 40%. (Part of this trend, of course, may be due to people growing fed up of being pestered by salespersons and 'market researchers'.) Researchers have responded with 'random digit dialling' which is presumably made easier by modern computerised equipment.

Matters are unlikely to improve for researchers in the future. The answering machine is often used as a barrier to unwanted calls and many residential lines connect to fax machines. Increasing deregulation and mobile phone use mean it will probably become more and more difficult to obtain a decent sampling frame for a proper survey.

Source: M. Collins, Sampling for UK telephone surveys, *J. Royal Statistical Society, Series A*, 1999, **162** (1), 1–4.

Even when these procedures are adhered to there can be various types of response bias. The first problem is of non-response, due to the subject not being at home when the interviewer calls. There might be a temptation to remove that person from the sample and call on someone else, but this should be resisted. There could well be important differences between those who are at home all day and those who are not, especially if the survey concerns employment or spending patterns, for example. Continued efforts should be made to contact the subject. One should be wary of surveys which have low response rates, particularly where it is suspected that the non-response is in some way systematic and related to the goal of the survey.

A second problem is that subjects may not answer the question truthfully for one reason or another, sometimes inadvertently. An interesting example of this occurred in the survey into sexual behaviour carried out in Britain in 1992 (see *Nature*, 3 December 1992). Among other things, this found the following

- The average number of heterosexual partners during a woman's lifetime is 3.4.
- The average number of heterosexual partners during a man's lifetime is 9.9.

This may be in line with one's beliefs about behaviour, but, in fact, the figures must be wrong. The *total* number of partners of all women must by definition equal the *total* number for all men. Since there are approximately equal numbers of males and females in the UK the averages must therefore be about the same. So how do the above figures come about?

It is too much to believe that international trade holds the answer. It seems unlikely that British men are so much more attractive to foreign women than British women are to foreign men. Nor is an unrepresentative sample likely. It was carefully chosen and quite large (around 20 000). The answer would appear to be that some people are lying. Either women are being excessively modest or (more likely?) men are boasting. Perhaps the answer is to divide by three whenever a man talks about his sexual exploits!

For an update on this story, see the article by J. Wadsworth *et al.*, What is a mean? An examination of the inconsistency between men and women in reporting sexual partnerships, *Journal of the Royal Statistical Society, Series A*, 1996, **159** (1), 111–123.

Case study: the UK Expenditure and Food Survey

Introduction

The Expenditure and Food Survey (EFS) is an example of a large government survey which examines households' expenditure patterns (with a particular focus on food expenditures) and income receipts. It is worth having a brief look at it, therefore, to see how the principles of sampling techniques outlined in this chapter are put into practice. The EFS succeeded the Family Expenditure Survey in 2001 and uses a similar design. The EFS is used for many different purposes, including the calculation of weights to be used in the UK Retail Price Index, and the assessment of the effects of changes in taxes and state benefits upon different households.

Choosing the sample

The sample design follows is known as a **three-stage**, **rotating**, **stratified**, **random sample**. This is obviously quite complex so will be examined stage by stage.

Stage 1

The country is first divided into around 150 strata, each stratum made up of a number of local authorities sharing similar characteristics. The characteristics used as stratification factors are

- geographic area;
- urban or rural character (based on a measure of population density);
- prosperity (based on a measure of property values).

A stratum might therefore be made up of local authorities in the South West region, of medium population density and high prosperity.

In each quarter of the year, one local authority from each stratum is chosen at random, the probability of selection being proportional to population. Once an authority has been chosen, it remains in the sample for one year (four quarters) before being replaced. Only a quarter of the authorities in the sample are replaced in any quarter, which gives the sample its 'rotating' characteristic. Each quarter some authorities are discarded, some kept and some new ones brought in.

Stage 2

From each local authority selected, four wards (smaller administrative units) are selected, one to be used in each of the four quarters for which the local authority appears in the sample.

Stage 3

Finally, within each ward, 16 addresses are chosen at random, and these constitute the sample.

The sampling frame

The Postcode Address File, a list of all postal delivery addresses, is used as the sampling frame. Previously the register of electors in each ward was used but had some drawbacks: it was under-representative of those who have no permanent home or who move frequently (e.g. tramps, students, etc.). The fact that many people took themselves off the register in the early 1990s in order to avoid paying the Community Charge could also have affected the sample. The addresses are chosen from the register by interval sampling from a random starting point.

About 12 000 addresses are targeted each year, but around 11% prove to be business addresses, leaving approximately 11 000 households. The response rate is about 60%, meaning that the actual sample consists of about 6500 households each year. Given the complexity of the information gathered, this is a remarkably good figure.

Collection of information

The data are collected by interview, and by asking participants to keep a diary in which they record everything they purchase over a two-week period. Highly skilled interviewers are required to ensure accuracy and compliance with the survey, and each participating family is visited serveral times. As a small inducement to cooperate, each member of the family is paid a small sum of money (£10, it is to be hoped that the anticipation of this does not distort their expenditure patterns!).

Sampling errors

Given the complicated survey design it is difficult to calculate sampling errors exactly. The multistage design of the sample actually tends to increase the sampling error relative to a simple random sample, but, of course, this is offset by cost savings which allow a greatly increased sample size. Overall, the results of the survey are of good quality, and can be verified by comparison with other statistics, such as retail sales, for example.

Summary

- A primary data source is one where you obtain the data yourself or have access to all the original observations.
- A secondary data source contains a summary of the original data, usually in the form of tables.
- When collecting data always keep detailed notes of the sources of all information, how it was collected, precise definitions of the variables, etc.

- Some data can be obtained electronically, which saves having to type them into a computer, but the data still need to be checked for errors.
- There are various types of random sample, including simple, stratified and clustered random samples. The methods are sometimes combined in multi-stage samples.
- The type of sampling affects the size of the standard errors of the sample statistics. The most precise sampling method is not necessarily the best if it costs more to collect (since the overall sample size that can be afforded will be smaller).
- Quota sampling is a non-random method of sampling which has the advantage of being extremely cheap. It is often used for opinion polls and surveys.
- The sampling frame is the list (or lists) from which the sample is drawn. If it omits important elements of the population its use could lead to biased results.
- Careful interviewing techniques are needed to ensure reliable answers are obtained from participants in a survey.

Key terms and concepts

cluster sampling
finite population correction
multistage sampling
online data sources
primary and secondary data
quota sampling
random sample
sampling frame
sampling methods
simple random sampling
spreadsheet
stratified sampling

References

C. A. Moser and G. Kalton, *Survey Methods in Social Investigations*, 1971, Heinemann.

Rand Corporation, *A Million Random Digits with 100 000 Normal Deviates*, 1955, The Glencoe Press.

Problems

Some of the more challenging problems are indicated by highlighting the problem number in colour.

9.1 What issues of definition arise in trying to measure 'output'?

9.2 What issues of definition arise in trying to measure 'unemployment'?

9.3 Find the gross domestic product for both the UK and the US for the period 1995–2003. Obtain both series in constant prices.

9.4 Find figures for the monetary aggregate M0 for the years 1995–2003 in the UK, in nominal terms.

9.5 A firm wishes to know the average weekly expenditure on food by households to within £2, with 95% confidence. If the variance of food expenditure is thought to be about 400, what sample size does the firm need to achieve its aim?

9.6 A firm has £10 000 to spend on a survey. It wishes to know the average expenditure on gas by businesses to within £30 with 99% confidence. The variance of expenditure is believed to be about 40 000. The survey costs £7000 to set up and then £15 to survey each firm. Can the firm achieve its aim with the budget available?

9.7 **(Project)** Visit your college library or online sources to collect data to answer the following question. Has women's remuneration risen relative to men's over the past 10 years? You should write a short report on your findings. This should include a section describing the data collection process, including any problems encountered and decisions you had to make. Compare your results with those of other students. It might be interesting to compare your experiences of using online and offline sources of data.

9.8 **(Project)** Do a survey to find the average age of cars parked on your college campus. (A letter or digit denoting the registration year can be found on the number plate – precise details can be obtained in various guides to used-car prices.) You might need stratified sampling (e.g. if administrators have newer cars than faculty and students). You could extend the analysis by comparing the results with a public car park. You should write a brief report outlining your survey methods and the results you obtain. If several students do such a survey you could compare results.

Section Title 3

Calculus

33 Gradients of curves

Objectives: This chapter:

- introduces a technique called differentiation for calculating the gradient of a curve at any point
- illustrates how to find the gradient function of several common functions using a table
- introduces some simple rules for finding gradient functions
- explains what is meant by the terms 'first derivative' and 'second derivative'
- explains the terms 'maximum' and 'minimum' when applied to functions
- applies the technique of differentiation to locating maximum and minimum values of a function

33.1 The gradient function

Suppose we have a function, $y = f(x)$, and are interested in its slope, or gradient, at several points. For example, Figure 33.1 shows a graph of the function $y = 2x^2 + 3x$. Imagine tracing the graph from the left to the right. At point A the graph is falling rapidly. At point B the graph is falling but less rapidly than at A. Point C lies at the bottom of the dip. At point D the graph is rising. At E it is rising but more quickly than at D. The important point is that the gradient of the curve changes from point to point. The following section describes a mathematical technique for measuring the gradient at different points. We introduce another function called the **gradient function**, which we write as $\frac{dy}{dx}$. This is read as 'dee y by dee x'. We sometimes simplify this notation to y', read as 'y dash'. For almost all of the functions you will meet this gradient function can be found by using a formula, applying a rule or checking in a table. Knowing the gradient function we can find the gradient of the curve at any point. The following sections show how to find the gradient function of a number of common functions.

y' stands for the gradient function of $y = f(x)$. We can also denote the gradient function by $f'(x)$.

Figure 33.1
Graph of $y = 2x^2 + 3x$

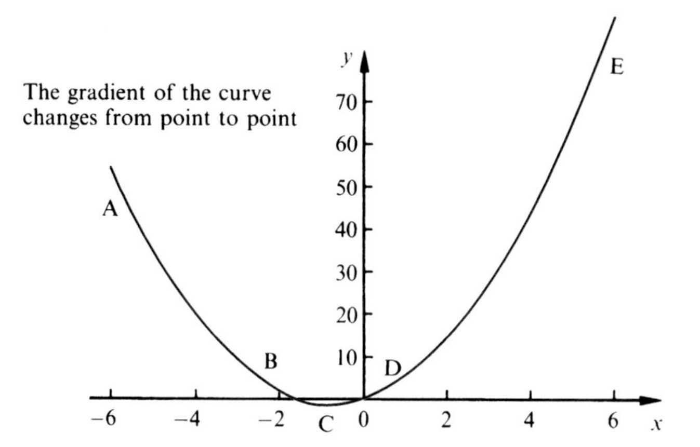

Key point

Given a function $y = f(x)$ we denote its gradient function by $\frac{dy}{dx}$ or simply by y'.

33.2 Gradient function of $y = x^n$

For any function of the form $y = x^n$ the gradient function is found from the following formula:

Key point

If $y = x^n$ then $y' = nx^{n-1}$.

WORKED EXAMPLES

33.1 Find the gradient function of (a) $y = x^3$, (b) $y = x^4$.

Solution

(a) Comparing $y = x^3$ with $y = x^n$ we see that $n = 3$. Then $y' = 3x^{3-1} = 3x^2$.

(b) Applying the formula with $n = 4$ we find that if $y = x^4$ then $y' = 4x^{4-1} = 4x^3$.

33.2 Find the gradient function of (a) $y = x^2$, (b) $y = x$.

Solution

(a) Applying the formula with $n = 2$ we find that if $y = x^2$ then $y' = 2x^{2-1} = 2x^1$. Because x^1 is simply x we find that the gradient function is $y' = 2x$.

(b) Applying the formula with $n = 1$ we find that if $y = x^1$ then $y' = 1x^{1-1} = 1x^0$. Because x^0 is simply 1 we find that the gradient function is $y' = 1$.

Finding the gradient of a graph is now a simple matter. Once the gradient function has been found, the gradient at any value of x is found by substituting that value into the gradient function. If, after carrying out this substitution, the result is negative, then the curve is falling. If the result is positive, the curve is rising. The size of the gradient function is a measure of how rapidly this fall or rise is taking place. We write $y'(x = 2)$ or simply $y'(2)$ to denote the value of the gradient function when $x = 2$.

WORKED EXAMPLES

33.3 Find the gradient of $y = x^2$ at the points where

(a) $x = -1$ (b) $x = 0$ (c) $x = 2$ (d) $x = 3$

Solution From Worked Example 33.2, or from the formula, we know that the gradient function of $y = x^2$ is given by $y' = 2x$.

(a) When $x = -1$ the gradient of the graph is then $y'(-1) = 2(-1) = -2$. The fact that the gradient is negative means that the curve is falling at the point.

(b) When $x = 0$ the gradient is $y'(0) = 2(0) = 0$. The gradient of the curve is zero at this point. This means that the curve is neither falling nor rising.

(c) When $x = 2$ the gradient is $y'(2) = 2(2) = 4$. The fact that the gradient is positive means that the curve is rising.

(d) When $x = 3$ the gradient is $y'(3) = 2(3) = 6$ and so the curve is rising here. Comparing this answer with that of part (c) we conclude that the curve is rising more rapidly at $x = 3$ than at $x = 2$, where the gradient was found to be 4.

The graph of $y = x^2$ is shown in Figure 33.2. If we compare our results with the graph we see that the curve is indeed falling when

Figure 33.2
Graph of $y = x^2$

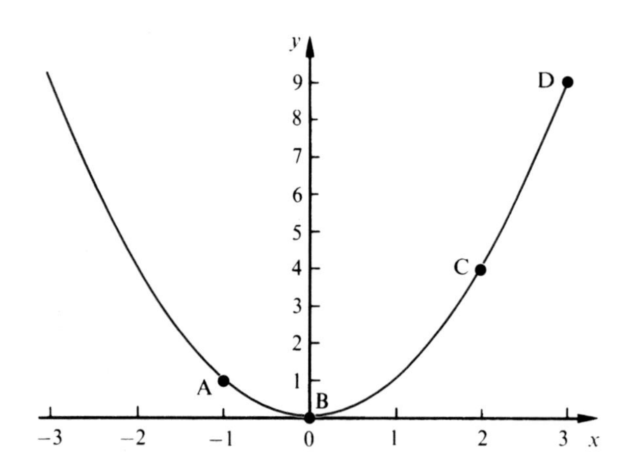

$x = -1$ (point A) and is rising when $x = 2$ (point C), and $x = 3$ (point D). At the point where $x = 0$ (point B) the curve is neither rising nor falling.

33.4 Find the gradient function of $y = x^{-3}$. Hence find the gradient of $y = x^{-3}$ when $x = 4$.

Solution Using the formula with $n = -3$ we find that if $y = x^{-3}$ then $y' = -3x^{-3-1} = -3x^{-4}$. Because x^{-4} can also be written as $1/x^4$ we could write $y' = -3(1/x^4)$ or $-3/x^4$. When $x = 4$ we find $y'(4) = -3/4^4 = -\frac{3}{256}$. This number is very small and negative, which means that when $x = 4$ the curve is falling, but only slowly.

33.5 Find the gradient function of $y = 1$.

Solution Before we use the formula to calculate the gradient function let us think about the graph of $y = 1$. Whatever the value of x, this function takes the value 1. Its graph must then be a horizontal line – it neither rises nor falls. We conclude that the gradient function must be zero, that is $y' = 0$. To obtain the same result using the formula we must rewrite 1 as x^0. Then, using the formula with $n = 0$, we find that if $y = x^0$ then $y' = 0x^{0-1} = 0$.

The previous worked example illustrates an important result. Any constant function has a gradient function equal to zero because its graph is a horizontal line, and is therefore neither rising nor falling.

The technique introduced in this section has a very long history and as a consequence a wide variety of alternative names for the gradient function

Table 33.1
The gradient function of some common functions

$y = f(x)$	$y' = f'(x)$	Notes
constant	0	
x	1	
x^2	$2x$	
x^n	nx^{n-1}	
e^x	e^x	
e^{kx}	ke^{kx}	k is a constant
$\sin x$	$\cos x$	
$\cos x$	$-\sin x$	
$\sin kx$	$k \cos kx$	k is a constant
$\cos kx$	$-k \sin kx$	k is a constant
$\ln kx$	$1/x$	k is a constant

have emerged. For example, the gradient function is also called the **first derivative**, or simply the **derivative**. The process of obtaining this is also known as **differentiation.** Being asked to **differentiate** $y = x^5$, say, is equivalent to being asked to find its gradient function, y'. Furthermore, because the gradient function measures how rapidly a graph is changing, it is also referred to as the **rate of change** of y. The area of study that is concerned with differentiation is known as *differential calculus.* When you meet these words you should realise that they are simply technical terms associated with finding the gradients of graphs.

It is possible to find the gradient function of a wide range of functions by simply referring to standard results. Some of these are given in Table 33.1. Note that, for technical reasons, when finding the gradient functions of trigonometrical functions the angle x must always be measured in radians.

WORKED EXAMPLES

33.6 Use Table 33.1 to find the gradient function y' when y is

(a) $\sin x$ (b) $\sin 2x$ (c) $\cos 3x$ (d) e^x

Solution

(a) Directly from the table we see that if $y = \sin x$ then its gradient function is given by $y' = \cos x$. This result occurs frequently and is worth remembering.

(b) Using the table and taking $k = 2$ we find that if $y = \sin 2x$ then $y' = 2 \cos 2x$.

(c) Using the table and taking $k = 3$ we find that if $y = \cos 3x$ then $y' = -3 \sin 3x$.

(d) Using the table we see directly that if $y = e^x$ then $y' = e^x$. Note that the exponential function e^x is very special because its gradient function y' is the same as y.

33.7 Find the gradient function of $y = e^{-x}$. Hence find the gradient of the graph of y at the point where $x = 1$.

Solution

Noting that $e^{-x} = e^{-1x}$ and using Table 33.1 with $k = -1$ we find that if $y = e^{-x}$ then $y' = -1e^{-x} = -e^{-x}$. Using a calculator to evaluate this when $x = 1$ we find $y'(1) = -e^{-1} = -0.368$.

33.8 Find the gradient function of $y = \sin 4x$ where $x = 0.3$.

Solution

Using Table 33.1 with $k = 4$ gives $y' = 4 \cos 4x$. Remembering to measure x in radians we evaluate this when $x = 0.3$:

$$y'(0.3) = 4 \cos 4(0.3) = 4 \cos 1.2 = 1.4494$$

Self-assessment questions 33.2

1. What is the purpose of the gradient function?
2. State the formula for finding the gradient function when $y = x^n$.
3. State three alternative names for the gradient function.
4. State an alternative notation for y'.
5. What is the derivative of any constant? What is the graphical explanation of this answer?
6. When evaluating the derivative of the sine and cosine functions, in which units must the angle x be measured – radians or degrees?
7. Which function is equal to its derivative?

Exercise 33.2

1. Find y' when y is given by
   (a) x^8 (b) x^7 (c) x^{-1} (d) x^{-5}
   (e) x^{13} (f) x^5 (g) x^{-2}
2. Find y' when y is given by
   (a) $x^{3/2}$ (b) $x^{5/2}$ (c) $x^{-1/2}$
   (d) $x^{1/2}$ (e) $\sqrt{x}$ (f) $x^{0.2}$
3. Sketch a graph of $y = 8$. State the gradient function, y'.
4. Find the first derivative of each of the following functions:
   (a) $y = x^{16}$ (b) $y = x^{0.5}$ (c) $y = x^{-3.5}$
   (d) $y = \dfrac{1}{x^{1/2}}$ (e) $y = \dfrac{1}{\sqrt{x}}$
5. Find the gradient of the graph of the function $y = x^4$ when
   (a) $x = -4$ (b) $x = -1$ (c) $x = 0$
   (d) $x = 1$ (e) $x = 4$
   Can you infer anything about the shape of the graph from this information?
6. Find the gradient of the graphs of each of the following functions at the points given:
   (a) $y = x^3$ at $x = -3$, at $x = 0$ and at $x = 3$
   (b) $y = x^4$ at $x = -2$ and at $x = 2$
   (c) $y = x^{1/2}$ at $x = 1$ and $x = 2$
   (d) $y = x^{-2}$ at $x = -2$ and $x = 2$
7. Sometimes a function is given in terms of variables other than x. This should pose no problems. Find the gradient functions of
   (a) $y = t^2$ (b) $y = t^{-3}$ (c) $y = t^{1/2}$
   (d) $y = t^{-1.5}$
8. Use Table 33.1 to find the gradient function y' when y is equal to
   (a) $\cos 7x$ (b) $\sin \frac{1}{2}x$ (c) $\cos \frac{x}{2}$
   (d) e^{4x} (e) e^{-3x}
9. Find the derivative of each of the following functions:
   (a) $y = \ln 3x$ (b) $y = \sin \frac{1}{3}x$
   (c) $y = e^{x/2}$
10. Find the gradient of the graph of the function $y = \sin x$ at the points where
    (a) $x = 0$ (b) $x = \pi/2$ (c) $x = \pi$
11. Table 33.1, although written in terms of the variable x, can still be applied when other variables are involved. Find y' when
    (a) $y = t^7$ (b) $y = \sin 3t$ (c) $y = e^{2t}$
    (d) $y = \ln 4t$ (e) $y = \cos \frac{t}{3}$

33.3 Some rules for finding gradient functions

Up to now the formulae we have used for finding gradient functions have applied to single terms. It is possible to apply these much more widely with the addition of three more rules. The first applies to the sum of two functions, the second to their difference and the third to a constant multiple of a function.

Key point

Rule 1: If $y = f(x) + g(x)$ then $y' = f'(x) + g'(x)$.

In words this says that to find the gradient function of a sum of two functions we simply find the two gradient functions separately and add these together.

WORKED EXAMPLE

33.9 Find the gradient function of $y = x^2 + x^4$.

Solution The gradient function of x^2 is $2x$. The gradient function of x^4 is $4x^3$. Therefore

$$\text{if } y = x^2 + x^4 \quad \text{then} \quad y' = 2x + 4x^3$$

The second rule is really an extension of the first, but we state it here for clarity.

Key point

Rule 2: If $y = f(x) - g(x)$ then $y' = f'(x) - g'(x)$.

WORKED EXAMPLE

33.10 Find the gradient function of $y = x^5 - x^7$.

Solution We find the gradient function of each term separately and subtract them. That is, $y' = 5x^4 - 7x^6$.

Key point

Rule 3: If $y = kf(x)$, where k is a number, then $y' = kf'(x)$.

WORKED EXAMPLE

33.11 Find the gradient function of $y = 3x^2$.

Solution This function is 3 times x^2. The gradient function of x^2 is $2x$. Therefore, using Rule 3, we find that

$$\text{if } y = 3x^2 \qquad \text{then } y' = 3(2x) = 6x$$

The rules can be applied at the same time. Consider the following examples.

WORKED EXAMPLES

33.12 Find the derivative of $y = 4x^2 + 3x^{-3}$.

Solution The derivative of $4x^2$ is $4(2x) = 8x$. The derivative of $3x^{-3}$ is $3(-3x^{-4}) = -9x^{-4}$. Therefore, if $y = 4x^2 + 3x^{-3}$ then $y' = 8x - 9x^{-4}$.

33.13 Find the derivative of

(a) $y = 4 \sin t - 3 \cos 2t$ (b) $y = \dfrac{e^{2t}}{3} + 6 + \dfrac{\ln(2t)}{5}$

Solution (a) We differentiate each quantity in turn using Table 33.1:

$$y' = 4 \cos t - 3(-2 \sin 2t) = 4 \cos t + 6 \sin 2t$$

(b) Writing y as $\frac{1}{3}e^{2t} + 6 + \frac{1}{5}\ln(2t)$ we find

$$y' = \frac{2}{3}e^{2t} + 0 + \frac{1}{5}\left(\frac{1}{t}\right) = \frac{2e^{2t}}{3} + \frac{1}{5t}$$

Self-assessment question 33.3

1. State the rules for finding the derivative of (a) $f(x) + g(x)$, (b) $f(x) - g(x)$, (c) $kf(x)$.

Exercise 33.3

MyMathLab Global

1. Find the gradient function of each of the following:
   (a) $y = x^6 - x^4$ (b) $y = 6x^2$
   (c) $y = 9x^{-2}$ (d) $y = \dfrac{1}{2}x$
   (e) $y = 2x^3 - 3x^2$
2. Find the gradient function of each of the following:
   (a) $y = 2 \sin x$ (b) $y = 3 \sin 4x$
   (c) $y = 7 \cos 9x + 3 \sin 4x$
   (d) $y = e^{3x} - 5e^{-2x}$
3. Find the gradient function of $y = 3x^3 - 9x + 2$. Find the value of the gradient function at $x = 1$, $x = 0$ and $x = -1$. At which of these points is the curve steepest?
4. Write down a function which when differentiated gives $2x$. Are there any other functions you can differentiate to give $2x$?

5. Find the gradient of $y = 3\sin 2t + 4\cos 2t$ when $t = 2$. Remember to use radians.

6. Find the gradient of $y = 2x - x^3 + e^{2x}$ when $x = 1$.

7. Find the gradient of $y = (2x - 1)^2$ when $x = 0$.

8. Show that the derivative of $f(x) = x^2 - 2x$ is 0 when $x = 1$.

9. Find the values of x such that the gradient of $y = \frac{x^3}{3} - x + 7$ is 0.

10. The function $y(x)$ is given by

$$y(x) = x + \cos x \qquad 0 \leqslant x \leqslant 2\pi$$

Find the value(s) of x where the gradient of y is 0.

33.4 Higher derivatives

In some applications, and in more advanced work, it is necessary to find the derivative of the gradient function itself. This is termed the **second derivative** and is written y'' and read 'y double dash'. Some books write

$$y'' \text{ as } \frac{d^2y}{dx^2}$$

Key point

y'' or $\frac{d^2y}{dx^2}$ is found by differentiating y'.

It is a simple matter to find the second derivative by differentiating the first derivative. No new techniques or tables are required.

WORKED EXAMPLES

33.14 Find the first and second derivatives of $y = x^4$.

Solution From Table 33.1, if $y = x^4$ then $y' = 4x^3$. The second derivative is found by differentiating the first derivative. Therefore

$$\text{if } y' = 4x^3 \qquad \text{then } y'' = 4(3x^2) = 12x^2$$

33.15 Find the first and second derivatives of $y = 3x^2 - 7x + 2$.

Solution Using Table 33.1 we find $y' = 6x - 7$. The derivative of the constant 2 equals 0. Differentiating again we find $y'' = 6$, because the derivative of the constant -7 equals 0.

33.16 If $y = \sin x$, find

(a) $\frac{dy}{dx}$ (b) $\frac{d^2y}{dx^2}$

Solution

(a) Recall that $\frac{dy}{dx}$ is the first derivative of y. From Table 33.1 this is given by

$$\frac{dy}{dx} = \cos x$$

(b) $\frac{d^2y}{dx^2}$ is the second derivative of y. This is found by differentiating the first derivative. From Table 33.1 we find that the derivative of $\cos x$ is $-\sin x$, and so

$$\frac{d^2y}{dx^2} = -\sin x$$

Self-assessment questions 33.4

1. Explain how the second derivative of y is found.
2. Give two alternative notations for the second derivative of y.

Exercise 33.4

MyMathLab Global

1. Find the first and second derivatives of the following functions:
(a) $y = x^2$ (b) $y = 3x^8$ (c) $y = 9x^5$
(d) $y = 3x^{1/2}$ (e) $y = 3x^4 + 5x^2$
(f) $y = 9x^3 - 14x + 11$
(g) $y = x^{1/2} + 4x^{-1/2}$
2. Find the first and second derivatives of
(a) $y = \sin 3x + \cos 2x$
(b) $y = e^x + e^{-x}$
3. Find the second derivative of $y = \sin 4x$. Show that the sum of $16y$ and y'' is zero, that is $y'' + 16y = 0$.
4. If $y = x^4 - 2x^3 - 36x^2$, find the values of x for which $y'' = 0$.
5. Determine y'' given y is
(a) $2 \sin 3x + 4 \cos 2x$
(b) $\sin kt$, k constant
(c) $\cos kt$, k constant
(d) $A \sin kt + B \cos kt$
A, B, k constants
6. Find y'' when y is given by
(a) $6 + e^{3x}$ (b) $2 + 3x + 2e^{4x}$
(c) $e^{2x} - e^{-2x}$ (d) $\frac{1}{e^x}$ (e) $e^x(e^x + 3)$
7. Given $y = 1000 + 200x + 6x^2 - x^3$, find the value(s) of x for which $y'' = 0$.
8. Given $y = 2e^x + \sin 2x$, evaluate y'' when $x = 1$.
9. Given $y = 2 \sin 4t - 5 \cos 2t$, evaluate y'' when $t = 1.2$.

33.5 Finding maximum and minimum points of a curve

Consider the graph sketched in Figure 33.3. There are a number of important points marked on this graph, all of which have one thing in

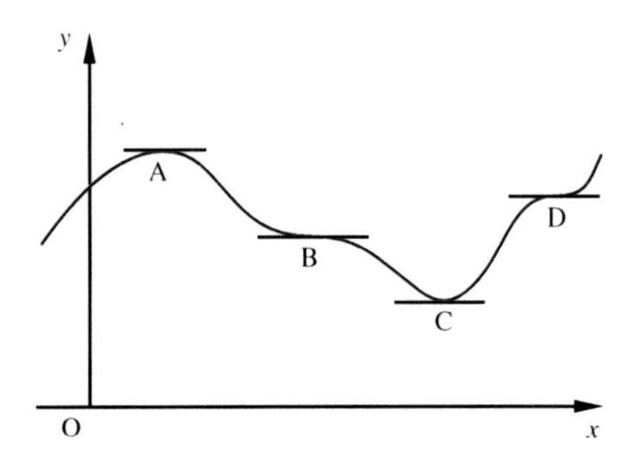

Figure 33.3
The curve has a maximum at A, a minimum at C and points of inflexion at B and D

common. At each point A, B, C and D the gradient of the graph is zero. If you were travelling from the left to the right then at these points the graph would appear to be flat. Points where the gradient is zero are known as **stationary points**. To the left of point A the curve is rising; to its right the curve is falling. A point such as A is called a **maximum turning point** or simply a **maximum**. To the left of point C the curve is falling; to its right the curve is rising. A point such as C is called a **minimum turning point**, or simply a **minimum**. Points B and D are known as **points of inflexion**. At these points the slope of the curve is momentarily zero but then the curve continues rising or falling as before.

Because at all these points the gradient is zero, they can be located by looking for values of x that make the gradient function zero.

Key point

Stationary points are located by setting the gradient function equal to zero, that is $y' = 0$.

WORKED EXAMPLE

33.17 Find the stationary points of $y = 3x^2 - 6x + 8$.

Solution We first determine the gradient function y' by differentiating y. This is found to be $y' = 6x - 6$. Stationary points occur when the gradient is zero, that is when $y' = 6x - 6 = 0$. From this

$$6x - 6 = 0$$
$$6x = 6$$
$$x = 1$$

When $x = 1$ the gradient is zero. When $x = 1$, $y = 5$. Therefore $(1, 5)$ is a stationary point. At this stage we cannot tell whether to expect a maximum, minimum or point of inflexion; all we know is that one of these occurs when $x = 1$. However, a sketch of the graph of $y = 3x^2 - 6x + 8$ is shown in Figure 33.4, which reveals that the point is a minimum.

Figure 33.4
There is a minimum turning point at $x = 1$

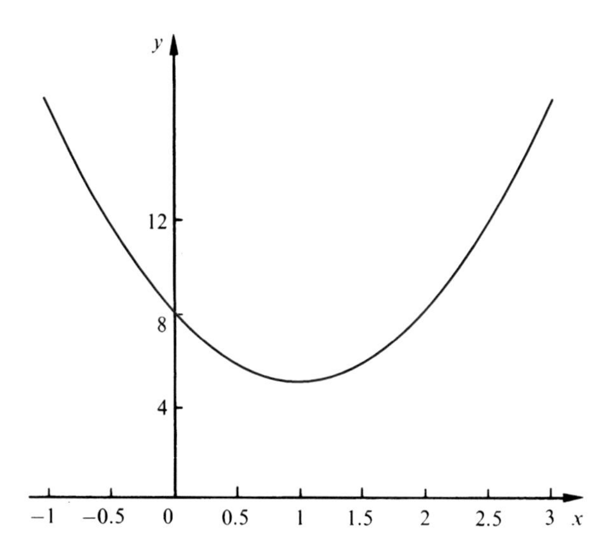

There are a number of ways to determine the nature of a stationary point once its location has been found. One way, as we have seen, is to sketch a graph of the function. Two alternative methods are now described.

Finding the nature of a stationary point by looking at the gradient on either side

At a stationary point we know that the gradient is zero. If this stationary point is a minimum, then we see from Figure 33.5 that, as we move from left to right, the gradient changes from negative to zero to positive. Therefore a little to the left of a minimum the gradient function will be negative; a little to the right it will be positive. Conversely, if the stationary point is a maximum, as we move from left to right, the gradient changes from positive to zero to negative. A little to the left of a maximum the gradient function will be positive; a little to the right it will be negative.

Figure 33.5
The sign of the gradient function close to a stationary point

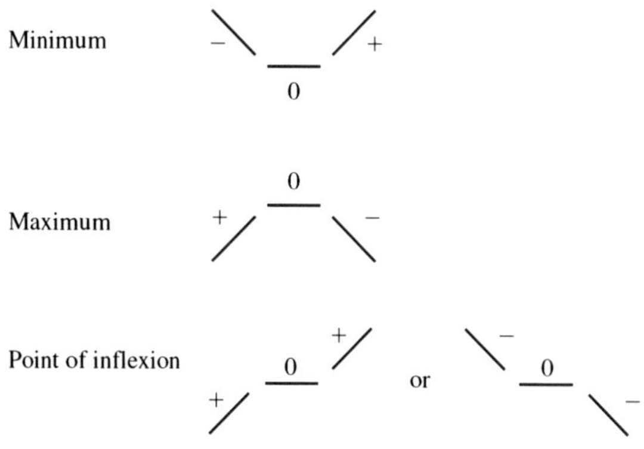

Minimum − 0 +

Maximum + 0 −

Point of inflexion + 0 + or − 0 −

The behaviour of the gradient function close to a point of inflexion is also shown. The sign of the gradient function is the same on both sides of a point of inflexion. With this information we can determine the nature of a stationary point without first plotting a graph.

WORKED EXAMPLES

33.18 Find the location and nature of the stationary points of $y = 2x^3 - 6x^2 - 18x$.

Solution In order to find the location of the stationary points we must calculate the gradient function. This is $y' = 6x^2 - 12x - 18$. At a stationary point $y' = 0$ and so the equation that must be solved is $6x^2 - 12x - 18 = 0$. This quadratic equation is solved as follows:

$$6x^2 - 12x - 18 = 0$$

Factorising:

$$6(x^2 - 2x - 3) = 0$$

$$6(x + 1)(x - 3) = 0$$

so that $x = -1$ and 3

The stationary points are located at $x = -1$ and $x = 3$. At these values, $y = 10$ and $y = -54$ respectively. We now find their nature:

When $x = -1$ A little to the left of $x = -1$, say at $x = -2$, we calculate the gradient of the graph using the gradient function. That is, $y'(-2) = 6(-2)^2 - 12(-2) - 18 = 24 + 24 - 18 = 30$. Therefore the graph is rising when $x = -2$. A little to the right of $x = -1$, say at $x = 0$, we calculate the gradient of the graph using the gradient function. That is, $y'(0) = 6(0)^2 - 12(0) - 18 = -18$. Therefore the graph is falling when $x = 0$. From this information we conclude that the turning point at $x = -1$ must be a maximum.

When $x = 3$ A little to the left of $x = 3$, say at $x = 2$, we calculate the gradient of the graph using the gradient function. That is, $y'(2) = 6(2)^2 - 12(2) - 18 = 24 - 24 - 18 = -18$. Therefore the graph is falling when $x = 2$. A little to the right of $x = 3$, say at $x = 4$, we calculate the gradient of the graph using the gradient function. That is, $y'(4) = 6(4)^2 - 12(4) - 18 = 96 - 48 - 18 = 30$. Therefore the graph is rising when $x = 4$. From this information we conclude that the turning point at $x = 3$ must be a minimum.

To show this behaviour a graph of $y = 2x^3 - 6x^2 - 18x$ is shown in Figure 33.6 where these turning points can be clearly seen.

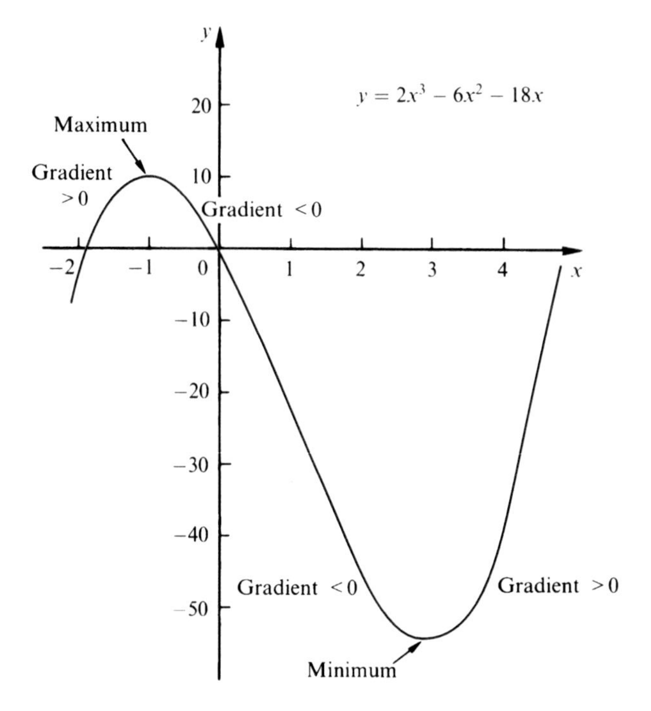

Figure 33.6
Graph of
$y = 2x^3 - 6x^2 - 18x$

33.19 Find the location and nature of the stationary points of $y = x^3 - 3x^2 + 3x - 1$.

Solution First we find the gradient function: $y' = 3x^2 - 6x + 3$. At a stationary point $y' = 0$ and so

$$3x^2 - 6x + 3 = 0$$
$$3(x^2 - 2x + 1) = 0$$
$$3(x - 1)(x - 1) = 0$$
$$\text{and so } x = 1$$

When $x = 1$, $y = 0$.

We conclude that there is only one stationary point, and this is at $(1, 0)$. To determine its nature we look at the gradient function on either side of

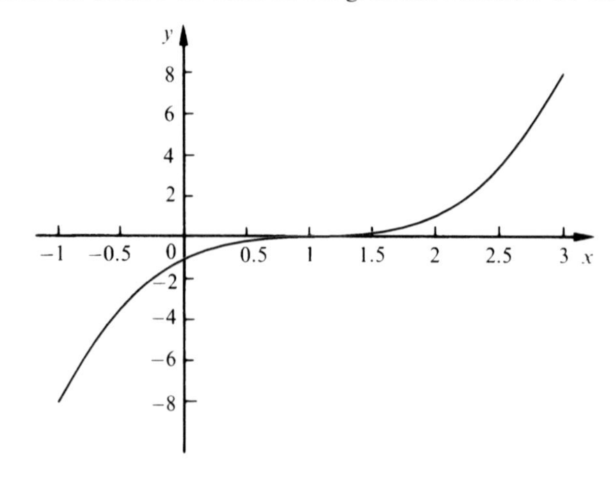

Figure 33.7
Graph of
$y = x^3 - 3x^2 + 3x - 1$
showing the point of inflexion

this point. At $x = 0$, say, $y'(0) = 3$. At $x = 2$, say, $y'(2) = 3(2^2) - 6(2) + 3 = 3$. The gradient function changes from positive to zero to positive as we move through the stationary point. We conclude that the stationary point is a point of inflexion. For completeness a graph is sketched in Figure 33.7.

The second-derivative test for maximum and minimum points

A simple test exists that uses the second derivative for determining the nature of a stationary point once it has been located.

Key point

If y'' is positive at the stationary point, the point is a minimum.
If y'' is negative at the stationary point, the point is a maximum.
If y'' is equal to zero, this test does not tell us anything and the previous method should be used.

WORKED EXAMPLE

33.20 Locate the stationary points of

$$y = \frac{x^3}{3} + \frac{x^2}{2} - 12x + 5$$

and determine their nature using the second-derivative test.

Solution We first find the gradient function: $y' = x^2 + x - 12$. Setting this equal to zero to locate the stationary points we find

$$x^2 + x - 12 = 0$$

$$(x + 4)(x - 3) = 0$$

$$x = -4 \text{ and } 3$$

When $x = -4$, $y = \frac{119}{3}$. When $x = 3$, $y = -\frac{35}{2}$.

There are two stationary points: at $(-4, \frac{119}{3})$ and $(3, -\frac{35}{2})$. To apply the second-derivative test we need to find y''. This is found by differentiating y' to give $y'' = 2x + 1$. We now evaluate this at each stationary point in turn:

$x = -4$ When $x = -4$, $y''(-4) = 2(-4) + 1 = -7$. This is negative and we conclude from the second-derivative test that the point is a maximum point.

$x = 3$ When $x = 3$, $y''(3) = 2(3) + 1 = 7$. This is positive and we conclude from the second-derivative test that the point is a minimum point. Remember, if in doubt, that these results could be confirmed by sketching a graph of the function.

Self-assessment questions 33.5

1. State the three types of stationary point.
2. State the second-derivative test for determining the nature of a stationary point.
3. Under what condition will the second-derivative test fail?

Exercise 33.5

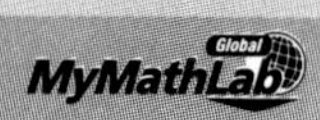

1. Determine the location and nature of any stationary point of the following functions:
   (a) $y = x^2 + 1$ (b) $y = -x^2$
   (c) $y = 2x^3 + 9x^2$
   (d) $y = -2x^3 + 27x^2$
   (e) $y = x^3 - 3x^2 + 3x + 1$
2. Determine the location of the maximum and minimum points of $y = \sin x$, $0 \leqslant x \leqslant 2\pi$.
3. Find the maximum and minimum points of $y = e^x - x$.
4. Find the maximum and minimum points of $y = 2x^3 - 3x^2 - 12x + 1$.
5. Determine the location and nature of the stationary points of $y = -\frac{x^3}{3} + x$.
6. Determine the location and nature of the stationary points of $y = x^5 - 5x + 1$.

Test and assignment exercises 33

1. Find y' when y equals
   (a) $7x$ (b) x^{15} (c) $x^{3/2}$ (d) $3x^{-1}$ (e) $5x^4$ (f) $16x^{-3}$
2. Find y' when y equals
   (a) $\dfrac{x+1}{x}$ (b) $3x^2 - 7x + 14$ (c) $\sin 3x$ (d) $e^{5x} + e^{4x}$ (e) $3e^{3x}$
3. Find the first and second derivatives of
   (a) $f(x) = e^{-3x}$ (b) $f(x) = \ln 2x$ (c) $f(x) = x + \dfrac{x^2}{2} + \dfrac{x^3}{3}$ (d) $f(x) = -15$
   (e) $f(x) = 3x$
4. Find y' when
   (a) $y = 3t - 7$ (b) $y = 16t^2 + 7t + 9$
5. Find the first derivative of each of the following functions:
   (a) $f(x) = x(x+1)$ (b) $f(x) = x(x-2)$ (c) $f(x) = x^2(x+1)$
   (d) $f(x) = (x-2)(x+3)$ (e) $f(t) = 3t^2 + 7t + 11$ (f) $f(t) = 7(t-1)(t+1)$
6. Find the first and second derivatives of each of the following:
   (a) $2x^2$ (b) $4x^3$ (c) $8x^7 - 5x^2$ (d) $\dfrac{1}{x}$

7. Find the location and nature of the maxima and minima of $y = 2x^3 + 15x^2 - 84x$

8. Find a function which when differentiated gives $\frac{1}{2}x^3$.

9. Find the gradient of y at the specified points:
(a) $y = 3x + x^2$ at $x = -1$ (b) $y = 4\cos 3x + \dfrac{\sin 4x}{2}$ at $x = 0.5$
(c) $y = 3e^{2x} - 2e^{-3x}$ at $x = 0.1$

10. Evaluate the second derivative of y at the specified point:
(a) $y = \frac{1}{2}\sin 4x$ $x = 1$ (b) $y = e^{-2x} + x^2$ $x = 0$
(c) $y = 3\ln 5x$ $x = 1$

11. Find the value(s) of t for which y' is zero:
(a) $y = t^3 - 12t^2 + 1$ (b) $y = \ln(2t) - 2t^2$

12. Determine the location and nature of the stationary points of
(a) $y = x^4 - 108x$ (b) $y = \ln 2x + \frac{1}{x}$

34 Techniques of differentiation

Objectives: This chapter:

- explains and illustrates the product, quotient and chain rules of differentiation

34.1 Introduction

In this chapter we study three rules of differentiation: the product, quotient and chain rules. These rules extend the range of functions that we are able to differentiate. They build upon the list of functions and their derivatives as given in Table 33.1 on page 408. This chapter assumes that you are familiar with this table; if you are not then please go back and study Chapter 33.

Using the product rule we are able to differentiate products of functions. Products of functions occur when one function is multiplied by another. For example, the product of the functions x^3 and $\sin 2x$ is $x^3 \sin 2x$. Using the product rule enables us to differentiate the product $y = x^3 \sin 2x$.

Using the quotient rule we can differentiate quotients of functions. A quotient is formed when one function is divided by another. For example, when x^2 is divided by $x^3 + 1$ we obtain the quotient $\frac{x^2}{x^3+1}$. The quotient rule enables us to differentiate functions of the form $y = \frac{x^2}{x^3+1}$.

Sometimes we will come across a function $y(x)$, say, where the variable x is itself a function of another variable, t say. So we have $y = y(x)$ and $x = x(t)$. For example, suppose $y(x) = x^3$ and $x(t) = \sin t$. We say that y is a **function of a function** and we can write

$$y = x^3 = (\sin t)^3$$

The chain rule enables us to differentiate a function of a function.

We now study each rule in turn.

34.2 The product rule

Suppose we are able to differentiate individually expressions $u(x)$ and $v(x)$. For example, we may have $u(x) = x^3$ and $v(x) = \sin 2x$. Using Table 33.1 we are able to find the derivatives $\frac{du}{dx}$ and $\frac{dv}{dx}$. The product rule then allows us to differentiate the product of u and v; that is, we can use the product rule to differentiate $y = u \times v$, in this case $y = x^3 \sin 2x$.

Key point

The product rule: if $y = uv$ then

$$\frac{dy}{dx} = \frac{du}{dx}v + u\frac{dv}{dx}$$

WORKED EXAMPLES

34.1 Use the product rule to differentiate $y = x^3 \sin 2x$.

Solution We let $u(x) = x^3$ and $v(x) = \sin 2x$.

Then clearly

$$y = uv$$

and using Table 33.1 we have

$$\frac{du}{dx} = 3x^2 \qquad \frac{dv}{dx} = 2\cos 2x$$

Using the product rule we have

$$\begin{aligned}\frac{dy}{dx} &= \frac{du}{dx}v + u\frac{dv}{dx}\\ &= 3x^2 \sin 2x + x^3(2\cos 2x)\\ &= x^2(3\sin 2x + 2x\cos 2x)\end{aligned}$$

34.2 Use the product rule to differentiate $y = e^{2x} \cos x$.

Solution We let

$$u(x) = e^{2x} \qquad v(x) = \cos x$$

so that

$$y = e^{2x}\cos x = uv$$

Using Table 33.1 we have

$$\frac{du}{dx} = 2e^{2x} \qquad \frac{dv}{dx} = -\sin x$$

Using the product rule we have

$$\frac{dy}{dx} = \frac{du}{dx}v + u\frac{dv}{dx}$$

$$= 2e^{2x}\cos x + e^{2x}(-\sin x)$$

$$= e^{2x}(2\cos x - \sin x)$$

34.3 Use the product rule to differentiate $y = x \ln x$.

Solution Let $u = x$, $v = \ln x$. Then

$$y = x \quad \ln x = uv$$

and

$$\frac{du}{dx} = 1 \qquad \frac{dv}{dx} = \frac{1}{x}$$

So using the product rule

$$\frac{dy}{dx} = \frac{du}{dx}v + u\frac{dv}{dx}$$

$$= 1\ \ln x + x\left(\frac{1}{x}\right)$$

$$= \ln x + 1$$

34.4 Find the gradient of $y = 2x^2e^{-x}$ when $x = 1$.

Solution Let $u = 2x^2$, $v = e^{-x}$ so that

$$y = 2x^2e^{-x} = uv$$

and

$$\frac{du}{dx} = 4x \qquad \frac{dv}{dx} = -e^{-x}$$

Using the product rule we see

$$\frac{dy}{dx} = \frac{du}{dx}v + u\frac{dv}{dx}$$

$$= 4xe^{-x} + 2x^2(-e^{-x})$$

$$= 2xe^{-x}(2 - x)$$

Recall that the derivative, $\frac{dy}{dx}$, is the gradient of y. When $x = 1$

$$\frac{dy}{dx} = 2e^{-1}(2 - 1) = 0.7358$$

so that the gradient of $y = 2x^2e^{-x}$ when $x = 1$ is 0.7358.

Self-assessment question 34.2

1. State the product rule for differentiating $y(x) = u(x)v(x)$.

Exercise 34.2

1. Use the product rule to differentiate
(a) $x \sin x$ (b) $x^2 \cos x$
(c) $\sin x \cos x$ (d) xe^x (e) xe^{-x}

2. Find the derivative of the following expressions:
(a) $2x \ln(3x)$ (b) $e^{3x} \sin x$
(c) $e^{-\frac{x}{2}} \cos 4x$ (d) $\sin 2x \cos 3x$
(e) $e^x \ln(2x)$

3. By writing $\sin x/e^x$ in the form $e^{-x} \sin x$ find the gradient of
$$y = \frac{\sin x}{e^x}$$
when $x = 1$.

4. Find the derivative of
$$y = \frac{\cos 2x}{x^2}$$
(Hint: See the technique of Question 3.)

5. Differentiate
$$y = (\sin x + x)^2$$

6. Differentiate
(a) $e^{-x}x^2$ (b) $e^{-x}x^2 \sin x$

7. Given that
$$y = x^3e^x$$
calculate the value(s) of x for which $\frac{dy}{dx} = 0$.

8. If u, v and w are functions of x and
$$y = uvw$$
show, using the product rule twice, that
$$\frac{dy}{dx} = \frac{du}{dx} vw + u \frac{dv}{dx} w + uv \frac{dw}{dx}$$

34.3 The quotient rule

The quotient rule is the partner of the product rule. Whereas the product rule allows us to differentiate a product, that is uv, the quotient rule allows us to differentiate a quotient, that is u/v.

Suppose we are able to differentiate individually the expressions $u(x)$ and $v(x)$. The quotient rule then allows us to differentiate the quotient of u and v: that is, the quotient rule allows us to differentiate $y = u/v$.

Key point

The quotient rule: if

$$y = \frac{u}{v}$$

then

$$\frac{dy}{dx} = \frac{v\frac{du}{dx} - u\frac{dv}{dx}}{v^2}$$

WORKED EXAMPLES

34.5 Use the quotient rule to differentiate

$$y = \frac{\sin x}{x^2}$$

Solution We let $u(x) = \sin x$ and $v(x) = x^2$. Then

$$y = \frac{u}{v} \quad \text{and} \quad \frac{du}{dx} = \cos x \quad \frac{dv}{dx} = 2x$$

Using the quotient rule we have

$$\frac{dy}{dx} = \frac{v\frac{du}{dx} - u\frac{dv}{dx}}{v^2}$$

$$= \frac{x^2 \cos x - \sin x(2x)}{(x^2)^2}$$

$$= \frac{x(x \cos x - 2 \sin x)}{x^4}$$

$$= \frac{x \cos x - 2 \sin x}{x^3}$$

34.6 Use the quotient rule to differentiate $y = \dfrac{x+1}{x^2+1}$.

Solution We let

$$u(x) = x + 1 \qquad v(x) = x^2 + 1$$

and so clearly

$$y = \frac{u}{v}$$

Now

$$\frac{du}{dx} = 1 \qquad \frac{dv}{dx} = 2x$$

Applying the quotient rule we have

$$\frac{dy}{dx} = \frac{v\dfrac{du}{dx} - u\dfrac{dv}{dx}}{v^2}$$

$$= \frac{(x^2+1)1 - (x+1)2x}{(x^2+1)^2}$$

$$= \frac{x^2+1-2x^2-2x}{(x^2+1)^2}$$

$$= \frac{1-2x-x^2}{(x^2+1)^2}$$

34.7 Use the quotient rule to differentiate $y = \dfrac{e^x + x}{\sin x}$.

Solution We let

$$u(x) = e^x + x \qquad v(x) = \sin x$$

and so

$$\frac{du}{dx} = e^x + 1 \qquad \frac{dv}{dx} = \cos x$$

Clearly

$$y = \frac{u}{v}$$

and so

$$\frac{dy}{dx} = \frac{v\dfrac{du}{dx} - u\dfrac{dv}{dx}}{v^2}$$

$$= \frac{\sin x(e^x+1) - (e^x+x)\cos x}{(\sin x)^2}$$

Self-assessment question 34.3

1. State the quotient rule for differentiating $y(x) = \dfrac{u(x)}{v(x)}$.

Exercise 34.3

1. Use the quotient rule to differentiate the following functions:

   (a) $y = \dfrac{e^x}{x+1}$ (b) $y = \dfrac{\ln x}{x}$

   (c) $y = \dfrac{x}{\ln x}$ (d) $y = \dfrac{\cos t}{\sin t}$

   (e) $y = \dfrac{t^2 + t}{2t + 1}$

2. Evaluate the derivative of the following functions when $x = 2$:

   (a) $y = \dfrac{x+1}{x+2}$ (b) $y = \dfrac{2\ln(3x)}{x+1}$

   (c) $y = \dfrac{\sin 2x}{\cos 3x}$ (d) $y = \dfrac{e^{2x}+1}{e^x+1}$

   (e) $y = \dfrac{x^3+1}{x^2+1}$

3. Noting that

   $$\tan\theta = \frac{\sin\theta}{\cos\theta}$$

   find the derivative of $y = \tan\theta$.

4. Noting that

   $$\tan(k\theta) = \frac{\sin(k\theta)}{\cos(k\theta)} \qquad k \text{ constant}$$

   find the derivative of $y = \tan(k\theta)$.

34.4 The chain rule

Suppose we are given a function $y(x)$ where the variable x is itself a function of another variable, t say. We say that y is a **function of a function**. For example, suppose

$$y(x) = \cos x \qquad \text{and} \qquad x(t) = t^2$$

Then we can write

$$y = \cos(t^2)$$

There will be occasions when it is necessary to calculate $\frac{dy}{dt}$. This can be done by first finding $\frac{dy}{dx}$ and $\frac{dx}{dt}$ and then using the chain rule.

Key point

The **chain rule** states that if $y = y(x)$ and $x = x(t)$, then

$$\frac{dy}{dt} = \frac{dy}{dx} \times \frac{dx}{dt}$$

WORKED EXAMPLE

34.8 Use the chain rule to find $\frac{dy}{dt}$ when $y = \cos x$ and $x = t^2$.

Solution When $y = \cos x$ our previous knowledge of differentiation (or use of Table 33.1) tells us that $\frac{dy}{dx} = -\sin x$. Also, when $x = t^2$, $\frac{dx}{dt} = 2t$. Using the chain rule,

$$\frac{dy}{dt} = \frac{dy}{dx} \times \frac{dx}{dt}$$

$$= (-\sin x) \times 2t$$

Using the fact that $x = t^2$ enables us to write this as

$$\frac{dy}{dt} = -2t \sin t^2$$

The chain rule is more useful when we are given a complicated function of a function and we are able to break it down into two simpler functions. This is usually done by making a suitable substitution. Consider carefully the following examples.

WORKED EXAMPLES

34.9 If $y = (7t + 3)^4$ find $\dfrac{dy}{dt}$.

Solution Note that if we introduce a new variable x and make the substitution $7t + 3 = x$ then y takes the much simpler form $y = x^4$. Then, from $y = x^4$,

$$\frac{dy}{dx} = 4x^3$$

and from $x = 7t + 3$,

$$\frac{dx}{dt} = 7$$

Using the chain rule

$$\frac{dy}{dt} = \frac{dy}{dx} \times \frac{dx}{dt}$$

$$= 4x^3 \times 7$$

$$= 28x^3$$

$$= 28(7t + 3)^3$$

34.10 If $y = (3t^2 + 4)^{1/2}$, find the derivative $\frac{dy}{dt}$.

Solution If we let $3t^2 + 4 = x$, then y looks much simpler: $y = x^{1/2}$. Then

$$\frac{dy}{dx} = \frac{1}{2}x^{-1/2} \quad \text{and} \quad \frac{dx}{dt} = 6t$$

Using the chain rule,

$$\begin{aligned} \frac{dy}{dt} &= \frac{dy}{dx} \times \frac{dx}{dt} \\ &= \frac{1}{2}x^{-1/2} \times 6t \\ &= 3t(3t^2 + 4)^{-1/2} \end{aligned}$$

34.11 Find the gradient of the function $y = e^{(t^2)}$ when $t = 0.5$.

Solution Note that by writing $t^2 = x$ then $y = e^x$. Using the chain rule,

$$\begin{aligned} \frac{dy}{dt} &= \frac{dy}{dx} \times \frac{dx}{dt} \\ &= e^x \times 2t \\ &= 2te^{(t^2)} \end{aligned}$$

This is the gradient function of $y = e^{(t^2)}$. So, when $t = 0.5$ the value of the gradient is $(2)(0.5)e^{(0.5^2)} = e^{0.25} = 1.284$.

Self-assessment questions 34.4

1. State the chain rule for differentiating $y(t)$ when $y = y(x)$ and $x = x(t)$.
2. How would the chain rule be written if we wanted to differentiate $y(x)$ when y is given as $y = y(z)$ and $z = z(x)$?
3. Explain what is meant by a function of a function. Give an example.
4. Decide which rule – product, quotient or chain – it might be appropriate to use in order to differentiate the following functions:

(a) $\frac{1}{x}\cos x$ (b) $x \sin x$ (c) $\sin\left(\frac{1}{x}\right)$

Exercise 34.4

1. Use the chain rule to find $\frac{dy}{dt}$ when
   (a) $y = \sin(x)$ and $x = t^2$
   (b) $y = x^{1/2}$ and $x = (t + 1)$
   (c) $y = e^x$ and $x = 2t^2$
   (d) $y = 3x^{5/2}$ and $x = (t^3 + 2t)$
2. Use the chain rule to differentiate
   (a) $y = \cos 3t^2$
   (b) $y = \sin(2t + 1)$
   (c) $y = \ln(3t - 2)$
   (d) $y = \sin(t^2 + 3t + 2)$
3. Use the chain rule to differentiate
   (a) $y = \sin(3x + 1)$
   (b) $y = e^{4x-3}$
   (c) $y = \cos(2x - 4)$
   (d) $y = (x^2 + 5x - 4)^{1/2}$
   (e) $y = \tan\frac{1}{x}$ (Hint: See Exercise 34.3, Question 4.)
4. In the special case of a function of a function when $y = \ln f(x)$, it can be shown from the chain rule that

$$\frac{dy}{dx} = \frac{f'(x)}{f(x)}$$

   Verify this result using the chain rule in each of the following cases:
   (a) $y = \ln(4x - 3)$
   (b) $y = \ln(3x^2)$
   (c) $y = \ln(\sin x)$
   (d) $y = \ln(x^2 + 3x)$

Test and assignment exercises 34

1. Use the product rule to differentiate the following:
   (a) $e^x \sin 3x$ (b) $x^3 \ln x$ (c) $2x^2 \cos 3x$ (d) $(e^{2x} + 1)(x^2 + 3)$ (e) $\ln 2x \ln 3x$
2. Use the quotient rule to differentiate the following:
   (a) $\frac{x^2 + 3}{x^2 - 3}$ (b) $\frac{1}{x + 1}$ (c) $\frac{\cos x}{\sin x}$ (d) $\frac{e^x + 1}{e^x + 2}$ (e) $\frac{\ln 2x}{2x}$
3. Evaluate the derivative, $\frac{dy}{dx}$, when $x = 1$ given
   (a) $y = x^2 \sin 3x$ (b) $y = \frac{\sin 3x}{x^2}$ (c) $y = \frac{x^2}{\sin 3x}$
4. Differentiate
   (a) $x \ln x$ (b) $e^{2x} x \ln x$ (c) $\frac{e^{2x}}{x \ln x}$
5. (a) Differentiate $y = e^{-3x}x$ using the product rule.
   (b) Differentiate $y = x/e^{3x}$ using the quotient rule.
   (c) What do you notice about your answers? Can you explain this?

6. Calculate the value(s) of x for which the function

$$y = (x^2 - 3)e^{-x}$$

has a zero gradient.

7. Use the chain rule to find $\frac{dy}{dt}$ when
   (a) $y = (t^2 + 1)^{1/3}$
   (b) $y = \sin(3t^2 - 4t + 9)$
   (c) $y = e^{-2t+3}$
   (d) $y = (5t + 3)^4$
8. Evaluate the derivative of $P = (4t + 3)^5$ when $t = 1$.
9. Find the gradient of $y = 2\ln(x^2 + 1)$ when $x = 1$
10. Differentiate, with respect to x,
    (a) $y = x^2 \sin x$
    (b) $y = \dfrac{\sin x}{x}$
    (c) $y = x\sin(x^2)$
    (d) $y = \dfrac{\sin(x^2)}{x}$

Functions of more than one variable and partial differentiation

37

Objectives: This chapter:

- introduces functions that have two (or more) inputs
- shows how functions with two inputs can be represented graphically
- introduces partial differentiation

37.1 Functions of two independent variables

In Chapter 16 a function was defined as a mathematical rule that operates upon an input to produce a single output. We saw that the input is referred to as the independent variable because we are free, within reason, to choose its value. The output is called the dependent variable because its value *depends* upon the value of the input. Commonly we use the letter x to represent the input, y the output and f the function, in which case we write $y = f(x)$. Examples include $y = 5x^2$, $y = 1/x$ and so on. In such cases, there is a single independent variable, x.

We now move on to consider examples in which there are two independent variables. This means that there will be two inputs to a function, each of which can be chosen independently. Once these values have been chosen, the function rule will be used to process them in order to produce a single output – the dependent variable. Notice that although there is now more than one input, there is still a single output.

WORKED EXAMPLES

37.1 A function f is defined by $f(x, y) = 3x + 4y$. Calculate the output when the input values are $x = 5$ and $y = 6$.

Solution In this example, the two independent variables are labelled x and y. The function is labelled f. When the input x takes the value 5, and the input y takes the value 6, the output is $3 \times 5 + 4 \times 6 = 15 + 24 = 39$. We could write this as $f(5, 6) = 39$.

37.2 A function f is defined by $f(x, y) = 11x^2 - 7y + 2$. Calculate the output when the inputs are $x = -2$ and $y = 3$.

Solution In this example we are required to find $f(-2, 3)$. Substitution into the function rule produces

$$f(-2, 3) = 11 \times (-2)^2 - 7 \times (3) + 2 = 44 - 21 + 2 = 25$$

37.3 A function is defined by $f(x, y) = x/y$. Calculate the output when the inputs are $x = 9$ and $y = 3$.

Solution $f(9, 3) = 9/3 = 3$

37.4 A function is defined by $f(x, y) = e^{x+y}$.

(a) Calculate the output when the inputs are $x = -1$ and $y = 2$.
(b) Show that this function can be written in the equivalent form $f(x, y) = e^x e^y$.

Solution (a) $f(-1, 2) = e^{-1+2} = e^1 = e = 2.718$ (3 d.p.)
(b) Using the first law of indices we can write e^{x+y} as $e^x e^y$.

Note that in the above worked examples both x and y are independent variables. We can choose a value for y quite independently of the value we have chosen for x. This is quite different from the case of a function of a single variable when, for example, if we write $y = 5x^2$, choosing x automatically determines y.

It is common to introduce another symbol to stand for the output. So, in Worked Example 37.1 we may write $z = f(x, y) = 3x + 4y$ or simply $z = 3x + 4y$. In Worked Example 37.2 we can write simply $z = 11x^2 - 7y + 2$. Here the value of z depends upon the values chosen for x and y, and so z is referred to as the dependent variable.

Key point A function of two variables is a rule that produces a single output when values of two independent variables are chosen.

Functions of two variables are introduced because they arise naturally in applications in business, economics, physical sciences, engineering etc. Consider the following examples, which illustrate this.

WORKED EXAMPLES

37.5 The volume V of the cylinder shown in Figure 37.1 is given by the formula $V = \pi r^2 h$, where r is the radius and h is the height of the cylinder. Suppose

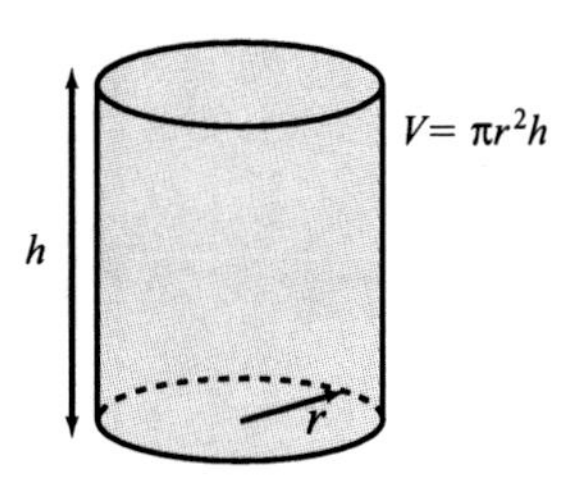

Figure 37.1
The volume V depends upon the two independent variables r and h, and is given by $V = \pi r^2 h$

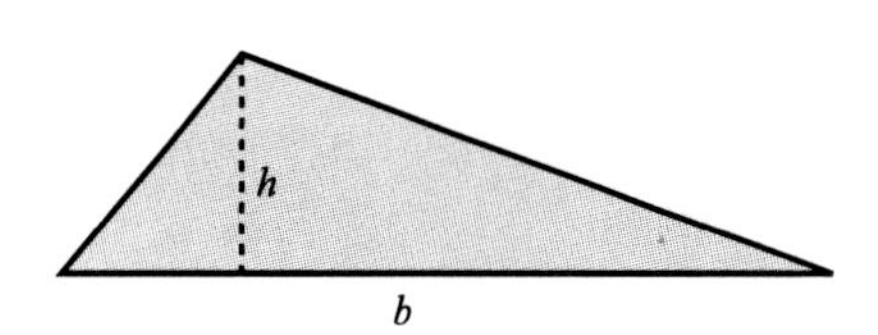

Figure 37.2
The area A depends upon the two independent variables b and h, and is given by $A = \frac{1}{2}bh$

we choose a value for the radius and a value for the height. Note that we can choose these values independently. We can then use the formula to determine the volume of the cylinder. We can regard V as depending upon the two independent variables r and h and write $V = f(r, h) = \pi r^2 h$. Here, the dependent variable is V since the value of the volume depends upon the values chosen for the radius and the height.

37.6 The area A of the triangle shown in Figure 37.2 is given by the formula $A = \frac{1}{2}bh$, where b is the base length and h is the height. We can choose values for b and h independently, and we can use the formula to find the area. We see that the area depends upon the two independent variables b and h, and so $A = f(b, h)$. For example, if we choose $b = 7$ cm and $h = 3$ cm then the function rule tells us $A = f(7, 3) = \frac{1}{2} \times 7 \times 3 = 10.5$. The area of the triangle is 10.5 cm^2.

37.7 When a sum of money is invested for one year the simple interest earned, I, depends upon the two independent variables P, the amount invested, and i, the interest rate per year expressed as a decimal, that is $I = f(P, i)$. The function rule is $I = Pi$. Calculate the interest earned when £500 is invested for one year at an interest rate of 0.06 (that is, 6%).

Solution We seek $f(500, 0.06)$, which equals $500 \times 0.06 = 30$. That is, the simple interest earned is £30.

Self-assessment questions 37.1

1. Explain what is meant by a function of two variables.
2. Explain the different roles played by the variable y when (a) f is a function of two variables such that $z = f(x, y)$, and (b) f is a function of a single variable such that $y = f(x)$.
3. Suppose $P = f(V, T) = RT/V$ where R is a constant. State which variables are dependent and which are independent.
4. Suppose $y = f(x, t)$. State which variables are dependent and which are independent.

Exercise 37.1

MyMathLab Global

1. Given $z = f(x, y) = 7x + 2y$ find the output when $x = 8$ and $y = 2$.

2. If $z = f(x, y) = -11x + y$ find (a) $f(2, 3)$, (b) $f(11, 1)$.

3. If $z = f(x, y) = 3e^x - 2e^y + x^2y^3$ find $z(1, 1)$.

4. If $w = g(x, y) = 7 - xy$ find the value of the dependent variable w when $x = -3$ and $y = -9$.

5. If $z = f(x, y) = \sin(x + y)$ find $f(20°, 30°)$ where the inputs are angles measured in degrees.

6. If $f(x, t) = e^{2xt}$ find $f(0.5, 3)$.

37.2 Representing a function of two independent variables graphically

By now you will be familiar with the way in which a function of one variable is represented graphically. For example, you have seen in Chapter 17 how the graph of the function $y = f(x)$ is drawn by plotting the independent variable x on the horizontal axis, and the dependent variable y on the vertical axis. Given a value of x, the function rule enables us to calculate a value for y, and the point with coordinates (x, y) is then plotted. Joining all such points produces a graph of the function.

When two independent variables are involved, as in $z = f(x, y)$, the plotting of the graph is rather more difficult because we now need two axes for the independent variables and a third axis for the dependent variable. This means that the graph is drawn in three dimensions – a task that is particularly difficult to do in the two-dimensional plane of the paper. Easily usable computer software is readily available for plotting graphs in three dimensions, so rather than attempt this manually we shall be content with understanding the process involved in producing the graph, and illustrating this with several examples.

Three axes are drawn at right angles, and these are labelled x, y and z as shown in Figure 37.3. There is more than one way of doing this, but the usual convention is shown here. When labelled in this way the axes are said to form a **right-handed set**. The axes intersect at the origin O.

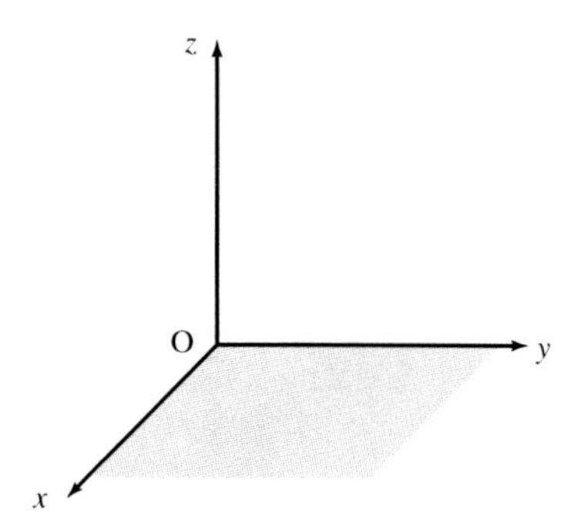

Figure 37.3
Three perpendicular axes, labelled x, y and z

WORKED EXAMPLE

37.8 Given the function of two variables $z = f(x, y) = 3x + 4y$ calculate the value of the function when x and y take the following values. Represent each point graphically.

(a) $x = 3$, $y = 2$ (b) $x = 5$, $y = 0$ (c) $x = 0$, $y = 5$

Solution In each case we use the function rule $z = f(x, y) = 3x + 4y$ to calculate the corresponding value of z:

(a) $z = 3(3) + 4(2) = 17$ (b) $z = 3(5) + 4(0) = 15$
(c) $z = 3(0) + 4(5) = 20$

It is conventional to write the coordinates of the points in the form (x, y, z) so that the three points in this example are $(3, 2, 17)$, $(5, 0, 15)$ and $(0, 5, 20)$.

Each of the points can then be drawn as shown in Figure 37.4, where they have been labelled A, B and C respectively. Notice that the z coordinate of point A, which is 17, gives the height of A above the point $(3, 2)$ in the x–y plane. Similar comments apply to points B and C.

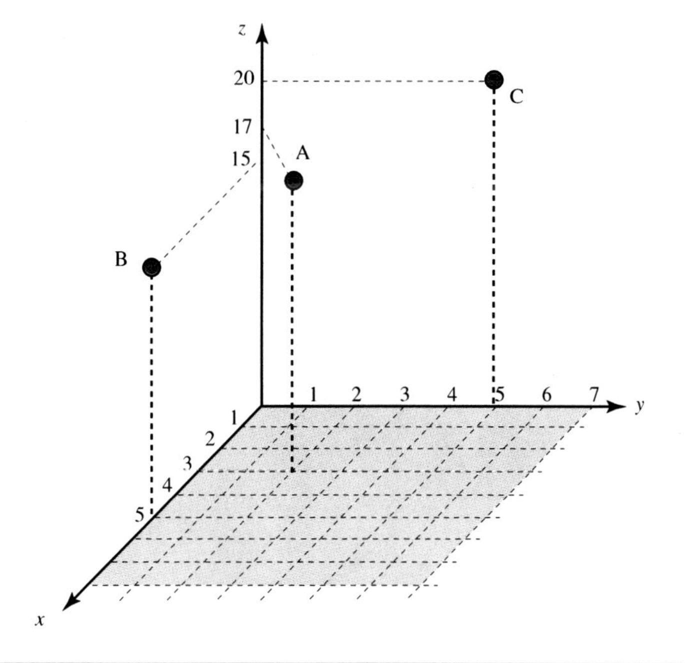

Figure 37.4
Three points in three-dimensional space with coordinates $A(3, 2, 17)$, $B(5, 0, 15)$ and $C(0, 5, 20)$

If we were to continue selecting more points and plotting them we would find that all the points lie in a plane. This plane is shown in Figure 37.5. Notice that in each case the z coordinate is the height of the plane above the point (x, y) in the x–y plane. For example, the point C is 20 units above the point $(0, 5)$ in the x–y plane.

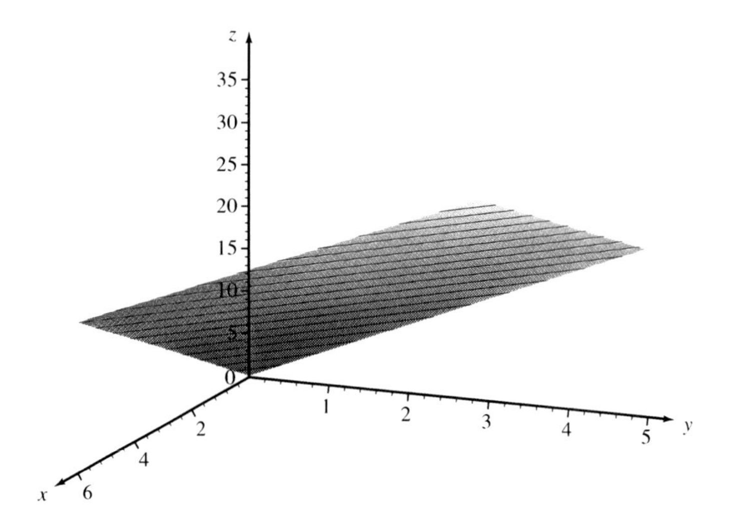

Figure 37.5
The plane $z = 3x + 4y$

In the previous worked example we saw that when we drew a graph of the function of two variables $z = 3x + 4y$ we obtained a plane. In more general cases the graph of $z = f(x, y)$ will be a curved surface, and the z coordinate is the height of the surface above the point (x, y). Some more examples of functions of two variables and their graphs are shown in Figures 37.6 and 37.7. It would be a useful exercise for you to try to reproduce these graphs for yourself using computer software available in your college or university.

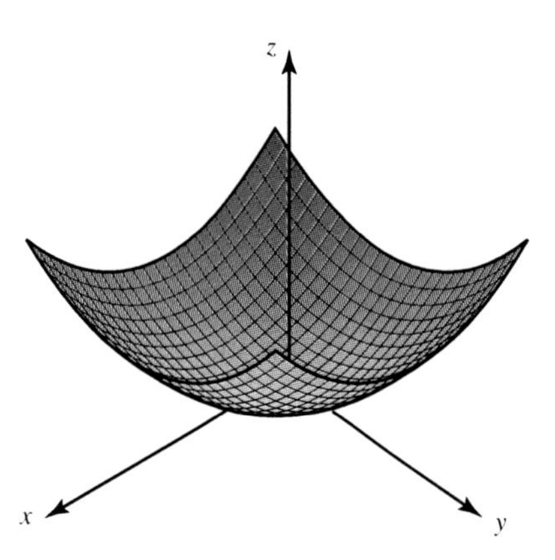

Figure 37.6
The function $z = x^2 + y^2$

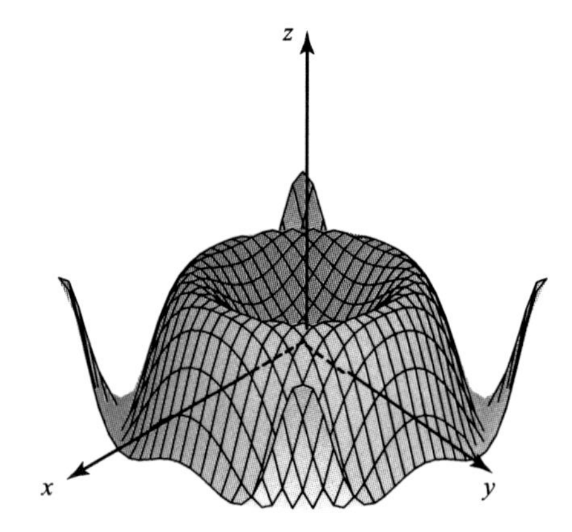

Figure 37.7
The function $z = \sin(x^2 + y^2)$

Self-assessment question 37.2

1. Draw three perpendicular axes and label the axes in the correct order according to the convention stated in this section.

Exercise 37.2 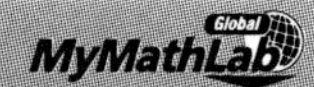

1. Given the function $z = 3x - 6y$, find the z coordinate corresponding to each of the following points, which lie in the $x - y$ plane:
(a) $(4, 4)$ (b) $(0, 3)$ (c) $(5, 0)$
Plot the points on a three-dimensional graph.

37.3 Partial differentiation

Before attempting this section it is essential that you have a thorough understanding of differentiation of functions of one variable, and you should revise Chapters 33 and 34 if necessary. In particular, you may need to refer frequently to Table 33.1 on page 408, which gives the derivatives of some common functions.

In Chapter 33 we introduced differentiation of a function of a single variable and showed how the derivative of such a function can be calculated. Recall that this gives useful information about the gradient of the graph of the function at different points. In this section we explain how functions of two variables are differentiated.

In Chapter 33 we differentiated y, which was a function of x, to obtain the derivative $\frac{dy}{dx}$. Note that the dependent variable (in this case y) is differentiated with respect to the independent variable (in this case x).

Consider now z, which depends upon two variables x and y. Recall that z is then the dependent variable, and x and y are the independent variables. Hence z can be differentiated with respect to x to produce a derivative, and it can also be differentiated with respect to y to produce another, different derivative. So for functions of two variables there are two derivatives: we can no longer talk about *the* derivative of z. This is a fundamental difference between functions of one variable and functions of two variables.

When differentiating functions of two variables we refer to this process as **partial differentiation**, and instead of using a normal letter d, as in $\frac{dy}{dx}$, we use a curly d instead and write ∂. Do not be put off by this notation – you will soon get used to it.

When we differentiate z with respect to (w.r.t.) x we denote the derivative produced by $\frac{\partial z}{\partial x}$. When we differentiate z w.r.t. y we denote the derivative produced by $\frac{\partial z}{\partial y}$.

Key point

If $z = f(x, y)$, then the derivatives $\frac{\partial z}{\partial x}$ and $\frac{\partial z}{\partial y}$ are called the (first) partial derivatives of z.

We now explain how these derivatives are calculated.

Partial differentiation with respect to *x*

Suppose we have a function $z = 5x + 11$. You will recall that its derivative $\frac{dz}{dx} = 5$. Note that the derivative of the constant 11 is zero. Similarly if $z = 6x + 12$, $\frac{dz}{dx} = 6$. Here the derivative of the constant 12 is zero. We are now ready to introduce partial differentiation with respect to x.

Consider the function $z = 5x + y$. When we differentiate with respect to x we treat any occurrence of the variable y *as though it were a constant.* Hence in this case the derivative of y is zero, and we write

$$\text{if } z = 5x + y \quad \text{then} \quad \frac{\partial z}{\partial x} = 5 + 0 = 5$$

Similarly,

$$\text{if } z = 7x^2 - y \quad \text{then} \quad \frac{\partial z}{\partial x} = 14x - 0 = 14x$$

If y is treated as a constant, then so too will be multiples of y, such as $7y$ and $-3y$. Furthermore, any functions of y, such as y^2 and e^y, will also be regarded as constants.

WORKED EXAMPLES

37.9 Calculate $\frac{\partial z}{\partial x}$ when $z = 9x + 2$.

Solution We must differentiate $z = 9x + 2$ with respect to x. This function is particularly simple because y does not appear at all. The derivative of $9x$ is 9, and the derivative of the constant 2 is zero. Hence $\frac{\partial z}{\partial x} = 9$.

37.10 Calculate $\frac{\partial z}{\partial x}$ when $z = 9x + y$.

Solution To find $\frac{\partial z}{\partial x}$ we treat y as though it were a constant. Imagine it were just a number like the '2' in the previous example. Then $\frac{\partial z}{\partial x} = 9$.

37.11 Calculate $\frac{\partial z}{\partial x}$ when $z = y - 19x$.

Solution

$$\frac{\partial z}{\partial x} = -19$$

37.12 Calculate $\frac{\partial z}{\partial x}$ when $z = 3x + 4y + 11$.

Solution

$$\frac{\partial z}{\partial x} = 3$$

In this case 11 is a constant and we treat y, and hence $4y$, as a constant.

37.13 Calculate $\frac{\partial z}{\partial x}$ when $z = 4x^2 - 3y$.

Solution

$$\frac{\partial z}{\partial x} = 8x - 0 = 8x$$

The quantity $-3y$ is treated as a constant.

37.14 Calculate $\frac{\partial z}{\partial x}$ when $z = x^2 - y^2$.

Solution y, and hence y^2, is treated as a constant. When $-y^2$ is differentiated with respect to x the result will be zero. Hence if $z = x^2 - y^2$ then $\frac{\partial z}{\partial x} = 2x$.

37.15 Calculate $\frac{\partial z}{\partial x}$ when $z = 5y^3 - x^4$.

Solution y, and hence $5y^3$, is treated as a constant. If $z = 5y^3 - x^4$, then

$$\frac{\partial z}{\partial x} = -4x^3$$

37.16 Calculate $\frac{\partial z}{\partial x}$ when $z = 3x - 4y^3$.

Solution If $z = 3x - 4y^3$, then $\frac{\partial z}{\partial x} = 3$.

Recall that

if $4x$ is differentiated with respect to x the result is 4
if $5x$ is differentiated with respect to x the result is 5

Extending this to the following function of two variables,

if $z = yx$ is differentiated with respect to x the result is y

We see that because y is treated as if it were a constant then $\frac{\partial z}{\partial x} = y$.

WORKED EXAMPLES

37.17 Calculate $\frac{\partial z}{\partial x}$ when $z = yx^2$.

Solution To find $\frac{\partial z}{\partial x}$ imagine that you were trying to differentiate $3x^2$, say. The result would be $3(2x) = 6x$. Hence if $z = yx^2$ then $\frac{\partial z}{\partial x} = y(2x)$ or simply $2xy$.

37.18 Calculate $\frac{\partial z}{\partial x}$ when $z = 3yx^2$.

Solution Treating y as a constant we see that this function is of the form constant $\times\, x^2$, and so $\frac{\partial z}{\partial x} = (3y)(2x) = 6xy$.

37.19 Calculate $\frac{\partial z}{\partial x}$ when $z = 3x^2 + 4xy + 11$.

Solution

$$\frac{\partial z}{\partial x} = 6x + 4y$$

37.20 Calculate $\frac{\partial z}{\partial x}$ when

(a) $z = x^2 e^y$ (b) $z = x^2 + e^y$ (c) $z = 3x^2 \cos y$
(d) $z = 3x^2 + \cos y$ (e) $z = 3x^2 + 4x \sin y$

Solutions

(a) y and hence e^y are treated as constants. Then $\frac{\partial z}{\partial x} = 2xe^y$.

(b) $\frac{\partial z}{\partial x} = 2x$.

(c) y and hence $\cos y$ are treated as constants. Then $\frac{\partial z}{\partial x} = 6x \cos y$.

(d) $\frac{\partial z}{\partial x} = 6x$.
(e) $\frac{\partial z}{\partial x} = 6x + 4 \sin y$.

37.21 Calculate $\frac{\partial z}{\partial x}$ when (a) $z = \sin 3x$, (b) $z = \sin yx$.

Solution (a) If $z = \sin 3x$ then $\frac{\partial z}{\partial x} = 3 \cos 3x$.
(b) If $z = \sin yx$ then $\frac{\partial z}{\partial x} = y \cos yx$ because y is treated as a constant.

Key point The partial derivative with respect to x of a function $z = f(x, y)$ is denoted by $\frac{\partial z}{\partial x}$ and is calculated by differentiating the function with respect to x and treating y as though it were a constant.

Partial differentiation with respect to *y*

When we differentiate a function $f(x, y)$ with respect to y we treat any occurrence of the variable x *as though it were a constant*. The partial derivative with respect to y of a function $z = f(x, y)$ is denoted by $\frac{\partial z}{\partial y}$. Consider the following examples.

WORKED EXAMPLES

37.22 Find $\frac{\partial z}{\partial y}$ when $z = 3y^4 + 4x^2 + 8$.

Solution When calculating $\frac{\partial z}{\partial y}$ we treat any occurrence of x as if it were a constant. So the term $4x^2$ is treated as a constant, and its partial derivative with respect to y is zero. That is,

$$\text{if } z = 3y^4 + 4x^2 + 8 \quad \text{then} \quad \frac{\partial z}{\partial y} = 12y^3 + 0 + 0 = 12y^3$$

37.23 Find $\frac{\partial z}{\partial y}$ when $z = 3x^2 y$.

Solution Because x is treated as a constant, we are dealing with a function of the form $z = \text{constant} \times y$. The derivative with respect to y will be simply the constant factor. That is,

$$\text{if } z = 3x^2 y \quad \text{then} \quad \frac{\partial z}{\partial y} = 3x^2$$

37.24 Find $\frac{\partial z}{\partial y}$ when $z = 4xy^3$.

Solution Because x is treated as a constant, we are dealing with a function of the form $z = \text{constant} \times y^3$. The derivative with respect to y will be the constant $\times\ 3y^2$, that is

$$\text{if } z = 4xy^3 \quad \text{then} \quad \frac{\partial z}{\partial y} = (4x)(3y^2) = 12xy^2$$

Key point

The partial derivative with respect to y of a function $z = f(x, y)$ is denoted by $\frac{\partial z}{\partial y}$ and is calculated by differentiating the function with respect to y and treating x as though it were a constant.

It will be necessary to work with symbols other than z, x and y. Consider the following worked example.

WORKED EXAMPLE

37.25 Consider the function $w = f(p, t)$. Find $\frac{\partial w}{\partial p}$ and $\frac{\partial w}{\partial t}$ when $w = 3t^7 + 4pt + p^2$.

Solution

$$\frac{\partial w}{\partial p} = 4t + 2p \qquad \frac{\partial w}{\partial t} = 21t^6 + 4p$$

Self-assessment questions 37.3

1. State the derivative of (a) 7, (b) 11, (c) k, a constant.
2. Suppose $z = f(x, y)$. When y is differentiated partially with respect to x, state the result.
3. What is the symbol for the partial derivative of z with respect to x?
4. What is the symbol for the partial derivative of z with respect to y?
5. If y is a function of x and t, that is $y = f(x, t)$, write down the symbol for the partial derivative of y with respect to x and the symbol for the partial derivative of y with respect to t.

Exercise 37.3

1. In each case, given $z = f(x, y)$, find $\frac{\partial z}{\partial x}$ and $\frac{\partial z}{\partial y}$.
   (a) $z = 5x + 11y$ (b) $z = -7y - 14x$
   (c) $z = 8x$ (d) $z = -5y$
   (e) $z = 3x + 8y - 2$
   (f) $z = 17 - 3x + 2y$ (g) $z = 8$
   (h) $z = 8 - 3y$ (i) $z = 2x^2 - 7y$
   (j) $z = 9 - 3y^3 + 7x$
   (k) $z = 9 - 9(x - y)$
   (l) $z = 9(x + y + 3)$
2. In each case, given $z = f(x, y)$, find $\frac{\partial z}{\partial x}$ and $\frac{\partial z}{\partial y}$.
   (a) $z = xy$ (b) $z = 3xy$
   (c) $z = -9yx$ (d) $z = x^2y$
   (e) $z = 9x^2y$ (f) $z = 8xy^2$
3. If $z = 9x + y^2$ evaluate $\frac{\partial z}{\partial x}$ and $\frac{\partial z}{\partial y}$ at the point $(4, -2)$.
4. Find $\frac{\partial z}{\partial x}$ and $\frac{\partial z}{\partial y}$ when
   (a) $z = e^{2x}$ (b) $z = e^{5y}$
   (c) $z = e^{xy}$ (d) $z = 4e^{2y}$
5. If $y = x \sin t$ find $\frac{\partial y}{\partial x}$ and $\frac{\partial y}{\partial t}$.

37.4 Partial derivatives requiring the product or quotient rules

Consider the following more demanding worked examples, which use the rules developed in Chapter 34.

WORKED EXAMPLES

37.26 Find $\frac{\partial z}{\partial x}$ and $\frac{\partial z}{\partial y}$ when $z = yxe^{2x}$.

Solution To find $\frac{\partial z}{\partial x}$ we treat y as constant. We are dealing with a function of the form constant $\times$ xe^{2x}. Note that this function itself contains a product of the functions x and e^{2x}, and so we must use the product rule for differentiation. The derivative of xe^{2x} is $e^{2x}(1) + x(2e^{2x}) = e^{2x}(1 + 2x)$. Hence

$$\text{if } z = yxe^{2x} \quad \text{then} \quad \frac{\partial z}{\partial x} = y(e^{2x}(1 + 2x)) = ye^{2x}(1 + 2x)$$

To find $\frac{\partial z}{\partial y}$ we treat x as constant. In turn, this means that xe^{2x} is constant too. This time the calculation is much simpler because we are dealing with a function of the form $z = \text{constant} \times y$. So,

$$\text{if } z = yxe^{2x} \quad \text{then} \quad \frac{\partial z}{\partial y} = xe^{2x}$$

37.27 Find $\frac{\partial z}{\partial x}$ and $\frac{\partial z}{\partial y}$ when $z = ye^{x}/x$.

Solution To find $\frac{\partial z}{\partial x}$ we treat y as constant. We are dealing with a function of the form constant $\times\, e^{x}/x$. Note that this function itself contains a quotient of the functions e^{x} and x and so we must use the quotient rule for differentiation. The derivative of e^{x}/x is $(e^{x}x - e^{x}(1))/x^{2} = e^{x}(x-1)/x^{2}$. Hence

$$\text{if } z = \frac{ye^{x}}{x} \quad \text{then} \quad \frac{\partial z}{\partial x} = \frac{ye^{x}(x-1)}{x^{2}}$$

To find $\frac{\partial z}{\partial y}$ we treat x as constant. In turn, this means that e^{x}/x is constant too. This calculation is simple because we are dealing with a function of the form $z = \text{constant} \times y$. So,

$$\text{if } z = \frac{ye^{x}}{x} \quad \text{then} \quad \frac{\partial z}{\partial y} = \frac{e^{x}}{x}$$

Exercise 37.4

1. Find $\frac{\partial z}{\partial x}$ and $\frac{\partial z}{\partial y}$ when
(a) $z = yxe^x$ (b) $z = xye^y$ (c) $z = xe^{xy}$ (d) $z = ye^{xy}$ (e) $z = x^2 \sin(xy)$
(f) $z = y \cos(xy)$ (g) $z = x \ln(xy)$ (h) $z = 3xy^3e^x$

37.5 Higher-order derivatives

Just as functions of one variable have second derivatives found by differentiating the first derivative, so too do functions of two variables. If $z = f(x, y)$ the first partial derivatives are $\frac{\partial z}{\partial x}$ and $\frac{\partial z}{\partial y}$. The second partial derivatives are found by differentiating the first partial derivatives. We can differentiate either first partial derivative with respect to x or with respect to y to obtain various second partial derivatives as summarised below:

Key point

differentiating $\frac{\partial z}{\partial x}$ w.r.t. x produces $\frac{\partial}{\partial x}\left(\frac{\partial z}{\partial x}\right) = \frac{\partial^2 z}{\partial x^2}$

differentiating $\frac{\partial z}{\partial x}$ w.r.t. y produces $\frac{\partial}{\partial y}\left(\frac{\partial z}{\partial x}\right) = \frac{\partial^2 z}{\partial y \partial x}$

differentiating $\frac{\partial z}{\partial y}$ w.r.t. x produces $\frac{\partial}{\partial x}\left(\frac{\partial z}{\partial y}\right) = \frac{\partial^2 z}{\partial x \partial y}$

differentiating $\frac{\partial z}{\partial y}$ w.r.t. y produces $\frac{\partial}{\partial y}\left(\frac{\partial z}{\partial y}\right) = \frac{\partial^2 z}{\partial y^2}$

The second partial derivatives of z are

$$\frac{\partial^2 z}{\partial x^2} \qquad \frac{\partial^2 z}{\partial y \partial x} \qquad \frac{\partial^2 z}{\partial x \partial y} \qquad \frac{\partial^2 z}{\partial y^2}$$

WORKED EXAMPLES

37.28 Given $z = 3xy^3 - 2xy$ find all the second partial derivatives of z.

Solution First of all the first partial derivatives must be found:

$$\frac{\partial z}{\partial x} = 3y^3 - 2y \qquad \frac{\partial z}{\partial y} = 9xy^2 - 2x$$

Then each of these is differentiated with respect to x:

$$\frac{\partial^2 z}{\partial x^2} = \frac{\partial}{\partial x}\left(\frac{\partial z}{\partial x}\right) = 0$$

$$\frac{\partial^2 z}{\partial x \partial y} = \frac{\partial}{\partial x}\left(\frac{\partial z}{\partial y}\right) = 9y^2 - 2$$

Now, each of the first partial derivatives must be differentiated with respect to y:

$$\frac{\partial^2 z}{\partial y \partial x} = \frac{\partial}{\partial y}\left(\frac{\partial z}{\partial x}\right) = 9y^2 - 2$$

$$\frac{\partial^2 z}{\partial y^2} = \frac{\partial}{\partial y}\left(\frac{\partial z}{\partial y}\right) = 18xy$$

Note that

$$\frac{\partial^2 z}{\partial y \partial x} = \frac{\partial^2 z}{\partial x \partial y}$$

It is usually the case that the result is the same either way.

37.29 Find all second partial derivatives of $z = \sin(xy)$.

Solution First of all the first partial derivatives are found:

$$\frac{\partial z}{\partial x} = y\cos(xy) \qquad \frac{\partial z}{\partial y} = x\cos(xy)$$

Then each of these is differentiated with respect to x:

$$\frac{\partial^2 z}{\partial x^2} = \frac{\partial}{\partial x}\left(\frac{\partial z}{\partial x}\right) = -y^2\sin(xy)$$

$$\frac{\partial^2 z}{\partial x \partial y} = \frac{\partial}{\partial x}\left(\frac{\partial z}{\partial y}\right) = -xy\sin(xy) + \cos(xy)$$

Note here the need to use the product rule to differentiate $x\cos(xy)$ with respect to x.

Now each of the first partial derivatives must be differentiated with respect to y:

$$\frac{\partial^2 z}{\partial y \partial x} = \frac{\partial}{\partial y}\left(\frac{\partial z}{\partial x}\right) = -xy\ \sin(xy) + \cos(xy)$$

$$\frac{\partial^2 z}{\partial y^2} = \frac{\partial}{\partial y}\left(\frac{\partial z}{\partial y}\right) = -x^2\ \sin(xy)$$

Self-assessment question 37.5

1. Given $z = f(x, y)$, explain what is meant by

$$\frac{\partial^2 z}{\partial y \partial x} \quad \text{and} \quad \frac{\partial^2 z}{\partial x \partial y}$$

Exercise 37.5

1. Find all the second partial derivatives in each of the following cases:
   (a) $z = xy$ (b) $z = 7xy$
   (c) $z = 8x + 9y + 10$
   (d) $z = 8y^2x + 11$ (e) $z = -2y^3x^2$
   (f) $z = x + y$
2. Find all the second partial derivatives in each of the following cases:
   (a) $z = \frac{1}{x}$ (b) $z = \frac{y}{x}$ (c) $z = \frac{x}{y}$
   (d) $z = \frac{1}{x} + \frac{1}{y}$
3. Find all the second partial derivatives in each of the following cases:
   (a) $z = x \sin y$ (b) $z = y \cos x$
   (c) $z = ye^{2x}$ (d) $z = ye^{-x}$
4. Find all the second partial derivatives in each of the following cases:
   (a) $z = 8e^{xy}$ (b) $z = -3e^x \sin y$
   (c) $z = 4e^y \cos x$
5. Find all the second partial derivatives in each of the following cases:
   (a) $z = \ln x$ (b) $z = \ln y$ (c) $z = \ln xy$
   (d) $z = x \ln y$ (e) $z = y \ln x$

37.6 Functions of several variables

In more advanced applications, functions of several variables arise. For example, suppose w is a function of three independent variables x, y and z, that is $w = f(x, y, z)$. It is not possible to represent such a function graphically in the usual manner because this would require four dimensions. Nevertheless, partial derivatives can be calculated in exactly the same way as we have done for two independent variables.

WORKED EXAMPLES

37.30 If $w = 3x + 5y - 4z$ find $\frac{\partial w}{\partial x}$, $\frac{\partial w}{\partial y}$ and $\frac{\partial w}{\partial z}$.

Solution When finding $\frac{\partial w}{\partial x}$ the variables y and z are treated as constants. So $\frac{\partial w}{\partial x} = 3$. When finding $\frac{\partial w}{\partial y}$ the variables x and z are treated as constants. So $\frac{\partial w}{\partial y} = 5$. When finding $\frac{\partial w}{\partial z}$ the variables x and y are treated as constants. So $\frac{\partial w}{\partial z} = -4$.

37.31 Find all second derivatives of $w = f(x, y, z) = x + y + z + xy + yz + xz$.

Solution The first partial derivatives must be found first:

$$\frac{\partial w}{\partial x} = 1 + y + z \qquad \frac{\partial w}{\partial y} = 1 + x + z \qquad \frac{\partial w}{\partial z} = 1 + y + x$$

The second partial derivatives are found by differentiating the first partial derivatives w.r.t. x, w.r.t. y and w.r.t. z. This gives

$$\frac{\partial}{\partial x}\left(\frac{\partial w}{\partial x}\right) = 0 \qquad \frac{\partial}{\partial x}\left(\frac{\partial w}{\partial y}\right) = 1 \qquad \frac{\partial}{\partial x}\left(\frac{\partial w}{\partial z}\right) = 1$$

Similarly,

$$\frac{\partial}{\partial y}\left(\frac{\partial w}{\partial x}\right) = 1 \qquad \frac{\partial}{\partial y}\left(\frac{\partial w}{\partial y}\right) = 0 \qquad \frac{\partial}{\partial y}\left(\frac{\partial w}{\partial z}\right) = 1$$

And finally,

$$\frac{\partial}{\partial z}\left(\frac{\partial w}{\partial x}\right) = 1 \qquad \frac{\partial}{\partial z}\left(\frac{\partial w}{\partial y}\right) = 1 \qquad \frac{\partial}{\partial z}\left(\frac{\partial w}{\partial z}\right) = 0$$

Self-assessment question 37.6

1. If $w = f(x, y, z)$ explain how

$$\frac{\partial^3 w}{\partial x \partial y \partial z}$$

is calculated.

Exercise 37.6

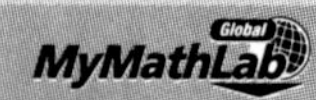

1. If $w = f(x, y, t)$ such that $w = 8x - 3y - 4t$, find $\frac{\partial w}{\partial x}$, $\frac{\partial w}{\partial y}$ and $\frac{\partial w}{\partial t}$.
2. If $w = f(x, y, t)$ such that $w = x^2 + y^3 - 5t^2$, find $\frac{\partial w}{\partial x}$, $\frac{\partial w}{\partial y}$ and $\frac{\partial w}{\partial t}$.
3. Find all the first partial derivatives of $w = xyz$.

Test and assignment exercises 37

1. If $z = 14x - 13y$ state $\frac{\partial z}{\partial x}$ and $\frac{\partial z}{\partial y}$.
2. If $w = 5y - 2x$ state $\frac{\partial^2 w}{\partial x^2}$ and $\frac{\partial^2 w}{\partial y^2}$.
3. If $z = 3x^2 + 7xy - y^2$ find $\frac{\partial^2 z}{\partial y \partial x}$ and $\frac{\partial^2 z}{\partial x \partial y}$.
4. If $z = 14 - 4xy$ evaluate $\frac{\partial z}{\partial x}$ and $\frac{\partial z}{\partial y}$ at the point $(1, 2)$.
5. If $z = 4e^{5xy}$ find $\frac{\partial z}{\partial x}$ and $\frac{\partial z}{\partial y}$.
6. If $y = x \cos t$ find $\frac{\partial y}{\partial x}$ and $\frac{\partial y}{\partial t}$.
7. If $w = 3xy^2 + 2yz^2$ find all first partial derivatives of w at the point with coordinates $(1, 2, 3)$.
8. If $V = D^{1/4}T^{-5/6}$ find $\frac{\partial V}{\partial T}$ and $\frac{\partial V}{\partial D}$.
9. If $p = RT/V$ where R is a constant, find $\frac{\partial p}{\partial V}$ and $\frac{\partial p}{\partial T}$.
10. Find $\frac{\partial f}{\partial x}$, $\frac{\partial f}{\partial y}$, $\frac{\partial^2 f}{\partial x^2}$, $\frac{\partial^2 f}{\partial y^2}$ and $\frac{\partial^2 f}{\partial x \partial y}$ if $f = (x - y)^2$.

Appendix 1

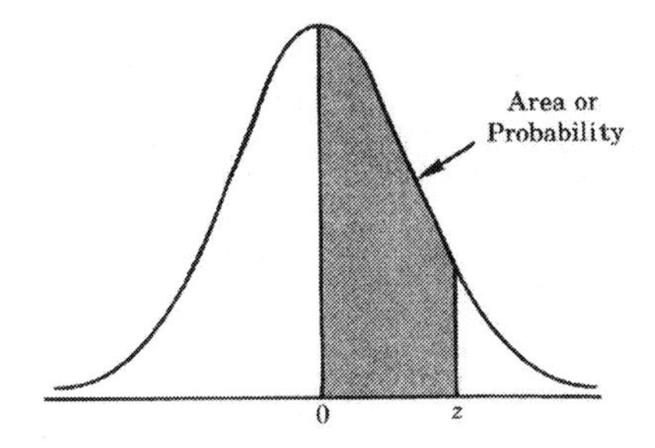

Entries in the table give the area under the standard Normal curve between the mean and z standard deviations above the mean. For example, for $z = 1.25$ the area under the curve between the mean and z is 0.3944 (underlined in the table).

z	0	0.01	0.02	0.03	0.04	0.05	0.06	0.07	0.08	0.09
0	0.0000	0.0040	0.0080	0.0120	0.0160	0.0199	0.0239	0.0279	0.0319	0.0359
0.1	0.0398	0.0438	0.0478	0.0517	0.0557	0.0596	0.0636	0.0675	0.0714	0.0753
0.2	0.0793	0.0832	0.0871	0.0910	0.0948	0.0987	0.1026	0.1064	0.1103	0.1141
0.3	0.1179	0.1217	0.1255	0.1293	0.1331	0.1368	0.1406	0.1443	0.1480	0.1517
0.4	0.1554	0.1591	0.1628	0.1664	0.1700	0.1736	0.1772	0.1808	0.1844	0.1879
0.5	0.1915	0.1950	0.1985	0.2019	0.2054	0.2088	0.2123	0.2157	0.2190	0.2224
0.6	0.2257	0.2291	0.2324	0.2357	0.2389	0.2422	0.2454	0.2486	0.2517	0.2549
0.7	0.2580	0.2611	0.2642	0.2673	0.2704	0.2734	0.2764	0.2794	0.2823	0.2852
0.8	0.2881	0.2910	0.2939	0.2967	0.2995	0.3023	0.3051	0.3078	0.3106	0.3133
0.9	0.3159	0.3186	0.3212	0.3238	0.3264	0.3289	0.3315	0.3340	0.3365	0.3389
1	0.3413	0.3438	0.3461	0.3485	0.3508	0.3531	0.3554	0.3577	0.3599	0.3621
1.1	0.3643	0.3665	0.3686	0.3708	0.3729	0.3749	0.3770	0.3790	0.3810	0.3830
1.2	0.3849	0.3869	0.3888	0.3907	0.3925	<u>0.3944</u>	0.3962	0.3980	0.3997	0.4015
1.3	0.4032	0.4049	0.4066	0.4082	0.4099	0.4115	0.4131	0.4147	0.4162	0.4177
1.4	0.4192	0.4207	0.4222	0.4236	0.4251	0.4265	0.4279	0.4292	0.4306	0.4319
1.5	0.4332	0.4345	0.4357	0.4370	0.4382	0.4394	0.4406	0.4418	0.4429	0.4441
1.6	0.4452	0.4463	0.4474	0.4484	0.4495	0.4505	0.4515	0.4525	0.4535	0.4545
1.7	0.4554	0.4564	0.4573	0.4582	0.4591	0.4599	0.4608	0.4616	0.4625	0.4633
1.8	0.4641	0.4649	0.4656	0.4664	0.4671	0.4678	0.4686	0.4693	0.4699	0.4706
1.9	0.4713	0.4719	0.4726	0.4732	0.4738	0.4744	0.4750	0.4756	0.4761	0.4767
2	0.4772	0.4778	0.4783	0.4788	0.4793	0.4798	0.4803	0.4808	0.4812	0.4817
2.1	0.4821	0.4826	0.4830	0.4834	0.4838	0.4842	0.4846	0.4850	0.4854	0.4857
2.2	0.4861	0.4864	0.4868	0.4871	0.4875	0.4878	0.4881	0.4884	0.4887	0.4890
2.3	0.4893	0.4896	0.4898	0.4901	0.4904	0.4906	0.4909	0.4911	0.4913	0.4916
2.4	0.4918	0.4920	0.4922	0.4925	0.4927	0.4929	0.4931	0.4932	0.4934	0.4936
2.5	0.4938	0.4940	0.4941	0.4943	0.4945	0.4946	0.4948	0.4949	0.4951	0.4952
2.6	0.4953	0.4955	0.4956	0.4957	0.4959	0.4960	0.4961	0.4962	0.4963	0.4964
2.7	0.4965	0.4966	0.4967	0.4968	0.4969	0.4970	0.4971	0.4972	0.4973	0.4974
2.8	0.4974	0.4975	0.4976	0.4977	0.4977	0.4978	0.4979	0.4979	0.4980	0.4981
2.9	0.4981	0.4982	0.4982	0.4983	0.4984	0.4984	0.4985	0.4985	0.4986	0.4986
3	0.4987	0.4987	0.4987	0.4988	0.4988	0.4989	0.4989	0.4989	0.4990	0.4990

Appendix 2

Solutions

Chapter 1

Self-assessment questions 1.1

1. An integer is a whole number, that is a number from the set

   ... −4, −3, −2, −1, 0, 1, 2, 3, 4 ...

   The positive integers are 1, 2, 3, 4 .... The negative integers are ...− 4, −3, −2, −1.
2. The sum is the result of adding numbers. The difference is found by subtracting one number from another. The product of numbers is found by multiplying the numbers. The quotient of two numbers is found by dividing one number by the other.
3. (a) positive (b) negative (c) negative (d) positive

Exercise 1.1

1. (a) 3 (b) 9 (c) 11 (d) 21 (e) 30 (f) 56 (g) −13 (h) −19 (i) −19 (j) −13 (k) 29 (l) −75 (m) −75 (n) 29
2. (a) −24 (b) −32 (c) −30 (d) 16 (e) −42
3. (a) −5 (b) 3 (c) −3 (d) 3 (e) −3 (f) −6 (g) 6 (h) −6
4. (a) Sum $= 3+6 = 9$
   Product $= 3 \times 6 = 18$
   (b) Sum = 17, Product = 70
   (c) Sum $= 2+3+6 = 11$
   Product $= 2 \times 3 \times 6 = 36$
5. (a) Difference $= 18 - 9 = 9$
   Quotient $= \frac{18}{9} = 2$
   (b) Difference $= 20 - 5 = 15$
   Quotient $= \frac{20}{5} = 4$
   (c) Difference $= 100 - 20 = 80$
   Quotient $= \frac{100}{20} = 5$

Self-assessment questions 1.2

1. BODMAS is a priority rule used when evaluating expressions: Brackets (do first), Of, Division, Multiplication (do secondly), Addition, Subtraction (do thirdly).
2. False. For example, $(12 - 4) - 3$ is not the same as $12 - (4 - 3)$. The former is equal to 5 whereas the latter is equal to 11. The position of the brackets is clearly important.

Exercise 1.2

1. (a) $6 - 2 \times 2 = 6 - 4 = 2$
   (b) $(6 - 2) \times 2 = 4 \times 2 = 8$
   (c) $6 \div 2 - 2 = 3 - 2 = 1$
   (d) $(6 \div 2) - 2 = 3 - 2 = 1$
   (e) $6 - 2 + 3 \times 2 = 6 - 2 + 6 = 10$
   (f) $6 - (2 + 3) \times 2 = 6 - 5 \times 2$
   $= 6 - 10 = -4$

(g) $(6-2)+3\times 2=4+3\times 2=4+6$
$=10$
(h) $\frac{16}{-2}=-8$ (i) $\frac{-24}{-3}=8$
(j) $(-6)\times(-2)=12$
(k) $(-2)(-3)(-4)=-24$

2. (a) $6\times(12-3)+1=6\times 9+1$
$=54+1=55$
(b) $6\times 12-(3+1)=72-4=68$
(c) $6\times(12-3+1)=6\times 10=60$
(d) $5\times(4-3)+2=5\times 1+2=5+2=7$
(e) $5\times 4-(3+2)=20-5=15$
or $5\times(4-3+2)=5\times 3=15$
(f) $5\times(4-(3+2))=5\times(4-5)$
$=5\times(-1)=-5$

Self-assessment questions 1.3

1. A prime number is a positive integer larger than 1 that cannot be expressed as the product of two smaller positive integers.
2. 2, 3, 5, 7, 11, 13, 17, 19, 23, 29
3. All even numbers have 2 as a factor and so can be expressed as the product of two smaller numbers. The exception to this is 2 itself, which can only be expressed as 1×2, and since these numbers are not both smaller than 2, then 2 is prime.

Exercise 1.3

1. 13, 2 and 29 are prime.
2. (a) 2×13 (b) $2\times 2\times 5\times 5$
(c) $3\times 3\times 3$ (d) 71
(e) $2\times 2\times 2\times 2\times 2\times 2$ (f) 3×29
(g) 19×23 (h) 29×31
3. $30=2\times 3\times 5$
$42=2\times 3\times 7$
2 and 3 are common prime factors.

Self-assessment questions 1.4

1. H.c.f. stands for 'highest common factor'. The h.c.f. is the largest number that is a factor of each of the numbers in the original set.
2. L.c.m. stands for 'lowest common multiple'. It is the smallest number that can be divided exactly by each of the numbers in the set.

Exercise 1.4

1. (a) $12=2\times 2\times 3$ $15=3\times 5$
$21=3\times 7$
Hence h.c.f. $=3$
(b) $16=2\times 2\times 2\times 2$
$24=2\times 2\times 2\times 3$
$40=2\times 2\times 2\times 5$
So h.c.f. $=2\times 2\times 2=8$
(c) $28=2\times 2\times 7$ $70=2\times 5\times 7$
$120=2\times 2\times 2\times 3\times 5$
$160=2\times 2\times 2\times 2\times 2\times 5$
So h.c.f. $=2$
(d) $35=5\times 7$ $38=2\times 19$
$42=2\times 3\times 7$
So h.c.f. $=1$
(e) $96=2\times 2\times 2\times 2\times 2\times 3$
$120=2\times 2\times 2\times 3\times 5$
$144=2\times 2\times 2\times 2\times 3\times 3$
So h.c.f. $=2\times 2\times 2\times 3=24$

2. (a) 5 $6=2\times 3$ $8=2\times 2\times 2$
So l.c.m. $=2\times 2\times 2\times 3\times 5$
$=120$
(b) $20=2\times 2\times 5$ $30=2\times 3\times 5$
So l.c.m. $=2\times 2\times 3\times 5=60$
(c) 7 $9=3\times 3$ $12=2\times 2\times 3$
So l.c.m. $=2\times 2\times 3\times 3\times 7$
$=252$
(d) $100=2\times 2\times 5\times 5$
$150=2\times 3\times 5\times 5$
$235=5\times 47$
So l.c.m. $=$
$2\times 2\times 3\times 5\times 5\times 47=14100$
(e) $96=2\times 2\times 2\times 2\times 2\times 3$
$120=2\times 2\times 2\times 3\times 5$
$144=2\times 2\times 2\times 2\times 3\times 3$
So l.c.m. $=$
$2\times 2\times 2\times 2\times 2\times 3\times 3\times 5$
$=1440$

Chapter 2

Self-assessment questions 2.1

1. (a) A fraction is formed by dividing a whole number by another whole number, for example $\frac{11}{3}$.
   (b) If the top number (the numerator) of a fraction is greater than or equal to the bottom number (the denominator) then the fraction is improper. For example, $\frac{101}{100}$ and $\frac{9}{9}$ are both improper fractions.
   (c) If the top number is less than the bottom number then the fraction is proper. For example, $\frac{99}{100}$ is a proper fraction.
2. (a) The numerator is the 'top number' of a fraction.
   (b) The denominator is the 'bottom number' of a fraction. For example, in $\frac{17}{14}$, the numerator is 17 and the denominator is 14.

Exercise 2.1

1. (a) Proper (b) Proper (c) Improper (d) Proper (e) Improper

Self-assessment questions 2.2

1. True. For example $7 = \frac{7}{1}$.
2. The numerator and denominator can both be divided by their h.c.f. This simplifies the fraction.
3. $\frac{3}{4}, \frac{9}{12}, \frac{75}{100}$.

Exercise 2.2

1. (a) $\frac{18}{27} = \frac{2}{3}$ (b) $\frac{12}{20} = \frac{3}{5}$ (c) $\frac{15}{45} = \frac{1}{3}$
   (d) $\frac{25}{80} = \frac{5}{16}$ (e) $\frac{15}{60} = \frac{1}{4}$ (f) $\frac{90}{200} = \frac{9}{20}$
   (g) $\frac{15}{20} = \frac{3}{4}$ (h) $\frac{2}{18} = \frac{1}{9}$ (i) $\frac{16}{24} = \frac{2}{3}$
   (j) $\frac{30}{65} = \frac{6}{13}$ (k) $\frac{12}{21} = \frac{4}{7}$ (l) $\frac{100}{45} = \frac{20}{9}$
   (m) $\frac{6}{9} = \frac{2}{3}$ (n) $\frac{12}{16} = \frac{3}{4}$ (o) $\frac{13}{42}$
   (p) $\frac{13}{39} = \frac{1}{3}$ (q) $\frac{11}{33} = \frac{1}{3}$ (r) $\frac{14}{30} = \frac{7}{15}$
   (s) $-\frac{12}{16} = -\frac{3}{4}$ (t) $\frac{11}{-33} = -\frac{1}{3}$
   (u) $\frac{-14}{-30} = \frac{7}{15}$
2. $\frac{3}{4} = \frac{21}{28}$
3. $4 = \frac{20}{5}$
4. $\frac{5}{12} = \frac{15}{36}$
5. $2 = \frac{8}{4}$
6. $6 = \frac{18}{3}$
7. $\frac{2}{3} = \frac{8}{12}, \frac{5}{4} = \frac{15}{12}, \frac{5}{6} = \frac{10}{12}$
8. $\frac{4}{9} = \frac{8}{18}, \frac{1}{2} = \frac{9}{18}, \frac{5}{6} = \frac{15}{18}$
9. (a) $\frac{1}{2} = \frac{6}{12}$ (b) $\frac{3}{4} = \frac{9}{12}$ (c) $\frac{5}{2} = \frac{30}{12}$
   (d) $5 = \frac{60}{12}$ (e) $4 = \frac{48}{12}$ (f) $12 = \frac{144}{12}$

Self-assessment question 2.3

1. The l.c.m. of the denominators is found. Each fraction is expressed in an equivalent form with this l.c.m. as denominator. Addition and subtraction can then take place.

Exercise 2.3

1. (a) $\frac{1}{4}+\frac{2}{3}=\frac{3}{12}+\frac{8}{12}=\frac{11}{12}$

(b) $\frac{3}{5}+\frac{5}{3}=\frac{9}{15}+\frac{25}{15}=\frac{34}{15}$

(c) $\frac{12}{14}-\frac{2}{7}=\frac{6}{7}-\frac{2}{7}=\frac{4}{7}$

(d) $\frac{3}{7}-\frac{1}{2}+\frac{2}{21}=\frac{18}{42}-\frac{21}{42}+\frac{4}{42}=\frac{1}{42}$

(e) $1\frac{1}{2}+\frac{4}{9}=\frac{3}{2}+\frac{4}{9}=\frac{27}{18}+\frac{8}{18}=\frac{35}{18}$

(f) $2\frac{1}{4}-1\frac{1}{3}+\frac{1}{2}=\frac{9}{4}-\frac{4}{3}+\frac{1}{2}$

$=\frac{27}{12}-\frac{16}{12}+\frac{6}{12}=\frac{17}{12}$

(g) $\frac{10}{15}-1\frac{2}{5}+\frac{8}{3}=\frac{10}{15}-\frac{7}{5}+\frac{8}{3}$

$=\frac{10}{15}-\frac{21}{15}+\frac{40}{15}=\frac{29}{15}$

(h) $\frac{9}{10}-\frac{7}{16}+\frac{1}{2}-\frac{2}{5}=\frac{72}{80}-\frac{35}{80}+\frac{40}{80}-\frac{32}{80}$

$=\frac{45}{80}=\frac{9}{16}$

2. (a) $\frac{7}{8}+\frac{1}{3}=\frac{21}{24}+\frac{8}{24}=\frac{29}{24}$

(b) $\frac{1}{2}-\frac{3}{4}=\frac{2}{4}-\frac{3}{4}=-\frac{1}{4}$

(c) $\frac{3}{5}+\frac{2}{3}+\frac{1}{2}=\frac{18}{30}+\frac{20}{30}+\frac{15}{30}=\frac{53}{30}=1\frac{23}{30}$

(d) $\frac{3}{8}+\frac{1}{3}+\frac{1}{4}=\frac{9}{24}+\frac{8}{24}+\frac{6}{24}=\frac{23}{24}$

(e) $\frac{2}{3}-\frac{4}{7}=\frac{14}{21}-\frac{12}{21}=\frac{2}{21}$

(f) $\frac{1}{11}-\frac{1}{2}=\frac{2}{22}-\frac{11}{22}=-\frac{9}{22}$

(g) $\frac{3}{11}-\frac{5}{8}=\frac{24}{88}-\frac{55}{88}=\frac{-31}{88}$

3. (a) $\frac{5}{2}$ (b) $\frac{11}{3}$ (c) $\frac{41}{4}$ (d) $\frac{37}{7}$ (e) $\frac{56}{9}$

(f) $\frac{34}{3}$ (g) $\frac{31}{2}$ (h) $\frac{55}{4}$ (i) $\frac{133}{11}$ (j) $\frac{41}{3}$

(k) $\frac{113}{2}$

4. (a) $3\frac{1}{3}$ (b) $3\frac{1}{2}$ (c) $3\frac{3}{4}$ (d) $4\frac{1}{6}$

Self-assessment question 2.4

1. The numerators are multiplied together to form the numerator of the product. The denominators are multiplied to form the denominator of the product.

Exercise 2.4

1. (a) $\frac{2}{3}\times\frac{6}{7}=\frac{2}{1}\times\frac{2}{7}=\frac{4}{7}$

(b) $\frac{8}{15}\times\frac{25}{32}=\frac{1}{15}\times\frac{25}{4}=\frac{1}{3}\times\frac{5}{4}=\frac{5}{12}$

(c) $\frac{1}{4}\times\frac{8}{9}=\frac{1}{1}\times\frac{2}{9}=\frac{2}{9}$

(d) $\frac{16}{17}\times\frac{34}{48}=\frac{1}{17}\times\frac{34}{3}=\frac{1}{1}\times\frac{2}{3}=\frac{2}{3}$

(e) $2\times\frac{3}{5}\times\frac{5}{12}=\frac{3}{5}\times\frac{5}{6}=\frac{3}{1}\times\frac{1}{6}=\frac{1}{2}$

(f) $2\frac{1}{3}\times1\frac{1}{4}=\frac{7}{3}\times\frac{5}{4}=\frac{35}{12}$

(g) $1\frac{3}{4}\times2\frac{1}{2}=\frac{7}{4}\times\frac{5}{2}=\frac{35}{8}$

(h) $\frac{3}{4}\times1\frac{1}{2}\times3\frac{1}{2}=\frac{3}{4}\times\frac{3}{2}\times\frac{7}{2}=\frac{63}{16}$

2. (a) $\frac{2}{3}\times\frac{3}{4}=\frac{1}{2}$ (b) $\frac{4}{7}\times\frac{21}{30}=\frac{6}{15}=\frac{2}{5}$

(c) $\frac{9}{10}\times80=72$ (d) $\frac{6}{7}\times42=36$

3. Yes, because $\frac{3}{4} \times \frac{12}{15} = \frac{12}{15} \times \frac{3}{4}$

4. (a) $-\frac{5}{21}$ (b) $-\frac{3}{8}$ (c) $-\frac{5}{11}$ (d) $\frac{10}{7}$

5. (a) $5\frac{1}{2} \times \frac{1}{2} = \frac{11}{2} \times \frac{1}{2} = \frac{11}{4}$

(b) $3\frac{3}{4} \times \frac{1}{3} = \frac{15}{4} \times \frac{1}{3} = \frac{5}{4}$

(c) $\frac{2}{3} \times 5\frac{1}{9} = \frac{2}{3} \times \frac{46}{9} = \frac{92}{27}$

(d) $\frac{3}{4} \times 11\frac{1}{2} = \frac{3}{4} \times \frac{23}{2} = \frac{69}{8}$

6. (a) $\frac{3}{5} \times 11\frac{1}{4} = \frac{3}{5} \times \frac{45}{4} = \frac{27}{4}$

(b) $\frac{2}{3} \times 15\frac{1}{2} = \frac{2}{3} \times \frac{31}{2} = \frac{31}{3}$

(c) $\frac{1}{4} \times \left(-8\frac{1}{3}\right) = \frac{1}{4} \times \left(-\frac{25}{3}\right) = -\frac{25}{12}$

Self-assessment question 2.5

1. To divide one fraction by a second fraction, the second fraction is inverted (that is, the numerator and denominator are interchanged) and then multiplication is performed.

Exercise 2.5

1. (a) $\frac{3}{4} \div \frac{1}{8} = \frac{3}{4} \times \frac{8}{1} = \frac{3}{1} \times \frac{2}{1} = 6$

(b) $\frac{8}{9} \div \frac{4}{3} = \frac{8}{9} \times \frac{3}{4} = \frac{2}{9} \times \frac{3}{1} = \frac{2}{3}$

(c) $-\frac{2}{7} \div \frac{4}{21} = -\frac{2}{7} \times \frac{21}{4} = -\frac{2}{1} \times \frac{3}{4} = -\frac{3}{2}$

(d) $\frac{9}{4} \div 1\frac{1}{2} = \frac{9}{4} \div \frac{3}{2} = \frac{9}{4} \times \frac{2}{3} = \frac{3}{4} \times \frac{2}{1} = \frac{3}{2}$

(e) $\frac{5}{6} \div \frac{5}{12} = \frac{5}{6} \times \frac{12}{5} = \frac{1}{6} \times \frac{12}{1} = 2$

(f) $\frac{99}{100} \div 1\frac{4}{5} = \frac{99}{100} \div \frac{9}{5} = \frac{99}{100} \times \frac{5}{9}$

$= \frac{11}{100} \times \frac{5}{1} = \frac{11}{20}$

(g) $3\frac{1}{4} \div 1\frac{1}{8} = \frac{13}{4} \div \frac{9}{8} = \frac{13}{4} \times \frac{8}{9}$

$= \frac{13}{1} \times \frac{2}{9} = \frac{26}{9}$

(h) $\left(2\frac{1}{4} \div \frac{3}{4}\right) \times 2 = \left(\frac{9}{4} \times \frac{4}{3}\right) \times 2$

$= \left(\frac{3}{4} \times 4\right) \times 2 = 3 \times 2 = 6$

(i) $2\frac{1}{4} \div \left(\frac{3}{4} \times 2\right) = \frac{9}{4} \div \left(\frac{3}{4} \times \frac{2}{1}\right)$

$= \frac{9}{4} \div \frac{3}{2} = \frac{9}{4} \times \frac{2}{3} = \frac{3}{4} \times \frac{2}{1} = \frac{3}{2}$

(j) $6\frac{1}{4} \div 2\frac{1}{2} + 5 = \frac{25}{4} \div \frac{5}{2} + 5$

$= \frac{25}{4} \times \frac{2}{5} + 5 = \frac{5}{2} + 5 = \frac{15}{2}$

(k) $6\frac{1}{4} \div \left(2\frac{1}{2} + 5\right) = \frac{25}{4} \div \left(\frac{5}{2} + 5\right)$

$= \frac{25}{4} \div \frac{15}{2} = \frac{25}{4} \times \frac{2}{15} = \frac{5}{4} \times \frac{2}{3} = \frac{5}{6}$

Chapter 3

Self-assessment questions 3.1

1. Largest is 23.01; smallest is 23.0.
2. 0.1

Exercise 3.1

1. (a) $\frac{7}{10}$ (b) $\frac{4}{5}$ (c) $\frac{9}{10}$
2. (a) $\frac{11}{20}$ (b) $\frac{79}{500}$ (c) $\frac{49}{50}$ (d) $\frac{99}{1000}$
3. (a) $4\frac{3}{5}$ (b) $5\frac{1}{5}$ (c) $8\frac{1}{20}$ (d) $11\frac{59}{100}$
   (e) $121\frac{9}{100}$
4. (a) 0.697 (b) 0.083 (c) 0.517

Self-assessment questions 3.2

1. Writing a number to 2 (or 3 or 4 etc.) significant figures is a way of approximating the number. The number of s.f. is the maximum number of non-zero digits in the approximation. The approximation is as close as possible to the original number.
2. The digits after the decimal point are considered. To write to 1 d.p. we consider the first two digits; to write to 2 d.p. we consider the first three digits, and so on. If the last digit considered is 5 or greater, we round up the previous digit; otherwise we round down.

Exercise 3.2

1. (a) 6960 (b) 70.4 (c) 0.0123 (d) 0.0110
   (e) 45.6 (f) 2350
2. (a) 66.00 (b) 66.0 (c) 66 (d) 70
   (e) 66.00 (f) 66.0
3. (a) 10 (b) 10.0
4. (a) 65.456 (b) 65.46 (c) 65.5
   (d) 65.456 (e) 65.46 (f) 65.5 (g) 65
   (h) 70

Chapter 4

Self-assessment question 4.1

1. Converting fractions to percentages allows for easy comparison of numbers.

Exercise 4.1

1. 23% of 124 $= \frac{23}{100} \times 124 = 28.52$
2. (a) $\frac{9}{11} = \frac{9}{11} \times 100\% = \frac{900}{11}\% = 81.82\%$
   (b) $\frac{15}{20} = \frac{15}{20} \times 100\% = 75\%$
   (c) $\frac{9}{10} = \frac{9}{10} \times 100\% = 90\%$
   (d) $\frac{45}{50} = \frac{45}{50} \times 100\% = 90\%$
   (e) $\frac{75}{90} = \frac{75}{90} \times 100\% = 83.33\%$
3. $\frac{13}{12} = \frac{13}{12} \times 100\% = 108.33\%$
4. 217% of 500 $= \frac{217}{100} \times 500 = 1085$
5. New weekly wage is 106% of £400,

   106% of 400 $= \frac{106}{100} \times 400 = 424$

   The new weekly wage is £424.

6. 17% of 1200 $= \frac{17}{100} \times 1200 = 204$
   The debt is decreased by £204 to £1200 − 204 = £996.
7. (a) $50\% = \frac{50}{100} = 0.5$
   (b) $36\% = \frac{36}{100} = 0.36$
   (c) $75\% = \frac{75}{100} = 0.75$
   (d) $100\% = \frac{100}{100} = 1$
   (e) $12.5\% = \frac{12.5}{100} = 0.125$
8. £204.80
9. £1125
10. percentage change
    $= \frac{\text{new value} - \text{original value}}{\text{original value}} \times 100$
    $= \frac{7495 - 6950}{6950} \times 100$
    $= 7.84\%$
11. percentage change
    $= \frac{\text{new value} - \text{original value}}{\text{original value}} \times 100$
    $= \frac{399 - 525}{525} \times 100$
    $= -24\%$
    Note that the percentage change is negative and this indicates a reduction in price. There has been a 24% reduction in the price of the washing machine.

Self-assessment question 4.2

1. True

Exercise 4.2

1. $8 + 1 + 3 = 12$. The first number is $\frac{8}{12}$ of 180, that is 120; the second number is $\frac{1}{12}$ of 180, that is 15; and the third number is $\frac{3}{12}$ of 180, that is 45. Hence 180 is divided into 120, 15 and 45.
2. $1 + 1 + 3 = 5$. We calculate $\frac{1}{5}$ of 930 to be 186 and $\frac{3}{5}$ of 930 to be 558. The length is divided into 186 cm, 186 cm and 558 cm.
3. $2 + 3 + 4 = 9$. The first piece is $\frac{2}{9}$ of 6 m, that is 1.33 m; the second piece is $\frac{3}{9}$ of 6 m, that is 2 m; the third piece is $\frac{4}{9}$ of 6 m, that is 2.67 m.
4. $1 + 2 + 3 + 4 = 10$: $\frac{1}{10}$ of $1200 = 120$; $\frac{2}{10}$ of $1200 = 240$; $\frac{3}{10}$ of $1200 = 360$; $\frac{4}{10}$ of $1200 = 480$. The number 1200 is divided into 120, 240, 360 and 480.
5. $2\frac{3}{4} : 1\frac{1}{2} : 2\frac{1}{4} = \frac{11}{4} : \frac{3}{2} : \frac{9}{4} = 11 : 6 : 9$
   Now, $11 + 6 + 9 = 26$, so
   $\frac{11}{26} \times 2600 = 1100$ $\quad \frac{6}{26} \times 2600 = 600$
   $\frac{9}{26} \times 2600 = 900$
   Alan receives £1100, Bill receives £600 and Claire receives £900.
6. 8 kg, 10.67 kg, 21.33 kg
7. (a) 1 : 2 (b) 1 : 2 (c) 1 : 2 : 4 (d) 1 : 21
8. 24, 84

Chapter 5

Self-assessment questions 5.1

1. 'Algebra' refers to the manipulation of symbols, as opposed to the manipulation of numbers.
2. The product of the two numbers is written as mn.
3. An algebraic fraction is formed by dividing one algebraic expression by another algebraic expression. The 'top' of the fraction is the numerator; the 'bottom' of the fraction is the denominator.

4. Superscripts and subscripts are located in different positions, relative to the symbol. Superscripts are placed high; subscripts are placed low.
5. A variable can have many different values; a constant has one, fixed value.

Self-assessment questions 5.2

1. In the expression a^x, a is the base and x is the power.
2. 'Index' is another word meaning 'power'.
3. $(xyz)^2$ means $(xyz)(xyz)$, which can be written as $x^2y^2z^2$. Clearly this is distinct from xyz^2, in which only the quantity z is squared.
4. $(-3)^4 = (-3)(-3)(-3)(-3) = 81$. $-3^4 = -(3)(3)(3)(3) = -81$. Here, the power (4) has the higher priority.

Exercise 5.2

1. $2^4 = 16$; $(\frac{1}{2})^2 = \frac{1}{4}$; $1^8 = 1$; $3^5 = 243$; $0^3 = 0$.
2. $10^4 = 10000$; $10^5 = 100000$; $10^6 = 1000000$.
3. $11^4 = 14641$; $16^8 = 4294967296$; $39^4 = 2313441$; $1.5^7 = 17.0859375$.
4. (a) $a^4b^2c = a \times a \times a \times a \times b \times b \times c$
   (b) $xy^2z^4 = x \times y \times y \times z \times z \times z \times z$
5. (a) x^4y^2 (b) $x^2y^2z^3$ (c) $x^2y^2z^2$
   (d) $a^2b^2c^2$
6. (a) $7^4 = 2401$ (b) $7^5 = 16807$
   (c) $7^4 \times 7^5 = 40353607$
   (d) $7^9 = 40353607$ (e) $8^3 = 512$
   (f) $8^7 = 2097152$
   (g) $8^3 \times 8^7 = 1073741824$
   (h) $8^{10} = 1073741824$
   The rule states that $a^m \times a^n = a^{m+n}$; the powers are added.
7. $(-3)^3 = -27$; $(-2)^2 = 4$; $(-1)^7 = -1$; $(-1)^4 = 1$.
8. -4492.125; 324; -0.03125.
9. (a) 36 (b) 9 (c) -64 (d) -8
   $-6^2 = -36$, $-3^2 = -9$,
   $-4^3 = -64$, $-2^3 = -8$

Self-assessment question 5.3

1. An algebraic expression is any quantity comprising symbols and operations (that is, $+$, $-$, $\times$, $\div$): for example, x, $2x^2$ and $3x + 2y$ are all algebraic expressions. An algebraic formula relates two or more quantities and must contain an '=' sign. For example, $A = \pi r^2$, $V = \pi r^2 h$ and $S = ut + \frac{1}{2}at^2$ are all algebraic formulae.

Exercise 5.3

1. 60
2. 69
3. (a) $314.2\ \text{cm}^2$ (b) $28.28\ \text{cm}^2$
   (c) $0.126\ \text{cm}^2$
4. $3x^2 = 3 \times 4^2 = 3 \times 16 = 48$;
   $(3x)^2 = 12^2 = 144$
5. $5x^2 = 5(-2)^2 = 20$; $(5x)^2 = (-10)^2 = 100$
6. (a) 33.95 (b) 23.5225 (c) 26.75
   (d) 109.234125
7. (a) $a + b + c = 25.5$ (b) $ab = 46.08$
   (c) $bc = 32.76$ (d) $abc = 419.328$
8. $C = \frac{5}{9}(100 - 32) = \frac{5}{9}(68) = 37.78$
9. (a) $x^2 = 49$ (b) $-x^2 = -49$
   (c) $(-x)^2 = (-7)^2 = 49$
10. (a) 4 (b) 4 (c) -4 (d) 12
    (e) -12 (f) 36
11. (a) 3 (b) 9 (c) -1 (d) 36
    (e) -36 (f) 144
12. $x^2 - 7x + 2 = (-9)^2 - 7(-9) + 2$
    $= 81 + 63 + 2 = 146$
13. $2x^2 + 3x - 11 = 2(-3)^2 + 3(-3) - 11$
    $= 18 - 9 - 11 = -2$
14. $-x^2 + 3x - 5 = -(-1)^2 + 3(-1) - 5$
    $= -1 - 3 - 5 = -9$
15. 0
16. (a) 49 (b) 43 (c) 1
    (d) 5
17. (a) 21 (b) 27 (c) 0
    (d) $\frac{1}{6}$
18. (a) 3 (b) $3\frac{4}{5}$ (c) 23 (d) 23

19. (a) $-\frac{1}{2}$ (b) 4 (c) 32
20. (a) 0 (b) 16 (c) $40\frac{1}{2}$
    (d) $2\frac{1}{2}$
21. (a) 17 (b) −0.5 (c) 7
22. (a) 6000 (b) 2812.5
23. (a) 17151 (b) 276951

Chapter 6

Self-assessment questions 6.1

1. $a^m \times a^n = a^{m+n}$
   $\frac{a^m}{a^n} = a^{m-n}$
   $(a^m)^n = a^{mn}$
2. a^0 is 1.
3. x^1 is simply x.

Exercise 6.1

1. (a) $5^7 \times 5^{13} = 5^{20}$ (b) $9^8 \times 9^5 = 9^{13}$
   (c) $11^2 \times 11^3 \times 11^4 = 11^9$
2. (a) $15^3/15^2 = 15^1 = 15$ (b) $4^{18}/4^9 = 4^9$
   (c) $5^{20}/5^{19} = 5^1 = 5$
3. (a) a^{10} (b) a^9 (c) b^{22}
4. (a) x^{15} (b) y^{21}
5. $19^8 \times 17^8$ cannot be simplified using the laws of indices because the two bases are not the same.
6. (a) $(7^3)^2 = 7^{3\times 2} = 7^6$ (b) $(4^2)^8 = 4^{16}$
   (c) $(7^9)^2 = 7^{18}$
7. $1/(5^3)^8 = 1/5^{24}$
8. (a) x^5y^5 (b) $a^3b^3c^3$
9. (a) $x^{10}y^{20}$ (b) $81x^6$ (c) $-27x^3$
   (d) x^8y^{12}
10. (a) z^3 (b) y^2 (c) 1

Self-assessment question 6.2

1. a^{-m} is the same as $\frac{1}{a^m}$. For example, 5^{-2} is the same as $\frac{1}{5^2}$.

Exercise 6.2

1. (a) $\frac{1}{4}$ (b) $\frac{1}{8}$ (c) $\frac{1}{9}$ (d) $\frac{1}{27}$ (e) $\frac{1}{25}$
   (f) $\frac{1}{16}$ (g) $\frac{1}{9}$ (h) $\frac{1}{121}$ (i) $\frac{1}{7}$
2. (a) 0.1 (b) 0.01 (c) 0.000001 (d) 0.01
   (e) 0.001 (f) 0.0001
3. (a) $\frac{1}{x^4}$ (b) x^5 (c) $\frac{1}{x^7}$ (d) $\frac{1}{y^2}$
   (e) $y^1 = y$ (f) $\frac{1}{y^1} = \frac{1}{y}$ (g) $\frac{1}{y^2}$
   (h) $\frac{1}{z^1} = \frac{1}{z}$ (i) $z^1 = z$
4. (a) $x^{-3} = \frac{1}{x^3}$ (b) $x^{-5} = \frac{1}{x^5}$ (c) $x^{-1} = \frac{1}{x}$
   (d) x^5 (e) $x^{-13} = \frac{1}{x^{13}}$ (f) $x^{-8} = \frac{1}{x^8}$
   (g) $x^{-9} = \frac{1}{x^9}$ (h) $x^{-4} = \frac{1}{x^4}$
5. (a) a^{11} (b) x^{-16} (c) x^{-18} (d) 4^{-6}
6. (a) 0.001 (b) 0.0001 (c) 0.00001
7. $4^{-8}/4^{-6} = 4^{-2} = 1/4^2 = \frac{1}{16}$
   $3^{-5}/3^{-8} = 3^3 = 27$

Self-assessment questions 6.3

1. $x^{\frac{1}{2}}$ is the square root of x. $x^{\frac{1}{3}}$ is the cube root of x.

2. 10, −10. Negative numbers, such as −100, do not have square roots, since squaring a number always gives a positive result.

Exercise 6.3

1. (a) $64^{1/3} = \sqrt[3]{64} = 4$ since $4^3 = 64$
   (b) $144^{1/2} = \sqrt{144} = \pm 12$
   (c) $16^{-1/4} = 1/16^{1/4} = 1/\sqrt[4]{16} = \pm\frac{1}{2}$
   (d) $25^{-1/2} = 1/25^{1/2} = 1/\sqrt{25} = \pm\frac{1}{5}$
   (e) $1/32^{-1/5} = 32^{1/5} = \sqrt[5]{32} = 2$ since $2^5 = 32$
2. (a) $(3^{-1/2})^4 = 3^{-2} = 1/3^2 = \frac{1}{9}$
   (b) $(8^{1/3})^{-1} = 8^{-1/3} = 1/8^{1/3} = 1/\sqrt[3]{8} = \frac{1}{2}$
3. (a) $8^{1/2}$ (b) $12^{1/3}$
   (c) $16^{1/4}$ (d) $13^{3/2}$ (e) $4^{7/3}$
4. (a) $x^{1/2}$ (b) $y^{1/3}$
   (c) $x^{5/2}$ (d) $5^{7/3}$

Exercise 6.4

1. (a) 743
   (b) 74300
   (c) 70
   (d) 0.0007
2. (a) 3×10^2
   (b) 3.56×10^2
   (c) 0.32×10^2
   (d) 0.0057×10^2

Self-assessment question 6.5

1. Scientific notation is useful for writing very large or extremely small numbers in a concise way. It is easier to manipulate such numbers when they are written using scientific notation.

Exercise 6.5

1. (a) $45 = 4.5 \times 10^1$
   (b) $45000 = 4.5 \times 10^4$
   (c) $-450 = -4.5 \times 10^2$
   (d) $90000000 = 9.0 \times 10^7$
   (e) $0.15 = 1.5 \times 10^{-1}$
   (f) $0.00036 = 3.6 \times 10^{-4}$
   (g) 3.5 is already in standard form.
   (h) $-13.2 = -1.32 \times 10^1$
   (i) $1000000 = 1 \times 10^6$
   (j) $0.0975 = 9.75 \times 10^{-2}$
   (k) $45.34 = 4.534 \times 10^1$
2. (a) $3.75 \times 10^2 = 375$
   (b) $3.97 \times 10^1 = 39.7$
   (c) $1.875 \times 10^{-1} = 0.1875$
   (d) $-8.75 \times 10^{-3} = -0.00875$
3. (a) 2.4×10^8 (b) 7.968×10^8
   (c) 1.044×10^{-4}
   (d) 1.526×10^{-1}
   (e) 5.293×10^2

Chapter 7

Exercise 7.1

1. (a) $-5p + 19q$ (b) $-5r - 13s + z$ (c) not possible to simplify
   (d) $8x^2 + 3y^2 - 2y$ (e) $4x^2 - x + 9$
2. (a) $-12y + 8p + 9q$ (b) $21x^2 - 11x^3 + y^3$
   (c) $7xy + y^2$ (d) $2xy$ (e) 0

Self-assessment questions 7.2

1. Positive
2. Negative

Exercise 7.2

1. (a) 84 (b) 84 (c) 84
2. (a) 40 (b) 40
3. (a) $14z$ (b) $30y$ (c) $6x$ (d) $27a$
   (e) $55a$ (f) $6x$
4. (a) $20x^2$ (b) $6y^3$ (c) $22u^2$ (d) $8u^2$
   (e) $26z^2$
5. (a) $21x^2$ (b) $21a^2$ (c) $14a^2$
6. (a) $15y^2$ (b) $8y$
7. (a) $a^3b^2c^2$ (b) x^3y^2 (c) x^2y^4
8. No difference; both equal x^2y^4.

9. $(xy^2)(xy^2) = x^2y^4$; $xy^2 + xy^2 = 2xy^2$
10. (a) $-21z^2$ (b) $-4z$
11. (a) $-3x^2$ (b) $2x$
12. (a) $2x^2$ (b) $-3x$

Exercise 7.3

1. (a) $4x + 4$ (b) $-4x - 4$ (c) $4x - 4$
   (d) $-4x + 4$
2. (a) $5x - 5y$ (b) $19x + 57y$ (c) $8a + 8b$
   (d) $5y + xy$ (e) $12x + 48$
   (f) $17x - 153$ (g) $-a + 2b$ (h) $x + \frac{1}{2}$
   (i) $6m - 12m^2 - 9mn$
3. (a) $18 - 13x - 26 = -8 - 13x$ (b) $x^2 + xy$ will not simplify any further
4. (a) $x^2 + 7x + 6$ (b) $x^2 + 9x + 20$
   (c) $x^2 + x - 6$ (d) $x^2 + 5x - 6$
   (e) $xm + ym + nx + yn$
   (f) $12 + 3y + 4x + yx$ (g) $25 - x^2$
   (h) $51x^2 - 79x - 10$
5. (a) $x^2 - 4x - 21$ (b) $6x^2 + 11x - 7$
   (c) $16x^2 - 1$ (d) $x^2 - 9$
   (e) $6 + x - 2x^2$
6. (a) $\frac{29}{2}x - \frac{5}{2}y$ (b) $\frac{5}{4}x + \frac{5}{4}$
7. (a) $-x + y$ (b) $-a - 2b$ (c) $-\frac{3}{2}p - \frac{1}{2}q$
8. $(x + 1)(x + 2) = x^2 + 3x + 2$. So
   $(x + 1)(x + 2)(x + 3) = (x^2 + 3x + 2)(x + 3)$
   $= (x^2 + 3x + 2)(x) + (x^2 + 3x + 2)(3)$
   $= x^3 + 3x^2 + 2x + 3x^2 + 9x + 6$
   $= x^3 + 6x^2 + 11x + 6$

Chapter 8

Self-assessment question 8.1

1. To factorise an expression means to write it as a product, usually of two or more simpler expressions.

Exercise 8.1

1. (a) $9x + 27$ (b) $-5x + 10$ (c) $\frac{1}{2}x + \frac{1}{2}$
   (d) $-a + 3b$ (e) $1/(2x + 2y)$
   (f) $x/(yx - y^2)$
2. (a) $4x^2$ has factors 1, 2, 4, x, $2x$, $4x$, x^2, $2x^2$, $4x^2$
   (b) $6x^3$ has factors 1, 2, 3, 6, x, $2x$, $3x$, $6x$, x^2, $2x^2$, $3x^2$, $6x^2$, x^3, $2x^3$, $3x^3$, $6x^3$
3. (a) $3(x + 6)$ (b) $3(y - 3)$ (c) $-3(y + 3)$
   (d) $-3(1 + 3y)$ (e) $5(4 + t)$ (f) $5(4 - t)$
   (g) $-5(t + 4)$ (h) $3(x + 4)$ (i) $17(t + 2)$
   (j) $4(t - 9)$
4. (a) $x(x^3 + 2)$ (b) $x(x^3 - 2)$
   (c) $x(3x^3 - 2)$ (d) $x(3x^3 + 2)$
   (e) $x^2(3x^2 + 2)$ (f) $x^3(3x + 2)$
   (g) $z(17 - z)$ (h) $x(3 - y)$ (i) $y(3 - x)$
   (j) $x(1 + 2y + 3yz)$
5. (a) $10(x + 2y)$ (b) $3(4a + b)$
   (c) $2x(2 - 3y)$ (d) $7(a + 2)$ (e) $5(2m - 3)$
   (f) $1/[5(a + 7b)]$ (g) $1/[5a(a + 7b)]$
6. (a) $3x(5x + 1)$ (b) $x(4x - 3)$
   (c) $4x(x - 2)$ (d) $3(5 - x^2)$
   (e) $5x^2(2x + 1 + 3y)$ (f) $6ab(a - 2b)$
   (g) $8b(2ac - ba + 3c)$

Self-assessment question 8.2

1. $2x^2 + x + 6$ cannot be factorised. On the other hand, $2x^2 + x - 6$ can be factorised as $(2x - 3)(x + 2)$.

Exercise 8.2

1. (a) $(x + 2)(x + 1)$ (b) $(x + 7)(x + 6)$
   (c) $(x + 5)(x - 3)$ (d) $(x + 10)(x - 1)$
   (e) $(x - 8)(x - 3)$ (f) $(x - 10)(x + 10)$
   (g) $(x + 2)(x + 2)$ or $(x + 2)^2$
   (h) $(x + 6)(x - 6)$ (i) $(x + 5)(x - 5)$
   (j) $(x + 1)(x + 9)$ (k) $(x + 9)(x - 1)$
   (l) $(x + 1)(x - 9)$ (m) $(x - 1)(x - 9)$
   (n) $x(x - 5)$

2. (a) $(2x+1)(x-3)$ (b) $(3x+1)(x-2)$
   (c) $(5x+3)(2x+1)$
   (d) $2x^2+12x+16$ has a common factor of 2, which should be written outside a bracket to give $2(x^2+6x+8)$. The bracket can then be factorised to give $2(x+4)(x+2)$.
   (e) $(2x+3)(x+1)$ (f) $(3s+2)(s+1)$
   (g) $(3z+2)(z+5)$
   (h) $9(x^2-4)=9(x+2)(x-2)$
   (i) $(2x+5)(2x-5)$
3. (a) $(x+y)(x-y)=x^2+yx-yx-y^2$
   $=x^2-y^2$
   (b) (i) $(4x+1)(4x-1)$
   (ii) $(4x+3)(4x-3)$
   (iii) $(5t-4r)(5t+4r)$
4. (a) $(x-2)(x+5)$ (b) $(2x+5)(x-4)$
   (c) $(3x+1)(3x-1)$ (d) $10x^2+14x-12$ has a common factor of 2, which is written outside a bracket to give $2(5x^2+7x-6)$. The bracket can then be factorised to give $2(5x-3)(x+2)$; (e) $(x+13)(x+2)$.
   (f) $(-x+1)(x+3)$
5. (a) $(10+7x)(10-7x)$
   (b) $(6x+5y)(6x-5y)$
   (c) $\left(\frac{1}{2}+3v\right)\left(\frac{1}{2}-3v\right)$ (d) $\left(\frac{x}{y}+2\right)\left(\frac{x}{y}-2\right)$

Chapter 9

Self-assessment questions 9.2

1. There are no factors common to both numerator and denominator and hence cancellation is not possible. In particular 3 is not a factor of the denominator.
2. Same as Q1 – there are no factors common to numerator and denominator. In particular x is a factor neither of the denominator nor of the numerator.
3. $\frac{x+1}{2x+2}=\frac{x+1}{2(x+1)}=\frac{1}{2}$

   The common factor, $x+1$, has been cancelled.

Exercise 9.2

1. (a) $\frac{3x}{y}$ (b) $\frac{9}{x}$ (c) $3y$ (d) $3x$ (e) 9
   (f) 3
2. (a) $\frac{5x}{y}$ (b) $\frac{3x}{y}$ (c) $15y$ (d) 15 (e) $-x^2$
   (f) $-\frac{1}{y^4}=-y^{-4}$ (g) $\frac{1}{y}=y^{-1}$ (h) y^{-7}
3. (a) $\frac{1}{3+2x}$ (b) $1+2x$ (c) $\frac{1}{2+7x}$
   (d) $\frac{x}{2+7x}$ (e) $\frac{x}{1+7x}$ (f) $\frac{1}{7x+y}$
   (g) $\frac{y}{7x+y}$ (h) $\frac{x}{7x+y}$
4. (a) $5x+1$ (b) $\frac{3(5x+1)}{3(x+2y)}=\frac{5x+1}{x+2y}$
   (c) $\frac{3}{x+2}$ (d) $\frac{3}{y+2}$ (e) $\frac{13}{x+5}$
   (f) $\frac{17}{9y+4}$
5. (a) $\frac{1}{3+2x}$ (b) $\frac{2}{x+7}$
   (c) $\frac{2(x+4)}{(x-2)(x+4)}=\frac{2}{x-2}$ (d) $\frac{7}{ab+9}$
   (e) $\frac{y}{y+1}$

6. (a) $\frac{1}{x-2}$ (b) $\frac{2(x-2)}{(x-2)(x+3)} = \frac{2}{x+3}$

(c) $\frac{1}{x+2}$ (d) $\frac{(x+1)(x+1)}{(x+1)(x-3)} = \frac{x+1}{x-3}$

(e) $\frac{2}{x-3}$ (f) $\frac{1}{x-3}$ (g) $\frac{1}{2(x-3)}$

(h) $\frac{2}{x-3}$ (i) $\frac{1}{2(x+4)}$ (j) $\frac{1}{2}$ (k) 2

(l) $x+4$ (m) $\frac{1}{x-3}$ (n) $\frac{1}{x+4}$ (o) $\frac{1}{2}$

(p) $\frac{x+4}{2x+9}$

Self-assessment question 9.3

1. True

Exercise 9.3

1. (a) $\frac{y}{6}$ (b) $\frac{z}{6}$ (c) $\frac{2}{5y}$ (d) $\frac{2}{5x}$ (e) $\frac{3x}{4y}$

(f) $\frac{3x^2}{5y}$ (g) $\frac{3x}{5y}$ (h) $\frac{7x}{16y}$ (i) $\frac{1}{4x}$ (j) $\frac{x}{4}$

(k) $\frac{1}{x}$ (l) $\frac{x}{9}$ (m) $\frac{1}{x}$ (n) $\frac{1}{9x}$ (o) $\frac{x}{2}$

2. (a) $\frac{1}{x}$ (b) $\frac{x}{4}$ (c) 1 (d) x (e) $\frac{1}{x}$

(f) $\frac{4}{x}$ (g) $\frac{6}{x}$

3. (a) $\frac{a}{20}$ (b) $\frac{5a}{4b}$ (c) $\frac{1}{2ab}$ (d) $\frac{6x^2}{y^3}$

(e) $\frac{3b}{5a^2}$ (f) $\frac{x}{4y}$ (g) $\frac{x}{3(x+y)}$ (h) $\frac{1}{3(x+4)}$

4. (a) $\frac{3y}{z^3}$ (b) $\frac{3+x}{y}$ (c) $\frac{x}{12}$ (d) $\frac{b}{c}$

5. (a) $\frac{1}{x+4}$ (b) $\frac{4(x-2)}{x}$

(c) $\frac{12ab}{5ef} \times \frac{f}{4ab^2} = \frac{3}{5eb}$

(d) $\frac{x+3y}{2x} \times \frac{4x^2}{y} = \frac{2x(x+3y)}{y}$ (e) $\frac{9}{xyz}$

6. $\frac{2}{x+3}$

7. $\frac{x+4}{x+3}$

Self-assessment question 9.4

1. The l.c.d. is the simplest expression that is divisible by all the given denominators. To find the l.c.d. first factorise all denominators. The l.c.d. is then formed by including the minimum number of factors from the denominators such that each denominator can divide into the l.c.d.

Exercise 9.4

1. (a) $\frac{5z}{6}$ (b) $\frac{7x}{12}$ (c) $\frac{6y}{25}$

2. (a) $\frac{x+2}{2x}$ (b) $\frac{1+2x}{2}$ (c) $\frac{1+3y}{3}$

(d) $\frac{y+3}{3y}$ (e) $\frac{8y+1}{y}$

3. (a) $\frac{10-x}{2x}$ (b) $\frac{5+2x}{x}$ (c) $\frac{9-x}{3x}$

(d) $\frac{2x-3}{6}$ (e) $\frac{9+x}{3x}$

4. (a) $\frac{3}{x} + \frac{4}{y} = \frac{3y}{xy} + \frac{4x}{xy} = \frac{3y+4x}{xy}$

(b) $\frac{3}{x^2} + \frac{4y}{x} = \frac{3}{x^2} + \frac{4xy}{x^2} = \frac{3+4xy}{x^2}$

(c) $\frac{4ab}{x} + \frac{3ab}{2y} = \frac{8aby}{2xy} + \frac{3abx}{2xy} = \frac{8aby+3abx}{2xy}$

(d) $\frac{4xy}{a} + \frac{3xy}{2b} = \frac{8xyb}{2ab} + \frac{3xya}{2ab}$

$= \frac{8xyb+3xya}{2ab}$

(e) $\frac{3}{x} - \frac{6}{2x} = \frac{3}{x} - \frac{3}{x} = 0$

(f) $\frac{3x}{2y} - \frac{7y}{4x} = \frac{6x^2}{4xy} - \frac{7y^2}{4xy} = \frac{6x^2 - 7y^2}{4xy}$

(g) $\frac{3}{x+y} - \frac{2}{y} = \frac{3y}{(x+y)y} - \frac{2(x+y)}{(x+y)y}$
$= \frac{3y - 2x - 2y}{(x+y)y} = \frac{y - 2x}{(x+y)y}$

(h) $\frac{1}{a+b} - \frac{1}{a-b}$
$= \frac{a-b}{(a+b)(a-b)} - \frac{a+b}{(a+b)(a-b)}$
$= \frac{a-b-a-b}{(a+b)(a-b)} = \frac{-2b}{(a+b)(a-b)}$

(i) $2x + \frac{1}{2x} = \frac{4x^2}{2x} + \frac{1}{2x} = \frac{4x^2+1}{2x}$

(j) $2x - \frac{1}{2x} = \frac{4x^2}{2x} - \frac{1}{2x} = \frac{4x^2-1}{2x}$

5. (a) $\frac{x}{y} + \frac{3x^2}{z} = \frac{xz}{yz} + \frac{3x^2y}{yz} = \frac{xz + 3x^2y}{yz}$

(b) $\frac{4}{a} + \frac{5}{b} = \frac{4b + 5a}{ab}$

(c) $\frac{6x}{y} - \frac{2y}{x} = \frac{6x^2 - 2y^2}{xy}$

(d) $3x - \frac{3x+1}{4} = \frac{12x}{4} - \frac{3x+1}{4}$
$= \frac{12x - 3x - 1}{4} = \frac{9x - 1}{4}$

(e) $\frac{5a}{12} + \frac{9a}{18} = \frac{15a + 18a}{36} = \frac{33a}{36} = \frac{11a}{12}$

(f) $\frac{x-3}{4} + \frac{3}{5} = \frac{5(x-3)+12}{20}$
$= \frac{5x - 15 + 12}{20} = \frac{5x-3}{20}$

6. (a) $\frac{2x+3}{(x+1)(x+2)}$ (b) $\frac{3x+1}{(x-1)(x+3)}$

(c) $\frac{4x+17}{(x+5)(x+4)}$ (d) $\frac{4x-10}{(x-2)(x-4)}$

(e) $\frac{5x+4}{(2x+1)(x+1)}$ (f) $\frac{x+1}{x(1-2x)}$

(g) $\frac{3x+7}{(x+1)^2}$ (h) $\frac{x}{(x-1)^2}$

Exercise 9.5

1. (a) $\frac{4}{x+2} + \frac{3}{x+3}$ (b) $\frac{5}{x+4} - \frac{3}{x+1}$

(c) $\frac{1}{x+6} - \frac{2}{2x+3}$ (d) $\frac{2}{x-1} + \frac{3}{x-4}$

(e) $\frac{5}{2x-1} - \frac{1}{x+2}$

2. $\frac{5}{2(3x-2)} - \frac{3}{2(2x+3)}$

3. (a) $\frac{4}{x+5} - \frac{3}{x-5}$ (b) $\frac{1}{x-3} - \frac{1}{(x-3)^2}$

(c) $\frac{3}{1-x} - \frac{2}{x+2}$ (d) $\frac{4}{3x-1} - \frac{1}{(3x-1)^2}$

Chapter 10

Self-assessment questions 10.1

1. The subject of a formula is a variable that appears by itself on one side of the formula. It appears nowhere else.
2. To change the subject of a formula you will have to transpose it.
3. Whatever is done to the left-hand side (l.h.s.) must be done to the right-hand side (r.h.s.). We can (a) add or subtract the same quantity

from both left- and right-hand sides, (b) multiply or divide both sides by the same quantity, (c) perform the same operation on both sides.

Exercise 10.1

1. (a) $x=\frac{y}{3}$ (b) $x=\frac{1}{y}$ (c) $x=\frac{5+y}{7}$

(d) $x=2y+14$ (e) $x=\frac{1}{2y}$

(f) $2x+1=\frac{1}{y}$ so $2x=\frac{1}{y}-1=\frac{1-y}{y}$

finally, $x=\frac{1-y}{2y}$

(g) $y-1=\frac{1}{2x}$ so $2x=\frac{1}{y-1}$

and then $x=\frac{1}{2(y-1)}$

(h) $x=\frac{y+21}{18}$ (i) $x=\frac{19-y}{8}$

2. (a) $m=\frac{y-c}{x}$ (b) $x=\frac{y-c}{m}$

(c) $c=y-mx$

3. (a) $y=13x-26$ and so $x=\frac{y+26}{13}$

(b) $y=x+1$ and so $x=y-1$

(c) $y=a+tx-3t$ and so $tx=y+3t-a$

and then $x=\frac{y+3t-a}{t}$

4. (a) $x=\frac{y-11}{7}$ (b) $I=\frac{V}{R}$ (c) $r^3=\frac{3V}{4\pi}$

and so $r=\sqrt[3]{\frac{3V}{4\pi}}$ (d) $m=\frac{F}{a}$

5. $n=(l-a)/d+1$
6. $\sqrt{x}=(m-n)/t$ and so $x=[(m-n)/t]^2$
7. (a) $x^2=1-y$ and so $x=\pm\sqrt{1-y}$

(b) $1-x^2=1/y$ and so $x^2=1-1/y$, finally, $x=\pm\sqrt{1-1/y}$

(c) Write $(1+x^2)y=1-x^2$, remove brackets to get $y+x^2y=1-x^2$, rearrange and factorise to give $x^2(y+1)=1-y$ from which $x^2=(1-y)/(1+y)$, finally $x=\pm\sqrt{(1-y)/(1+y)}$.

Chapter 11

Self-assessment questions 11.1

1. A root is a solution of an equation.
2. A linear equation can be written in the form $ax+b=0$, where a and b are constants. In a linear equation, the variable is always to the first power only.
3. Adding or subtracting the same quantity from both sides; multiplying or dividing both sides by the same quantity; performing the same operation on both sides of the equation.
4. Equation: The two sides of the equation are equal only for certain values of the variable involved. These values are the solutions of the equation. For all other values, the two sides are not equal. Formula: A formula is essentially a statement of how two, or more, variables relate to one another, as for example in $A=\pi r^2$. The area of a circle, A, is always πr^2, for any value of the radius r.

Exercise 11.1

2. (a) $x=\frac{9}{3}=3$ (b) $x=3\times 9=27$

(c) $3t=-6$, so $t=\frac{-6}{3}=-2$ (d) $x=22$

(e) $3x = 4$ and so $x = \frac{4}{3}$ (f) $x = 36$

(g) $5x = 3$ and so $x = \frac{3}{5}$ (h) $x + 3 = 6$ and so $x = 3$

(i) Adding the two terms on the left we obtain

$$\frac{3x+2}{2} + 3x = \frac{3x+2}{2} + \frac{6x}{2}$$

$$= \frac{3x+2+6x}{2} = \frac{9x+2}{2}$$

and the equation becomes $(9x + 2)/2 = 1$, therefore $9x + 2 = 2$, that is $9x = 0$ and so $x = 0$.

3. (a) $5x + 10 = 13$, so $5x = 3$ and so $x = \frac{3}{5}$
   (b) $3x - 21 = 2x + 2$ and so $x = 23$
   (c) $5 - 10x = 8 - 4x$ so that $-3 = 6x$ finally $x = \frac{-3}{6} = -\frac{1}{2}$
4. (a) $t = 9$ (b) $v = \frac{17}{7}$
   (c) $3s + 2 = 14s - 14$, so that $11s = 16$ and so $s = \frac{16}{11}$
5. (a) $t = 3$ (b) $t = 5$ (c) $t = -5$ (d) $t = 3$
   (e) $x = \dfrac{12}{5}$ (f) $x = -9$ (g) $x = 24$
   (h) $x = 15$ (i) $x = 23$ (j) $x = -\dfrac{43}{19}$
6. (a) $x = \dfrac{1}{5}$ (b) $x = \dfrac{2}{5}$ (c) $x = -\dfrac{3}{5}$
   (d) $x = 1\dfrac{2}{5}$ (e) $x = -1$ (f) $x = -3/5$
   (g) $x = -1/2$ (h) $x = -1/10$

Exercise 11.2

2. (a) Adding the two equations gives

$$\begin{aligned} 3x + y &= 1 \\ 2x - y &= 2 \\ \hline 5x \quad &= 3 \end{aligned}$$

from which $x = \frac{3}{5}$. Substitution into either equation gives $y = -\frac{4}{5}$.

(b) Subtracting the given equations eliminates y to give $x = 4$. From either equation $y = 1$.

(c) Multiplying the second equation by 2 and subtracting the first gives

$$\begin{aligned} 2x + 6y &= 24 \\ 2x - y \ &= 17 \quad - \\ \hline 7y &= 7 \end{aligned}$$

from which $y = 1$. Substitute into either equation to get $x = 9$.

(d) Multiplying the second equation by -2 and subtracting from the first gives

$$\begin{aligned} -2x + y \ &= -21 \\ -2x - 6y &= \ 28 \quad - \\ \hline 7y &= -49 \end{aligned}$$

from which $y = -7$. Substitution into either equation gives $x = 7$.

(e) Multiplying the first equation by 3 and adding the second gives

$$\begin{aligned} -3x + 3y &= -30 \\ 3x + 7y &= \ 20 \quad + \\ \hline 10y &= -10 \end{aligned}$$

from which $y = -1$. Substitution into either equation gives $x = 9$.

(f) Multiplying the second equation by 2 and subtracting this from the first gives

$$\begin{aligned} 4x - 2y &= \ 2 \\ 6x - 2y &= \ 8 \quad - \\ \hline -2x \quad &= -6 \end{aligned}$$

from which $x = 3$. Substitution into either equation gives $y = 5$.

3. (a) $x = 7, y = 1$ (b) $x = -7, y = 2$
   (c) $x = 10, y = 0$ (d) $x = 0, y = 5$
   (e) $x = -1, y = -1$

Self-assessment questions 11.3

1. If $b^2 - 4ac > 0$ a quadratic equation will have distinct real roots.
2. If $b^2 - 4ac = 0$ a quadratic equation will have a repeated root.
3. If $b^2 - 4ac < 0$ the quadratic equation has complex roots. When this case arises we are faced with finding the square root of a negative number. This is handled by introducing the symbol i to stand for the square root of -1.

Exercise 11.3

1. (a) $(x+2)(x-1)=0$ so that $x=-2$ and 1
   (b) $(x-3)(x-5)=0$ so that $x=3$ and 5
   (c) $2(2x+1)(x+1)=0$ so that $x=-1$ and $x=-\frac{1}{2}$
   (d) $(x-3)(x-3)=0$ so that $x=3$ twice
   (e) $(x+9)(x-9)=0$ so that $x=-9$ and 9
   (f) $-1, -3$ (g) $1, -3$ (h) $1, -4$
   (i) $-1, -5$ (j) $5, 7$ (k) $-5, -7$
   (l) $1, -\frac{3}{2}$ (m) $-\frac{3}{2}, 2$ (n) $-\frac{3}{2}, 5$
   (o) $1, -1/3$ (p) $-1/3, 5/3$ (q) $0, -\frac{1}{7}$
   (r) $-\frac{3}{2}$ twice

2. (a)
$$x = \frac{-(-6) \pm \sqrt{(-6)^2 - 4(3)(-5)}}{6} = \frac{6 \pm \sqrt{96}}{6}$$
$$= 2.633 \text{ and } -0.633$$

   (b)
$$x = \frac{-3 \pm \sqrt{9 - 4(1)(-77)}}{2} = \frac{-3 \pm \sqrt{317}}{2}$$
$$= 7.402 \text{ and } -10.402$$

   (c)
$$x = \frac{-(-9) \pm \sqrt{(-9)^2 - 4(2)(2)}}{4} = \frac{9 \pm \sqrt{65}}{4}$$
$$= 4.266 \text{ and } 0.234$$

   (d) $1, -4$ (e) $1.758, -0.758$
   (f) $0.390, -0.640$ (g) $7.405, -0.405$
   (h) $0.405, -7.405$
   (i) The roots are complex numbers:
$$x = \frac{-1 \pm \sqrt{1^2 - 4(11)(1)}}{22} = \frac{-1 \pm \sqrt{-43}}{22} = \frac{-1 \pm \sqrt{43}\mathrm{i}}{22}$$
   (j) $2.766, -1.266$

3. (a) $(3x+2)(2x+3)=0$
   from which $x=-2/3$ and $-3/2$
   (b) $(t+3)(3t+4)=0$
   from which $t=-3$ and $t=-4/3$
   (c) $t=(7 \pm \sqrt{37})/2 = 6.541$ and 0.459

Chapter 16

Self-assessment questions 16.2

1. A function is a rule that receives an input and produces a single output.
2. The input to a function is called the independent variable; the output is called the dependent variable.
3. False. For example, if $f(x) = 5x$, then $f(\frac{1}{x}) = 5 \times \frac{1}{x} = \frac{5}{x}$ whereas $\frac{1}{f(x)} = \frac{1}{5x}$.
4. $f(x) = x^2 - 5x$

Exercise 16.2

1. (a) Multiply the input by 10.
   (b) Multiply the input by -1 and then add 2; alternatively we could say subtract the input from 2.
   (c) Raise the input to the power 4 and then multiply the result by 3.
   (d) Divide 4 by the square of the input.
   (e) Take three times the square of the input, subtract twice the input from the result and finally add 9.
   (f) The output is always 5 whatever the value of the input.
   (g) The output is always zero whatever the value of the input.
2. (a) Take three times the square of the input and add to twice the input; (b) take three times the square of the input and add to twice the input. The functions in parts (a) and (b) are the same. Both instruct us to do the same thing.
3. (a) $f(x) = x^3/12$ – letters other than f and x are also valid.
   (b) $f(x) = (x + 3)^2$
   (c) $f(x) = x^2 + 4x - 10$
   (d) $f(x) = x/(x^2 + 5)$ (e) $f(x) = x^3 - 1$
   (f) $f(x) = (x - 1)^2$ (g) $f(x) = (7 - 2x)/4$
   (h) $f(x) = -13$

4. (a) $A(2) = 3$ (b) $A(3) = 7$ (c) $A(0) = 1$
   (d) $A(-1) = (-1)^2 - (-1) + 1 = 3$
5. (a) $y(1) = 1$
   (b) $y(-1) = [2(-1) - 1]^2 = (-3)^2 = 9$
   (c) $y(-3) = 49$ (d) $y(0.5) = 0$
   (e) $y(-0.5) = 4$
6. (a) $f(t+1) = 4(t+1) + 6 = 4t + 10$
   (b) $f(t+2) = 4(t+2) + 6 = 4t + 14$
   (c) $f(t+1) - f(t) = (4t + 10) - (4t + 6)$
   $= 4$
   (d) $f(t+2) - f(t) = (4t + 14) - (4t + 6)$
   $= 8$
7. (a) $f(n) = 2n^2 - 3$ (b) $f(z) = 2z^2 - 3$
   (c) $f(t) = 2t^2 - 3$
   (d) $f(2t) = 2(2t)^2 - 3 = 8t^2 - 3$
   (e) $f(1/z) = 2(1/z)^2 - 3 = 2/z^2 - 3$
   (f) $f(3/n) = 2(3/n)^2 - 3 = 18/n^2 - 3$
   (g) $f(-x) = 2(-x)^2 - 3 = 2x^2 - 3$
   (h) $f(-4x) = 2(-4x)^2 - 3 = 32x^2 - 3$
   (i) $f(x+1) = 2(x+1)^2 - 3$
   $= 2(x^2 + 2x + 1) - 3 = 2x^2 + 4x - 1$
   (j) $f(2x-1) = 2(2x-1)^2 - 3$
   $= 8x^2 - 8x - 1$
8. $a(p+1) = (p+1)^2 + 3(p+1) + 1$
   $= p^2 + 2p + 1 + 3p + 3 + 1 = p^2 + 5p + 5$
   $a(p+1) - a(p) = (p^2 + 5p + 5)$
   $-(p^2 + 3p + 1) = 2p + 4$
9. (a) $f(3) = 6$ (b) $h(2) = 3$
   (c) $f[h(2)] = f(3) = 6$
   (d) $h[f(3)] = h(6) = 7$
10. (a) $f[h(t)] = f(t+1) = 2(t+1) = 2t + 2$
    (b) $h[f(t)] = h(2t) = 2t + 1$. Note that $h[f(t)]$ is not equal to $f[h(t)]$.
11. (a) $f(0.5) = 0.5$ (first part)
    (b) $f(1.1) = 1$ (third part) (c) $f(1) = 2$ (second part)

Self-assessment questions 16.3

1. Suppose $f(x)$ and $g(x)$ are two functions. If the input to f is $g(x)$ then we have $f(g(x))$. This is a composite function.
2. Suppose $f(x) = x$, $g(x) = \frac{1}{x}$. Then both $f(g(x))$ and $g(f(x))$ equal $\frac{1}{x}$. Note that this is in contrast to the more general case in which $f(g(x))$ is not equal to $g(f(x))$.

Exercise 16.3

1. (a) $f(g(x)) = f(3x - 2) = 4(3x - 2)$
   $= 12x - 8$
   (b) $g(f(x)) = g(4x) = 3(4x) - 2 = 12x - 2$
2. (a) $y(x(t)) = y(t^3) = 2t^3$
   (b) $x(y(t)) = x(2t) = (2t)^3 = 8t^3$
3. (a) $r(s(x)) = r(3x) = \frac{1}{6x}$
   (b) $t(s(x)) = t(3x) = 3x - 2$
   (c) $t(r(s(x))) = t(\frac{1}{6x}) = \frac{1}{6x} - 2$
   (d) $r(t(s(x))) = r(3x - 2) = \frac{1}{2(3x-2)}$
   (e) $r(s(t(x))) = r(s(x - 2)) = r(3(x - 2))$
   $= \frac{1}{6(x-2)}$
4. (a) $v(v(t)) = v(2t + 1) = 2(2t + 1) + 1$
   $= 4t + 3$
   (b) $v(v(v(t))) = v(4t + 3) = 2(4t + 3) + 1$
   $= 8t + 7$
5. (a) $m(n(t)) = m(t^2 - 1) = (t^2)^3 = t^6$
   (b) $n(m(t)) = n((t+1)^3) = (t+1)^6 - 1$
   (c) $m(p(t)) = m(t^2) = (t^2 + 1)^3$
   (d) $p(m(t)) = p((t+1)^3) = (t+1)^6$
   (e) $n(p(t)) = n(t^2) = t^4 - 1$
   (f) $p(n(t)) = p(t^2 - 1) = (t^2 - 1)^2$
   (g) $m(n(p(t))) = m(t^4 - 1) = (t^4)^3 = t^{12}$
   (h) $p(p(t)) = p(t^2) = t^4$
   (i) $n(n(t)) = n(t^2 - 1) = (t^2 - 1)^2 - 1$
   (j) $m(m(t)) = m((t+1)^3) = [(t+1)^3 + 1]^3$

Self-assessment questions 16.4

1. The inverse of a function, say $f(x)$, is another function, denoted by $f^{-1}(x)$. When the input to f^{-1} is $f(x)$, the output is x, that is

   $$f^{-1}(f(x)) = x$$

   Thus f^{-1} reverses the process in f.
2. The function $f(x) = 4x^4$ produces the same output for various inputs, for example $f(1) = 4$, $f(-1) = 4$. If an inverse of f existed, then when the input is 4, there would have to be outputs of both 1 and -1. This cannot be the case, as a function must produce a unique output for every input.

Exercise 16.4

1. In each case let the inverse be denoted by g
   (a) $g(x) = x/3$ (b) $g(x) = 4x$
   (c) $g(x) = x - 1$ (d) $g(x) = x + 3$
   (e) $g(x) = 3 - x$ (f) $g(x) = (x - 6)/2$
   (g) $g(x) = (7 - x)/3$ (h) $g(x) = 1/x$
   (i) $g(x) = 3/x$ (j) $g(x) = -3/4x$
2. (a) $f^{-1}(x) = \frac{x}{6}$
   (b) $f^{-1}(x) = \frac{x-1}{6}$
   (c) $f^{-1}(x) = x - 6$
   (d) $f^{-1}(x) = 6x$
   (e) $f^{-1}(x) = \frac{6}{x}$
3. (a) $g^{-1}(t) = \frac{t-1}{3}$

   (b) We require $g^{-1}\left(\frac{1}{3t+1}\right) = t$

   Let $z = \dfrac{1}{3t+1}$

   so that $t = \dfrac{1}{3}\left(\dfrac{1}{z} - 1\right)$

   Then $g^{-1}(z) = \dfrac{1}{3}\left(\dfrac{1}{z} - 1\right)$

   so $g^{-1}(t) = \dfrac{1}{3}\left(\dfrac{1}{t} - 1\right)$

   (c) $g^{-1}(t) = t^{\frac{1}{3}}$
   (d) $g^{-1}(t) = (\frac{t}{3})^{\frac{1}{3}}$
   (e) $g^{-1}(t) = (\frac{t-1}{3})^{\frac{1}{3}}$

   (f) We require $g^{-1}\left(\frac{3}{t^3+1}\right) = t$

   Let $z = \dfrac{3}{t^3+1}$

   so that $t = \left(\dfrac{3}{z} - 1\right)^{\frac{1}{3}}$

   Then $g^{-1}(z) = \left(\dfrac{3}{z} - 1\right)^{\frac{1}{3}}$

   so $g^{-1}(t) = \left(\dfrac{3}{t} - 1\right)^{\frac{1}{3}}$

4. (a) $h^{-1}(t) = \frac{t-3}{4}$ (b) $g^{-1}(t) = \frac{t+1}{2}$

   (c) $g^{-1}(h^{-1}(t)) = g^{-1}\left(\dfrac{t-3}{4}\right)$

   $$= \frac{\frac{t-3}{4} + 1}{2}$$

   $$= \frac{t+1}{8}$$

   (d) $h(g(t)) = h(2t - 1) = 4(2t - 1) + 3$
   $= 8t - 1$
   (e) $h(g(t))^{-1} = \frac{t+1}{8}$. We note that $[h(g(t))]^{-1}$ is the same as $g^{-1}(h^{-1}(t))$.

Chapter 17

Self-assessment questions 17.1

1. The horizontal axis is a straight line with a scale marked on it. It represents the independent variable. The vertical axis is also a straight line with a scale marked on it; this scale may be different from that which is marked on the horizontal axis. These two axes are at 90° to each other. The point where they intersect is the origin.
2. x coordinate
3. (0, 0)

Exercise 17.1

1.

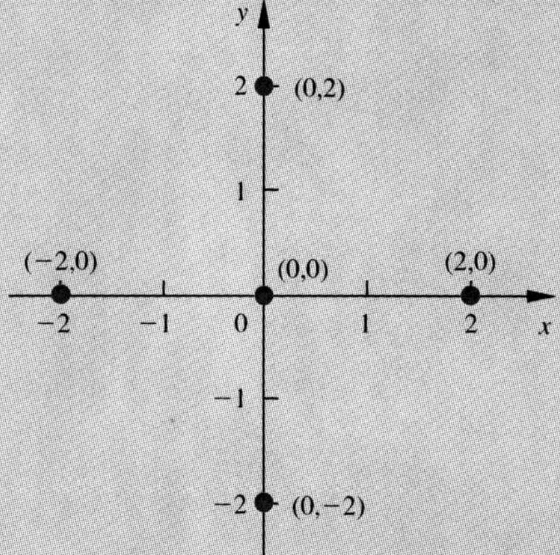

2. All y coordinates must be the same.
3. All x coordinates must be the same.

Self-assessment questions 17.2

1. (a) A closed interval is one that includes the two end-points.
   (b) An open interval is one that does not include the end-points.
   (c) A semi-closed interval is closed at one end and open at the other: that is, one end-point is included, the other is not.
2. (a) Square brackets, [ ], denote that the end-point is included. On a graph, a filled bullet, ●, is used to indicate this.
   (b) Round brackets, ( ), denote that the end-point is not included. On a graph, an open bullet, ○, is used to indicate this.

Exercise 17.2

1. (a) $\{x : x \in \mathbb{R}, 2 \leqslant x \leqslant 6\}$

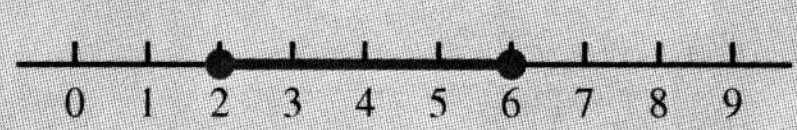

The interval [2, 6]

(b) $\{x : x \in \mathbb{R}, 6 < x \leqslant 8\}$

0 1 2 3 4 5 6 7 8 9

The interval (6, 8]

(c) $\{x : x \in \mathbb{R}, -2 < x < 0\}$

−3 −2 −1 0 1

The interval (−2, 0)

(d) $\{x : x \in \mathbb{R}, -3 \leqslant x < -1.5\}$

−4 −3 −2 −1 0 1

The interval [−3, −1.5)

2. (a) T (b) T (c) T (d) T (e) F (f) T (g) F (h) T

Self-assessment questions 17.3

1. Dependent
2. Horizontal axis

Exercise 17.3

1.

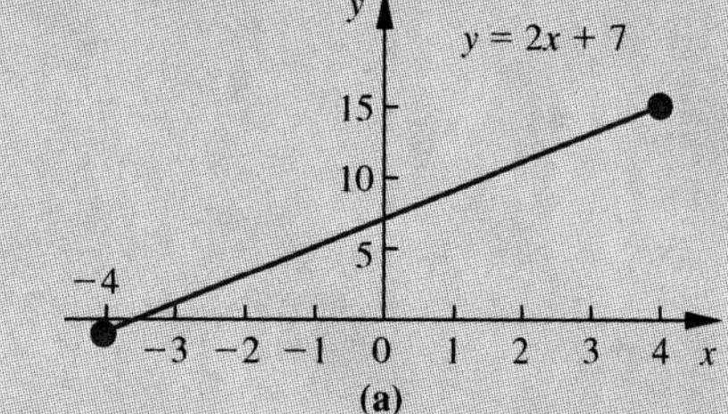

(a)

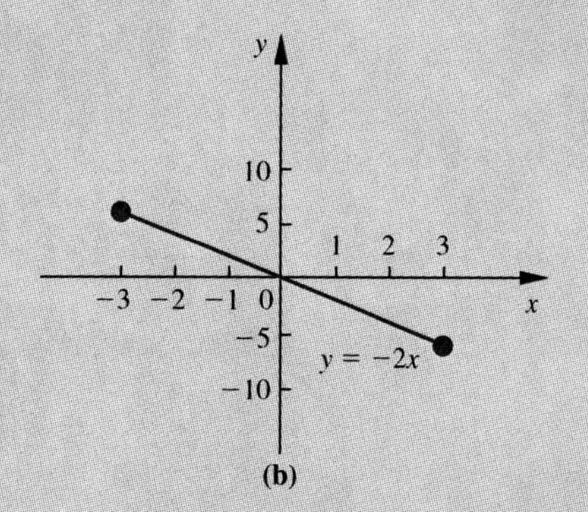

(b)

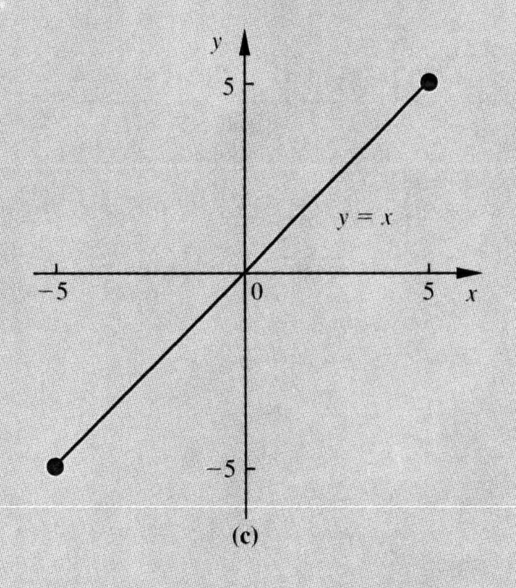

(c)

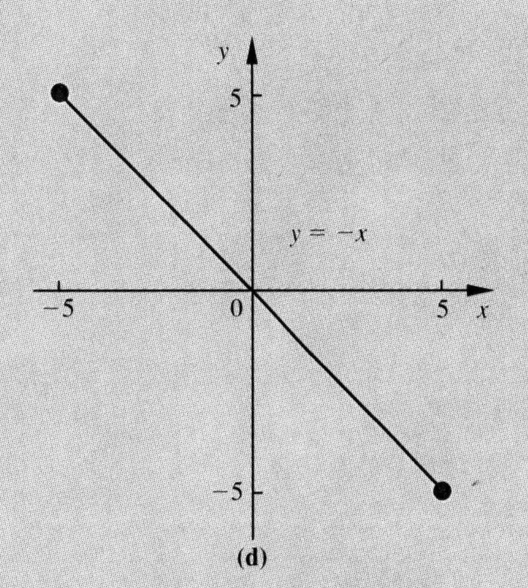

(d)

2.

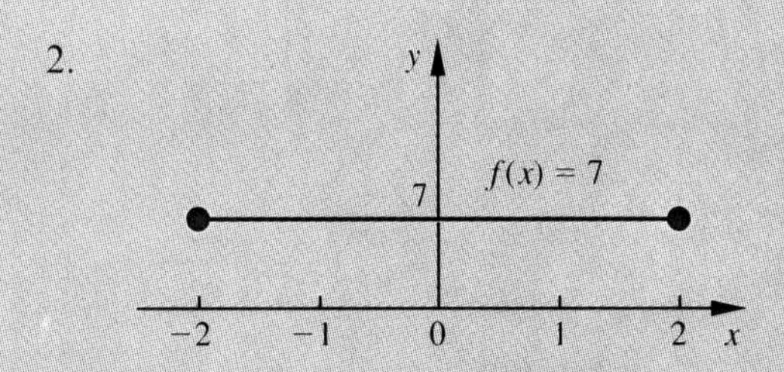

3. (a)

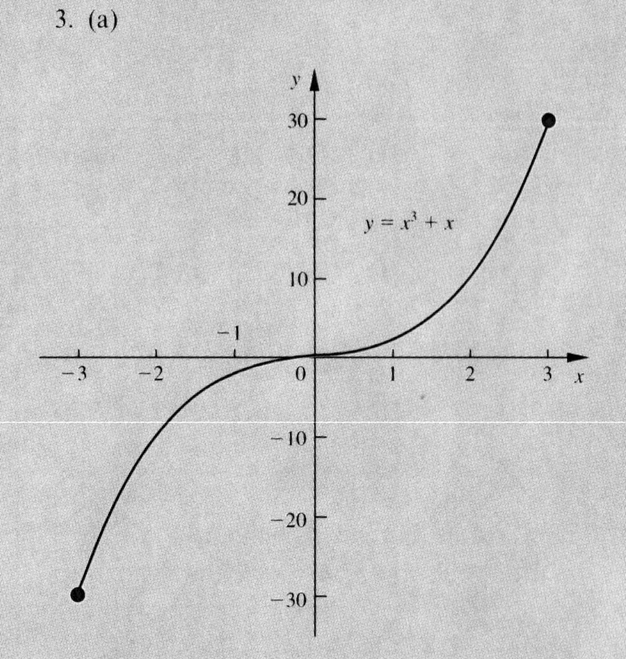

(b) (1, 2) and (−1, −2) lie on the curve.

4.

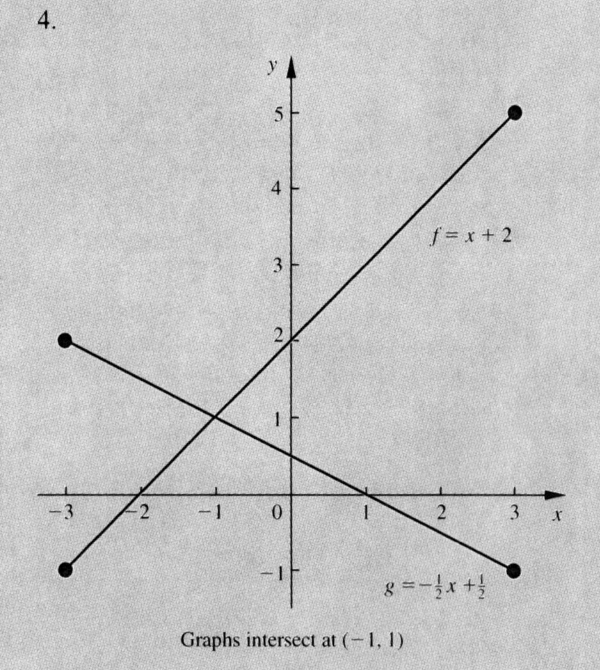

Graphs intersect at (−1, 1)

5.

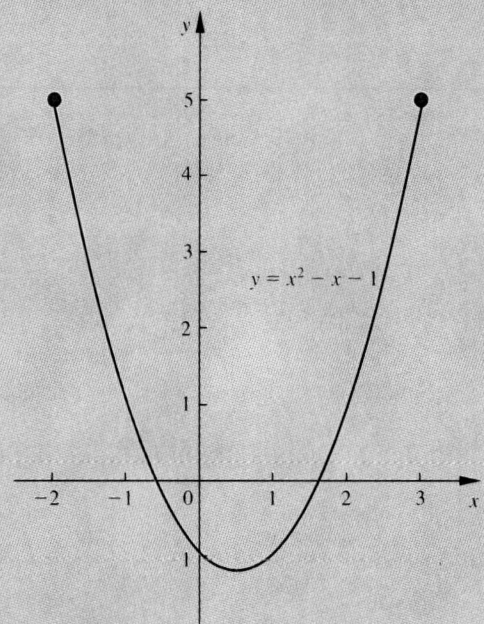

(a) The curve cuts the horizontal axis at $x = -0.6$ and $x = 1.6$.
(b) The curve cuts the vertical axis at $y = -1$.

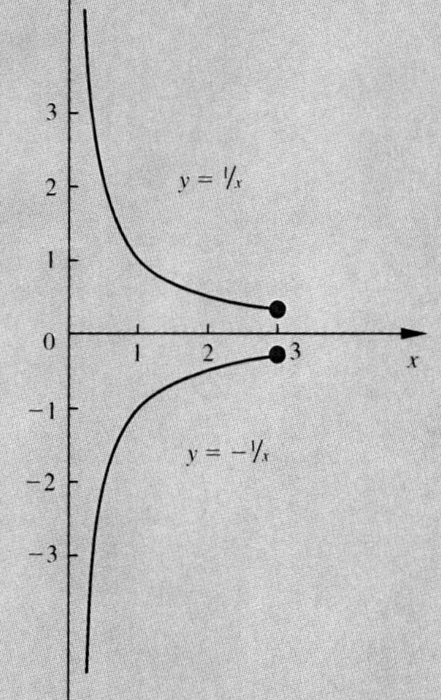

7.

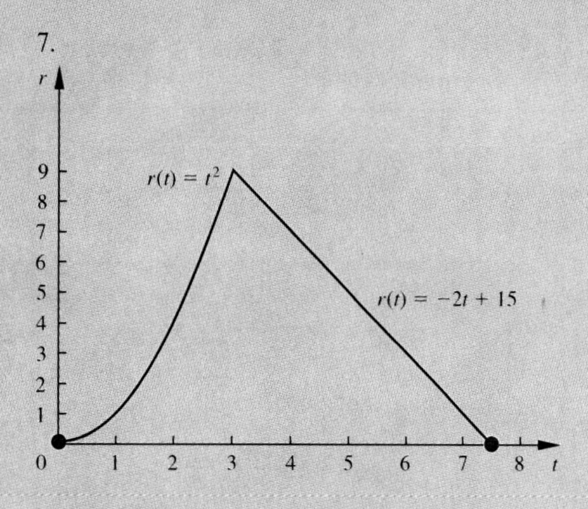

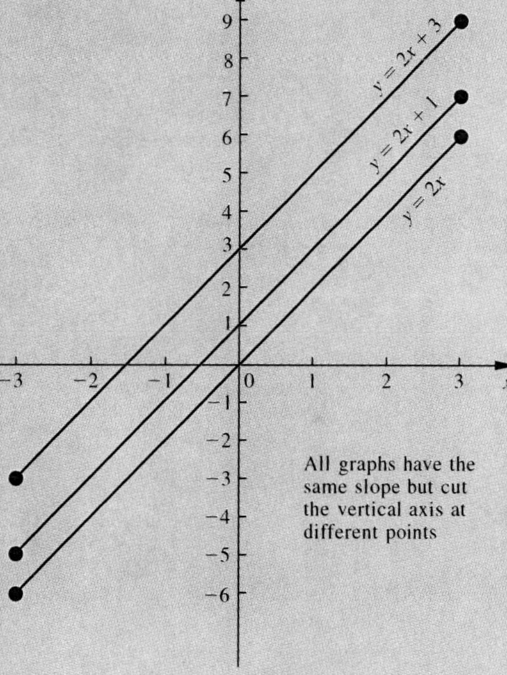

Self-assessment questions 17.4

1. The domain is the set of values that the independent variable can take. The range is the set of values that the dependent variable can take.
2. When $x = 3$, the denominator of the fraction is 0. Division by 0 is not defined and so the value $x = 3$ must be excluded.

3. $f(x) = \frac{1}{x+2}$. When $x = -2$, the denominator is zero and division by zero is not defined.

Exercise 17.4

1.

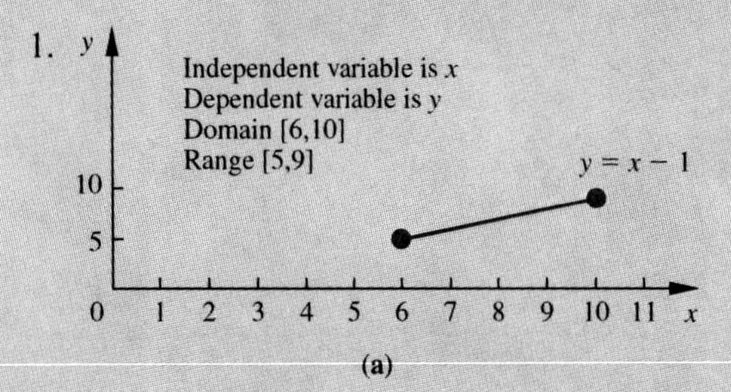

(a)

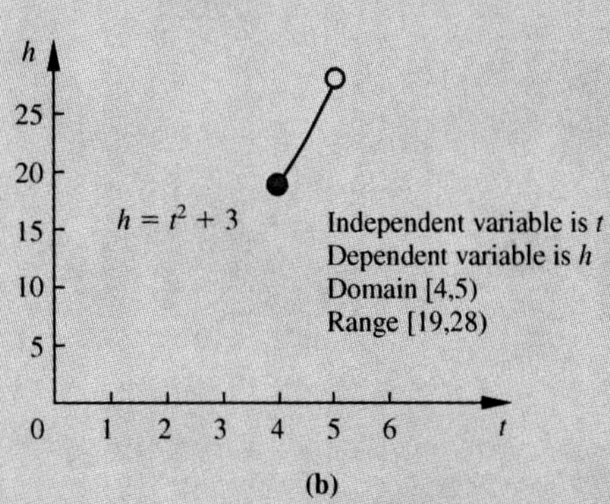

(b)

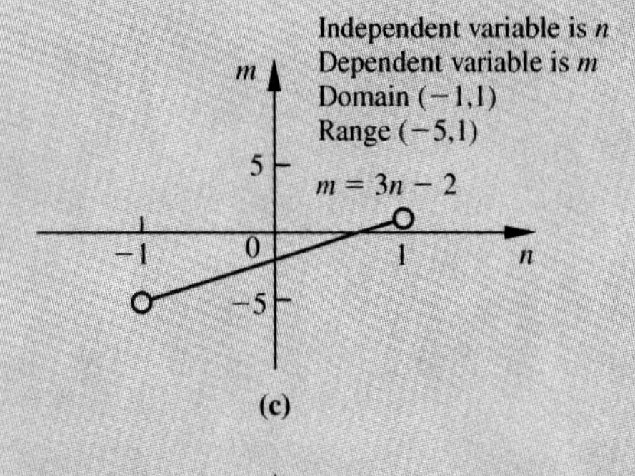

(c)

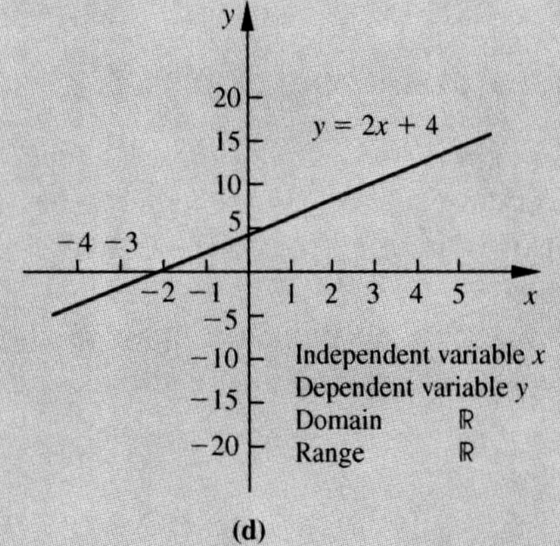

(d)

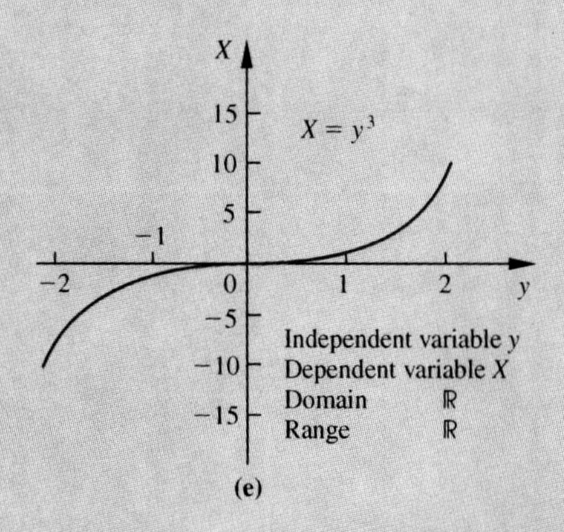

(e)

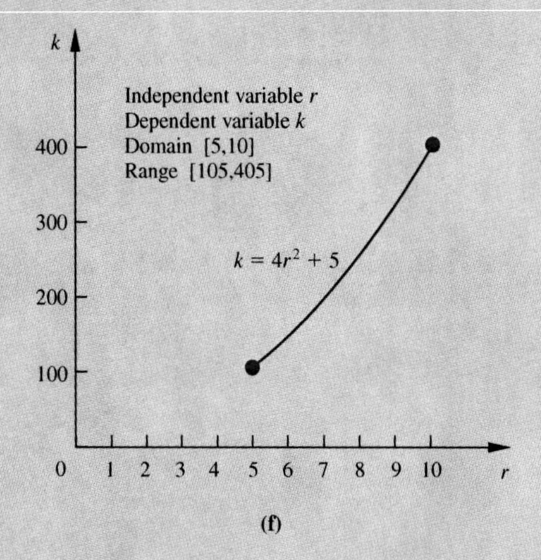

(f)

2.

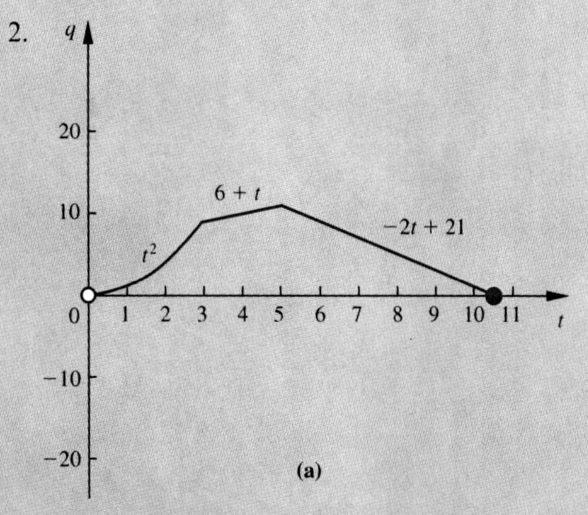

(a)

(b) Domain $(0, 10.5]$ (c) range $= [0, 11]$
(d) $q(1) = 1$, $q(3) = 9$, $q(5) = 11$, $q(7) = 7$

Self-assessment question 17.5

1. (a) The intersection of $f(x)$ with the x axis gives the solutions to $f(x) = 0$.
   (b) The intersection of $f(x)$ with the straight line $y = 1$ gives the solutions to $f(x) = 1$.

Exercise 17.5

1.

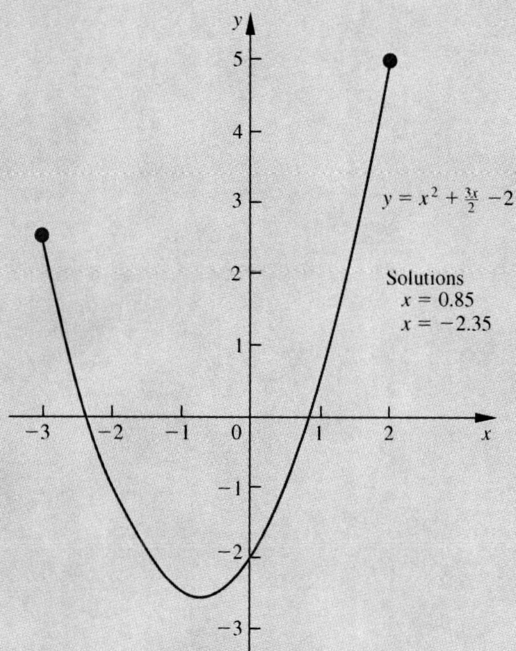

2.

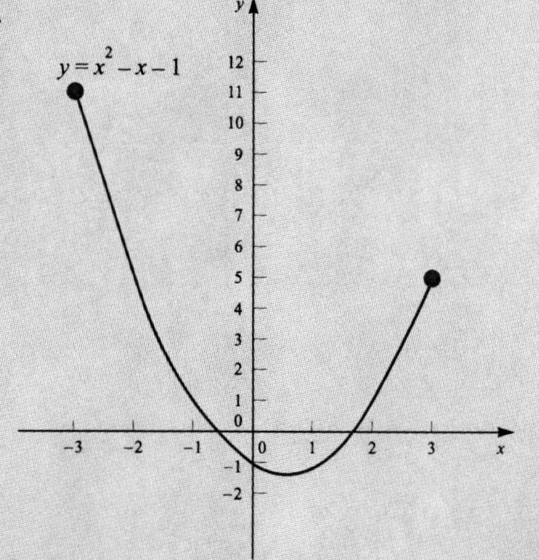

From the graph the roots of $x^2 - x - 1 = 0$ are $x = -0.6$ and $x = 1.6$.

3.

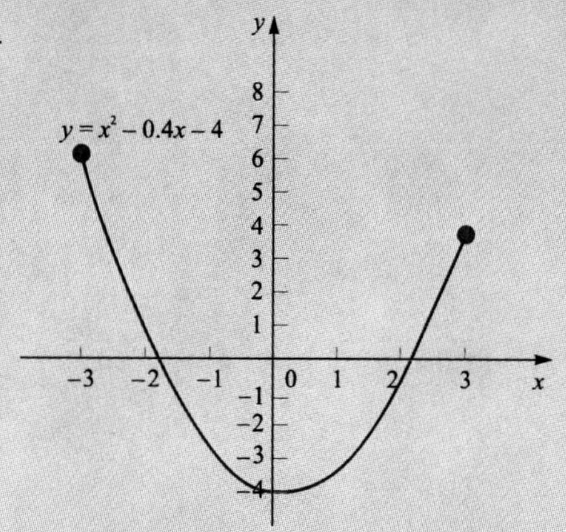

The required roots are $x = -1.8$ and $x = 2.2$.

4. (a)

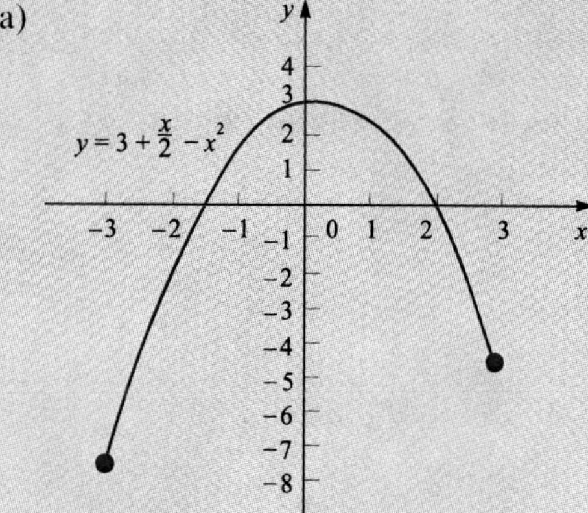

(b) From the graph the roots of $x^2 - \frac{x}{2} - 3 = 0$ are $x = -1.5$ and $x = 2$.

Self-assessment questions 17.6

1. Points of intersection provide the solutions to simultaneous equations.
2. Three.
3. There are no solutions.

Exercise 17.6

1. (a) We write the equations in the form

$$y = -\frac{3}{2}x + 2, \qquad y = x - 3$$

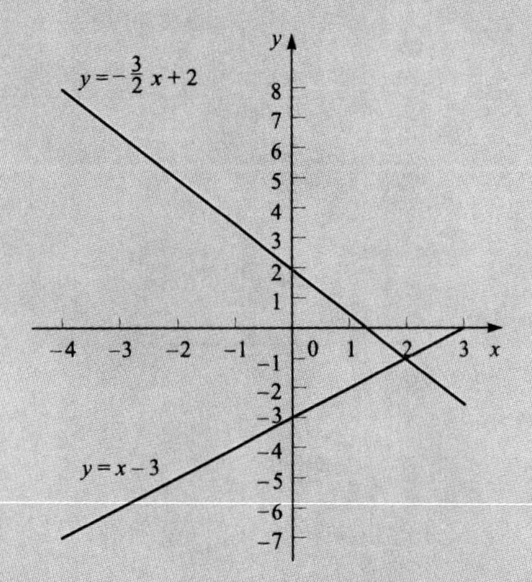

The point of intersection is $x=2, y=-1$, which is the solution to the simultaneous equations.

(b) The equations are written as

$$y=-2x-2, \qquad y=\frac{x}{4}+\frac{5}{2}$$

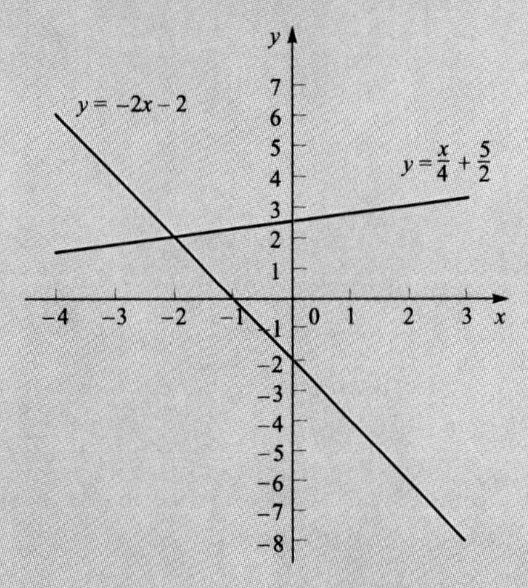

From the graph the solution is $x=-2, y=2$.

(c) The equations are written in the form

$$y=-2x+4, \qquad y=\frac{4}{3}x+\frac{7}{3}$$

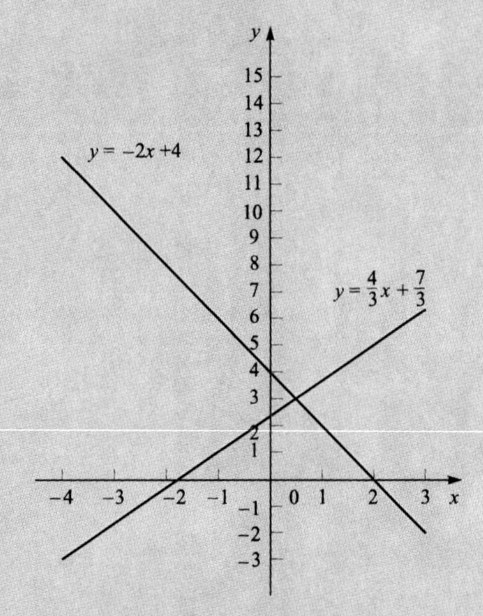

The solution is $x=\frac{1}{2}, y=3$.

(d) The equations are written

$$y=\frac{x}{4}+\frac{7}{4}, \qquad y=2x+7$$

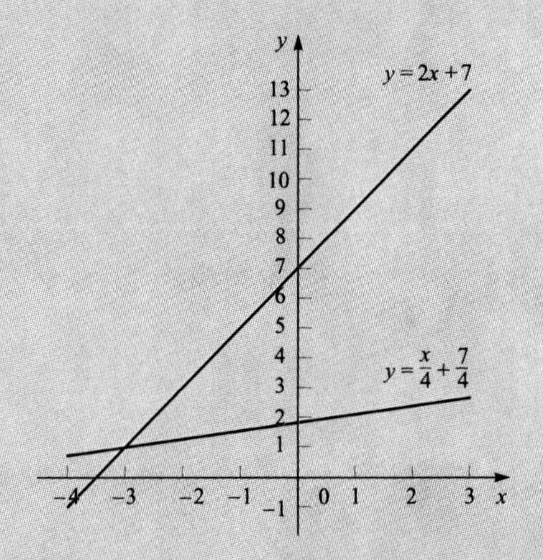

The root is $x=-3, y=1$.

(e) The equations are written

$$y=-\frac{x}{2}+\frac{1}{2}, \qquad y=-\frac{x}{10}-\frac{1}{10}$$

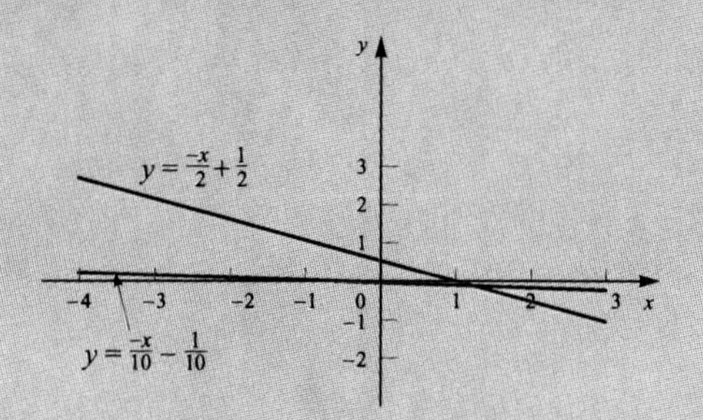

The root is $x = 1.5$, $y = -0.25$.

2.

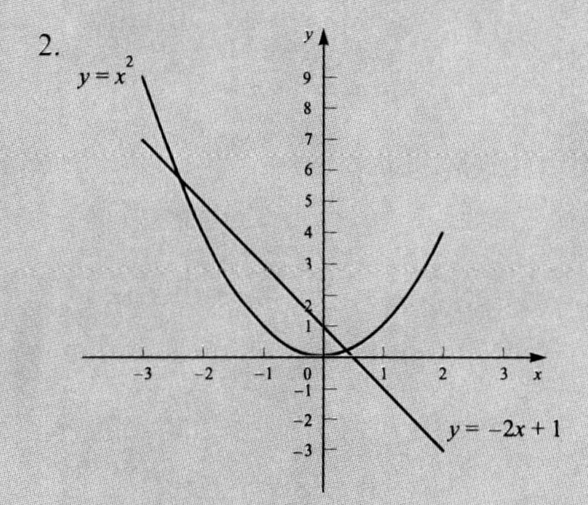

From the graph the required roots are approximately $x = -2.4$, $y = 5.8$ and $x = 0.4$, $y = 0.2$.

3.

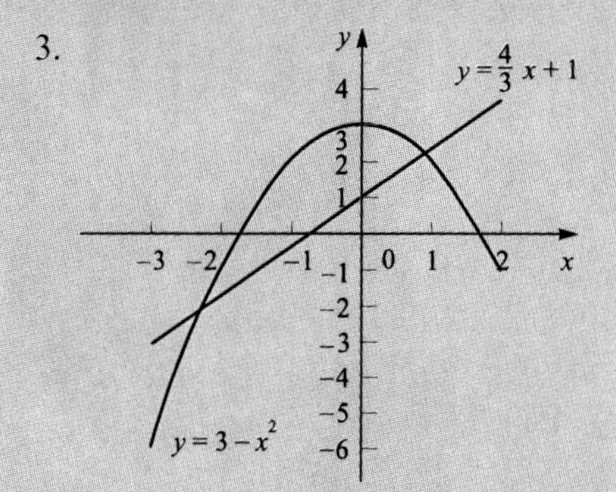

The roots of the simultaneous equations are approximately $x = -2.2$, $y = -2$ and $x = 0.9$, $y = 2.2$.

4.

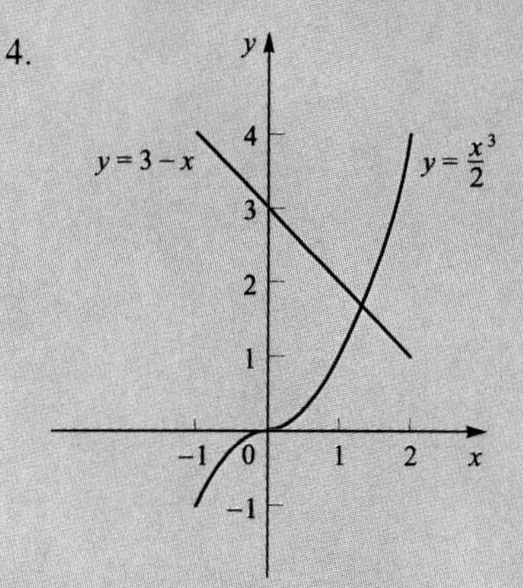

From the point of intersection we see the required root is approximately $x = 1.4$, $y = 1.5$.

5.

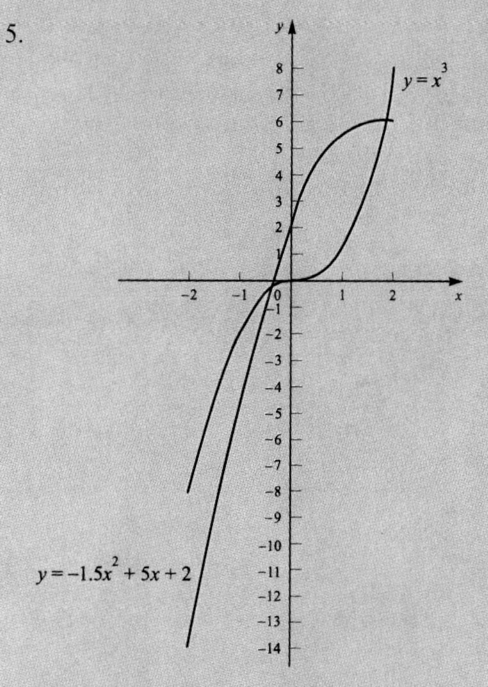

The points of intersection, and hence the required roots, are approximately $x = -0.4$, $y = -0.2$ and $x = 1.8$, $y = 6.1$.

Chapter 18

Self-assessment questions 18.1

1. $y = mx + c$
2. m is the gradient of the line. c is the intercept of the line with the vertical axis.

Exercise 18.1

1. (a), (b), (c) and (e) are straight lines.
2. (a) $m = 9$, $c = -11$ (b) $m = 8$, $c = 1.4$
   (c) $m = \frac{1}{2}$, $c = -11$ (d) $m = -2$, $c = 17$
   (e) $m = \frac{2}{3}$, $c = \frac{1}{3}$ (f) $m = -\frac{2}{5}$, $c = \frac{4}{5}$
   (g) $m = 3$, $c = -3$ (h) $m = 0$, $c = 4$
3. (a) (i) -1 (ii) 6 (b) (i) 2 (ii) -1
   (c) (i) 2 (ii) -1.5 (d) (i) $\frac{3}{4}$ (ii) 3
   (e) (i) -6 (ii) 18

Self-assessment questions 18.2

1. gradient $= m =$

$$\frac{\text{difference in } y \text{ (vertical) coordinates}}{\text{difference in } x \text{ (horizontal) coordinates}}$$

$$m = \frac{y_2 - y_1}{x_2 - x_1}$$

2. The value of c is given by the intersection of the straight line graph with the vertical axis.

Exercise 18.2

1. $y = 2x + 5$

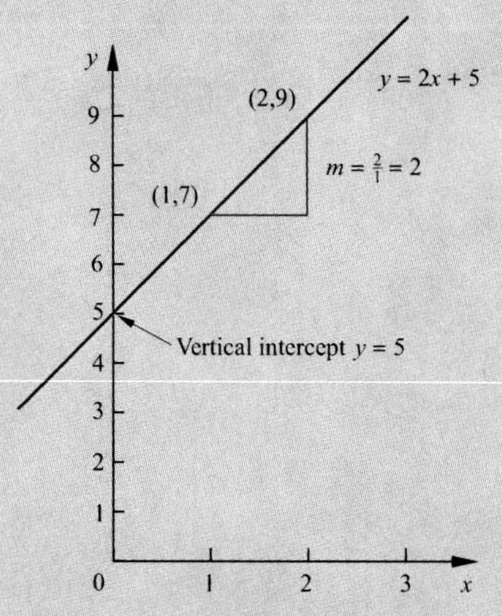

2. $y = 6x - 10$
3. $y = 2$
4. $y = x$
5. $y = -x$
6. $y = 2x + 8$
7. (b), (c) and (d) lie on the line.
8. The equation of the line is $y = mx + c$. The gradient is $m = -1$ and so

   $$y = -x + c$$

   When $x = -3$, $y = 7$ and so $c = 4$. Hence the required equation is $y = -x + 4$.
9. The gradient of $y = 3x + 17$ is 3. Let the required equation be $y = mx + c$.
   Since the lines are parallel then $m = 3$ and so $y = 3x + c$. When $x = -1$, $y = -6$ and so $c = -3$. Hence the equation is $y = 3x - 3$.
10. The equation is $y = mx + c$.
    The vertical intercept is $c = -2$ and so

    $$y = mx - 2$$

    When $x = 3$, $y = 10$ and so $m = 4$. Hence the required equation is $y = 4x - 2$.

Self-assessment questions 18.3

1. A tangent is a straight line that touches a curve at a single point.
2. The gradient of a curve at a particular point is the gradient of the tangent at that point.

Exercise 18.3

1.

$y = 2x^2 - 1$

Tangent at A(2,7)

A(2,7)

Tangent at (−1,1)

Gradient = −4

Gradient of tangent at (2,7) is $\frac{8}{1} = 8$

2.

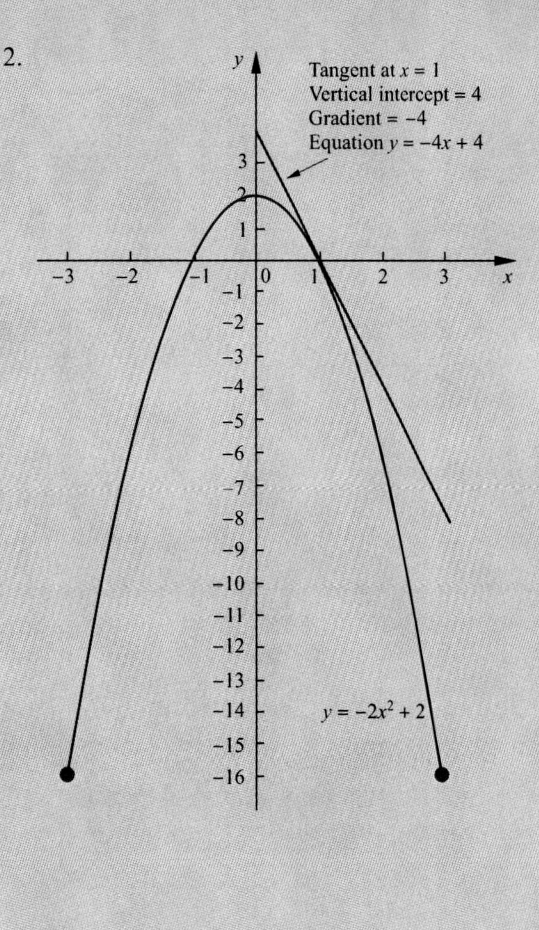

Chapter 19

Exercise 19.1

1. (a) 9.9742 (b) 6.6859 (c) 0.5488
   (d) 0.8808 (e) −8.4901
2. (a) $e^2.e^7 = e^{2+7} = e^9$ (b) $\frac{e^7}{e^4} = e^{7-4} = e^3$
   (c) $\frac{e^{-2}}{e^3} = e^{-2-3} = e^{-5}$ or $\frac{1}{e^5}$
   (d) $\frac{e^3}{e^{-1}} = e^{3-(-1)} = e^4$
   (e) $\frac{(4e)^2}{(2e)^3} = \frac{4^2e^2}{2^3e^3} = \frac{16e^{2-3}}{8} = 2e^{-1} = \frac{2}{e}$
   (f) $e^2(e + \frac{1}{e}) - e = e^2e + \frac{e^2}{e} - e$
   $= e^3 + e - e = e^3$
   (g) $e^{1.5}e^{2.7} = e^{1.5+2.7} = e^{4.2}$
   (h) $\sqrt{e}\sqrt{4e} = e^{0.5}(4e)^{0.5} = e^{0.5}4^{0.5}e^{0.5}$
   $= 2e^1 = 2e$
3. (a) $e^xe^{-3x} = e^{x-3x} = e^{-2x}$
   (b) $e^{3x}e^{-x} = e^{3x-x} = e^{2x}$
   (c) $e^{-3x}e^{-x} = e^{-3x-x} = e^{-4x}$

(d) $(e^{-x})^2 e^{2x} = e^{-2x}e^{2x} = e^0 = 1$

(e) $\frac{e^{3x}}{e^x} = e^{3x-x} = e^{2x}$

(f) $\frac{e^{-3x}}{e^{-x}} = e^{-3x-(-x)} = e^{-2x}$

4. (a) $\frac{e^t}{e^3} = e^{t-3}$ (b) $\frac{2e^{3x}}{4e^x} = \frac{1}{2}e^{3x-x} = \frac{e^{2x}}{2}$

(c) $\frac{(e^x)^3}{3e^x} = \frac{e^{3x}}{3e^x} = \frac{e^{3x-x}}{3} = \frac{e^{2x}}{3}$

(d) $e^x + 2e^{-x} + e^x = 2e^x + 2e^{-x}$
$= 2(e^x + e^{-x})$

(e) $\frac{e^x e^y e^z}{e^{(x/2)-y}} = e^{x+y+z-((x/2)-y)} = e^{(x/2)+2y+z}$

(f) $\frac{(e^{\frac{t}{3}})^6}{(e^t+e^t)} = \frac{e^{(\frac{t}{3})\times 6}}{2e^t} = \frac{e^{2t}}{2e^t} = \frac{e^t}{2}$

(g) $\frac{e^t}{e^{-t}} = e^{t-(-t)} = e^{2t}$

(h) $\frac{1+e^{-t}}{e^{-t}} = (1+e^{-t})e^t = e^t + e^{-t}e^t$
$= e^t + e^0 = e^t + 1$

(i) $(e^z e^{-z/2})^2 = (e^{z-z/2})^2 = (e^{z/2})^2 = e^z$

(j) $\frac{e^{-3+t}}{2e^t} = \frac{e^{-3+t-t}}{2} = \frac{e^{-3}}{2} = \frac{1}{2e^3}$

(k) $(\frac{e^{-1}}{e^{-x}})^{-1} = (e^{-1-(-x)})^{-1} = (e^{x-1})^{-1}$
$= e^{(x-1)(-1)} = e^{1-x}$

Self-assessment questions 19.2

1. The functions are never negative. When $x = 0$ the functions have a value of 1.
2. False; e^{-x} is positive for all values of x.

Exercise 19.2

1.

x	−1	−0.8	−0.6	−0.4	−0.2
e^{3x}	0.0498	0.0907	0.1653	0.3012	0.5488

x	0	0.2	0.4	0.6	0.8	1
e^{3x}	1	1.8221	3.3201	6.0496	11.0232	20.0855

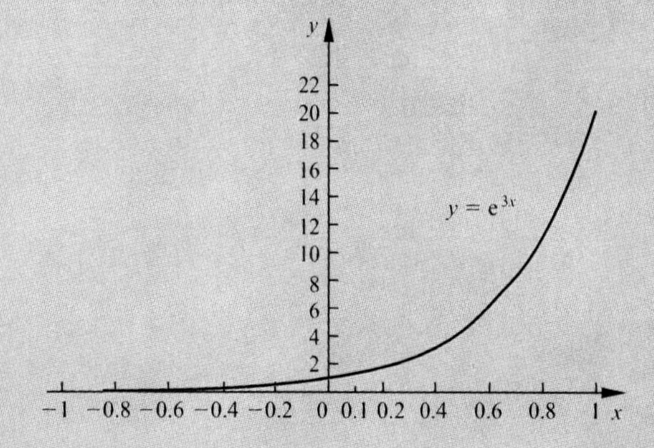

The graph has similar shape and properties to that of $y = e^x$.

2. (a)

t	0	1	2	3	4	5
P	5	8.1606	9.3233	9.7511	9.9084	9.9663

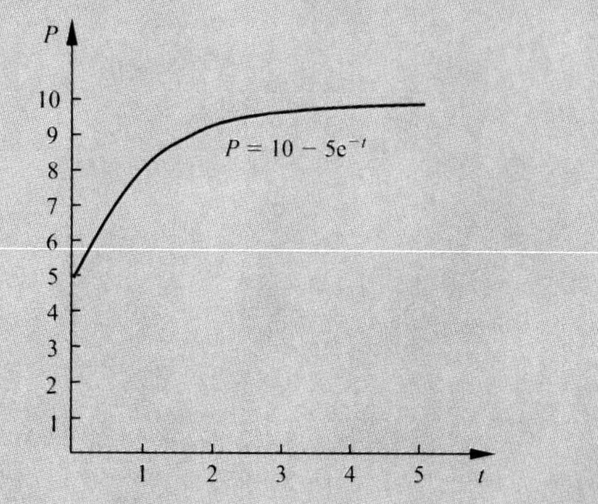

(b) Maximum population is 10, obtained for large values of t.

3. (a)

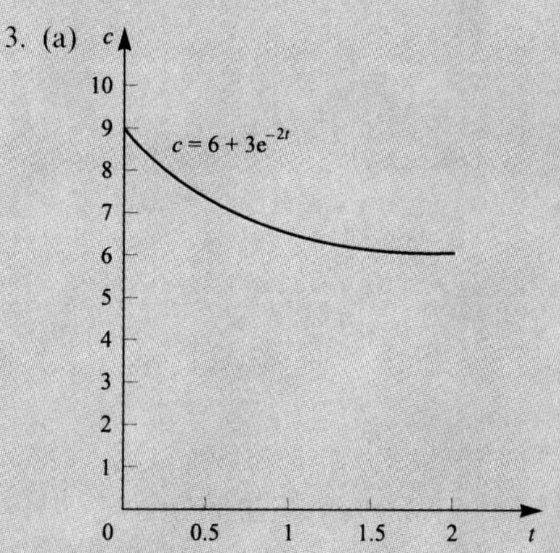

(b) 6

4. (a) $\cosh x + \sinh x$

$$= \frac{e^x + e^{-x}}{2} + \frac{e^x - e^{-x}}{2}$$

$$= \frac{e^x + e^{-x} + e^x - e^{-x}}{2} = e^x$$

(b) $\cosh x - \sinh x$

$$= \frac{e^x + e^{-x}}{2} - \frac{(e^x - e^{-x})}{2}$$

$$= \frac{e^x + e^{-x} - e^x + e^{-x}}{2} = e^{-x}$$

5. $3 \sinh x + 7 \cosh x$

$$= 3\left(\frac{e^x - e^{-x}}{2}\right) + 7\left(\frac{e^x + e^{-x}}{2}\right)$$

$$= 5e^x + 2e^{-x}$$

6. $6e^x - 9e^{-x} = 6(\cosh x + \sinh x)$

$$-9(\cosh x - \sinh x)$$

$$= 15 \sinh x - 3 \cosh x$$

Exercise 19.3

1. (a)

x	0	0.5	1	1.5	2	2.5	3
$15 - x^2$	15	14.75	14	12.75	11	8.75	6
e^x	1	1.65	2.72	4.48	7.39	12.18	20.09

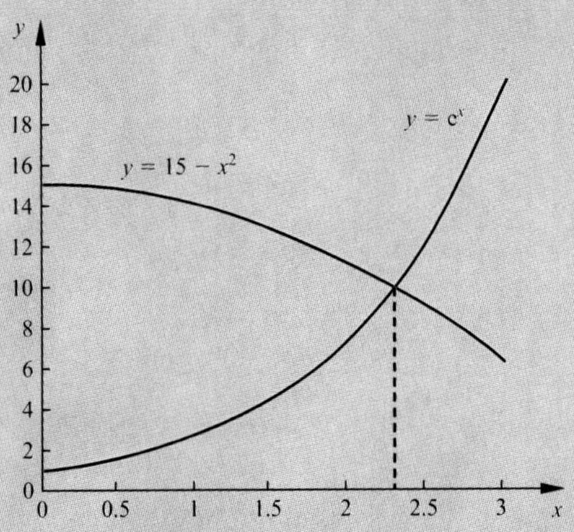

(b) Solving $e^x + x^2 = 15$ is equivalent to solving $e^x = 15 - x^2$. Approximate solution is $x = 2.3$.

2. (a)

x	−1	−0.75	−0.5	−0.25	0	0.25	0.5	0.75	1
$12x^2 - 1$	11	5.75	2	−0.25	−1	−0.25	2	5.75	11
$3e^{-x}$	8.15	6.35	4.95	3.85	3	2.34	1.82	1.42	1.10

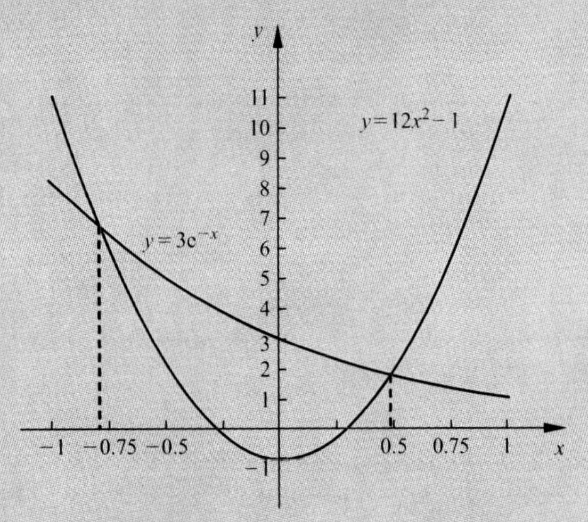

(b) Approximate solutions are $x = -0.80$ and $x = 0.49$.

3.

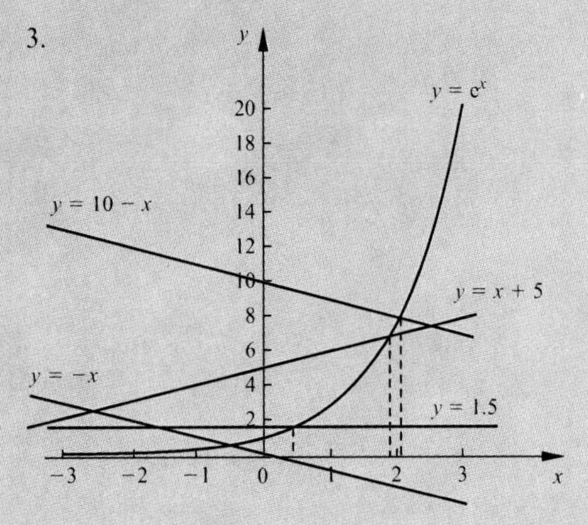

(a) The equation $e^x + x = 0$ is equivalent to $e^x = -x$. Hence draw $y = -x$. Intersection of $y = e^x$ and $y = -x$ will give solution to $e^x + x = 0$. Approximate solution: $x = -0.57$.

(b) The equation $e^x - 1.5 = 0$ is equivalent to $e^x = 1.5$. Hence also draw $y = 1.5$. This is a horizontal straight line. Intersection point gives solution to $e^x - 1.5 = 0$. Approximate solution: $x = 0.41$.

(c) The equation $e^x - x - 5 = 0$ can be written as $e^x = x + 5$. Hence draw $y = x + 5$. Approximate solution: $x = 1.94$.

(d) The equation

$$\frac{e^x}{2} + \frac{x}{2} - 5 = 0$$

can be written as

$$e^x = 10 - x$$

Hence also draw $y = 10 - x$. Approximate solution: $x = 2.07$.

(b) The t value corresponding to $y = 5$ on the graph is $t = 0.7$. Hence the root of $2 + 6e^{-t} = 5$ is $t = 0.7$.

(c) $y = t + 4$ is also shown. The point of intersection of $y = 2 + 6e^{-t}$ and $y = t + 4$ is a solution of

$$2 + 6e^{-t} = t + 4$$

$$6e^{-t} = t + 2$$

The solution is $t = 0.77$.

4. (a)

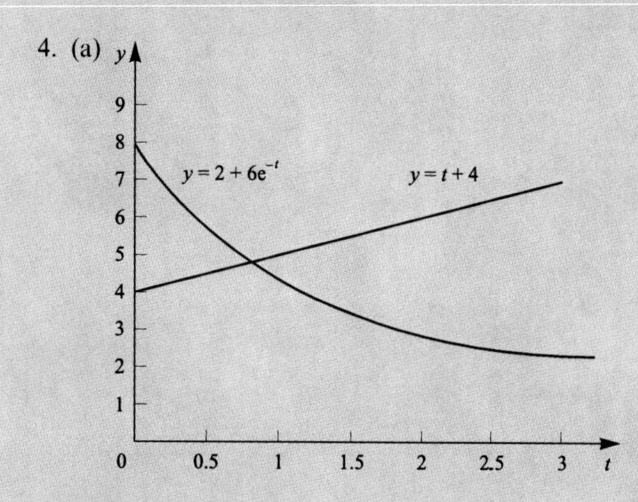

Chapter 20

Self-assessment question 20.1

1. False. For example, natural logarithms have base e (≈ 2.718).

Exercise 20.1

1. (a) 2.1761 (b) 5.0106 (c) −0.5003 (d) −2.3026
2. (a) $\log_3 6561 = 8$ (b) $\log_6 7776 = 5$ (c) $\log_2 1024 = 10$ (d) $\log_{10} 100000 = 5$ (e) $\log_4 16384 = 7$ (f) $\log_{0.5} 0.03125 = 5$ (g) $\log_{12} 1728 = 3$ (h) $\log_9 6561 = 4$
3. (a) $1296 = 6^4$ (b) $225 = 15^2$ (c) $512 = 8^3$ (d) $2401 = 7^4$ (e) $243 = 3^5$ (f) $216 = 6^3$ (g) $8000 = 20^3$ (h) $4096 = 16^3$ (i) $4096 = 2^{12}$
4. (a) 0.4771

(b)
$$3 = 10^{0.4771}$$

$$3 \times 10^2 = 10^2 \times 10^{0.4771}$$

$$300 = 10^{2.4771}$$

$$\log 300 = 2.4771$$

(c) $3 = 10^{0.4771}$

$3 \times 10^{-2} = 10^{-2} \times 10^{0.4771}$

$0.03 = 10^{-1.5229}$

$\log 0.03 = -1.5229$

5. (a) 7

(b) $\log 7 = 0.8451$

$7 = 10^{0.8451}$

$7 \times 10^2 = 10^2 \times 10^{0.8451}$

$700 = 10^{2.8451}$

$\log 700 = 2.8451$

(c) $7 = 10^{0.8451}$

$10^{-2} \times 7 = 10^{-2} \times 10^{0.8451}$

$0.07 = 10^{-1.1549}$

$\log 0.07 = -1.1549$

Exercise 20.2

1. (a) $\log_4 6 = \log 6/\log 4 = 1.2925$
   (b) $\log_3 10 = \log 10/\log 3 = 2.0959$
   (c) $\log_{20} 270 = \log 270/\log 20 = 1.8688$
   (d) $\log_5 0.65 = \log 0.65/\log 5 = -0.2677$
   (e) $\log_2 100 = \log 100/\log 2 = 6.6439$
   (f) $\log_2 0.03 = \log 0.03/\log 2 = -5.0589$
   (g) $\log_{100} 10 = \log 10/\log 100 = \frac{1}{2}$
   (h) $\log_7 7 = 1$

2.
$$\ln X = \frac{\log_{10} X}{\log_{10} \mathrm{e}} = \frac{\log_{10} X}{0.4343} = 2.3026 \log_{10} X$$

3. (a) $\log_3 7 + \log_4 7 + \log_5 7$

$$= \frac{\log 7}{\log 3} + \frac{\log 7}{\log 4} + \frac{\log 7}{\log 5} = 4.3840$$

(b) $\log_8 4 + \log_8 0.25 = \dfrac{\log 4}{\log 8} + \dfrac{\log 0.25}{\log 8}$

$= 0.6667 + (-0.6667)$

$= 0$

(c) $\log_{0.7} 2 = \dfrac{\log 2}{\log 0.7} = -1.9434$

(d) $\log_2 0.7 = \dfrac{\log 0.7}{\log 2} = -0.5146$

Self-assessment question 20.3

1. $\log A + \log B = \log AB$

$$\log A - \log B = \log\left(\frac{A}{B}\right)$$

$$\log A^n = n \log A$$

Exercise 20.3

1. (a) $\log 5 + \log 9 = \log(5 \times 9) = \log 45$
   (b) $\log 9 - \log 5 = \log(\frac{9}{5}) = \log 1.8$
   (c) $\log 5 - \log 9 = \log(\frac{5}{9})$
   (d) $2 \log 5 + \log 1 = 2 \log 5 = \log 5^2$
   $= \log 25$
   (e) $2 \log 4 - 3 \log 2 = \log 4^2 - \log 2^3$
   $= \log 16 - \log 8 = \log(\frac{16}{8}) = \log 2$
   (f) $\log 64 - 2 \log 2 = \log(64/2^2) = \log 16$
   (g) $3 \log 4 + 2 \log 1 + \log 27 - 3 \log 12$
   $= \log 4^3 + 2(0) + \log 27 - \log 12^3$
   $$= \log\left(\frac{4^3 \times 27}{12^3}\right)$$
   $= \log 1 = 0$
2. (a) $\log 3x$ (b) $\log 8x$ (c) $\log 1.5$
   (d) $\log T^2$ (e) $\log 10X^2$
3. (a) $3 \log X - \log X^2 = 3 \log X - 2 \log X$
   $= \log X$
   (b) $\log y - 2 \log \sqrt{y} = \log y - \log(\sqrt{y})^2$
   $= \log y - \log y = 0$
   (c) $5 \log x^2 + 3 \log \dfrac{1}{x} = 10 \log x - 3 \log x$
   $= 7 \log x$
   (d) $4 \log X - 3 \log X^2 + \log X^3$
   $= 4 \log X - 6 \log X + 3 \log X$
   $= \log X$

(e) $3\log y^{1.4} + 2\log y^{0.4} - \log y^{1.2}$
$= 4.2\log y + 0.8\log y - 1.2\log y$
$= 3.8\log y$

4. (a) $\log 4x - \log x = \log(4x/x) = \log 4$

(b) $\log t^3 + \log t^4 = \log(t^3 \times t^4) = \log t^7$

(c) $\log 2t - \log\left(\frac{t}{4}\right) = \log\left(2t \div \frac{t}{4}\right)$

$= \log\left(2t \times \frac{4}{t}\right) = \log 8$

(d) $\log 2 + \log\left(\frac{3}{x}\right) - \log\left(\frac{x}{2}\right)$

$= \log\left(2 \times \frac{3}{x}\right) - \log\left(\frac{x}{2}\right)$

$= \log\left(\frac{6}{x} \div \frac{x}{2}\right)$

$= \log\left(\frac{6}{x} \times \frac{2}{x}\right)$

$= \log\left(\frac{12}{x^2}\right)$

(e) $\log\left(\frac{t^2}{3}\right) + \log\left(\frac{6}{t}\right) - \log\left(\frac{1}{t}\right)$

$= \log\left(\frac{t^2}{3} \times \frac{6}{t}\right) - \log\left(\frac{1}{t}\right)$

$= \log(2t) - \log\left(\frac{1}{t}\right)$

$= \log\left(\frac{2t}{1/t}\right)$

$= \log 2t^2$

(f) $2\log y - \log y^2 = \log y^2 - \log y^2 = 0$

(g) $3\log\left(\frac{1}{t}\right) + \log t^2 = \log\left(\frac{1}{t}\right)^3 + \log t^2$

$= \log\left(\frac{1}{t^3}\right) + \log t^2$

$= \log\left(\frac{t^2}{t^3}\right)$

$= \log\left(\frac{1}{t}\right) = -\log t$

(h) $4\log\sqrt{x} + 2\log\left(\frac{1}{x}\right)$

$= 4\log x^{0.5} + \log\left(\frac{1}{x}\right)^2$

$= \log x^2 + \log\left(\frac{1}{x^2}\right)$

$= \log\left(x^2 \frac{1}{x^2}\right)$

$= \log 1 = 0$

(i) $2\log x + 3\log t = \log x^2 + \log t^3$
$= \log(x^2 t^3)$

(j) $\log A - \frac{1}{2}\log 4A = \log A - \log(4A)^{1/2}$

$= \log\left[\frac{A}{(4A)^{1/2}}\right]$

$= \log\left(\frac{A^{1/2}}{2}\right)$

(k) $$\frac{\log 9x + \log 3x^2}{3} = \frac{\log(9x.3x^2)}{3}$$

$$= \frac{\log(27x^3)}{3}$$

$$= \frac{\log(3x)^3}{3}$$

$$= \frac{3\log(3x)}{3}$$

$$= \log(3x)$$

(l) $$\log xy + 2\log\left(\frac{x}{y}\right) + 3\log\left(\frac{y}{x}\right)$$

$$= \log xy + \log\left(\frac{x^2}{y^2}\right) + \log\left(\frac{y^3}{x^3}\right)$$

$$= \log\left(xy\frac{x^2}{y^2}\frac{y^3}{x^3}\right)$$

$$= \log\left(\frac{x^3y^4}{x^3y^2}\right)$$

$$= \log y^2$$

(m) $$\log\left(\frac{A}{B}\right) - \log\left(\frac{B}{A}\right) = \log\left(\frac{A}{B} \div \frac{B}{A}\right)$$

$$= \log\left(\frac{A}{B} \times \frac{A}{B}\right) = \log\left(\frac{A^2}{B^2}\right)$$

(n) $$\log\left(\frac{2t}{3}\right) + \frac{1}{2}\log 9t - \log\left(\frac{1}{t}\right)$$

$$= \log\left(\frac{2t}{3}\right) + \log(9t)^{1/2} - \log t^{-1}$$

$$= \log\left[\frac{2t}{3}(9t)^{1/2}\right] + \log t$$

$$= \log\left[\frac{2t}{3}(9t)^{1/2}t\right]$$

$$= \log\left(\frac{2t}{3}3t^{1/2}t\right)$$

$$= \log(2t^{2.5})$$

5. $$\log_{10} X + \ln X = \log_{10} X + \frac{\log_{10} X}{\log_{10} \mathrm{e}}$$

$$= \log_{10} X + 2.3026\log_{10} X$$

$$= 3.3026\log_{10} X$$

$$= \log X^{3.3026}$$

6. (a) $\log(9x - 3) - \log(3x - 1)$

$$= \log\left(\frac{9x-3}{3x-1}\right) = \log 3$$

(b) $\log(x^2 - 1) - \log(x + 1)$

$$= \log\left(\frac{x^2-1}{x+1}\right) = \log(x-1)$$

(c) $\log(x^2 + 3x) - \log(x + 3)$

$$= \log\left(\frac{x^2+3x}{x+3}\right) = \log x$$

Exercise 20.4

1. (a) $\log x = 1.6000$

$$x = 10^{1.6000} = 39.8107$$

(b) $10^x = 75$

$$x = \log 75 = 1.8751$$

(c) $\ln x = 1.2350$

$$x = \mathrm{e}^{1.2350} = 3.4384$$

(d) $\mathrm{e}^x = 36$

$$x = \ln 36 = 3.5835$$

2. (a) $\log(3t) = 1.8$

$$3t = 10^{1.8}$$

$$t = \frac{10^{1.8}}{3} = 21.0319$$

(b) $10^{2t} = 150$

$$2t = \log 150$$

$$t = \frac{1}{2}\log 150 = 1.0880$$

(c) $\ln(4t) = 2.8$

$$4t = e^{2.8}$$

$$t = \frac{1}{4}e^{2.8} = 4.1112$$

(d) $e^{3t} = 90$

$$3t = \ln 90$$

$$t = \frac{1}{3}\ln 90 = 1.4999$$

3. (a) $\log x = 0.3940$

$$x = 10^{0.3940} = 2.4774$$

(b) $\ln x = 0.3940$

$$x = e^{0.3940} = 1.4829$$

(c) $10^{y} = 5.5$

$$y = \log 5.5 = 0.7404$$

(d) $e^{z} = 500$

$$z = \ln 500 = 6.2146$$

(e) $\log(3v) = 1.6512$

$$3v = 10^{1.6512}$$

$$v = \frac{10^{1.6512}}{3} = 14.9307$$

(f) $\ln\left(\frac{t}{6}\right) = 1$

$$\frac{t}{6} = e^{1} = e$$

$$t = 6e = 16.3097$$

(g) $10^{2r+1} = 25$

$$2r + 1 = \log 25$$

$$2r = \log 25 - 1$$

$$r = \frac{\log 25 - 1}{2} = 0.1990$$

(h) $e^{(2t-1)/3} = 7.6700$

$$\frac{2t-1}{3} = \ln 7.6700$$

$$2t - 1 = 3\ln 7.6700$$

$$2t = 3\ln 7.6700 + 1$$

$$t = \frac{3\ln 7.6700 + 1}{2} = 3.5560$$

(i) $\log(4b^2) = 2.6987$

$$4b^2 = 10^{2.6987}$$

$$b^2 = \frac{10^{2.6987}}{4} = 124.9223$$

$$b = \sqrt{124.9223} = \pm 11.1769$$

(j) $\log\left(\frac{6}{2+t}\right) = 1.5$

$$\frac{6}{2+t} = 10^{1.5}$$

$$6 = 10^{1.5}(2+t)$$

$$\frac{6}{10^{1.5}} = 2 + t$$

$$t = \frac{6}{10^{1.5}} - 2 = -1.8103$$

(k) $\ln(2r^3 + 1) = 3.0572$

$$2r^3 + 1 = e^{3.0572}$$

$$2r^3 = e^{3.0572} - 1$$

$$r^3 = \frac{e^{3.0572} - 1}{2} = 10.1340$$

$$r = 2.1640$$

(l) $\ln(\log t) = -0.3$

$$\log t = e^{-0.3} = 0.7408$$

$$t = 10^{0.7408} = 5.5058$$

(m) $10^{t^2-1} = 180$

$$t^2 - 1 = \log 180$$

$$t^2 = \log 180 + 1 = 3.2553$$

$$t = \sqrt{3.2553} = \pm 1.8042$$

(n) $10^{3r^4} = 170000$

$$3r^4 = \log 170000$$

$$r^4 = \frac{\log 170000}{3} = 1.7435$$

$$r = \pm 1.1491$$

(o) $\log 10^t = 1.6$

$$t \log 10 = 1.6$$

$$t = 1.6$$

(p) $\ln(e^x) = 20000$

$$x \ln e = 20000$$

$$x = 20000$$

4. (a) $e^{3x}e^{2x} = 59$

$$e^{5x} = 59$$

$$5x = \ln 59$$

$$x = \frac{\ln 59}{5} = 0.8155$$

(b) $10^{3t}.10^{4-t} = 27$

$$10^{4+2t} = 27$$

$$4 + 2t = \log 27$$

$$2t = \log 27 - 4$$

$$t = \frac{\log 27 - 4}{2} = -1.2843$$

(c) $\log(5 - t) + \log(5 + t) = 1.2$

$$\log(5 - t)(5 + t) = 1.2$$

$$\log(25 - t^2) = 1.2$$

$$25 - t^2 = 10^{1.2}$$

$$t^2 = 25 - 10^{1.2}$$

$$= 9.1511$$

$$t = \pm 3.0251$$

(d) $\log x + \ln x = 4$

$$\log x + \frac{\log x}{\log e} = 4$$

$$\log x \left(1 + \frac{1}{\log e}\right) = 4$$

$$3.3026 \log x = 4$$

$$\log x = \frac{4}{3.3026} = 1.2112$$

$$x = 10^{1.2112} = 16.2619$$

Self-assessment questions 20.5

1. Both $\log x$ and $\ln x$ tend to infinity as x tends to infinity. Both have a value of 0 when $x = 1$.
2. True. The graph of the logarithm function increases without bound, albeit very slowly.

Exercise 20.5

1. (a)

x	0.5	1	2	3	4
$\log x^2$	−0.6021	0	0.6021	0.9542	1.2041

x	5	6	7	8	9	10
$\log x^2$	1.3979	1.5563	1.6902	1.8062	1.9085	2

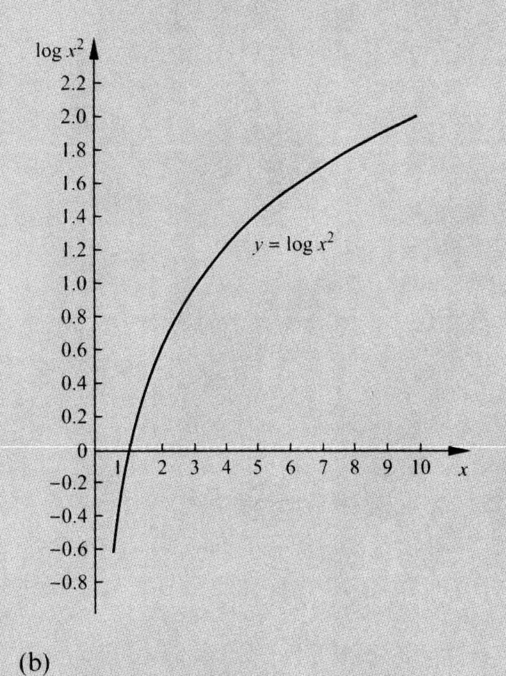

(b)

x	0.1	0.2	0.5	0.75	1
$\log(1/x)$	1	0.6990	0.3010	0.1249	0

x	2	3	4	5
$\log(1/x)$	−0.3010	−0.4771	−0.6021	−0.6990

x	6	7	8	9	10
$\log(1/x)$	−0.7782	−0.8451	−0.9031	−0.9542	−1

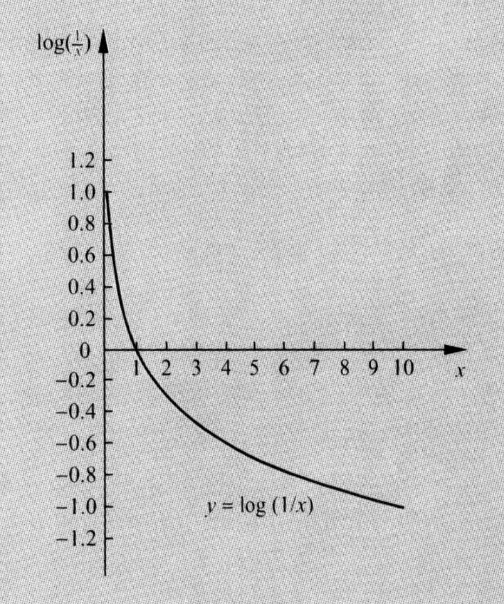

2. (a)

x	0.1	0.5	1	1.5	2	2.5	3
$\log x$	−1	−0.3010	0	0.1761	0.3010	0.3979	0.4771

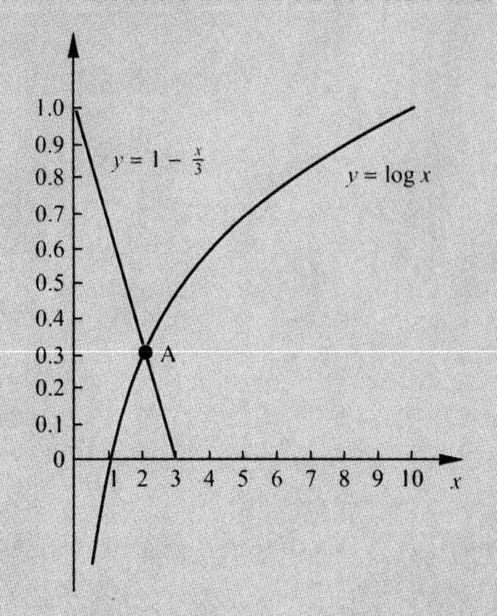

(b) $y = 1 - \frac{x}{3}$ is a straight line passing through (0, 1) and (3, 0).

(c) The graphs intersect at A. Approximate solution: $x = 2.06$.

3. (a)

(b) The line $y = 0.5 - 0.1x$ is also shown on the figure.

(c) Solution $x = 2$.

Chapter 33

Self-assessment questions 33.2

1. It is a function that provides a formula for calculating the gradient of another function.
2. $y' = nx^{n-1}$.
3. Derivative, derived function, first derivative, gradient function, rate of change.
4. $\frac{dy}{dx}$.
5. Zero. A constant function has a horizontal graph. Consequently its gradient is zero.
6. Radians.
7. The exponential function e^x.

Exercise 33.2

1. (a) $8x^7$ (b) $7x^6$ (c) $-x^{-2} = -1/x^2$ (d) $-5x^{-6} = -5/x^6$ (e) $13x^{12}$ (f) $5x^4$ (g) $-2x^{-3} = -2/x^3$
2. (a) $\frac{3}{2}x^{1/2}$ (b) $\frac{5}{2}x^{3/2}$ (c) $-\frac{1}{2}x^{-3/2}$
   (d) $(1/2)x^{-1/2} = 1/2\sqrt{x}$
   (e) $y = \sqrt{x} = x^{1/2}$ and so $y' = (1/2)x^{-1/2} = 1/2\sqrt{x}$
   (f) $0.2x^{-0.8}$
3. $y = 8$, $y' = 0$, the line is horizontal and so its gradient is zero.

4. (a) $16x^{15}$ (b) $0.5x^{-0.5}$ (c) $-3.5x^{-4.5}$
   (d) $y = 1/x^{1/2} = x^{-1/2}$ and so $y' = -\frac{1}{2}x^{-3/2}$

(e) $y = 1/\sqrt{x} = 1/x^{1/2} = x^{-1/2}$ and so $y' = -\frac{1}{2}x^{-3/2}$

5. $y' = 4x^3$:
(a) at $x = -4$, $y' = -256$
(b) at $x = -1$, $y' = -4$
(c) at $x = 0$, $y' = 0$ (d) at $x = 1$, $y' = 4$
(e) at $x = 4$, $y' = 256$. The graph is falling at the given negative x values and is rising at the given positive x values. At $x = 0$ the graph is neither rising nor falling.

6. (a) $y' = 3x^2$, $y'(-3) = 27$, $y'(0) = 0$, $y'(3) = 27$
(b) $y' = 4x^3$, $y'(-2) = -32$, $y'(2) = 32$
(c) $y' = \frac{1}{2}x^{-1/2} = 1/2\sqrt{x}$, $y'(1) = \frac{1}{2}$, $y'(2) = \frac{1}{2\sqrt{2}} = 0.354$
(d) $y' = -2x^{-3}$, $y'(-2) = \frac{1}{4}$, $y'(2) = -\frac{1}{4}$

7. (a) $2t$ (b) $-3t^{-4}$ (c) $\frac{1}{2}t^{-1/2}$
(d) $-1.5t^{-2.5}$

8. (a) $-7 \sin 7x$ (b) $\frac{1}{2} \cos(x/2)$
(c) $-\frac{1}{2} \sin(x/2)$ (d) $4e^{4x}$ (e) $-3e^{-3x}$

9. (a) $1/x$ (b) $\frac{1}{3} \cos(x/3)$ (c) $\frac{1}{2}e^{x/2}$

10. $y' = \cos x$, (a) $y'(0) = 1$, (b) $y'(\pi/2) = 0$, (c) $y'(\pi) = -1$

11. (a) $7t^6$ (b) $3 \cos 3t$ (c) $2e^{2t}$ (d) $1/t$
(e) $-\frac{1}{3} \sin(t/3)$

Self-assessment question 33.3

1. (a) The derivative of the sum of two functions is the sum of the separate derivatives.
(b) The derivative of the difference of two functions is the difference of the separate derivatives. This means we can differentiate sums and differences term by term.
(c) The derivative of $k f(x)$ is equal to k times the derivative of $f(x)$.

Exercise 33.3

1. (a) $y' = 6x^5 - 4x^3$
(b) $y' = 6 \times (2x) = 12x$
(c) $y' = 9(-2x^{-3}) = -18x^{-3}$ (d) $y' = \frac{1}{2}$
(e) $y' = 6x^2 - 6x$

2. (a) $y' = 2 \cos x$
(b) $y' = 3(4 \cos 4x) = 12 \cos 4x$
(c) $y' = -63 \sin 9x + 12 \cos 4x$
(d) $y' = 3e^{3x} + 10e^{-2x}$

3. The gradient function is $y' = 9x^2 - 9$. $y'(1) = 0$, $y'(0) = -9$, $y'(-1) = 0$. Of the three given points, the curve is steepest at $x = 0$ where it is falling.

4. When x^2 is differentiated we obtain $2x$. Also, because the derivative of any constant is zero, the derivative of $x^2 + c$ will be $2x$ where c can take any value.

5. $y' = 6 \cos 2t - 8 \sin 2t$
$y'(2) = 6 \cos 4 - 8 \sin 4 = 2.1326$

6. $y' = 2 - 3x^2 + 2e^{2x}$
$y'(1) = 2 - 3 + 2e^2 = 13.7781$

7. $y = (2x - 1)^2 = 4x^2 - 4x + 1$
$y' = 8x - 4$
$y'(0) = -4$

8. $f' = 2x - 2$
$f'(1) = 0$

9. $y' = x^2 - 1$. We see that $y' = 0$ when $x^2 - 1 = 0$, that is when $x = -1, 1$.

10. $y' = 1 - \sin x$. When $y' = 0$ then $\sin x = 1$, and hence $x = \frac{\pi}{2}$.

Self-assessment questions 33.4

1. The second derivative of y is found by differentiating the first derivative of y.
2. y'' and, if y is a function of x, $\frac{d^2y}{dx^2}$.

Exercise 33.4

1. (a) $y' = 2x$, $y'' = 2$
(b) $y' = 24x^7$, $y'' = 168x^6$
(c) $y' = 45x^4$, $y'' = 180x^3$
(d) $y' = \frac{3}{2}x^{-1/2}$, $y'' = -\frac{3}{4}x^{-3/2}$
(e) $y' = 12x^3 + 10x$, $y'' = 36x^2 + 10$
(f) $y' = 27x^2 - 14$, $y'' = 54x$
(g) $y' = \frac{1}{2}x^{-1/2} - 2x^{-3/2}$, $y'' = -\frac{1}{4}x^{-3/2} + 3x^{-5/2}$

2. (a) $y' = 3 \cos 3x - 2 \sin 2x$, $y'' = -9 \sin 3x - 4 \cos 2x$
(b) $y' = e^x - e^{-x}$, $y'' = e^x + e^{-x}$

3. $y' = 4\cos 4x$, $y'' = -16\sin 4x$, and so $16y + y'' = 16\sin 4x + (-16\sin 4x) = 0$ as required.
4. $y' = 4x^3 - 6x^2 - 72x$, $y'' = 12x^2 - 12x - 72$. The values of x where $y'' = 0$ are found from $12x^2 - 12x - 72 = 0$, that is $12(x^2 - x - 6) = 12(x+2)(x-3) = 0$ so that $x = -2$ and $x = 3$.
5. (a) $y' = 6\cos 3x - 8\sin 2x$, $y'' = -18\sin 3x - 16\cos 2x$
   (b) $y' = k\cos kt$, $y'' = -k^2 \sin kt$
   (c) $y' = -k\sin kt$, $y'' = -k^2\cos kt$
   (d) $y' = Ak\cos kt - Bk\sin kt$, $y'' = -Ak^2\sin kt - Bk^2\cos kt$
6. (a) $y' = 3e^{3x}$, $y'' = 9e^{3x}$
   (b) $y' = 3 + 8e^{4x}$, $y'' = 32e^{4x}$
   (c) $y' = 2e^{2x} + 2e^{-2x}$, $y'' = 4e^{2x} - 4e^{-2x}$
   (d) $y = e^{-x}$, $y' = -e^{-x}$, $y'' = e^{-x}$
   (e) $y = e^{2x} + 3e^{x}$, $y' = 2e^{2x} + 3e^{x}$, $y'' = 4e^{2x} + 3e^{x}$
7. $y' = 200 + 12x - 3x^2$, $y'' = 12 - 6x$
   $y'' = 0$ when $x = 2$
8. $$y' = 2e^x + 2\cos 2x$$ $$y'' = 2e^x - 4\sin 2x$$ $$y''(1) = 2e - 4\sin 2 = 1.7994$$
9. $$y' = 8\cos 4t + 10\sin 2t$$ $$y'' = -32\sin 4t + 20\cos 2t$$ $$y''(1.2) = -32\sin 4.8 + 20\cos 2.4 = 17.1294$$

Self-assessment questions 33.5

1. Maximum, minimum and point of inflexion.
2. Suppose $x = a$ is a stationary point of $y(x)$. If y'' evaluated at $x = a$ is positive, the point is a minimum. If y'' evaluated at $x = a$ is negative that point is a maximum. If $y'' = 0$ this test does not enable us to distinguish between the possible types of stationary point.
3. If $y'' = 0$.

Exercise 33.5

1. (a) $y' = 2x$, $y'' = 2$, (0, 1) is a minimum;
   (b) $y' = -2x$, $y'' = -2$, (0, 0) is a maximum; (c) $y' = 6x^2 + 18x$, $y'' = 12x + 18$, (0, 0) is a minimum and (−3, 27) is a maximum;
   (d) $y' = -6x^2 + 54x$, $y'' = -12x + 54$, (0, 0) is a minimum and (9, 729) is a maximum;
   (e) $y' = 3x^2 - 6x + 3$, $y'' = 6x - 6$, stationary point by solving $3x^2 - 6x + 3 = 0$, at (1, 2), second-derivative test fails, inspection of gradient on either side of $x = 1$ reveals a point of inflexion.
2. $y' = \cos x$. Solving $\cos x = 0$ gives $x = \frac{\pi}{2}, \frac{3\pi}{2}$. The maximum and minimum points occur at $x = \frac{\pi}{2}, \frac{3\pi}{2}$.
   We examine y''
   $$y'' = -\sin x$$
   $$y''\left(\frac{\pi}{2}\right) = -\sin\frac{\pi}{2} < 0 \quad \text{a maximum}$$
   $$y''\left(\frac{3\pi}{2}\right) = -\sin\frac{3\pi}{2} > 0 \quad \text{a minimum}$$
   When $x = \frac{\pi}{2}$, $y = 1$; when $x = \frac{3\pi}{2}$, $y = -1$. In summary, maximum at $(\frac{\pi}{2}, 1)$, minimum at $(\frac{3\pi}{2}, -1)$.
3. $y' = e^x - 1$. Solving $y' = 0$ gives $x = 0$. Now, $y'' = e^x$ and $y''(0) = 1 > 0$. When $x = 0$, $y = 1$, so there is a minimum at (0, 1).
4. $y' = 6x^2 - 6x - 12 = 6(x+1)(x-2)$. Solving $y' = 0$ gives $x = -1, 2$. $y'' = 12x - 6$, $y''(-1) < 0$, a maximum, $y''(2) > 0$, a minimum.
   When $x = -1$, $y = 8$; when $x = 2$, $y = -19$. Hence there is a maximum at (−1, 8), a minimum at (2, −19).
5. $y' = -x^2 + 1$. Solving $y' = 0$ gives $x = -1, 1$. $y'' = -2x$, $y''(-1) > 0$, a minimum, $y''(1) < 0$, a maximum.
   When $x = -1$, $y = -\frac{2}{3}$, when $x = 1$, $y = \frac{2}{3}$. So there is a minimum at $(-1, -\frac{2}{3})$ and a maximum at $(1, \frac{2}{3})$.

6. $y' = 5x^4 - 5$
   $= 5(x^4 - 1)$
   $= 5(x^2 - 1)(x^2 + 1)$
   $= 5(x - 1)(x + 1)(x^2 + 1)$
   Solving $y' = 0$ gives $x = -1, 1$.
   $y'' = 20x^3$, $y''(-1) < 0$, a maximum, $y''(1) > 0$, a minimum.
   When $x = -1$, $y = 5$, when $x = 1$, $y = -3$.
   There is a maximum at $(-1, 5)$; a minimum at $(1, -3)$.

Chapter 34

Self-assessment question 34.2

1. The product rule states that if $y = uv$ then

$$\frac{dy}{dx} = \frac{du}{dx}v + u\frac{dv}{dx}$$

Exercise 34.2

1. (a) $\sin x + x\cos x$ (b) $2x\cos x - x^2\sin x$
   (c) $\cos^2 x - \sin^2 x$ (d) $e^x + xe^x$
   (e) $e^{-x} - xe^{-x}$
2. (a) $2\ln(3x) + 2$ (b) $3e^{3x}\sin x + e^{3x}\cos x$
   (c) $-e^{-\frac{x}{2}}(\frac{\cos 4x}{2} + 4\sin 4x)$
   (d) $2\cos 2x\cos 3x - 3\sin 2x\sin 3x$
   (e) $e^x\ln(2x) + \frac{e^x}{x}$
3. $\frac{dy}{dx} = e^{-x}(\cos x - \sin x)$, $\frac{dy}{dx}\ (x = 1) = -0.1108$
4. $\frac{dy}{dx} = -2(\frac{\cos 2x + x\sin 2x}{x^3})$
5. $y' = 2(\sin x + x)(\cos x + 1)$
6. (a) $e^{-x}x(2 - x)$
   (b) $e^{-x}x(-x\sin x + 2\sin x + x\cos x)$
7. $y' = x^2e^x(3 + x)$, $y' = 0$ when $x = 0$ and $x = -3$

Self-assessment question 34.3

1. The quotient rule states that if $y = \frac{u}{v}$ then

$$\frac{dy}{dx} = \frac{v\frac{du}{dx} - u\frac{dv}{dx}}{v^2}$$

Exercise 34.3

1. (a) $\frac{xe^x}{(x+1)^2}$ (b) $\frac{1 - \ln x}{x^2}$ (c) $\frac{\ln x - 1}{(\ln x)^2}$
   (d) $-\frac{1}{\sin^2 t}$ (e) $\frac{2t^2 + 2t + 1}{(2t+1)^2}$
2. (a) $y' = \frac{1}{(x+2)^2}$, $y'(2) = \frac{1}{16}$
   (b) $y' = \frac{2(x+1)/x - 2\ln(3x)}{(x+1)^2}$, $y'(2) = -0.0648$
   (c) $y' = \frac{2\cos 2x\cos 3x + 3\sin 2x\sin 3x}{(\cos 3x)^2}$,
   $y'(2) = -0.6734$
   (d) $y' = \frac{e^{3x} + 2e^{2x} - e^x}{(e^x + 1)^2}$, $y'(2) = 7.1791$
   (e) $y' = \frac{x^4 + 3x^2 - 2x}{(x^2+1)^2}$, $y'(2) = \frac{24}{25}$
3. $\frac{1}{\cos^2\theta}$
4. $\frac{k}{\cos^2 k\theta}$

Self-assessment questions 34.4

1. $\frac{dy}{dt} = \frac{dy}{dx} \times \frac{dx}{dt}$
2. $\frac{dy}{dx} = \frac{dy}{dz} \times \frac{dz}{dx}$
3. When the input to a function is itself a function, we have a function of a function. So if, for example, the function $\theta = 7x^2$ is the input to the function $\sin\theta$, we obtain the function of a function $\sin(7x^2)$.
4. (a) Note that $\frac{1}{x}\cos x$ could be written as $\frac{\cos x}{x}$. So the function $\frac{1}{x}\cos x$ can be thought of as either a product of the functions $\frac{1}{x}$ and $\cos x$, or a quotient of $\cos x$ and x. Either the product rule or the quotient rule can be used.
   (b) $x\sin x$ is a product of the function x and $\sin x$.
   (c) $\sin(\frac{1}{x})$ is a function of a function. Note that if $y = \sin t$ and $t = \frac{1}{x}$ then $y = \sin\frac{1}{x}$.

Exercise 34.4

1. (a) $\frac{dy}{dt} = \cos x \times 2t = 2t \cos t^2$

   (b) $\frac{dy}{dt} = \frac{1}{2}x^{-1/2} \times 1$
   $= \frac{1}{2}x^{-1/2} = \frac{1}{2}(t+1)^{-1/2} = \frac{1}{2(t+1)^{1/2}}$

   (c) $\frac{dy}{dt} = e^x \times 4t = 4te^{2t^2}$

   (d) $\frac{dy}{dt} = \frac{15}{2}x^{3/2} \times (3t^2 + 2)$
   $= \frac{15}{2}(3t^2 + 2)(t^3 + 2t)^{3/2}$

2. (a) $\frac{dy}{dt} = -6t \sin(3t^2)$

   (b) $\frac{dy}{dt} = 2\cos(2t + 1)$

   (c) $\frac{dy}{dt} = \frac{3}{3t-2}$

   (d) $\frac{dy}{dt} = (2t + 3)\cos(t^2 + 3t + 2)$

3. (a) $\frac{dy}{dx} = 3\cos(3x + 1)$

   (b) $\frac{dy}{dx} = 4e^{4x-3}$

   (c) $\frac{dy}{dx} = -2\sin(2x - 4)$

   (d) $\frac{dy}{dx} = \frac{1}{2}(x^2 + 5x - 4)^{-1/2}(2x + 5)$

   (e) $\frac{dy}{dx} = -\frac{1}{x^2 \cos^2(1/x)}$

4. (a) $\frac{dy}{dx} = \frac{4}{4x-3}$

   (b) $\frac{dy}{dx} = \frac{6x}{3x^2} = \frac{2}{x}$

   (c) $\frac{dy}{dx} = \frac{\cos x}{\sin x}$

   (d) $\frac{dy}{dx} = \frac{2x+3}{x^2+3x}$

Chapter 37

Self-assessment questions 37.1

1. A function of two variables is a rule that operates upon two variables, chosen independently, and produces a single output value known as the dependent variable.
2. (a) In $z = f(x, y)$, y is an independent variable. The user is free to choose a value for y.
   (b) In $y = f(x)$, y is the dependent variable. Once a value for x has been chosen, then y is determined automatically – we have no choice.
3. Given $P = f(V, T) = \frac{RT}{V}$, P is the dependent variable. V and T are independent variables.
4. In $y = f(x, t)$, x and t are independent variables and y is the dependent variable.

Exercise 37.1

1. $f(8, 2) = 7(8) + 2(2) = 56 + 4 = 60$
2. (a) $f(2, 3) = -11(2) + 3 = -19$
   (b) $f(11, 1) = -11(11) + 1 = -121 + 1 = -120$
3. $z(1, 1) = 3e^1 - 2e^1 + 1 = 3.718$
4. $w = 7 - (-3)(-9) = 7 - 27 = -20$
5. $\sin(50°) = 0.7660$
6. $f(0.5, 3) = e^{(2)(0.5)(3)} = e^3 = 20.086$

Self-assessment question 37.2

1. Your sketch should look like Figure 37.3 in the main text, with the axes ordered in the same way.

Exercise 37.2

1. (a) $z = -12$ (b) -18 (c) 15

Self-assessment questions 37.3

1. In each case, zero. The derivative of a constant is always zero.
2. Zero. When performing partial differentiation with respect to x, y is treated as a constant and consequently its derivative is zero.
3. $\frac{\partial z}{\partial x}$. Always remember to use the curly d.
4. $\frac{\partial z}{\partial y}$. Always remember to use the curly d.
5. $\frac{\partial y}{\partial x}, \frac{\partial y}{\partial t}$.

Exercise 37.3

1. (a) $\frac{\partial z}{\partial x} = 5$, $\frac{\partial z}{\partial y} = 11$ (b) $-14, -7$ (c) $8, 0$ (d) $0, -5$ (e) $3, 8$ (f) $-3, 2$ (g) $0, 0$ (h) $0, -3$ (i) $4x, -7$ (j) $7, -9y^2$ (k) $-9, 9$ (l) $9, 9$
2. (a) $\frac{\partial z}{\partial x} = y$, $\frac{\partial z}{\partial y} = x$ (b) $3y, 3x$ (c) $-9y, -9x$ (d) $2xy, x^2$ (e) $18xy, 9x^2$ (f) $8y^2, 16xy$
3. $\frac{\partial z}{\partial x} = 9$ and $\frac{\partial z}{\partial y} = 2y$. Hence at $(4, -2)$ these derivatives are, respectively, 9 and -4.
4. (a) $\frac{\partial z}{\partial x} = 2e^{2x}$ and $\frac{\partial z}{\partial y} = 0$
   (b) $\frac{\partial z}{\partial x} = 0$ and $\frac{\partial z}{\partial y} = 5e^{5y}$
   (c) $\frac{\partial z}{\partial x} = ye^{xy}$ and $\frac{\partial z}{\partial y} = xe^{xy}$
   (d) $\frac{\partial z}{\partial x} = 0$ and $\frac{\partial z}{\partial y} = 8e^{2y}$
5. $\frac{\partial y}{\partial x} = \sin t$, $\frac{\partial y}{\partial t} = x \cos t$

Exercise 37.4

1. (a) $\frac{\partial z}{\partial x} = ye^x(x+1)$, $\frac{\partial z}{\partial y} = xe^x$
   (b) $ye^y, xe^y(y+1)$ (c) $e^{xy}(xy+1), x^2e^{xy}$
   (d) $y^2e^{xy}, e^{xy}(xy+1)$
   (e) $x^2y\cos(xy) + 2x\sin(xy), x^3\cos(xy)$
   (f) $-y^2\sin(xy), -yx\sin(xy) + \cos(xy)$
   (g) $1 + \ln(xy), \frac{x}{y}$ (h) $3y^3e^x(x+1), 9xy^2e^x$

Self-assessment question 37.5

1. $\frac{\partial^2 z}{\partial y \partial x}$ means differentiate $\frac{\partial z}{\partial x}$ with respect to y. $\frac{\partial^2 z}{\partial x \partial y}$ means differentiate $\frac{\partial z}{\partial y}$ with respect to x. It is usually the case that the result is the same either way.

Exercise 37.5

In all solutions $\frac{\partial^2 z}{\partial x \partial y}$ and $\frac{\partial^2 z}{\partial y \partial x}$ are identical.

1. (a) $\frac{\partial^2 z}{\partial x^2} = 0$, $\frac{\partial^2 z}{\partial y \partial x} = 1$, $\frac{\partial^2 z}{\partial y^2} = 0$
   (b) $\frac{\partial^2 z}{\partial x^2} = 0$, $\frac{\partial^2 z}{\partial y \partial x} = 7$, $\frac{\partial^2 z}{\partial y^2} = 0$
   (c) $\frac{\partial^2 z}{\partial x^2} = 0$, $\frac{\partial^2 z}{\partial y \partial x} = 0$, $\frac{\partial^2 z}{\partial y^2} = 0$
   (d) $\frac{\partial^2 z}{\partial x^2} = 0$, $\frac{\partial^2 z}{\partial y \partial x} = 16y$, $\frac{\partial^2 z}{\partial y^2} = 16x$
   (e) $\frac{\partial^2 z}{\partial x^2} = -4y^3$, $\frac{\partial^2 z}{\partial y \partial x} = -12y^2x$, $\frac{\partial^2 z}{\partial y^2} = -12yx^2$
   (f) $\frac{\partial^2 z}{\partial x^2} = 0$, $\frac{\partial^2 z}{\partial y \partial x} = 0$, $\frac{\partial^2 z}{\partial y^2} = 0$
2. (a) $\frac{\partial^2 z}{\partial x^2} = \frac{2}{x^3}$, $\frac{\partial^2 z}{\partial y \partial x} = 0$, $\frac{\partial^2 z}{\partial y^2} = 0$
   (b) $\frac{\partial^2 z}{\partial x^2} = \frac{2y}{x^3}$, $\frac{\partial^2 z}{\partial y \partial x} = -\frac{1}{x^2}$, $\frac{\partial^2 z}{\partial y^2} = 0$
   (c) $\frac{\partial^2 z}{\partial x^2} = 0$, $\frac{\partial^2 z}{\partial y \partial x} = -\frac{1}{y^2}$, $\frac{\partial^2 z}{\partial y^2} = \frac{2x}{y^3}$
   (d) $\frac{\partial^2 z}{\partial x^2} = \frac{2}{x^3}$, $\frac{\partial^2 z}{\partial y \partial x} = 0$, $\frac{\partial^2 z}{\partial y^2} = \frac{2}{y^3}$
3. (a) $\frac{\partial^2 z}{\partial x^2} = 0$, $\frac{\partial^2 z}{\partial y \partial x} = \cos y$, $\frac{\partial^2 z}{\partial y^2} = -x \sin y$
   (b) $\frac{\partial^2 z}{\partial x^2} = -y \cos x$, $\frac{\partial^2 z}{\partial y \partial x} = -\sin x$, $\frac{\partial^2 z}{\partial y^2} = 0$
   (c) $\frac{\partial^2 z}{\partial x^2} = 4ye^{2x}$, $\frac{\partial^2 z}{\partial y \partial x} = 2e^{2x}$, $\frac{\partial^2 z}{\partial y^2} = 0$
   (d) $\frac{\partial^2 z}{\partial x^2} = ye^{-x}$, $\frac{\partial^2 z}{\partial y \partial x} = -e^{-x}$, $\frac{\partial^2 z}{\partial y^2} = 0$
4. (a) $\frac{\partial^2 z}{\partial x^2} = 8y^2e^{xy}$, $\frac{\partial^2 z}{\partial y \partial x} = 8e^{xy}(xy+1)$, $\frac{\partial^2 z}{\partial y^2} = 8x^2e^{xy}$
   (b) $\frac{\partial^2 z}{\partial x^2} = -3e^x \sin y$, $\frac{\partial^2 z}{\partial y \partial x} = -3e^x \cos y$, $\frac{\partial^2 z}{\partial y^2} = 3e^x \sin y$

(c) $\dfrac{\partial^2 z}{\partial x^2} = -4e^y \cos x$, $\dfrac{\partial^2 z}{\partial y \partial x} = -4e^y \sin x$,

$\dfrac{\partial^2 z}{\partial y^2} = 4e^y \cos x$

5. (a) $\dfrac{\partial^2 z}{\partial x^2} = -\dfrac{1}{x^2}$, $\dfrac{\partial^2 z}{\partial y \partial x} = 0$, $\dfrac{\partial^2 z}{\partial y^2} = 0$

(b) $\dfrac{\partial^2 z}{\partial x^2} = 0$, $\dfrac{\partial^2 z}{\partial y \partial x} = 0$, $\dfrac{\partial^2 z}{\partial y^2} = -\dfrac{1}{y^2}$

(c) $\dfrac{\partial^2 z}{\partial x^2} = -\dfrac{1}{x^2}$, $\dfrac{\partial^2 z}{\partial y \partial x} = 0$, $\dfrac{\partial^2 z}{\partial y^2} = -\dfrac{1}{y^2}$

(d) $\dfrac{\partial^2 z}{\partial x^2} = 0$, $\dfrac{\partial^2 z}{\partial y \partial x} = \dfrac{1}{y}$, $\dfrac{\partial^2 z}{\partial y^2} = -\dfrac{x}{y^2}$

(e) $\dfrac{\partial^2 z}{\partial x^2} = -\dfrac{y}{x^2}$, $\dfrac{\partial^2 z}{\partial y \partial x} = \dfrac{1}{x}$, $\dfrac{\partial^2 z}{\partial y^2} = 0$

Self-assessment question 37.6

1. First calculate $\dfrac{\partial w}{\partial z}$. Then partially differentiate this with respect to y. Finally, partially differentiate the result with respect to x.

Exercise 37.6

1. 8, −3, −4
2. $2x$, $3y^2$, $-10t$
3. $\dfrac{\partial w}{\partial x} = yz$, $\dfrac{\partial w}{\partial y} = xz$, $\dfrac{\partial w}{\partial z} = xy$

Answers to Problems

Chapter 1

Problem 1.1

(a) Comparison is complicated by different numbers, but there are relatively more women in the 'Other' education category, relatively less in the other three.

(b) The biggest difference is the larger number of 'inactive' women, mostly engaged in child-rearing, one would suspect.

(c) The 'inactive' women show up again.

(d) Relatively fewer women have no qualifications, A levels and higher education, but relatively more have other qualifications.

Problem 1.3

(a) Higher education, 88%.

(b) Those in work, 20%.

Problem 1.5

The difference between bar chart and histogram should be similar to those for the 2003 distribution. Overall shape similar (heavily skewed to right). Comparison difficult because of different wealth levels (due to inflation) and because grouping into classes can affect precise shape of graph.

Problem 1.7

(a) Mean 16.399 (£000); median 8.92; mode 0–1 (£000) group has the greatest frequency density. They differ because of skewness in the distribution.

(b) Q1 = 3.295, Q3 = 18.339, IQR = 15.044; variance = 653.88; s.d. = 25.552; cv = 1.56.

(c) $95\,469.32/25.55^3 = 5.72 > 0$ as expected.

(d) Comparison in text.

(e) This would increase the mean substantially (to 31.12), but the median and mode would be unaffected.

Problem 1.9

57.62 pence/litre.

Problem 1.11

(a) $z = 1.5$ and -1.5 respectively.

(b) Using Chebyshev's theorem with $k = 1.5$, we have that at least $(1 - 1/1.5^2) = 0.56$ (56%) lies within 1.5 standard deviations of the mean, so at most 0.44 (44 students) lies outside the range.

(c) Chebyshev's theorem applies to *both* tails, so we cannot answer this part. You cannot halve the figure of 0.44 because the distribution may be skewed.

Problem 1.13

(a) Strong upwards trend for 1974–1989, but substantial falls in 1972–1974 and 1989–1991. The market appears quite volatile, therefore. Note that this shows the *volume* of car registrations. *If* the 1972–1976 pattern is repeated, one might expect car registrations to turn up again in 1992–1993.

(b) This figure shows the substantial volatility of the series, around an average not much above zero. 1989–1991 looks less like 1971–1973 on this graph. The log graphs are very similar to the levels graphs. There is not always an advantage to drawing these. Here, the time period is relatively short and the growth rate rather small.

Problem 1.15

(a) 1.81% p.a. (but 4.88% p.a. between 1975 and 1989).

(b) 0.187 (around the arithmetic mean).

(c) Registrations appear more volatile, as measured by the cv (10.33 for car registrations, 0.816 for investment). Possible reasons: investment covers several categories and fluctuations in one may offset those in another; the registrations series is shorter, so a big random fluctuation has a larger effect; investment is nominal and the price influence may help to smooth out the series.

Problem 1.17

(a) Non-linear, upwards trend. Likely to be positively autocorrelated. Variation around the trend is likely to grow over time (heteroscedasticity).

(b) Similar to (a), except trend would be shallower after deflating. Probably less heteroscedasticity because price variability has been removed, which may also increase the autocorrelation of the series.

(c) Unlikely to show a trend in the *very* long run, but there might be one over, say, five years, if inflation is increasing. Likely to be homoscedastic, with some degree of correlation.

Problem 1.19

(a) Using $S_t = S_0(1 - r)^t$, hence $S_0 = S_t/(1 + r)^t$. Setting $S_t = 1000$, $r = 0.07$ gives $S_0 = 712.99$. Price after two years: £816.30. If r rose to 10% the bond would fall to $1000/1.1^3 = 751.31$.

(b) The income stream should be discounted to the present using

$$\frac{200}{1+r} + \frac{200}{(1+r)^2} + \ldots + \frac{200}{(1+r)^5} = 820.04$$

so the bond should sell for £820.04. It is worth more than the previous bond because the return is obtained earlier.

Problem 1.21

(a) 17.9% p.a. for BMW, 14% p.a. for Mercedes.

(b) Depreciated values are

BMW 525i	22 275	18 284	15 008	12 319	10 112	8300
Merc 200E	21 900	18 833	16 196	13 928	11 977	10 300

which are close to actual values. Depreciation is initially slower than the average, then speeds up, for both cars.

Problem 1.23

$$E(x + k) = \frac{\Sigma(x + k)}{n} = \frac{\Sigma x + nk}{n} = \frac{\Sigma x}{n} + k = E(x) + k$$

Problem 1.25

The mistake is comparing non-comparable averages. A first-time buyer would have an above-average mortgage and purchase a below-average priced house, hence the amount of buyer's equity would be small.

Answers to problems on Σ notation

Problem A1

20, 90, 400, 5, 17, 11.

Problem A3

88, 372, 16, 85.

Problem A5

113, 14, 110.

Problem A7

$$\frac{\Sigma f(x - k)}{\Sigma f} = \frac{\Sigma fx - k\Sigma f}{\Sigma f} = \frac{\Sigma fx}{\Sigma f} - k$$

Answers to Problems on logarithms

Problem C1

–0.8239, 0.17609, 1.17609, 2.17609, 3.17609, 1.92284, 0.96142, impossible!

Problem C3

–1.89712, 0.40547, 2.70705, 5.41610, impossible!

Problem C5

0.15, 12.58925, 125.8925, 1,258.925, 10^{12}.

Problem C7

15, 40.77422, 2.71828, 22 026.4658.

Problem C9

3.16228, 1.38692, 1.41421, 0.0005787, 0.008.

Chapter 2

Problem 2.1

(a) 4/52 or 1/13.

(b) 12/52 or 3/13.

(c) 1/2.

(d) $4/52 \times 3/52 \times 2/52 = 3/17\,576$ (0.017%).

(e) $(4/52)^3 = 0.000455$.

Problem 2.3

(a) 0.25, 0.4, 5/9.

(b) 'Probabilities' are 0.33, 0.4, 0.5, which sum to 1.23. These cannot be real probabilities, therefore. The difference leads to an (expected) gain to the bookmaker.

(c) Suppose the true probabilities of winning are proportional to the odds, i.e. 0.33/1.23, 0.4/1.23, 0.5/1.23, or 0.268, 0.325, 0.407. If £1 were bet on each horse, then the bookie would expect to pay out $0.268 \times 2 + 0.325 \times 1.5 + 0.407 \times 1 = 1.43$, plus one of the £1 stakes, £2.43 in total. He would thus gain 57 pence on every £3 bet, or about 19%.

Problem 2.5

A number of factors might help: statistical ones such as the ratio of exports to debt interest, the ratio of GDP to external debt, the public sector deficit, etc., and political factors such as the policy stance of the government.

Problem 2.7

(a) is the more probable, since it encompasses her being active *or* not active in the feminist movement. Many people get this wrong, which shows how one's preconceptions can mislead.

Problem 2.9

The advertiser is a trickster and guesses at random. Every correct guess ($P = 0.5$) nets a fee, every wrong one costs nothing except reimbursing the fee. The trickster would thus keep half the money sent in. You should be wary of such advertisements!

Problem 2.11

(a) E(winnings) $= 0.5^{20} \times$ £1bn $+ (1 - 0.5^{20}) \times -$£100 = £853.67.

(b) Despite the positive expected value, most would not play because of their aversion to risk. Would you?

Problem 2.13

(a), (b) and (d) are independent, although legend says that rain on St Swithins day means rain for the next 40!

Problem 2.15

(a) There are 15 ways in which a 4–2 score could be arrived at, of which this is one. Hence the probability is 1/15.

(b) Six of the routes through the tree diagram involve a 2–2 score at some stage, so the probability is 6/15.

Problem 2.17

Pr(guessing all six) = 6/50 × 5/49 × . . . × 1/45 = 1/15 890 700. Pr(six from 10 guesses) = 10/50 × 9/49 × . . . × 5/45 = 151 200/11 441 304 000. This is exactly 210 times the first answer, so there is no discount for bulk gambling!

Problem 2.19

	Prior	Likelihood	Prior × likelihood	Posterior
Fair coin	0.5	0.25	0.125	0.2
Two heads	0.5	1.00	0.500	0.8
Total			0.625	

Problem 2.21

(a) Using Bayes' theorem the probability is $p/(p + (1 - p)) = p$.

(b) Again using Bayes' theorem we obtain $p^2/(p^2 + (1 - p)^2)$.

(c) If $p < 0.5$ then (b) < (a). The agreement of the second witness *reduces* the probability that the defendant is guilty. Intuitively this seems unlikely. The fallacy is that they can lie in many different ways, so Bayes' theorem is not applicable here.

Problem 2.23

(a) EVs are 142, 148.75 and 146 respectively. Hence B is chosen.

(b) The minima are 100, 130, 110, so B has the greatest minimum. The maxima are 180, 170, 200, so C is chosen.

(c) The regret table is

	Low	Middle	High	Max.
A	30	5	20	30
B	0	0	30	30
C	20	15	0	20

so C has the minimax regret figure.

(d) The EV assuming perfect information is 157.75, against an EV of 148.75 for project B, so the value of information is 9.

Problem 2.25

The probability of *no* common birthday is 365/365 × 364/365 × 363/365 × . . . × 341/365 = 0.43. Hence the probability of at least one birthday in common is 0.57, or greater than one-half. Most people underestimate this probability by a large amount. (This result could form the basis of a useful source of income at parties . . .)

Problem 2.27

(a) Choose low confidence. The expected score is 0.6 × 1 + 0.4 × 0 = 0.6. For medium confidence the score would be 0.6 × 2 + 0.4 × −2 = 0.4, and for high confidence 0.6 × 3 + 0.4 × −6 = −0.6.

(b) We require $p \times 2 + (1 - p) \times -2 > p$ (the payoff to admitting low confidence). Hence $4p - 2 > p \Rightarrow p > 2/3$. Similarly, we also require $p \times 2 + (1 - p) \times -2 > p \times 3 + (1 - p) \times -6$. This implies $p < 4/5$.

(c) The payoff to the correct choice (high confidence) is 0.85 × 3 + (1 − 0.85) × −6 = 1.65. Similar calculations show the payoffs to medium and low confidence to be 1.4 and 0.85, respectively. The losses are therefore 0.25 and 0.8.

Chapter 3

Problem 3.1

The graph looks like a pyramid, centred on the value of 7, which is the mean of the distribution. The probabilities of scores of 2, 3, . . . , 12 are (out of 36): 1, 2, 3, 4, 5, 6, 5, 4, 3, 2, 1 respectively. The probability that the sum is nine or greater is therefore 10/36. The variance is 5.83.

Problem 3.3

The distribution should be sharply peaked (at or just after the timetabled departure time) and should be skewed to the right.

Problem 3.5

Similar to the train departure time, except that it is a discrete distribution. The mode would be 0 accidents, and the probability above 1 accident per day very low indeed.

Problem 3.7

The probabilities are 0.33, 0.40, 0.20, 0.05, 0.008, 0.000, 0.000 of 0–6 sixes respectively.

Problem 3.9

(a) $\Pr(0) = 0.9^{15} = 0.21$, $\Pr(1) = 15 \times 0.9^{14} \times 0.1 = 0.34$, hence $\Pr(0 \text{ or } 1) = 0.55$.

(b) By taking a larger sample or tightening the acceptance criteria, e.g. only accepting if the sample is defect free.

(c) 7%.

(d) The assumption of a large batch means that the probability of a defective component being selected does not alter significantly as the sample is drawn.

Problem 3.11

The *y*-axis coordinates are:

0.05 0.13 0.24 0.35 0.40 0.35 0.24 0.13 0.05

This gives the outline of the central part of the Normal distribution.

Problem 3.13

(a) 5%.

(b) 30.85%.

(c) 93.32%.

(d) 91.04%.

(e) Zero! (You must have an *area* for a probability.)

Problem 3.15

(a) $z = 0.67$, area = 25%.

(b) $z = -1$, area = 15.87%.

(c) $z_L = -0.67$, $z_U = 1.67$, area = 70.11%.

(d) Zero again!

Problem 3.17

(a) $I \sim N(100, 16^2/10)$.

(b) $z = 1.98$, Pr = 2.39%.

(c) 2.39% (same as (b)).

(d) 95.22%. This is much greater than the previous answer. This question refers to the distribution of sample means, which is less dispersed than the population.

(e) Since the marginal student has an IQ of 108 (see previous question) nearly all university students will have an IQ above 110 and so, to an even greater extent, the sample mean will be above 110. Note that the distribution of students' IQ is not Normal but skewed to the right, since it is taken from the upper tail of a Normal distribution. The small sample size means we cannot safely use the Central Limit Theorem here.

(f) 105, *not* 100. The expected value of the last 9 is 100, so the average is 105.

Problem 3.19

(a) $r \sim B(10, \frac{1}{2})$.

(b) $r \sim N(5, 2.5)$.

(c) Binomial: Pr = 82.8%; Normal: 73.57% (82.9% using the continuity correction).

Problem 3.21

(a) By the Binomial, Pr(no errors) = 0.99^{100} = 36.6%. By the Poisson, $nP = 1$, so $\Pr(x = 0) = 1^0 3 \times {}^{-1}/0! = 36.8\%$.

(b) $\Pr(r = 1) = 100 \times 0.99^{99} \times 0.01 = 0.370$; $\Pr(r = 2) = 100C2 \times 0.99^{98} \times 0.01^2 = 0.185$. Hence $\Pr(r \leqslant 2) = 0.921$. Poisson method: $\Pr(x = 1) = 1^1 \times e^{-1}/1! = 0.368$; $\Pr(x = 2) = 1^2 \times e^{-1}/2! = 0.184$. Hence $\Pr(x \leqslant 2) = 0.920$. Hence the probability of more than two errors is about 8% using either method. Using the Normal method we would have $x \sim N(1, 0.99)$. So the probability of $x > 2.5$ (taking account of the continuity correction) is given by $z = (2.5 - 1)/\sqrt{0.99} = 1.51$, giving an answer of 6.55%, a significant underestimate of the true value.

Problem 3.23

(a) Normal.

(b) Uniform distribution between 0 and 1 (look up the =*RAND*() function in your software documentation).

(c) Mean = 0.5, variance = 5/12 = 0.42 for parent, mean = 0.5, variance = 0.42/5 for sample means (Normal distribution).

Chapter 4

Problem 4.1

(a) It gives the reader some idea of the reliability of an estimate.

(b) The population variance (or its sample estimate) and the sample size.

Problem 4.3

An estimator is the rule used to find the estimate or a parameter. A good estimator does not *guarantee* a good estimate, only that it is correct *on average* (if the estimator is unbiased) and close to the true value (if precise).

Problem 4.5

$E(w_1x_1 + w_2x_2) = w_1E(x_1) + w_2E(x_2) = w_1\mu + w_2\mu = \mu$ if $w_1 + w_2 = 1$.

Problem 4.7

$40 \pm 2.57 \times \sqrt{10^2/36} = [35.71, 44.28]$.

If $n = 20$, the t distribution should be used, giving

$40 \pm 2.861 \times \sqrt{10^2/20} = [33.60, 46.40]$.

Problem 4.9

$0.40 \pm 2.57 \times \sqrt{0.4 \times 0.6/50} = [0.22, 0.58]$.

Problem 4.11

$(25 - 22) \pm 1.96 \times \sqrt{12^2/80 + 18^2/100} = [-1.40, 7.40]$.

Problem 4.13

$$(0.67 - 0.62) \pm 2.57\sqrt{\frac{0.67 \times 0.33}{150} + \frac{0.62 \times 0.38}{120}} = [-0.10, 0.20].$$

Problem 4.15

$30 \pm 2.131 \times \sqrt{5^2/16} = [27.34, 32.66]$.

Problem 4.17

$(45 - 52) \pm \sqrt{40.32/12 + 40.32/18} = [-2.15, -11.85]$. 40.32 is the pooled variance.

Chapter 5

Problem 5.1

(a) False, you can alter the sample size.

(b) True.

(c) False, you need to consider the Type II error probability also.

(d) True.

(e) False, it's the probability of a Type I error.

(f) False, the confidence level is 1 – Pr(Type I error) or the probability of accepting H_0 when true.

Problem 5.3

H_0: fair coin, ($Pr(H) = \frac{1}{2}$), H_1: two heads ($Pr(H) = 1$). Pr(Type I error) $= (\frac{1}{2})^2 = \frac{1}{4}$; Pr(Type II error) = 0.

Problem 5.5

(a) Rejecting a good batch or accepting a bad batch.

(b) H_0: $\mu = 0.01$ and H_1: $\mu = 0.10$ are the hypotheses. Under H_0, Pr(0 or 1 defective in sample) = 0.911, hence 8.9% chance of rejecting a good batch. Under H_1, Pr(0 or 1) = 0.034, hence 3.4% chance of accepting a bad batch. One could also use the Normal approximation to the Binomial, giving probabilities of 7.78% and 4.95%.

(c) Pr(Type I error) = 26%; Pr(Type II error) = 4.2%.

(d) (i) Try to avoid faulty batches, hence increase the risk of rejecting good batches, the significance level of the test.

(ii) Since there are alternative suppliers it can increase the risk of rejecting good batches (which upsets its supplier).

(iii) Avoid accepting bad batches.

Problem 5.7

$z = 1.83$, hence Prob-value is 3.36%.

Problem 5.9

100%. You will always reject H_0 when false.

Problem 5.11

$z = 1 < 1.96$, the critical value at the 95% confidence level.

Problem 5.13

$z = 0.59$, not significant, do not reject.

Problem 5.15

$z = (115 - 105)/\sqrt{21^2/49 + 23^2/63} = 2.4 > 1.64$, the critical value, hence reject with 95% confidence.

Problem 5.17

$$z = \frac{0.57 - 0.47}{\sqrt{\frac{0.51 \times (1 - 0.51)}{180} + \frac{0.51 \times (1 - 0.51)}{225}}} = 2.12.$$

This is significant using either a one- or a two-tail test. Whether you used a one- or a two-tail test reveals something about your prejudices! The proportions passing are the actual outcomes for 1992 in the UK, based on 1.85 million tests altogether. You might have an interesting class discussion about what these statistics prove! To redress the balance you might investigate the relationship between gender and road accidents.

Problem 5.19

(a) $t = -5/\sqrt{10^2/20} = -2.24 < -2.093$, the critical value, hence reject H_0.

(b) The parent distribution is Normal.

Problem 5.21

$S^2 = 38.8$, and $t = 8.29$, so H_0 is rejected.

Problem 5.23

(a) $t_{20} = 1.18$.

(b) $t_{10} = 1.63$. Neither is significant at the 5% level, though the latter is closer. Note that only one worker performs worse, but this one does substantially worse, perhaps due to other factors.

Problem 5.25

(a) It would be important to check *all* the predictions of the astrologer. Too often, correct predictions are highlighted ex-post and incorrect ones ignored.

(b) Like astrology, a fair test is important, where it is possible to pass or fail, with known probabilities. Then performance can be judged.

(c) Samples of both taken-over companies and independent companies should be compared, with as little difference between samples as possible.

Chapter 7

Problem 7.1

(b) We would expect similar slopes to those using Todaro's data, but the graphs for growth and the income ratio don't look promising.

(c) There seems to exist a psychological propensity to overestimate the degree of correlation. See (d) to see if you did.

(d) $r = -0.73, -0.25, -0.22$ for GNP, growth, the income ratio respectively. Note that $r < 0$ for the income ratio, in contrast to the result in the text.

(e) $t = -3.7, -0.89, -0.78$, so only the first is significant. The critical value is 1.78 (for a one-tail test).

Problem 7.3

(a) Very high and positive.

(b) A medium degree of negative correlation (bigger countries can provide more for themselves).

(c) Theoretically negative, but empirically the association tends to be weak, especially using the real interest rate. There might be lags in (a) and (c). (b) would best be estimated in cross-section.

Problem 7.5

Rank correlations are:

	Birth rate
GNP	−0.770
Growth	−0.274
Inc. ratio	−0.371

These are similar to the ordinary r values. Only the first is significant at 5%.

Problem 7.7

(a)

	GNP	Growth	Income ratio
a	47.18	42.88	45.46
b	−0.006	−1.77	−1.40
R^2	0.53	0.061	0.047

These results are quite different from what was found before! GNP appears the best, not worst, explanatory variable.

(b)–(c)

	GNP	Growth	Income ratio
s_e	7.83	11.11	11.20
s_b	0.0017	1.99	1.82
t	−3.71	−0.89	−0.77
F	13.74	0.78	0.59

Only in the case of GNP is the *t*-ratio significant. The same is true for the *F*-statistic.

(d) You should be starting to have serious doubts! Two samples produce quite different results. We should think more carefully about how to model the birth rate and what data to apply it to.

(e) Using all 26 observations gives:

	GNP	Growth	IR
a	42.40	43.06	36.76
b	−0.01	−2.77	−0.20
R^2	0.31	0.28	0.002
F	11.02	9.52	0.04
S_b	0.00	0.90	1.01
t	−3.32	−3.08	−0.20

The results seem quite sensitive to the data employed. We should try to ensure that we obtain a *representative* sample for estimation.

Problem 7.9

(a) 27.91 (34.71).

(b) 37.58 (32.61).

(c) 35.67 (33.76).

(Predictions in brackets are obtained from Todaro's data.) Again, different samples, different results, which does not inspire confidence.

Chapter 9

Problem 9.1

GNP versus GDP; gross or net national product; factor cost or market prices; coverage (UK, GB, England and Wales); current or constant prices are some of the issues.

Problem 9.3

The following are measures of UK and US GDP, both at year 2000 prices. The UK figures are £bn, the US figures are $bn. Your own figures may be slightly different, but should be highly correlated with these numbers.

	1995	1996	1997	1998	1999	2000	2001	2002	2003
UK	821.4	843.6	875.0	897.7	929.7	961.9	979.2	997.5	1023.2
US	8031.7	8328.9	8703.5	9066.9	9470.3	9817.0	9890.7	10 074.8	10 381.3

Problem 9.5

$n = 1.96^2 \times 400/2^2 = 385.$

Chapter 10

Problem 10.1

(a)

	1987	1988	1989	1990	1991	1992
Exports	100	100.5	105.1	110.5	109.5	112.4
Imports	100	112.5	120.9	121.5	114.8	121.5

(b) No. Using the indices, information about the *levels* of imports and exports is lost.

Problem 10.3

(a)–(c)

Year	E	P_L	P_P	Q_L	Q_P
1995	100	100	100	100	100
1996	89.7	86.3	85.7	104.3	103.9
1997	90.3	85.5	85.1	105.6	105.6
1998	93.7	88.1	87.8	105.9	106.3
1999	97.3	88.1	87.3	110.5	110.5

Problem 10.5

(a)

Year	Coal	Petroleum	Electricity	Gas
1995	100	100	100	100
1996	95.0	105.8	97.7	68.5
1997	92.4	97.8	92.0	75.2
1998	94.3	93.9	91.5	82.7
1999	93.3	112.9	90.4	80.6
Shares	0.013	0.073	0.505	0.409

(b) Answer as in Problem 10.3(a).

Problem 10.7

The chain index is 100, 110, 115, 123.1, 127.7, 136.9, 139.2, using 2000 as the common year. Using one of the other years to chain yields a slightly different index. There is no definitive right answer.

Problem 10.9

Expenditure on energy in 2003 was £8863.52 m. The Laspeyres index increased from 101.68 to 115.22 between 2003 and 2004, an increase of 13.3%. Hence industry should be compensated 8.63% of 8863.52 = £737.12 m. A similar calculation using the Paasche index yields compensation of £692.35 m. The choice of index makes a difference of about £45 m, a substantial sum.

Problem 10.11

The index number series are as follows:

Year	Cash expenditure (a)	Real expenditure (b)	Volume of expenditure (c)	Real expenditure per capita (d)	Volume of expenditure per capita (d)	Needs index (e)	Spending deflated by need (e)
1987	100.0	100.0	100.0	100.0	100.0	100.0	100.0
1988	109.8	102.6	99.4	102.4	99.2	100.2	99.2
1989	120.5	105.7	101.9	105.1	101.3	100.5	101.4
1990	132.7	107.5	104.5	106.6	103.6	100.8	103.6
1991	150.4	113.6	108.8	111.9	107.1	101.6	107.1

(a) $109.8 = 23\,601/21\,495 \times 100$; $120.5 = 25\,906/21\,495 \times 100$; etc.

(b) This series is obtained by dividing column 1 by column 2 (and setting 1987 as the reference year). Clearly, much of the increase in column 1 is due to inflation.

(c) This series is column 1 divided by column 3. Since the NHS price index rose faster than the GDP deflator, the volume of expenditure rises more slowly than the real figure.

(d) Per capita figures are obtained by dividing by the population, column 4.

(f) Needs index could be improved by finding the true cost of treating people of different ages.

Problem 10.13

(a) 1702.20.

(b) Yes, 102.20.

Problem 10.15

18.3%.

Problem 10.17

(a) Area A = 0.291, B = 0.209, Gini = 0.581.

(b) The old have had a lifetime to accumulate wealth whereas the young have not. This does not apply to income.

Problem 10.19

(a) The Gini coefficients are 0.45, 0.33, 0.33, 0.36 respectively.

(b) These differ from the values given in Table 2.24 (from *Statbase*), substantially so in the case of original income. The figures based on quintiles are all lower than the figures from *Statbase*, as expected, although the bias is large in some cases. This finding suggests the method based on quintiles may not be very accurate when the Gini is around 0.5 or higher.

Problem 10.21

79.3%.